AF389084

Design and Application of Electrical Machines

Design and Application of Electrical Machines

Editors

Ryszard Palka
Marcin Wardach

MDPI • Basel • Beijing • Wuhan • Barcelona • Belgrade • Manchester • Tokyo • Cluj • Tianjin

Editors
Ryszard Palka
West Pomeranian University of Technology
Poland

Marcin Wardach
West Pomeranian University of Technology
Poland

Editorial Office
MDPI
St. Alban-Anlage 66
4052 Basel, Switzerland

This is a reprint of articles from the Special Issue published online in the open access journal *Energies* (ISSN 1996-1073) (available at: https://www.mdpi.com/journal/energies/special_issues/ design_application_electrical_machines).

For citation purposes, cite each article independently as indicated on the article page online and as indicated below:

LastName, A.A.; LastName, B.B.; LastName, C.C. Article Title. *Journal Name* **Year**, *Volume Number*, Page Range.

ISBN 978-3-0365-4323-9 (Hbk)
ISBN 978-3-0365-4324-6 (PDF)

Contents

About the Editors

Ryszard Palka

Ryszard Palka, D.Sc. Ph.D. Eng.; Head of Department of Electrical Machines and Drives, West Pomeranian University of Technology, Szczecin, Poland. In 1987-2005 he was with the Institute of Electrical Machines, Traction and Drives, TU Braunschweig, Germany. Areas of research: electromagnetic field theory, numerical field calculations, optimization of electromagnetic fields, electrical machines, high-temperature superconductivity. He is the author of about 320 refereed journal articles, conference papers, technical reports, and the co-author of 4 books. He is a member of IEEE, International Compumag Society, Polish Society of Theoretical and Applied Electrical Engineering, International Maglev Board and Committee on Electrical Engineering, Polish Academy of Sciences.

Marcin Wardach

Marcin Wardach, D.Sc. Ph.D. Eng.; was born in Poland in 1980. He graduated and received a Ph.D. degree from the Electrical Department, Szczecin University of Technology, Szczecin, Poland, in 2006 and 2011, respectively. From 2020, he has been an Associate Professor with the Faculty of Electrical Engineering, West Pomeranian University of Technology, Szczecin. His research interests include the design of electrical machines and drives, especially unconventional and hybrid excited machines and drives. He is the author of over 100 scientific papers and post-conference publications. He is a member of the Association of Polish Electrical Engineers and Polish Society of Theoretical and Applied Electrical Engineering.

Preface to "Design and Application of Electrical Machines"

Electrical machines are one of the most important components of the industrial world. They are at the heart of the new industrial revolution, brought forth by the development of electromobility and renewable energy systems. Electric motors must meet the most stringent requirements of reliability, availability, and high efficiency in order, among other things, to match the useful lifetime of power electronics in complex system applications and compete in the market under ever-increasing pressure to deliver the highest performance criteria.

Today, thanks to the application of highly efficient numerical algorithms running on high-performance computers, it is possible to design electric machines and very complex drive systems faster and at a lower cost. At the same time, progress in the field of material science and technology enables the development of increasingly complex motor designs and topologies. The purpose of this Special Issue is to contribute to the development of electric machines.

We hope that the publication of this collection of scientific articles, dedicated to the topic of electric machine design and application, contributes to the dissemination of the above information among professionals dealing with electrical machines.

Ryszard Palka, Marcin Wardach
Editors

Editorial

Design and Application of Electrical Machines

Ryszard Palka * and Marcin Wardach

Faculty of Electrical Engineering, West Pomeranian University of Technology, Sikorskiego 37, 70-313 Szczecin, Poland; marcin.wardach@zut.edu.pl
* Correspondence: ryszard.palka@zut.edu.pl

Citation: Palka, R.; Wardach, M. Design and Application of Electrical Machines. *Energies* **2022**, *15*, 523. https://doi.org/10.3390/en15020523

Received: 31 December 2021
Accepted: 4 January 2022
Published: 12 January 2022

Publisher's Note: MDPI stays neutral with regard to jurisdictional claims in published maps and institutional affiliations.

Design and Application of Electrical Machines is a Special Issue of *Energies*. In this Special Issue, authors from various research centers present the results of their scientific research on electrical machines. In eighteen papers, they presented simulation studies, design works, and experimental tests on built prototypes.

Electrical machines are the most important components of both the industrial and commercial worlds. They are at the heart of the new industrial revolution brought forth by the development of electromobility and renewable energy systems. Today's electric motors must meet the most stringent requirements of reliability, availability, and high efficiency to match (and exceed) the useful lifetime of power electronics in complex system applications and to compete in a market under ever-increasing pressure to deliver the highest performance criteria.

Today, thanks to the application of highly efficient numerical algorithms running on high-performance computers, it is possible to design electric machines and entire drive systems that are faster and at a lower cost. At the same time, progress in the field of material science and technology enables the development of ever-more complex motor designs and topologies.

The purpose of this Special Issue, "Design and Application of Electrical Machines", is to contribute to the development of electric machines. The aim is to present a variety of problems related to widely understood electrical machines. Unlike other very monothematic issues (available mostly to specialized readers), this allows us to reach a much wider group of scientists and engineers.

Here, scientists report results of their research in the field of design and application of many types of electric machines.

Topics of interest for publication include:

- Simulation tools, modeling, and analysis of electrical machines;
- New design methods of electrical machines;
- Optimization of electrical machines;
- Electrical machines for EVs and HEVs;
- Power electronics used for electrical machines;
- Supply and control of electrical machines;
- New technologies, materials, devices, and systems for electrical machines;
- Linear drives for transportation.

This Special Issue, "Design and Application of Electrical Machines", was established to provide original results of research, simulation, design, and the build of and testing of modern electrical machines. In total, 18 papers related to the main topic were accepted and published in this Special Issue. Most of the papers are concerned with research on cylindrical electrical machines, but there are also papers dealing with axial-flux machines and magnetic clutches.

The first paper, presented by Ullah et al. [1] presents linear flux switching machines (LFSMs) possess the capability of generating adhesive thrust force; thus, problems associated with conventional rotary electric machines and mechanical conversion assemblies can be eliminated. Additionally, the unique features of high force/power density, efficiency, and a robust secondary structure make LFSMs a suitable candidate for linear motion applications. However, deficiencies in controllable air-gap flux, risk of PM demagnetization, and the increasing cost of rare earth PM materials in case of PMLFSMs, as well as the inherent low thrust force capability of field excited LFSMs compel researchers to investigate new hybrid topologies. In this paper, a novel double-sided hybrid excited LFSM (DSHELFSM) with three excitation sources, i.e., PMs, DC, and AC windings confined to short moving primary and segmented secondary providing short flux paths is designed, investigated, and optimized. Second, unequal primary tooth width optimization and additional end-teeth at all four corners of the primary equipment's proposed design with a balanced magnetic circuit and a reduced end-effect and thrust force ripples. Third, the measured experimental results of the proposed manufactured machine prototype were compared with corresponding simulated model results and showed good agreement, thus validating the theoretical study.

The next paper by Kluszczyński and Pilch [2] focuses on magnetorheological clutches (MR clutches) with a disc structure that can be designed as one-disc or multi-disc clutches (number of discs: $n = 1$, $n > 2$). The main goal of the paper was to compare the overall dimensions (lengths and radii), masses, volumes, and characteristic factors—torque per mass ratio and torque per volume ratio for MR clutches that develop the same given clutching torque (Tc) but differ in the number of discs (it was assumed that the number of discs of the primary member varied from one to four). This analysis developed charts and guidelines that allow designers to choose the appropriate number of discs from the view point of various criteria, and with various limitations regarding geometry, geometric proportions, mass, and volume, or restrictions on the amounts of active materials used in the manufacturing process. The limitations on the used active materials are of particular importance in the case of mass production. The methodology used a comparative study, which can also be used when comparing design solutions of other electromechanical converters.

The article by Kocan et al. [3] is applied to high-speed electric machines. Currently, one of the most used motor types for high-speed applications is the permanent magnet synchronous motor. However, this type of machine has high costs and rare earth elements are needed for its production. For these reasons, permanent-magnet-free alternatives are sought. An overview of high-speed electrical machines shows that the switched reluctance motor is a possible alternative. The paper deals with the design and optimization of this motor, which could achieve the same output power as existing high-speed permanent-magnet synchronous motors while maintaining the same motor volume. The paper presents the initial design of the motor and the procedure for analyses performed using analytical and finite element methods. During electromagnetic analysis, the influence of motor geometric parameters on parameters, such as maximum current, average torque, torque ripple, output power, and losses, was analyzed. The analysis of windage losses was performed using analytical calculations. Based on the results, it was necessary to create a cylindrical rotor shape. The rotor modification method was chosen based on mechanical analysis. Using thermal analysis, the design was modified to meet thermal limits. The results of the work were a design that met all requirements and limits.

The paper by Ciurys and Fiebig [4] presents an innovative design solution of a balanced vane pump integrated with an electric motor that was developed by the authors. The designed and constructed bench, which enables testing of the system: power supply, converter, integrated motor–pump assembly and hydraulic load at different motor speeds and different pressures in the hydraulic system, is described. The electromagnetic and hydraulic processes in the motor–pump unit were investigated, and new, previously unpublished, results of experimental studies under steady and dynamic states are presented. The results of the study showed good dynamics of the integrated motor–pump assembly and proved its suitability to control the pump flow rate, and thus, the speed of the hydraulic cylinder or the speed of the hydraulic motor.

Wang et al. [5] discuss the subject of prediction of electromagnetic characteristics in stator end parts. In order to study the multiple restricted factors and parameters of the eddy current loss of generator end structures, both multi-layer perceptron (MLP) and support vector regression (SVR) were used to study and predict the mechanisms of the synergistic effect of metal shield conductivity, relative permeability of clamping plates, and structural characteristics of eddy current losses. Based on the eddy current losses of generator end structures under different metal shielding thicknesses and electromagnetic properties, the calculation accuracies of MLP and SVR were compared. The prediction method gave an effective means for the complex design of the end region of the generator, which reduced effort by the designers. It also promoted the design efficiency of the electrical generator.

In the paper by Radwan-Pragłowska et al. [6], an application of the harmonic balance method (HBM) for analysis of an axial flux permanent magnet generator (AFPMG) is presented. Particular attention was paid to the development of mathematical model equations, allowing to estimate machine properties without having to use quantitative solutions. The methodology used allowed for precise determination of the Fourier spectra with respect to the winding currents and electromagnetic torque (both quantitatively and qualitatively) in a steady state operation. Analyses of space harmonic interaction in steady states were presented for the three-phase AFPMG. Satisfactory convergence between the results of calculations and measurements confirmed the initial assumption that the developed circuit models of AFPMG and showed that they were sufficiently accurate and can be useful in diagnostic analyses, tests, and the final stages of the design process.

Ullah et al. [7] propose research on a single-sided variable flux permanent magnet linear machine with a flux bridge in mover core. The flux bridge prevents leakage flux from the mover and converts it into flux linkage, which greatly influences the performance of the machine. First, a lumped parameter model was used to find a suitable coil combination and a no-load flux linkage of the proposed machine, which greatly reduced the computational time and drive storage. Secondly, the proposed machine replaced the expensive rare earth permanent magnets with ferrite magnets and provided improved flux controlling capability under variable excitation currents. Multivariable geometric optimization was utilized to optimize the leading design parameters, such as split ratio, stator pole width, width and height of permanent magnet, flux bridge width, the width of mover's tooth, and stator slot depth at constant electric and magnetic loading. The optimized design increases the flux linkage by 44.11%, average thrust force by 35%, thrust force density by 35.02%, minimizes ripples in thrust force by 23%, and detent force by 87.5%. Furthermore, the results obtained by 2D analysis were verified using 3D analysis. Thermal analysis was done to set the operating limit of the proposed machine.

In the paper by Choi [8], analyses and experimental verification of the demagnetization vulnerability in PM synchronous machine are presented. Safety is a critical feature for all passenger vehicles, ensuring fail-safe operation of the traction drive system is highly important. Increasing demands for traction drives that can operate in challenging environments over wide constant power speed ranges expose permanent magnet (PM) machines to conditions that can cause irreversible demagnetization of the rotor magnets. In this paper, a comprehensive analysis of the demagnetization vulnerability in PM machines for an electric vehicle (EV) application is presented. The first half of the paper presents rotor demagnetization characteristics of several different PM machines to investigate the impact of different design configurations on demagnetization and to identify promising machine geometries that have higher demagnetization resistance. Experimental verification results of rotor demagnetization in an interior PM (IPM) machine are presented in the latter half of the paper. The experimental tests were carried out on a specially designed locked-rotor test setup combined with a closed-loop magnet temperature control. Experimental results confirm that both local and global demagnetization damage can be accurately predicted by time-stepped finite element (FE) analysis.

Bi et al. [9] show research on electromagnetic fields, eddy current losses, and heat transfer in synchronous condensers. Aiming at the problem of end structure heating caused by the excessive eddy current loss of large synchronous condensers used in ultra-high voltage (UHV) power transmission, combined with the actual operation characteristics of the synchronous condenser, a three-dimensional transient electromagnetic field physical model was established, and three schemes for adjusting the end structure of the condenser under rated condition were researched. The original structure had a copper shield and a steel clamping plate. During the research, two modifications were made relative to the original design; in the first, the copper shield was removed, while in the second, a steel clamping plate was additionally replaced with aluminum. By constructing a three-dimensional fluid–solid coupling heat transfer model at the end of the synchronous condenser, and giving the basic assumptions and boundary conditions, the eddy current loss of the structure calculated by the three schemes was applied to the end region of the synchronous condenser as the heat source, and the velocity distribution of the cooling medium and the temperature distribution of each structure under the three different schemes were obtained. In order to verify the rationality of the numerical analysis model and the effectiveness of the calculation method, the temperature of the inner edge of the copper shield in the end of the synchronous condenser was measured, and the temperature calculation results are consistent with the temperature measurement results, which provides a theoretical basis for the electromagnetic design, structural optimization, ventilation, and cooling of the synchronous condenser.

Franck et al. [10] refer to analysis of structural dynamic interactions of electric drivetrains. In electric drivetrains, the traction machines are often coupled to a gear transmission. For the noise and vibration analysis of such systems, linearized system models in the frequency domain are commonly used. In this paper, a system approach in the time domain is introduced, which gives the advantage of analyzing the transient behavior of an electric drivetrain. The focus of this paper is on the dynamic gear model. Finally, the modeling approach is applied to an exemplary drivetrain, and the results are discussed.

In the article by Groschup et al. [11], the influence of a preformed coil design on the thermal behavior of electric machines is discussed. Preformed coils are used in electrical machines to improve the copper slot fill factor. A higher utilization of the machine can be realized. The improvement is a result of both low copper losses due to the increased slot fill factor and an improved heat transition out of the slot. In this study, the influence of these two aspects on the operational improvement of the machine were studied. Detailed simulation models allowed a separation of the two effects. A preformed wound winding in comparison to a round wire winding was studied. Full machine prototypes as well as motorettes of the two designs were built. Thermal finite element models of the stator slot were developed and parameterized with the help of motorette microsections. The resulting thermal lumped parameter model is enlarged to represent the entire electric machine. Electromagnetic finite element models for loss calculation and the thermal lumped parameter models are parameterized using test bench measurements. The developed models show very good agreement in comparison to the test bench evaluation. The study indicates that both the improvements in the heat transition path and the advantages of the reduced losses in the slot contribute to the improved operational range in dependency of the studied operational point.

The paper by Dukalski and Krok [12] discusses the problem of decreasing the mass of a wheel hub motor by improving the design of a motor's electromagnetic circuit. The authors propose increasing the number of magnetic pole pairs. They present possibilities of mass reduction obtained through these means. They also analyzed the impact of design changes on losses and the temperature distribution in motor elements. Lab tests of a constructed prototype, as well as elaborated conjugate thermal-electromagnetic models of the prototype motor and modified motor (i.e., motor with increased number of magnetic poles) were used in the investigations. Simulation models were verified by tests on the prototype. Results of calculations for two motors, differing by the number of pair poles, were compared over a wide operational range specific to the motor application in the electric traction. A detailed analysis of the operational range for these motors was also performed.

Wang et al. [13] deal with nonlinear simulation models of a drive consisting of the four-phase 8/6 doubly salient switched reluctance motor (SRM), the four-phase dissymmetric bridge power converter and the closed-cycle rotor speed control strategy carried out by the pulse width modulation (PWM) with a variable angle and combined control scheme with the PI algorithm. All presented considerations are based on a MATLAB-SIMULINK platform. The nonlinear mathematical model of the analyzed SRM drive was obtained as a combination of the two dimensional (2D) finite element model of the motor and the nonlinear model of the electrical network of the power supply circuit. The main model and its seven sub-modules, the controller module, one phase simulation module, rotor position angle transformation module, power system module, phase current operation module, "subsystem" module, and electromagnetic torque of one phase operation module are described. MATLAB functions store the magnetization curves data of the motor obtained by the 2D FEM electromagnetic field calculations, as well as the data of magnetic co-energy curves of the motor calculated from the magnetization curves. The 2D specimen insert method is adopted in MATLAB functions for operating the flux linkage and the magnetic co-energy at the given phase current and rotor position. The phase current waveforms obtained during simulations match with the tested experimentally phases current waveforms at basically the same rotor speed and under the same load. The simulated rotor speed curves also agree with the experimental rotor speed curves. This means that the method of suggested nonlinear simulation models of the analyzed SRM drive is correct and that the model is accurate.

Palomo and Gwozdziewicz [14] present the design and analysis of a permanent magnet synchronous generator. The interior permanent magnet rotor was designed asymmetrically and with the consequent pole approach. The basis for the design was a series-produced three-phase induction motor and neodymium iron boron (Nd-Fe-B) cuboid magnets that were used for the design. For the partial demagnetization analysis, some of the magnets were extracted and the results are compared with the finite element analysis.

Gozdowiak [15] presents simulation results of hydro generator faulty synchronization during connection to the grid for various voltage phase shift changes in a full range ($-180°$; $180°$). A field-circuit model of salient pole synchronous hydro generator was used to perform the calculation results and was verified using the measured no-load and three-phase short-circuit characteristics. This model allowed observing the physical phenomena existing in the investigated machine, especially in the rotor which was hardly accessible for measurement. The presented analysis shows the influence of faulty synchronization on the power system stability and the construction components that are the most vulnerable to damage. From a mechanical point of view, the most dangerous case was for the voltage phase shift equal to $120°$, and this case was analyzed in detail. Great emphasis was placed on the following physical quantities: electromagnetic torque, stator current, stator voltage, rotor current, current in rotor bars, and active and reactive power. The physical quantities existing during faulty synchronization were compared with a three-phase sudden short-circuit state. From this comparison, the values of physical quantities that should be taken into account during design of new hydro generators to withstand the greatest possible threats during long-term work were taken into account.

In the paper by Korkosz et al. [16] tests on a brushless permanent magnet DC motor with different winding configurations were carried out. Three configurations were compared: star, delta, and combined star–delta. A mathematical model was constructed for the motor considering the different winding configurations. An analysis of the operation of the motor in the different configurations was performed, based on numerical calculations. The use of different winding configurations affects the properties of the motor. This is significant in the case of the occurrence of various fault states. Based on numerical calculations, an analysis of an open-circuit fault in one of the phases of the motor was performed. Fast Fourier transform (FFT) analysis of the artificial neutral-point voltage was used for the detection of fault states. The results were verified by tests carried out under laboratory conditions. It was shown that the winding configuration had an impact on the behavior of the motor in the case of an open circuit in one of the phases. The classical star configuration is the worst of the possible arrangements. The most favorable in this respect is the delta configuration. In the case of the combined star–delta configuration, the consequences of the fault depend on the location of the open circuit.

The paper by Wardach et al. [17] deals with the overview of different designs of hybrid excited electrical machines, i.e., those with conventional permanent magnets excitation and additional DC-powered electromagnetic systems in the excitation circuit. The paper presents the most common topologies for this type of machine found in the literature—they were divided according to their electrical, mechanical, and thermal properties. Against this background, the designs of hybrid excited machines, that were the subject of scientific research of the authors, are presented.

The final paper [18] by Baek et al. develops a simulation model of a 120-kW class electric all-wheel-drive (AWD) tractor and verify the model by comparing the measurement and simulation results. The platform was developed based on the power transmission system, including batteries, electric motors, reducers, wheels, and a charging system, composed of a generator, an AC/DC converter, and chargers on each axle. The data measurement system was installed on the platform, and consisted of an analog (current) and a digital part (rotational speed of electric motors and voltage and SOC (state of charge) level of batteries) by a CAN (controller area network) bus. The axle torque was calculated using the current and torque curves of the electric motor. The simulation model was developed using 1D simulation software and used axle torque and vehicle velocity data to create the simulation conditions. To compare the results of the simulation, a driving test using the platform was performed at a ground speed of 10 km/h in off- and on-road conditions. The similarities between the results were analyzed using statistical software and we found no significant difference in axle torque data. The simulation model was considered to be highly reliable given the change rate and average value of the SOC level. Using the simulation model, the workable time of driving operation was estimated to be about six hours and the workable time of plow tillage was estimated to be about 2.4 h. The results showed that the capacity of the battery was slightly low for plow tillage. However, in future studies, electric AWD tractor performance could be improved through battery optimization through simulations under various conditions.

The subject of the "Design and Application of Electrical Machines" Special Issue turned out to be very topical, as evidenced by 18 scientific papers included in it. Currently, more and more scientists are taking up the topic of electric vehicle drives. A huge role in this respect is played by the significant development in the field of modern electrotechnical materials (such as permanent magnets, electrical steel sheets, insulation materials, etc.). In addition, faster and more efficient computational software that supports the process of designing electrical machines prove that the results obtained using them are more and more similar to the results obtained during experimental research. All these problems were also discussed in this Special Issue of *Energies*. Similar research about design, simulation and electrical machine optimization are also discussed in some previous publications [19–22] not included in this Special Issue.

Author Contributions: The contributions of the two authors are the same. All authors have read and agreed to the published version of the manuscript.

Funding: This research received no external funding.

Institutional Review Board Statement: Not applicable.

Informed Consent Statement: Not applicable.

Acknowledgments: Guest Editors would like to thank all authors who prepared very interesting and professional papers. In addition, we would like to thank reviewers, thanks to whom manuscripts have been improved. A special thanks Guest Editors would like to send to Sherwin Chen for help and support during whole editorial process of Special Issue *Design and Application of Electrical Machines*.

Conflicts of Interest: The authors declare no conflict of interest.

References

1. Ullah, N.; Khan, F.; Basit, A.; Shahzad, M. Experimental Validations of Hybrid Excited Linear Flux Switching Machine. *Energies* **2021**, *14*, 7274. [CrossRef]
2. Kluszczyński, K.; Pilch, Z. The Choice of the Optimal Number of Discs in an MR Clutch from the Viewpoint of Different Criteria and Constraints. *Energies* **2021**, *14*, 6888. [CrossRef]
3. Kocan, S.; Rafajdus, P.; Bastovansky, R.; Lenhard, R.; Stano, M. Design and Optimization of a High-Speed Switched Reluctance Motor. *Energies* **2021**, *14*, 6733. [CrossRef]
4. Ciurys, M.P.; Fiebig, W. Experimental Investigation of a Double-Acting Vane Pump with Integrated Electric Drive. *Energies* **2021**, *14*, 5949. [CrossRef]
5. Wang, L.; Sun, Y.; Kou, B.; Bi, X.; Guo, H.; Marignetti, F.; Zhang, H. Prediction of Electromagnetic Characteristics in Stator End Parts of a Turbo-Generator Based on MLP and SVR. *Energies* **2021**, *14*, 5908. [CrossRef]
6. Radwan-Pragłowska, N.; Węgiel, T.; Borkowski, D. Application of the Harmonic Balance Method for Spatial Harmonic Interactions Analysis in Axial Flux PM Generators. *Energies* **2021**, *14*, 5570. [CrossRef]
7. Ullah, B.; Khan, F.; Qasim, M.; Khan, B.; Milyani, A.H.; Cheema, K.M.; Din, Z. Lumped Parameter Model and Electromagnetic Performance Analysis of a Single-Sided Variable Flux Permanent Magnet Linear Machine. *Energies* **2021**, *14*, 5494. [CrossRef]
8. Choi, G. Analysis and Experimental Verification of the Demagnetization Vulnerability in Various PM Synchronous Machine Configurations for an EV Application. *Energies* **2021**, *14*, 5447. [CrossRef]
9. Bi, X.; Wang, L.; Marignetti, F.; Zhou, M. Research on Electromagnetic Field, Eddy Current Loss and Heat Transfer in the End Region of Synchronous Condenser with Different End Structures and Material Properties. *Energies* **2021**, *14*, 4636. [CrossRef]
10. Franck, M.; Rickwärtz, J.; Butterweck, D.; Nell, M.; Hameyer, K. Transient System Model for the Analysis of Structural Dynamic Interactions of Electric Drivetrains. *Energies* **2021**, *14*, 1108. [CrossRef]
11. Groschup, B.; Pauli, F.; Hameyer, K. Influence of the Preformed Coil Design on the Thermal Behavior of Electric Traction Machines. *Energies* **2021**, *14*, 959. [CrossRef]
12. Dukalski, P.; Krok, R. Selected Aspects of Decreasing Weight of Motor Dedicated to Wheel Hub Assembly by Increasing Number of Magnetic Poles. *Energies* **2021**, *14*, 917. [CrossRef]
13. Wang, X.; Palka, R.; Wardach, M. Nonlinear Digital Simulation Models of Switched Reluctance Motor Drive. *Energies* **2020**, *13*, 6715. [CrossRef]
14. Palomo, R.E.Q.; Gwozdziewicz, M. Effect of Demagnetization on a Consequent Pole IPM Synchronous Generator. *Energies* **2020**, *13*, 6371. [CrossRef]
15. Gozdowiak, A. Faulty Synchronization of Salient Pole Synchronous Hydro Generator. *Energies* **2020**, *13*, 5491. [CrossRef]
16. Korkosz, M.; Prokop, J.; Pakla, B.; Podskarbi, G.; Bogusz, P. Analysis of Open-Circuit Fault in Fault-Tolerant BLDC Motors with Different Winding Configurations. *Energies* **2020**, *13*, 5321. [CrossRef]
17. Wardach, M.; Palka, R.; Paplicki, P.; Prajzendanc, P.; Zarebski, T. Modern Hybrid Excited Electric Machines. *Energies* **2020**, *13*, 5910. [CrossRef]
18. Baek, S.-Y.; Kim, Y.-S.; Kim, W.-S.; Kim, Y.-J. Development and Verification of a Simulation Model for 120 kW Class Electric AWD (All-Wheel-Drive) Tractor during Driving Operation. *Energies* **2020**, *13*, 2422. [CrossRef]
19. Palka, R.; Woronowicz, K. Linear Induction Motors in Transportation Systems. *Energies* **2021**, *14*, 2549. [CrossRef]
20. Wardach, M.; Bonislawski, M.; Palka, R.; Paplicki, P.; Prajzendanc, P. Hybrid Excited Synchronous Machine with Wireless Supply Control System. *Energies* **2019**, *12*, 3153. [CrossRef]
21. Yan, W.; Chen, H.; Liu, X.; Zhongwei, L.; Xiaoping, M.; Xing, W.X.; Palka, R.; Chen, L.; Wang, K. Design and multi-objective optimisation of switched reluctance machine with iron loss. *IET Electr. Power Appl.* **2019**, *13*, 435–444. [CrossRef]
22. Wardach, M.; Paplicki, P.; Palka, R. A Hybrid Excited Machine with Flux Barriers and Magnetic Bridges. *Energies* **2018**, *11*, 676. [CrossRef]

Article

Experimental Validations of Hybrid Excited Linear Flux Switching Machine

Noman Ullah [1,2,*], Faisal Khan [1], Abdul Basit [2] and Mohsin Shahzad [1]

[1] Department of Electrical and Computer Engineering, COMSATS University Islamabad (Abbottabad Campus), Abbottabad 22060, Pakistan; faisalkhan@cuiatd.edu.pk (F.K.); mohsinshahzad@cuiatd.edu.pk (M.S.)
[2] Pakistan Center for Advanced Studies in Energy, University of Engineering & Technology, Peshawar 25000, Pakistan; abdul.basit@uetpeshawar.edu.pk
[*] Correspondence: nomanullah@cuiatd.edu.pk; Tel.: +92-336-564-2442

Abstract: Linear Flux Switching Machines (LFSMs) possess the capability to generate adhesive thrust force, thus problems associated with conventional rotary electric machines and mechanical conversion assemblies can be eliminated. Additionally, the unique features of high force/power density, efficiency, and a robust secondary structure make LFSMs a suitable candidate for linear motion applications. However, deficiency of controllable air-gap flux, risk of PM demagnetization, and increasing cost of rare earth PM materials in case of PMLFSMs, and inherent low thrust force capability of Field Excited LFSMs compels researchers to investigate new hybrid topologies. In this paper, a novel Double-Sided Hybrid Excited LFSM (DSHELFSM) with all three excitation sources, i.e., PMs, DC, and AC windings confined to short moving primary and segmented secondary providing short flux paths is designed, investigated, and optimized. Secondly, unequal primary tooth width optimization and additional end-teeth at all four corners of the primary equip proposed design with balanced magnetic circuit and reduced end-effect and thrust force ripples. Thirdly, the measured experimental results of the manufactured proposed machine prototype are compared with corresponding simulated model results and shows good agreements, thus validating the theoretical study.

Keywords: electric train; finite element analysis; hybrid excited linear flux switching machine; rope-less elevator

Citation: Ullah, N.; Khan, F.; Basit, A.; Shahzad, M. Experimental Validations of Hybrid Excited Linear Flux Switching Machine. *Energies* **2021**, *14*, 7274. https://doi.org/10.3390/en14217274

Academic Editor: Ryszard Palka and Marcin Wardach

Received: 25 August 2021
Accepted: 18 September 2021
Published: 3 November 2021

Publisher's Note: MDPI stays neutral with regard to jurisdictional claims in published maps and institutional affiliations.

1. Introduction

Existing long stroke linear motion applications such as electric train and vertical lifting utilize rotatory machines plus Mechanical Conversion System (MCS). Two important problems of existing traction system can be highlighted as; (1) the installed rotatory motors are either DC series traction motor, induction motor, or newest traction system utilizes synchronous machines. DC series traction motors are excited by DC rotor bars and brushes are utilized to deliver electric power. This commutator section results in increased maintenance cost and sparking may lead to fire/faults [1]. Induction motor shows demerits of complex speed control, low starting torque, and low efficiency at low loads due to low power factor [2]. Synchronous machines result in high efficiency, constant speed, high power factor, and high torque density. However, placement of field system (PMs or electromagnets) at the rotor reduces its mechanical integrity at high speeds [3]. Furthermore, increased cost of rare earth material in case of Permanent Magnet Synchronous Machine (PMSM) and due to increased maintenance cost of collector rings and brushes in case of field excited synchronous machines forced electrical machine designers to investigate new topologies [4]. Second, MCS is required to convert rotating torque into linear thrust force, as linear translational motion is expected in case of electric train. MCS is meshing engagement of motion type conversion mechanical devices such as ball screw, lead screw and rack, and pinion, etc. This combination results in noise, vibration, mechanical power transfer loss, gearbox faults, and regular maintenance problems [5].

9

Linear Flux Switching Machine (LFSM) directly obtained from corresponding rotatory machine wiped out MCS requirements due to capability of generating direct thrust force, thus solving noise, vibration, and regular maintenance problems. Additionally, high force/power density, power factor, efficiency, and a unique inherent structural property of simple and robust secondary structure makes LFSMs suitable for long-distanced linear motion applications [6]. Linear motor directly obtained from corresponding rotatory machine is termed as single sided linear motor and shows a demerit of high normal or attraction forces [7]. High normal force values result in additional frictional force and reliability of the linear machine [8]. Double sided LFSM is a new introduction in the linear machines' family that successfully solved large undesired normal/attraction force problem [9,10].

Geometric structure and excitation sources are the two parameters that helps researchers to investigate new topologies of LFSMs. Based on geometric structure, LFSMs can be divided into (a) single sided and (b) double sided designs [11]. Based on excitation sources, LFSMs can be excited by: (a) PM plus AC known as Permanent Magnet LFSM (PMLFSM), (b) DC plus AC named as Field Excited LFSM (FELFSM), and (c) PM plus DC and AC termed as Hybrid Excited LFSM (HELFSM). PMLFSM possess drawbacks of fixed and uncontrollable air-gap magnetic field density, risk of PM demagnetization, and high manufacturing cost due to rapid increase in rare-earth PM materials [12]. PMs can be replaced by DC electromagnets to design FELFSM. However, FELFSM exhibit reduced thrust force density and high copper losses. HELFSM requires serious attention of researchers due to its unique features of combining PMLFSM and FELFSM advantages such as variable air-gap field density, reduced manufacturing cost, and flux strengthening/weakening capability.

In this paper, a novel HELFSM with modular stator and complementary coil connection mover is proposed for linear motion applications. Additional assistant teeth at all four end points of the moving primary is installed to balance magnetic circuit and reduce end-effect. Modular stator design reduces secondary material consumption and provides short paths [13]. Complementary coil design enables reduced Thrust Force Ripple Ratio (TFRR) by engaging symmetrical and sinusoidal flux linkages. Disadvantages of existing design schemes installed for targeted applications along with their proposed solution are discussed in Section 2. Design topology, development guidelines, structure variables, and operation principle of proposed machine is explained in Section 3. Section 4 of this paper presents single variable geometry based deterministic optimization (SVGBDO) approach applied to uplift thrust force performance and reduce TFRR of the proposed HELFSM. During optimization process, unequal primary tooth width [14] is enabled to provide appropriate low reluctance path where high flux is recorded during initial tests. Detailed comparison of initial and optimized HELFSM, thrust-force/power versus armature current density plot, thrust-force/power versus velocity characteristics, and efficiency at eight different points of thrust force-vs-velocity graph is also presented in this section. Experimental test bed and measured results are explained in Section 5. Finally, some conclusions are summarized in Section 6.

2. Targeted Applications

Multiple methods to decrease humans' and goods' delivery suspension down time are experienced such as dedicated auto-mobiles, sea ships, and air cargo. However, harmful emissions due to increased use of Internal Combustion Engines (ICEs) and dependency on continuously decaying fossil fuel reservoirs make these options uneconomical [15]. Furthermore, congested road conditions and chances of fatal accidents in case of dedicated auto-mobiles, delivery time ranging from weeks to months in case of sea ships, and high costs of air travel are few reasons that compels transport industry to move towards electric trains. Similarly, due to dramatic increase of human population and consequently land prices, skyscrapers and high-rise buildings are mushrooming at a very high speed. Thus, safe and fast vertical lifting system is an essential need of the current time.

Comparison of existing and proposed traction scheme for electric train and elevator system is shown in Figure 1. Both of these proposed configurations have the ability to wipe out meshing of rotatory machines and MCS. Besides these, two disadvantages of conventional elevator system are hoist cables and counterweights. Hoist cables may suffer strength and stability failures whereas counterweights absorb significant accommodation space throughout the building height [16]. Proposed ropeless vertical elevator scheme is capable to increase stability by avoiding hoist cables and helps in better utilization of building accommodation by removing counterweights requirements.

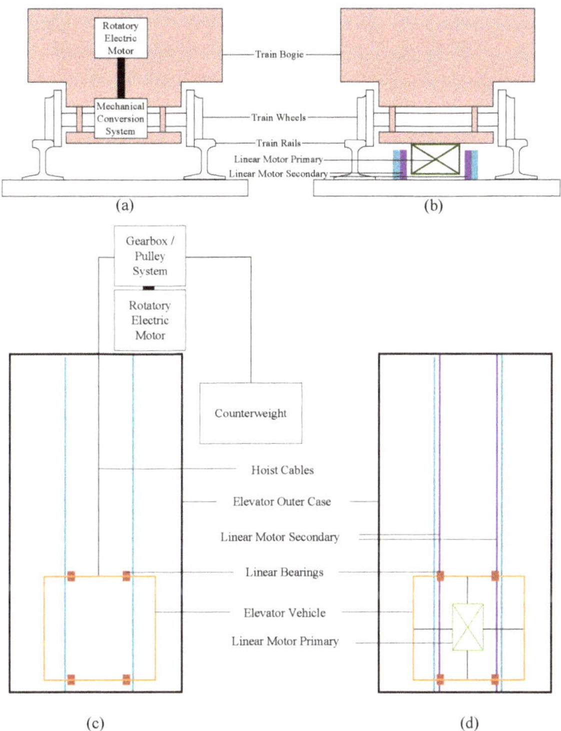

Figure 1. Existing and proposed schemes of targeted applications: (**a**) existing electric train, (**b**) proposed electric train, (**c**) existing vertical lifting, and (**d**) proposed rope-less elevator.

The schematic diagram presenting solutions of existing electric train and elevator system problems identifies replacement of rotatory electric motor and MCS with a linear motor. As shown in Figure 1, secondary of the linear motor in case of electric train will be stretched along with the train rails and primary will be attached with the train bogie. This solution reflects an additional advantage of easy power supply to the linear motor primary. Regarding elevator system, primary of the linear motor is to be attached with the elevator vehicle and its secondary will be stretched along with frame of the elevator.

3. Topology and Working Principle

3.1. Topology

Three dimensional illustration and two dimensional diagram of proposed machine is shown in Figures 2 and 3, respectively. Mover teeth (M_t), number of DC windings or PMs

$(W_{PM/DC})$, AC windings (W_{AC}), and stator to mover pole pitch (τ_s/τ_m) of the proposed machine are derived utilizing following design guidelines equations [17]:

$$M_t = 4pq + 1 \tag{1}$$
$$W_{PM/DC} = 2pq + 1 \tag{2}$$
$$W_{AC} = 2pq \tag{3}$$
$$\tau_s/\tau_m = 4pq/(2pq + 2) \tag{4}$$

Here, $p = 3$ is used to represent number of AC phases and $q = 2$ that reflects AC winding coil pair repetition in the machine. Aforementioned guidelines resulted in $M_t = 25, W_{PM/DC} = 13$, and $W_{AC} = 12$ leading to $\tau_s/\tau_m = 24/14$. Initial values of structure parameters are tabulated in Table 1 and their definitions are provided in Figure 4.

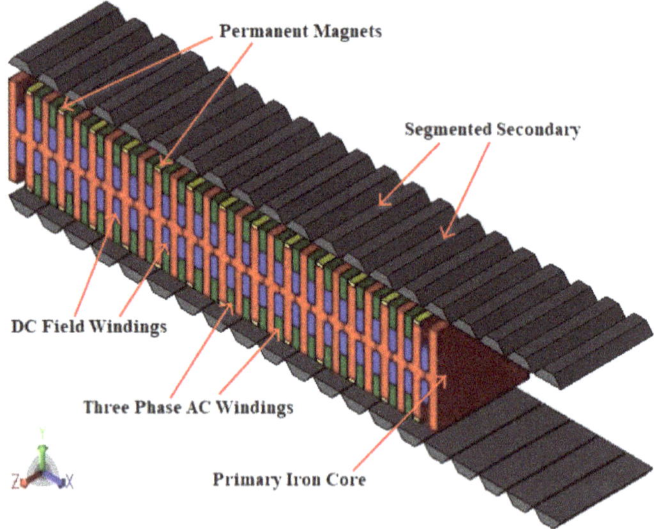

Figure 2. The 3-D structure of the proposed HELFSM.

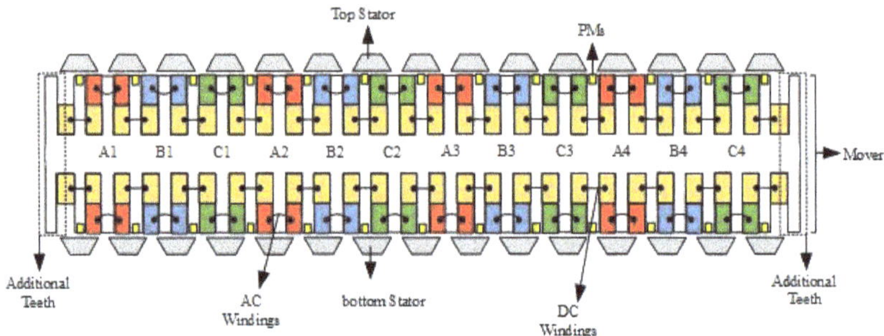

Figure 3. The 2-D schematic diagram of proposed HELFSM.

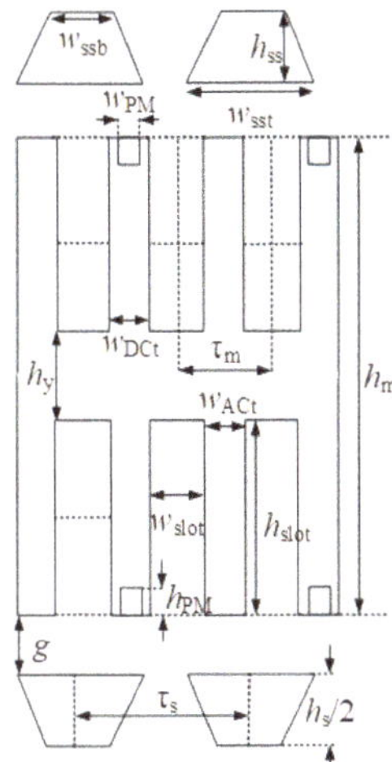

Figure 4. Design variables.

Table 1. Structure design parameters.

Symbol	Parameter (Unit)	Initial Value	Optimized Value
τ_s	Stator pole pitch (mm)	30	
τ_m	Mover pole pitch (mm)	35	
h_m	Mover height (mm)	85	91
w_{DCt}	Mover DC tooth width (mm)	7.5	9.5
h_s	Stator height (mm)	25	19
w_{slot}	Slot width (mm)	10	8.5
w_{ACt}	Mover AC tooth width (mm)	7.5	8.5
h_{slot}	Slot height (mm)	17.5	20.6
h_y	Mover yoke height (mm)	15	8.6
w_{PM}	PM width (mm)	5	7
h_{PM}	PM height (mm)	5	3.5
V_{PM}	PM volume (grams)	45.5	45.5
w_{sst}	Stator segment tip width (mm)	24	28.5
h_{ss}	Stator segment height (mm)	12.5	9.5
w_{ssb}	Stator segment base width (mm)	12	12.825
L	Stack length (mm)	10	
g	Air-gap height (mm)	2	
v	Mover velocity (m/s)	1.5	
J_{DC}	DC current density (A/mm^2)	4.52	
J_{AC}	AC current density (A/mm^2)	4.57	
$N_{AC/DC}$	Number of AC and DC coil turns	40	

3.2. Working Principle

Two types of explanation techniques (air-gap field modulation theory [18,19] or magnetic equivalent circuit) can be utilized to understand operation principle of proposed machine. Magnetic Equivalent Circuit (MEC) methodology is adopted in this paper to reduce complexity. Positive max. and negative max. of no-load flux linkage obtained during linear displacement of one stator pole pitch are shown in Figure 5a,b, respectively. Red lines indicate flux flow generated due to PMs and makes series magnetic circuit encompassing two stators and complete mover. Flux represented by green lines is due to DC electromagnets and make combination of two parallel magnetic circuits and also follow PM flux flow paths. Both PM and DC electromagnets' flux follow same paths to ensure philosophy of hybrid excitation.

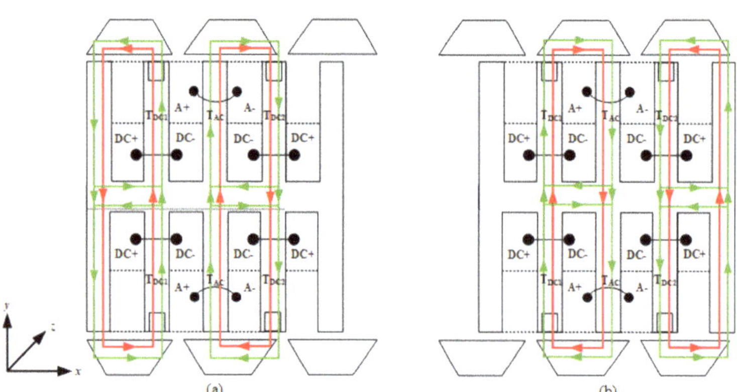

Figure 5. Magnetic circuit based working principle, (**a**) Positive max. flux linkage, and (**b**) Negative max. flux linkage.

4. Optimization and Comparisons

4.1. Optimization of Proposed HELFSM

SVGBDO is a sequential optimization approach that modify geometry variables and is aimed at improving two important performance indices. SVGBDO is conducted in two steps: (1) mover optimization and (2) stator optimization. During step 1, machine configuration with highest average thrust force is selected as optimized model. A TFRR of less than 10% is considered as optimization goal during step 2, while maintaining average thrust force of 10 kN. In other words, mover optimization is done to increase average thrust force and stator optimization is done to reduce TFRR while maintaining specific defined average thrust force limit. Stator pole pitch, mover pole pitch, stack length, air-gap height, whole machine height, primary length (x-direction), armature and field excitation current densities, PM volume, slot area, number of AC and DC coil turns, and mover velocity are kept constant during whole optimization process.

Unequal flux flow in the primary is due to presence of PMs in the DC tooth and is normalized by broadening PM+DC tooth width and shrinking Armature Winding (AW) tooth width in order to keep mover pole pitch constant. To optimize split ratio, mover height, mover yoke height, stator height, AW and PM+DC tooth width, slot area dimensions, PM dimensions, and stator segment dimensions, following six optimization coefficients are defined. Definitions of optimization coefficients, sequence of optimization process, initial values, constraints, and their optimized values are listed in Table 2.

Due to sequential nature of optimization approach, only one optimization coefficient is considered at an instance and investigation is performed until its optimization is completed. In the upcoming optimization process and comparisons, different colour encircling schemes are adopted to differentiate initial and optimized machine configurations. Black colour encircling conveys performance of initial or base machine configuration whereas green rectangle represents optimized value of optimization coefficients. Keeping optimization targets in mind, three Key Performance Indicators (KPIs), i.e., no-load detent force, on-load average thrust force, and TFRR are analysed during upcoming optimization process comparisons. Once, the optimization of an optimization coefficient is completed, the geometrically updated machine configuration is subjected to next optimization coefficient analysis. Detailed procedure of optimization approach is given in Figure 6a–f.

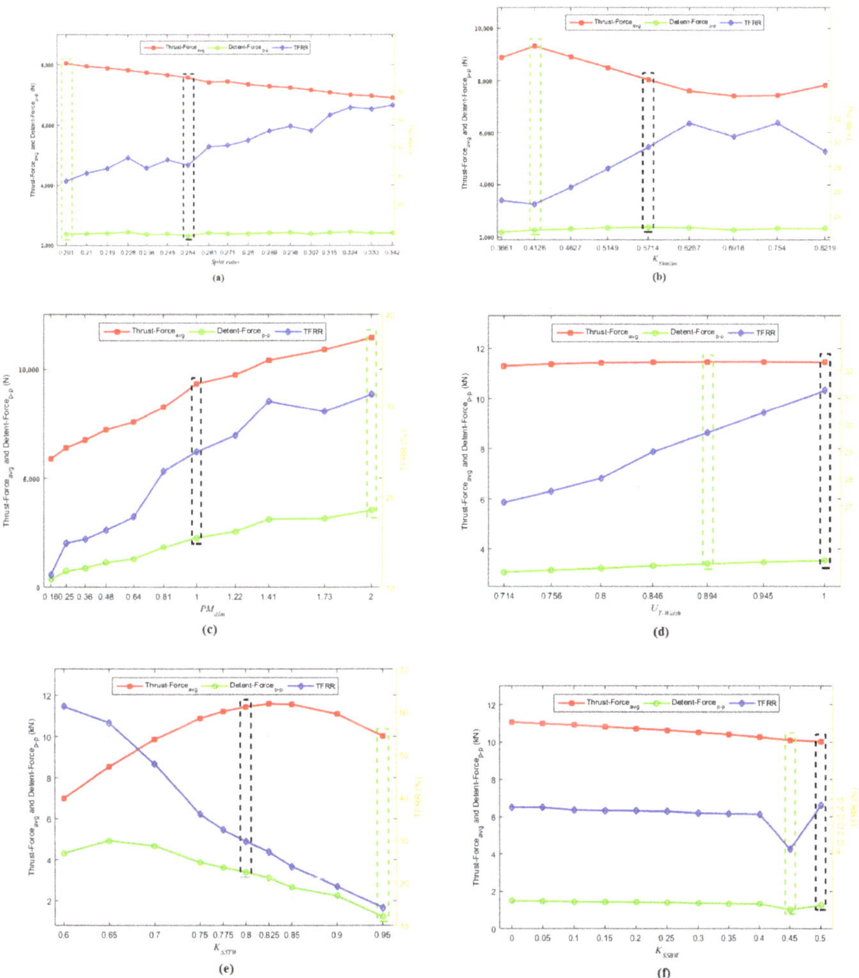

Figure 6. Optimization of HELFSM: (**a**) split ratio, (**b**) slot area dimensions, (**c**) PM dimensions, (**d**) unequal tooth width, (**e**) stator segment tip width, and (**f**) stator segment base width.

Table 2. Optimization process.

Coefficients (Symbol)	Definition	Initial Value	Constraints	Optimized Value
Split Ratio $(S.R)$	$\frac{h_s+2\cdot g}{h_s+2\cdot g+h_m}$	0.254	(0.20–0.34)	0.201
Slot area dimensions $(K_{slotdim})$	$\frac{w_{slot}}{h_{slot}}$	0.571	(0.36–0.82)	0.412
PM dimensions (PM_{dim})	$\frac{w_{PM}}{h_{PM}}$	1.0	(0.16–2.0)	2.0
Unequal tooth width $(U_{T-Width})$	$\frac{w_{ACt}}{w_{DCt}}$	1.0	(0.71–1.0)	0.894
Stator segment tip width (K_{SSTW})	$\frac{w_{sst}}{\tau_s}$	0.8	(0.60–0.95)	0.95
Stator segment base width (K_{SSBW})	$\frac{w_{ssb}}{w_{sst}}$	0.5	(0–0.50)	0.45

4.1.1. Split Ratio Optimization

Split ratio can be defined as ratio of stator to whole machine volume. As discussed in previous section, depth or stack length, whole machine height, and x-direction length of proposed HELFSM are kept constant during optimization process. Hence only stator, mover, and air-gap heights are incorporated in the split ratio coefficient defined in Table 2. As whole machine height is constant, any variation either increase or decrease in the mover height will be reflected in the same amount of decrement or increment of stator segment height. Two geometric variables, which are supposed to be altered are (1) mover yoke height and (2) stator segment height. Smaller value of split ratio coefficient will result in higher mover and lower stator volume. This condition reduces secondary material consumption and provides large cross sectional area to flux flow at the mover yoke height. In order to avoid saturation effect at stator segments, minimum value of split ratio coefficient is defined as 0.2. Maximum value of split ratio coefficient is limited by mechanical stability and manufacturing constraints of mover, as higher value of mentioned coefficient reduces mover yoke height.

Analysing Figure 6a, it can be seen that influence of split ratio coefficient is directly proportional to TFRR, shows the inverse relation with average thrust force, and does not shows strong influence on detent force profile. Minimum value of the split ratio coefficient results in average thrust force, peak-to-peak detent force, and TFRR of 8048.43 N, 2386.5 N, and 29.58%. Whereas maximum value of the split ratio coefficient reflects average thrust force, peak-to-peak detent force, and TFRR of 6910.55 N, 2426.07 N, and 35%. Observing KPIs trends with respect to split ratio coefficient and keeping in mind about selection of machine configuration with maximum average thrust force during mover optimization, HELFSM having split ratio coefficient of 0.201 is selected as optimized machine configuration and it is subjected for further analysis.

4.1.2. Slot Area Dimensions Optimization

Dimensions of AW and Field Excitation Coils (FECs) are optimized while keeping mover pole pitch and winding slot area constant. Under constant slot area constraint, increase in slot area width will result in decrease of slot area height and also reduces primary tooth width. This ultimately affects mover yoke height. Similarly, decrement of slot area width will directly influence machine's geometry by increasing mover tooth width and reducing mover yoke height. So, the geometric variables prone to variations are (1) primary tooth width and (2) mover yoke height. Slot area dimension coefficient declared in Table 2 have direct relation with width of slot area and inverse relation with height of slot area. Hence, minimum value of the slot area coefficient will result in reduced slot opening and mover yoke height.

It should be noted that the optimized machine configuration obtained from split ratio optimization step is considered as base machine configuration in this subsection and is subjected to slot area dimensions optimization. In order to avoid saturation effect at mover yoke height, minimum value of slot area coefficient is limited as 0.3661 reflecting mover yoke height of 3.6 mm. Maximum value of the mentioned coefficient will reduce mover tooth width and must be limited to avoid saturation. Hence, maximum value of slot area coefficient is selected as 0.8219 representing mover tooth width of 5.5 mm.

It can be seen that influence of slot area coefficient is non-linear in case of average thrust force and TFRR (Figure 6b). However, its influence is not that much significant in case of detent force profile. Regarding average thrust force, an increase at initial step is observed and then it goes on decreasing. Inversely, a decrease in TFRR profile is recorded at start and then it goes on increasing. Investigation of KPIs' numerical values and search for machine configuration with maximum average thrust force compelled the author to select slot area coefficient of 0.4126. Machine configuration with slot area width of 8.5 mm, slot area height of 20.60 mm resulted in slot area coefficient of 0.4126 and it is subjected to next stage of optimization process.

4.1.3. PM Dimensions Optimization

PM's width and length are optimized while keeping mover tooth width and total PM volume constant. As can be seen in the topology of the proposed machine, mechanical integrity of the linear machine is enhanced by burying PMs in the tip of primary teeth and primary core portion on both sides of the PM is used to firmly hold the external body. This additional primary core area is capable to avoid PM loss/drop and PM demagnetization during flux weakening operation, as FEC flux will find dedicated low reluctance path to flow. Increase in PM width will result in decrease of PM height and primary core area that is used for clamping of PMs and bypass path for FEC flux. Similarly, decrement of PM width will influence machine's geometry by increasing PM height and width of primary core area. In addition to PM width and height, the only geometric variable that is altered during this optimization stage is primary core area at both sides of PM. PM dimensions coefficient declared in Table 2 have direct relation with PM width and inverse relation with height of PM. Minimum value of the PM dimensions coefficient is limited due to manufacturing constraints of PM and it is selected as 0.16 resulting in a PM width of 2 mm. Maximum value of PM dimensions coefficient is selected as 2.00 that allows primary core width of 1 mm at each side of PM, any further increment in the mentioned coefficient may cause saturation at the specific location.

As practised in previous subsection, optimized machine configuration obtained from slot area optimization stage is considered as base machine configuration in this subsection and is subjected to PM dimensions optimization. All three KPIs of Figure 6c are almost linearly increasing with increment of PM dimensions coefficient and may shows same behaviour after the maximum limit of constraints. However, constraints of the optimization coefficient must be followed for realization of mechanically stable prototype. As PM is one of the major excitation source and strongly influence the performance, maximum value of the average thrust force (i.e., 11,441.09 N) is achieved when PM dimensions coefficient is 2.00, PM width is 7 mm, PM height is 3.5 mm, and each primary core area at both sides of the PM is 1 mm.

4.1.4. Unequal Primary Tooth Width Optimization

Figure 7a shows the on-load magnetic flux lines of HEI FSM optimized during PM dimensions optimization. It can be seen that, both AW and PM+DC tooth shows same width dimensions. On-load flux lines are displayed in this optimization stage to evaluate quantity of magnetic flux density in the AW and PM+DC tooth at rated armature current density. In depth analysis revealed that primary tooth with PM+DC is highly populated when compared to that of AW tooth. In order to convert highly populated primary tooth into low reluctance path and increase primary core utilization ratio, Unequal Tooth Width Optimization (UTWO) coefficient is defined in Table 2. Reduction in reluctance of highly populated primary tooth is possible by increasing its width. The purpose of UTWO coefficient is to increase PM+DC tooth width and reduce AW tooth width while keeping mover pole pitch and slot area dimensions constant. The fraction of millimetres added to PM+DC tooth width must be equal to the fraction of millimetres subtracted from AW tooth width. An increment of 0.25 mm in PM+DCC tooth width is utilized, and seven different machine configurations were examined. UTWO coefficient shows direct relation with AW tooth, hence maximum value of the coefficient is selected as 1.00 when PM+DC tooth width is equal to AW tooth width. Minimum value of the mentioned coefficient is decided as 0.714, when PM+DC tooth width is equal to 10.5 mm and AW tooth width is 7.5 mm. Electromagnetic performance of HELFSM optimized in PM dimensions optimization section under different UTWO coefficient values are compared and presented in Figure 6d.

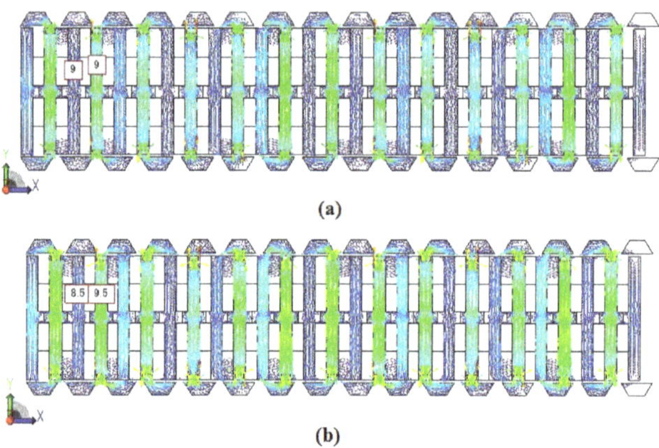

Figure 7. On-load magnetic flux lines: (**a**) equal primary tooth width and (**b**) unequal primary tooth width.

It can be seen that UTWO coefficient shows strong relation with TFRR, linearly lower gradient increasing plot of detent force, and merely influence average thrust force. Base machine configuration's results are shown in black encircled plot of Figure 6d and by reducing AW tooth width, maximum average thrust force of 11,464.85 N is achieved at UTWO coefficient of 0.894. The maximum average thrust force machine configuration results in PM+DC tooth width and AW tooth width of 9.5 mm and 8.5 mm, respectively. A HELFSM with the novelty of unequal primary tooth width is successfully realized and its on-load magnetic flux lines are shown in Figure 7b. Mover optimization is completed at this stage. Three KPIs, i.e., average thrust force, peak-to-peak detent force, and TFRR of 11,464.85 N, 3407.63 N, and 29.68%, respectively, are achieved.

4.1.5. Stator Segment Tip Width Optimization

Stator Segment Tip Width (SSTW) can be defined as the portion of stator segment facing primary teeth and it is the gateway of flux linkage to link in between stator and mover. Wider SSTW will allow flux linkage for major portion of the electrical cycle or x-direction displacement. In this subsection, SSTW is optimized with respect to constant stator pole pitch. Initial value of SSTW simulated for HELFSM optimized during UTWO subsection is 24 mm reflecting SSTW optimization coefficient described in Table 2 of 0.8. Minimum value of the SSTW optimization coefficient is set at 0.6, when SSTW is equal to 18 mm. Further reduction is also possible; however, it degrades the proposed machine performance. Maximum value of the SSTW optimization coefficient is set as 0.95. Mentioned optimization coefficient can be further increased to 1.00; however, at that point SSTW will be equal to stator pole pitch, the secondary will become long uniform rectangle, and concept of segmented secondary will not be maintained.

Figure 6e indicates that increase of SSTW optimization coefficient shows strong and negative slope effect on peak-to-peak detent force and TFRR. Thrust force profile show increasing behaviour up to SSTW optimization coefficient of 0.825 and then decreasing. Observing KPIs trends with respect to SSTW optimization coefficient and keeping in mind about selection of machine configuration with minimum TFRR while maintaining average thrust force of 10 kN, HELFSM having SSTW optimization coefficient of 0.95 is selected as optimized machine configuration and it is subjected for further analysis. Mentioned optimization coefficient of 0.95 results is SSTW of 28.5 mm.

4.1.6. Stator Segment Base Width Optimization

Stator Segment Base Width (SSBW) defines the dimension of stator segment facing opposite to SSTW. Initial value of SSBW is fraction (half) of SSTW optimized in previous section (SSTW Optimization). Optimized machine configuration of SSTW optimization coefficient resulted in SSTW of 28.5 mm and according to calculations SSBW became 14.25 mm. Hence, the initial value of SSBW optimization coefficient defined in Table 2 will result in 0.5 while utilizing SSBW of 14.25 mm and SSTW of 28.5 mm. Value of 0.5 is considered as maximum value of the SSBW optimization coefficient and results in one stator segment volume of 2030 mm^3. In order to bring TFRR below 10% and reduce secondary manufacturing cost by reducing stator segment volume, minimum value of the SSBW is considered as zero.

It can be verified from Figure 6f that all three KPIs shows lower gradient negative slope graph with increase of SSBW optimization coefficient. However, a sudden dip in the peak-to-peak detent force and TFRR profile is witnessed at SSBW optimization coefficient of 0.45. Numerical values of average thrust force, peak-to-peak detent force, and TFRR at SSBW optimization coefficient of 0.45 and SSBW of 12.825 mm are 10,111.13 N, 1028.83 N, and 9.15%.

Targets defined in the start of optimization process (i.e., average thrust force of 10 kN with TFRR of less than 10%) are successfully achieved. Detailed comparisons of important KPIs based on numerical values and corresponding waveforms of initial HELFSM and geometrically optimized HELFSM is done in next section. All KPIs due to three excitation types, i.e., PMs, DC, and PM+DC are analysed separately for in depth investigation. Furthermore, initial and updated geometry based parameters are compared and presented for reproduction.

4.2. Comparison of Initial and Optimized HELFSM

No-load and on-load KPIs recorded under PMs, DC electromagnets, and combined excitations of initial and optimized HELFSM are compared in Figures 8–10, respectively. Numerical values of optimized structure design variables are depicted in Table 1, whereas detailed and quantitative electromagnetic performance comparison of initial and optimized HELFSM is illustrated in Table 3. Average thrust force and power versus variable armature current density plot of optimized HELFSM at fixed field excitation is presented in Figure 11. Furthermore, average thrust force and power versus velocity graph of optimized HELFSM is shown in Figure 12. Furthermore, efficiency of optimized HELFSM at eight different points considering core and copper losses is computed and presented.

4.2.1. No-Load Flux Linkage

Open-circuit flux linkage waveform comparison of the initial HELFSM and geometrically optimized HELFSM is presented in Figure 8a. All three types of excitation sources are separately analysed and combined in one plot to understand their true behaviour. In order to increase ease of understanding, only centre phase (C-Phase) flux linkage is presented. Frequency spectrum up to tenth order harmonic and THD of corresponding no-load flux waveform is also shown in Figure 8b. It can be seen that peak-to-peak values of geometrically optimized HELFSM are greater in magnitude, more sinusoidal, and more symmetrical. Magnitude of peak-to-peak no-load flux linkage is increased from 8.10 mWb to 10.90 mWb. Detailed analysis of PM+DC excited machine's frequency spectrum revealed that the dominant third and fifth order harmonics resulting in THD of 4.76% are effectively curtailed during optimization process and THD is reduced to 1.17%.

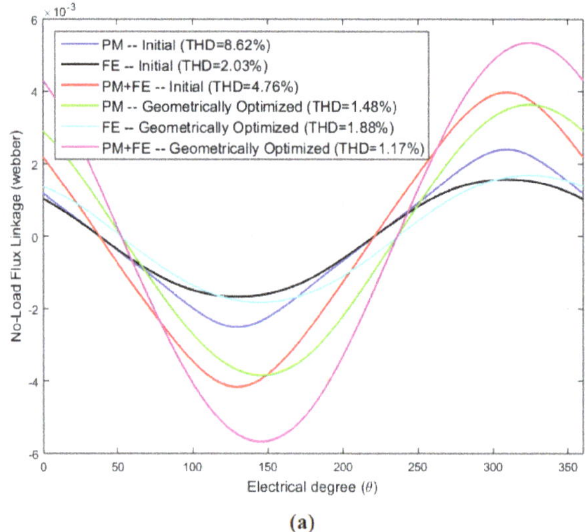

(a)

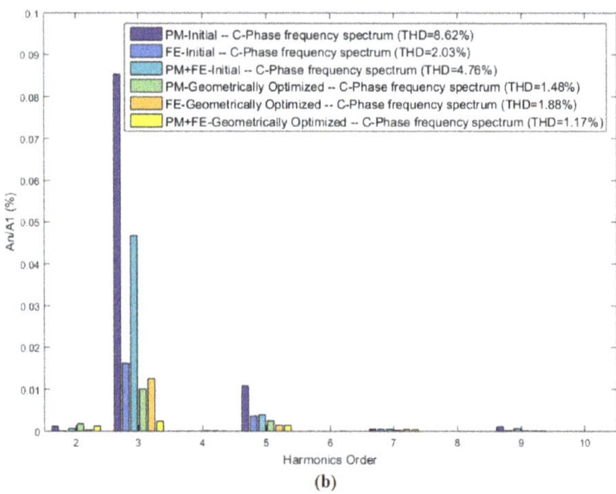

(b)

Figure 8. No-load flux linkage comparison: (**a**) waveforms and (**b**) frequency spectrum.

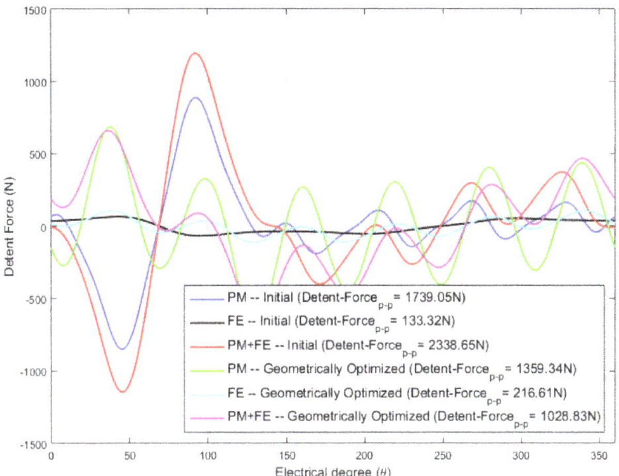

Figure 9. Detent force comparison.

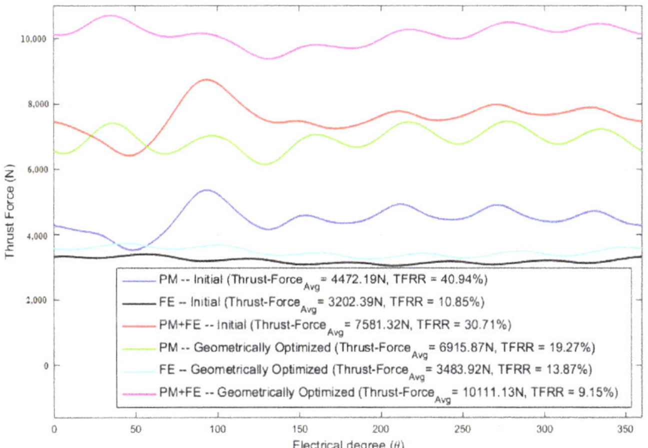

Figure 10. Thrust force comparison.

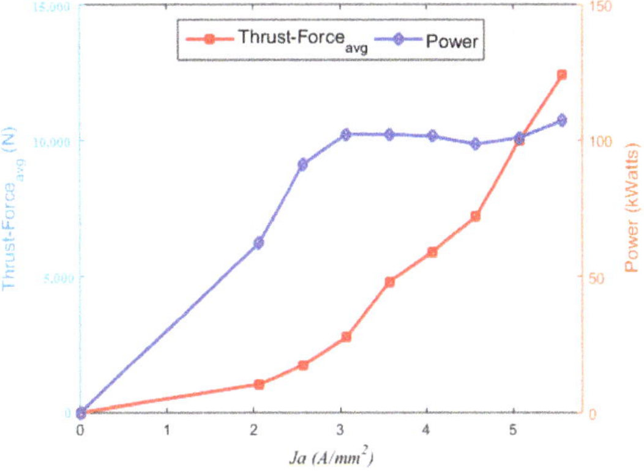

Figure 11. Average thrust force and power versus variable armature current density.

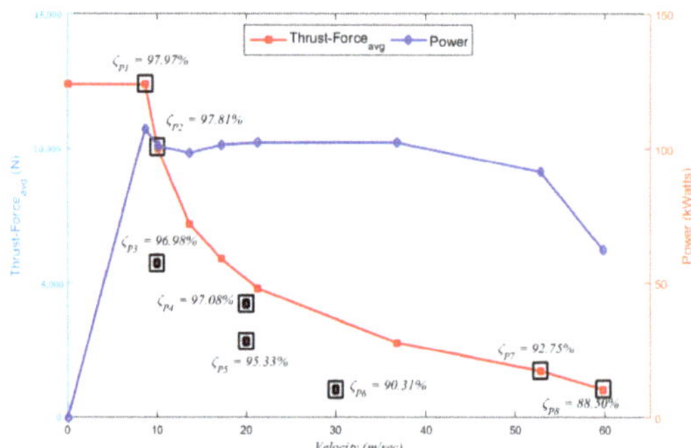

Figure 12. Average thrust force and power versus velocity.

4.2.2. Detent Force

Comparison of detent force profile under all three excitation sources combinations for initial HELFSM and geometrically optimized HELFSM is presented in Figure 9. As detent force is an undesired property, it causes ripples in the thrust force profile, and affects machine's positioning precision [20], hence its peak-to-peak value must be reduced after optimization process. While comparing initial and optimized machine configurations, peak-to-peak values of detent force due to only PMs excitation is reduced from 1739.05 N to 1359.34 N and due to PMs + DC is reduced from 2338.35 N to 1028.83 N. However, a slight increase in the peak-to-peak detent force values due to only DC excitation is observed, its value is increased from 133.32 N to 216.61 N.

4.2.3. Thrust Force and TFRR

Unidirectional thrust force profile of the initial HELFSM and geometrically optimized HELFSM under three different excitation sources methodology is compared in Figure 10. Average thrust force under PMs excitation is improved from 4472.19 N to 6915.87 N whereas its TFRR is reduced from 40.94% to 19.27%. Similarly, average thrust force under PMs + DC excitation is improved from 7581.32 N to 10,111.13 N whereas its TFRR is reduced from 30.71% to 9.15%. Average thrust force under DC excitation is improved from 3202.39 N to 3483.92 N whereas its TFRR is increased, i.e., from 10.85% to 13.87%.

Table 3. Optimization results and comparisons.

Performance Indicator (Unit)	Excitation	Initial Value	Optimized Value
Flux Linkage$_{p-p}$ (mWb)		4.88	7.41
THD (%)		8.62	1.48
Detent Force$_{p-p}$ (N)	PM	1739.05	1359.34
Thrust Force$_{avg}$ (N)		4472.19	6915.87
TFRR (%)		40.94	19.27

Table 3. *Cont.*

Performance Indicator (Unit)	Excitation	Initial Value	Optimized Value
Flux Linkage$_{p-p}$ (mWb)	FEC	3.20	3.45
THD (%)		2.03	1.88
Detent Force$_{p-p}$ (N)		133.32	216.61
Thrust Force$_{avg}$ (N)		3202.39	3483.92
TFRR (%)		10.85	13.87
Flux Linkage$_{p-p}$ (mWb)	PM+FEC	8.10	10.90
THD (%)		4.76	1.17
Detent Force$_{p-p}$ (N)		2338.65	1028.83
Thrust Force$_{avg}$ (N)		7581.32	10111.13
TFRR (%)		30.71	9.15

5. Experimental Validation

After recognition of geometric parameters delivering maximum average thrust force and lowest TFRR, a prototype proposed machine having stroke length of two meters is manufactured (as shown in Figure 13). Electrical steel (35H210) for mover and stator core, NdFeB (Neomax-35AH) for PMs, and SWG 18 copper conductor for windings Measured resistance and inductance of each AC phase is 0.7 ohm and 0.99 mH, respectively, whereas that of DC coil is 2.1 ohm and 7.5 mH, respectively.

Comparison of measured and theoretical results for centre phase no-load B-EMF and detent force are presented in Figures 14 and 15. Under no-load condition, HELFSM was driven by servo motor at the rated speed of 1500 mm/s resulting in a B-EMF frequency of 50 Hz. It can be seen that the results obtained by experiment show a good agreement with corresponding FE Analysis. At constant speed, average of detent force is equal to that of friction. Thus, by subtracting friction force from tested no-load force, detent force can be obtained.

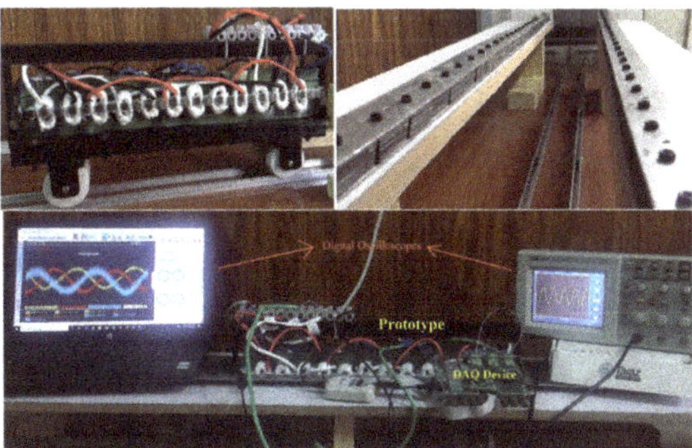

Figure 13. HELFSM prototype and test bed.

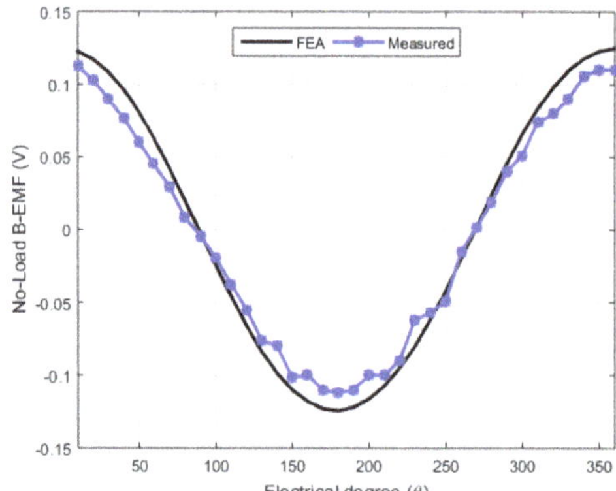

Figure 14. No-load induced B-EMF at 1.5 m/s.

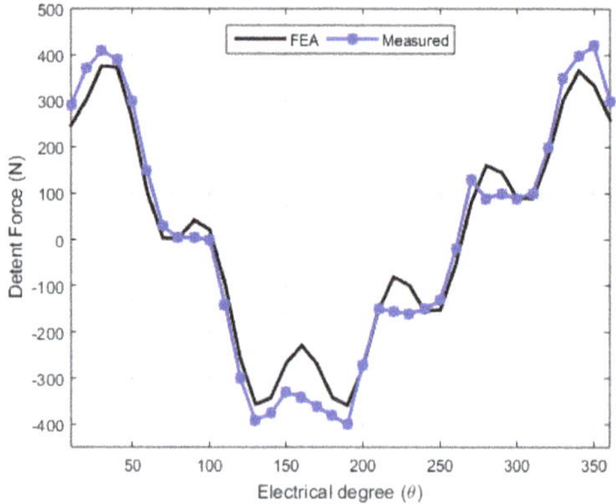

Figure 15. Detent force comparison.

6. Conclusions

In this paper, a novel LFSM combining the advantages of PMLFSM and FELFSM is realized by developing HELFSM for long stroke linear motion applications. The proposed machine shows the advantages of reduced PM volume and secondary material consumption, controllable air-gap magnetic flux density, high thrust force and power density, high efficiency, reduced thrust force ripple ratio, more symmetrical and sinusoidal flux linkages, and balanced magnetic circuit. Design guidelines, operation principle, and methodology of the proposed model are provided in detail for reproduction and further advancements. Segmented secondary design is involved to provide low reluctance short paths for flux linkage and also to reduce manufacturing cost by using less material. Advantages of complementary coil design and combination of series/parallel magnetic circuit such as more symmetrical and sinusoidal flux linkages and reduced TFRR are also incorporated in the proposed design. Unequal primary tooth width is another contribution of the paper that curtailed thrust force ripples by effective utilization of primary core area and it helped in provision of low reluctance path where high flux is recorded during initial tests. The SVGBDO approach is utilized to uplift thrust force performance and reduce TFRR of the proposed HELFSM. Optimization goals of 10 kN average thrust force with TFRR < 10% are successfully achieved. While comparing FEA and measured results, a maximum error of 0.02 V in B-EMF and 53 N in detent force was observed. Experimental results indicate that the proposed linear machine is suitable for direct drive long stroke applications.

Author Contributions: Conceptualization, N.U.; Software, N.U.; Investigation, N.U.; Validation, N.U.; Writing—Original Draft, M.S.; Writing—Review and Editing, M.S.; Supervision, F.K.; Project Administration, A.B.; Funding Acquisition, F.K. All authors have read and agreed to the published version of the manuscript.

Funding: This work was supported by Technology Development Fund-Higher Education Commission, Pakistan (Grant No. HEC-TDF 03-067).

Institutional Review Board Statement: Not applicable.

Informed Consent Statement: Not applicable.

Data Availability Statement: Data will be provided on demand.

Conflicts of Interest: The authors declare no conflict of interest.

References

1. Tang, Y.; Ilhan, E.; Paulides, J.; Lomonova, E. Design considerations of flux-switching machines with permanent magnet or DC excitation. In Proceedings of the 2013 15th European Conference on Power Electronics and Applications (EPE), Lille, France, 2–6 September 2013; pp. 1–10.
2. Barrero, F.; Duran, M.J. Recent advances in the design, modeling, and control of multiphase machines—Part I. *IEEE Trans. Ind. Electron.* **2015**, *63*, 449–458. [CrossRef]
3. Cao, R.; Mi, C.; Cheng, M. Quantitative comparison of flux-switching permanent-magnet motors with interior permanent magnet motor for EV, HEV, and PHEV applications. *IEEE Trans. Magn.* **2012**, *48*, 2374–2384. [CrossRef]
4. Thomas, A.; Zhu, Z.; Wu, L. Novel modular-rotor switched-flux permanent magnet machines. *IEEE Trans. Ind. Appl.* **2012**, *48*, 2249–2258. [CrossRef]
5. Li, W.; Chau, K.; Jiang, J. Application of linear magnetic gears for pseudo-direct-drive oceanic wave energy harvesting. *IEEE Trans. Magn.* **2011**, *47*, 2624–2627. [CrossRef]
6. Zeng, Z.; Lu, Q. A Novel Hybrid-Excitation Switched-Flux Linear Machine With Partitioned-Excitations. *IEEE Trans. Magn.* **2019**, *55*, 1–4. [CrossRef]
7. Xiao, F.; Du, Y.; Sun, Y.; Zhu, H.; Zhao, W.; Li, W.; Ching, T.; Qiu, C. A novel double-sided flux-switching permanent magnet linear motor. *J. Appl. Phys.* **2015**, *117*, 17B530. [CrossRef]
8. Hao, W.; Wang, Y.; Deng, Z. Study of two kinds of double-sided yokeless linear flux-switching permanent magnet machines. In Proceedings of the 2016 IEEE Vehicle Power and Propulsion Conference (VPPC), Hangzhou, China, 17–20 October 2016; pp. 1–6.
9. Cao, R.; Jin, Y.; Zhang, Z.; Cheng, M. A new double-sided linear flux-switching permanent magnet motor with yokeless mover for electromagnetic launch system. *IEEE Trans. Energy Convers.* **2018**, *34*, 680–690. [CrossRef]
10. Hao, W.; Wang, Y. Analysis of double-sided sandwiched linear flux-switching permanent-magnet machines with staggered stator teeth for urban rail transit. *IET Electr. Syst. Transp.* **2018**, *8*, 175–181. [CrossRef]
11. Cao, R.; Huang, W.; Cheng, M. A new modular and complementary double-sided linear flux-switching permanent magnet motor with yokeless secondary. In Proceedings of the 2014 17th International Conference on Electrical Machines and Systems (ICEMS), Hangzhou, China, 22–25 October 2014; pp. 3648–3652.
12. Liu, C.T.; Hwang, C.C.; Li, P.L.; Hung, S.S.; Wendling, P. Design optimization of a double-sided hybrid excited linear flux switching PM motor with low force ripple. *IEEE Trans. Magn.* **2014**, *50*, 1–4. [CrossRef]
13. Cao, R.; Jin, Y.; Zhang, Y.; Cheng, M. A new double-sided HTS flux-switching linear motor with series magnet circuit. *IEEE Trans. Appl. Supercond.* **2016**, *26*, 1–5. [CrossRef]
14. Petrov, I.; Ponomarev, P.; Alexandrova, Y.; Pyrhönen, J. Unequal teeth widths for torque ripple reduction in permanent magnet synchronous machines with fractional-slot non-overlapping windings. *IEEE Trans. Magn.* **2014**, *51*, 1–9. [CrossRef]
15. Chau, K.; Chan, C.C.; Liu, C. Overview of permanent-magnet brushless drives for electric and hybrid electric vehicles. *IEEE Trans. Ind. Electron.* **2008**, *55*, 2246–2257. [CrossRef]
16. Fan, H.; Chau, K.; Liu, C.; Cao, L.; Ching, T. Quantitative comparison of novel dual-PM linear motors for ropeless elevator system. *IEEE Trans. Magn.* **2018**, *54*, 1–6.
17. Ullah, N.; Khan, F.; Basit, A.; Ullah, W.; Haseeb, I. Analytical Airgap Field Model and Experimental Validation of Double Sided Hybrid Excited Linear Flux Switching Machine. *IEEE Access* **2021**, *9*, 117120–117131. [CrossRef]
18. Wang, Y.; Xu, W.; Zhang, X.; Ma, W. Harmonic analysis of air gap magnetic field in flux-modulation double-stator electrical-excitation synchronous machine. *IEEE Trans. Ind. Electron.* **2019**, *67*, 5302–5312. [CrossRef]
19. Shi, Y.; Ching, T.W.; Jian, L.; Li, W. A new dual-permanent-magnet-excited motor with hybrid stator configuration for direct-drive applications. In Proceedings of the 2019 22nd International Conference on Electrical Machines and Systems (ICEMS), Harbin, China, 11–14 August 2019; pp. 1–6.
20. Liu, Q.; Yu, H.; Hu, M.; Liu, C.; Zhang, J.; Huang, L.; Zhou, S. Cogging force reduction of double-sided linear flux-switching permanent magnet machine for direct drives. *IEEE Trans. Magn.* **2013**, *49*, 2275–2278. [CrossRef]

Article

The Choice of the Optimal Number of Discs in an MR Clutch from the Viewpoint of Different Criteria and Constraints

Krzysztof Kluszczyński [†] and Zbigniew Pilch [*,†]

Faculty of Electrical and Computer Engineering, Cracow University of Technology, Warszawska Str. 24, 31155 Cracow, Poland; krzysztof.kluszczynski@pk.edu.pl
* Correspondence: zbigniew.pilch@pk.edu.pl
† These authors contributed equally to this work.

Abstract: This paper focuses on magnetorheological clutches (MR clutches) with a disc structure that can be designed as one-disc or multi-disc clutches (number of discs: N = 1, N > 2). The main goal of the paper is to compare their overall dimensions (lengths and radii), masses, volumes, and characteristic factors—torque per mass ratio and torque per volume ratio for MR clutches that develop the same given clutching torque T_c but differ in the number of discs (it is assumed that the number of discs of the primary member varies from one to four). This analysis develops charts and guidelines that will allow designers to choose the appropriate number of discs from the view point of various criteria, and with various limitations regarding geometry, geometric proportions, mass, volume, or restrictions on the amount of active materials used in the manufacturing process. The limitations on the active materials used are of particular importance in the case of mass production. Our methodology uses a comparative study, which can also be used when comparing design solutions of other electromechanical converters.

Keywords: electromechanical convertor; drive system component; electromagnetic calculation; MR fluids; MR multi-disc clutch; clutch design

Citation: Kluszczyński K.; Pilch Z. The Choice of the Optimal Number of Discs in an MR Clutch from the Viewpoint of Different Criteria and Constraints. *Energies* **2021**, *14*, 6888. https://doi.org/10.3390/en14216888

Academic Editors: Ryszard Palka and Marcin Wardach

Received: 25 August 2021
Accepted: 13 October 2021
Published: 20 October 2021

1. Introduction

Magnetorheological (MR) fluids belong to the group of Smart Materials [1–3]. The feature of MR fluids that can be controlled with the help of an external magnetic field (generated by an excitation winding) is viscosity: the stronger the magnetic field, the greater the viscosity. Because of this property, MR fluids have found a number of applications in dampers [4] and brakes [1,5,6]. MR clutches are a very promising application, in which the coupling between two mechanical parts is brought about through magnetic means and not by mechanical contact. The structure of MR clutches can be divided into two basic varieties: a cylindrical structure or a disc type [7,8]. The latter can include one disc of the primary member (number of discs N = 1), or many discs (multi-disc construction with N ≥ 2) [9]. This paper focuses on a multi-disc MR clutch. The many possibilities of varying the geometrical proportions of MR clutches by selecting different numbers of discs appears to be a key advantage. They play a particularly important role in embedded drive systems applied in the automotive [10–13], aerospace, shipbuilding and robotics industries [14].

An example of a three-disc MR clutch is given in Figures 1 and 2: the main view with the description of consecutive constructional parts is presented in Figure 1 and an axis-symmetrical cross-section (due to symmetry, it is sufficient to consider only half of the cross-section) is presented in Figure 2.

The coil wound on a carcass is presented separately in Figure 3. It consists of z turns. The diameter of a bare conductor is denoted by d_{Cu} and the diameter of an insulated conductor by d_{Cu0}.

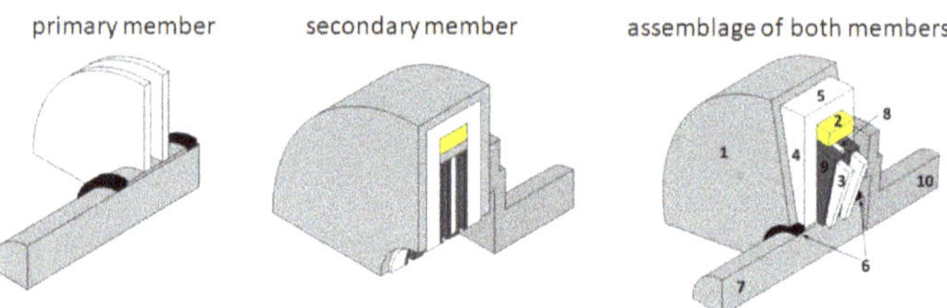

Figure 1. Main view of a two-disc (N = 2) MR clutch: 1—non-magnetic housing, 2—coil, 3—discs of primary member, 4—cover yoke, 5—cylinder yoke, 6—bearings, 7—non-magnetic shaft—primary member, 8—discs of secondary member, 9—MR fluid, 10—non-magnetic shaft—secondary member.

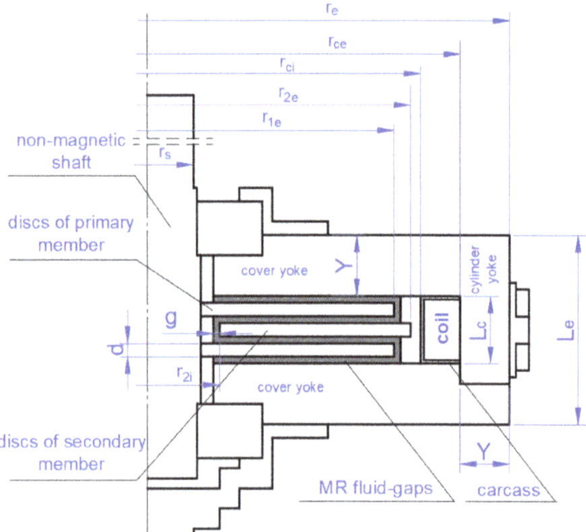

Figure 2. Axisymmetrical cross-section of a two-disc MR clutch (N = 2): L_e—external length of magnetic circuit, r_e—external radius of magnetic circuit, r_s—radius of non-magnetic shaft, r_{1e}—external radius of primary member discs, r_{2e}—external radius of secondary member discs, r_{2i}—inner radius of secondary member discs, d—thickness of primary and secondary member discs, g—thickness of MR fluid gap, L_c—external length of coil, r_{ce}—external radius of coil, r_{ci}—inner radius of coil, Y—thickness of cover and cylinder yoke.

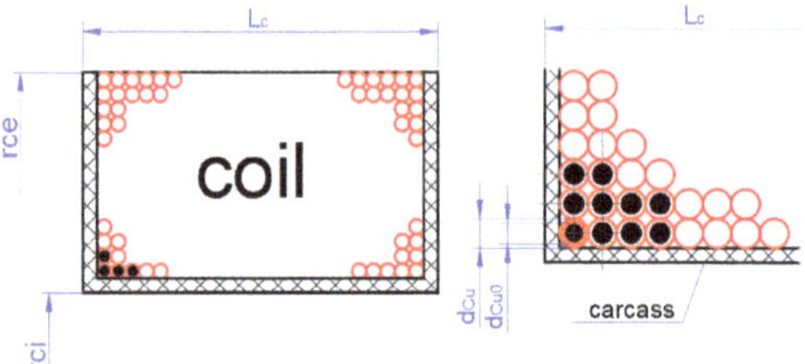

Figure 3. Dimensions of coil and conductors.

The value of the excitation current I feeding the coil is usually given; its value results from the properties of the supply system. The current density in the coil must be selected such that a satisfactory efficiency can be obtained without an excessive temperature increase. Assuming that the allowable current density for the given insulation class of wound wires is $j_{cu} = 4.5$ A/mm^2, one can calculate the values of the cross-sectional area S_{cu}, and the diameter of a bare conductor d_{cu}.

$$S_{cu} = \frac{I}{j_{cu}},\tag{1}$$

$$d_{cu} = \sqrt[2]{\frac{4 \cdot S_{cu}}{\pi}}.\tag{2}$$

The final nominal cross-sections and diameters of the bare conductors S_{cu0} and d_{cu0}, and the insulated conductors S_{cu} and d_{cu}, that are used to wind the coil can be found in the AWG International Standard Specification [15].

Note that the terminology used for the description of the MR clutch in Figures 1–3 is inspired by that used in the theory of transformers and electrical machines.

2. Materials and Methods

2.1. Assumptions that Must Be Fulfilled by the Compared Clutches with the Number of Discs N = 1,2,3,4

In the framework of a comparative study, we decided to compare clutches with a number of discs in the primary member varying from one to four (N = 1, 2, 3, 4) and developing a clutching torque equal to: $T_c = 20, 35, 50$ Nm. The comparison was made under the following assumptions:

- The thickness of the MR fluid gap g is assumed the same ($g = 1$ mm);
- Current density j_{Cu} is assumed the same (it is assumed that $j_{Cu} = 4.5$ A/mm^2 Equations (1) and (2));
- Excitation current I is assumed the same ($I = 0.6$ A);
- Allowable stress k_s is assumed the same $k_s = 75 \times 10^6$ Pa. Usually the designer introduces an additional safety factor k_{safe} by which the given value of the clutching torque is multiplied to determine the shaft radius r_s ensuring the appropriate level of safety;
- Safety factor for shaft k_{safe} is assumed the same $k_{safe} = 1.2$;
- The B-H curve for magnetic steel forming a ferromagnetic core is depicted in Figure 4a and the B-H curve for MR fluid MRF-140CG is depicted in Figure 4b [16];
- It is assumed that discs and yokes are made of the same magnetic steel;
- Magnetic flux density in MR fluid-gaps B_0 is kept the same in spite of variations in the geometries of clutches (the recommended value resulting from [12,17] is equal to $B_0 = 0.7$ T). More precisely, B_0 is the magnetic flux density in the middle of the MR fluid gap, and can be regarded as "its average value";
- Maximum magnetic field density B_{mx} is kept the same in spite of variations in the geometries of clutches. The most saturated point lies within the cover yoke at a length approximately equal to the external radius of the primary member discs [18,19] (we assumed a value of $B_{mx} = 1.2$ T located at the knee of the B-H curve for magnetic steel, which is remarkably less than the saturation point 1.6 T);

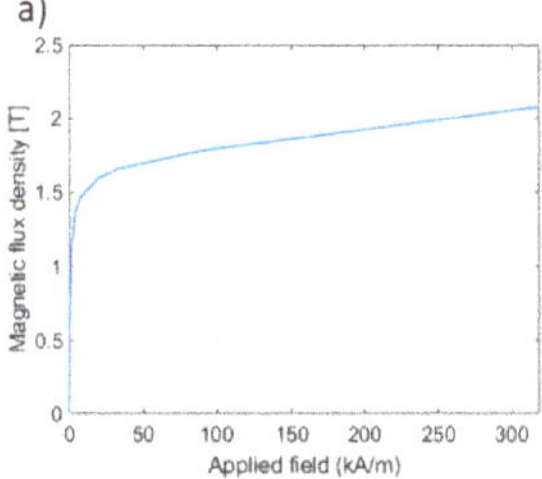
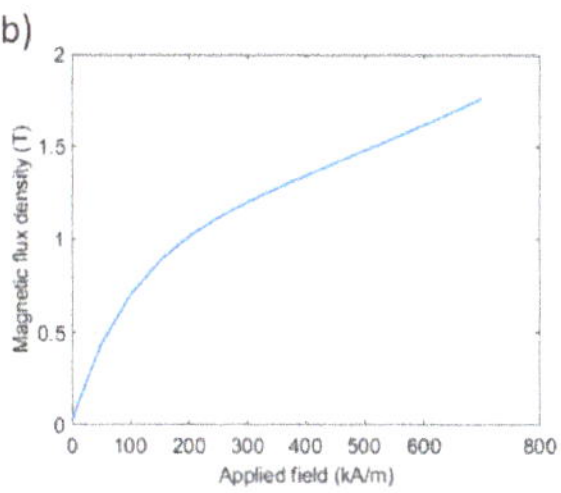

Figure 4. B-H curves for magnetic steel and MR fluid MRF-140CG. (**a**) B-H curves for ferromagnetic steel [20]. (**b**) B-H curves for MR fluid MRF-140CG [16].

2.2. Brief Description of the Applied Analytical-Field Design Method and Obtained Constructional Data for the Considered Variants

The design method that we used in the comparative study includes two stages: an analytical stage (composed of 36 algebraic formulas) and a field stage based on the finite element method (FEM). The analytical stage allows for a preliminary, quick determination of the approximate geometric dimensions and data of the excitation coil for the designed clutch, which develops the given torque T_c and meets all the assumptions listed in m. The final determination of the value of the clutching torque, the distribution of the magnetic field, the total magnetomotive force, the geometrical dimensions of the clutch and the data of the excitation coil are made in the field stage. From the point of view of the field stage, the analytical stage is a preliminary stage that allows us to determine a starting point close to the solution sought, which allows a significant reduction in the number of iterations and the computation time.

The clutching torque occurs in the space between the 2N overlapping surfaces of: N primary member discs, N − 1 secondary member discs and 2 internal surfaces of cover yokes facing the discs. In the analytical stage, the approximate value of the clutching torque is calculated according to the following formula:

$$T_c = 2N \cdot \int_{r_{2i}}^{r_{1e}} \int_0^{2\pi} dT = 2N \cdot \int_{r_{2i}}^{r_{1e}} \int_0^{2\pi} \tau_y(B_0) \cdot r^2 \cdot d\phi dr = \frac{4\pi}{3} \cdot N \cdot \tau_y(B_0) \cdot \left(r_{1e}^3 - r_{2i}^3\right), \tag{3}$$

where, $\tau_y(B)$ is a shear stress vs. magnetic flux density B curve, resulting from the Bingham model for plastics depicted in Figure 5 (in a clutch functioning in shear mode, the primary and secondary member discs do not move relative to each other, so $\dot{\gamma} = 0$). As seen, for the assumed value of magnetic flux density in the MR fluid-gaps $B_0 = 0.7$ T, shear stress is equal to: $\tau_y(B_0 = 0.7 \text{ T}) = 45.7$ kPa (Figure 5b).

In the field stage, the exact value of clutching torque is determined on the basis of the real spatial distribution of magnetic flux density resulting from the FEM model, according to the following formula (Figure 6):

$$T_c = \sum_{i=1}^{2N} T_{ci} = \sum_{i=1}^{2N} \int_{r_{2i}}^{r_{1e}} \int_0^{2\pi} \tau_y[B_i(r, \varphi)]dr d\varphi, \tag{4}$$

where i is the number of the considered surface, r is the current radius, $d\varphi$ is the angle increment and dr is the radius increment.

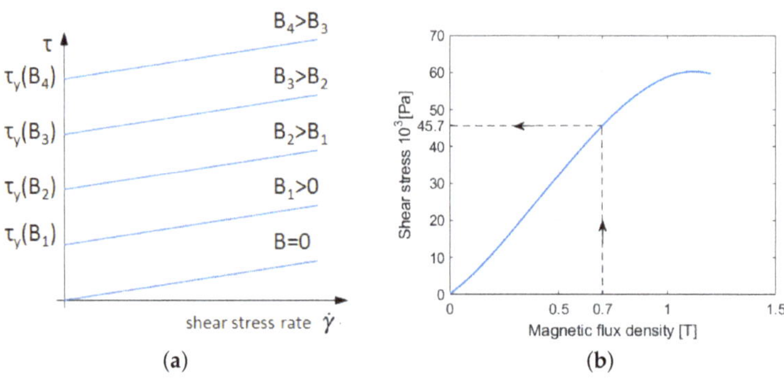

Figure 5. Graphical representation of Bingham model (**a**). Yield stress characteristics $\tau(B)$ for MRF-140CG [16] (**b**).

The FEM model is based on an open-access program Agros2D that was elaborated at the Pilzen University of Technology [20] (all of the scripts were written in PythonLAB language).

In the elaborated analytical-field method, the crucial role is played by the so-called yoke factor k_Y, the concept of which is based on the theory of induction machines. To define the yoke factor, it is necessary to introduce the term of the movement region. The movement region consists of the overlapping fragments of N primary member discs, N − 1 secondary member discs, 2N MR fluid-gaps and 2 internal surfaces of cover yokes facing the discs. The exact boundaries of the movement region on the example of a magnetorheological clutch with 2 primary member discs (2-disc MR clutch) are shown in Figure 7.

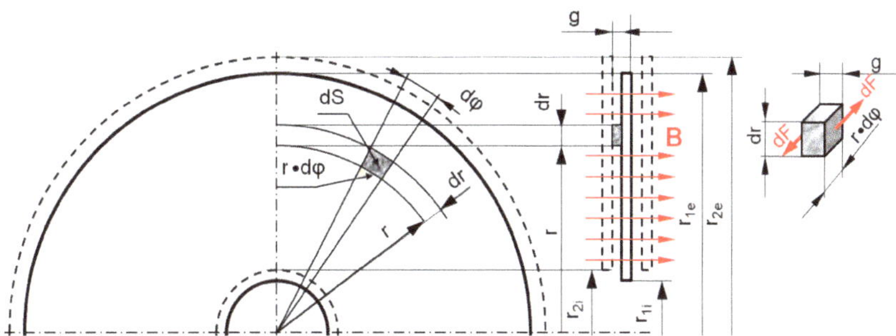

Figure 6. Geometrical denotations for calculating clutching torque acting on a single disc surface.

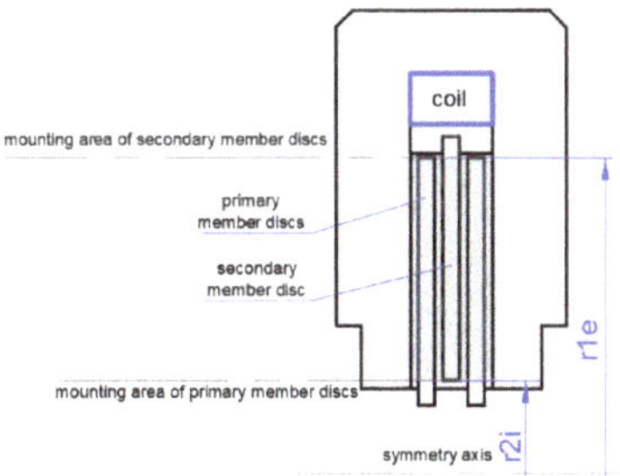

Figure 7. Exact boundaries of the movement region in the example of a magnetorheological clutch with 2 primary member discs (2-disc MR clutch).

The yoke factor determines the ratio of the total magnetomotive force required to magnetise the entire magnetic circuit of the clutch, $\Theta = \Theta_{mr} + \Theta_Y$ (where: Θ—the total magnetomotive force, Θ_{mr}—the magnetomotive force required to magnetise the movement region, Θ_Y—the magnetomotive force required to magnetise the yoke region composed of two covers and a cylinder) to the magnetomotive force Θ_{mr}:

$$k_Y = \frac{\Theta}{\Theta_{mr}} = \frac{\Theta_{mr} + \Theta_Y}{\Theta_{mr}}. \tag{5}$$

In the analytical stage, the approximate value of the total magnetomotive force is calculated according to the following formula:

$$\Theta = k_Y \cdot \Theta_{mr} = k_Y \cdot \left[\underbrace{2Ng \cdot \frac{B_0}{\mu_0 \mu_{MR}(B_0)}}_{\Theta_{mr}(MR)} + \underbrace{(2N-1)d \cdot \frac{B_0}{\mu_0 \mu_{Fe}(B_0)}}_{\Theta_{mr}(Fe)} \right], \tag{6}$$

where $\Theta_{mr(MR)}$ is the magnetomotive force associated with the MR fluid gaps, $\Theta_{mr(Fe)}$ is the magnetomotive force associated with the discs, $\mu_{MR}(B_0)$ is determined from the curve in Figure 4b, and $\mu_{Fe}(B_0)$ is determined from the curve in Figure 4a.

In the field stage, the exact value of the total magnetomotive force is determined on the basis of the real spatial distribution of magnetic flux density and the real value of the yoke factor k_Y determined using loop calculations. As shown in Figure 8, the applied analytical field-design method works according to the block diagram with iteration loop factor k_Y.

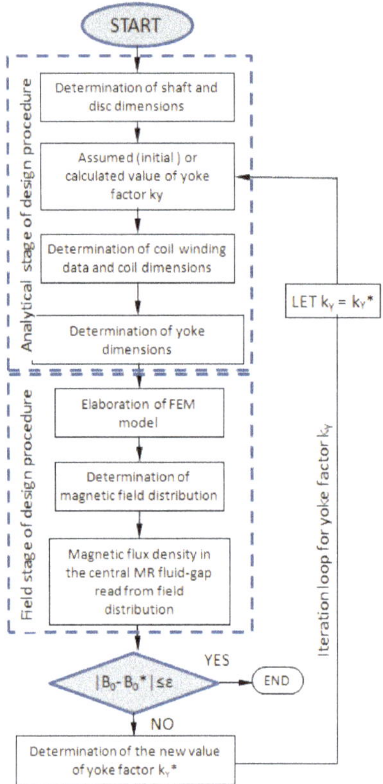

Figure 8. The applied analytical field-design method: block diagram.

The results of design calculations: magnetic field distributions and constructional data for the selected variants with the greatest considered values and the lowest considered values of the clutching torque and the number of discs:

- variant 1: $T_c = 20$ Nm, N = 1 (Figure 9a);
- variant 2: $T_c = 20$ Nm, N = 4 (Figure 9b);
- variant 3: $T_c = 50$ Nm, N = 1 (Figure 9c);
- variant 4: $T_c = 50$ Nm, N = 4 (Figure 9d);

are given in Figure 9 and in Table 1, respectively. In Figure 10 these variants are located in the corners of the table and are marked by green line contours.

Table 1 shows how the geometrical (constructional) parameters specified in Figure 2, as well as the total magnetomotive force Θ, the number of coil turns z and the yoke factor k_Y change for the four selected clutches with a different number of discs and different ratings, while maintaining the assumptions listed in Section 2.1.

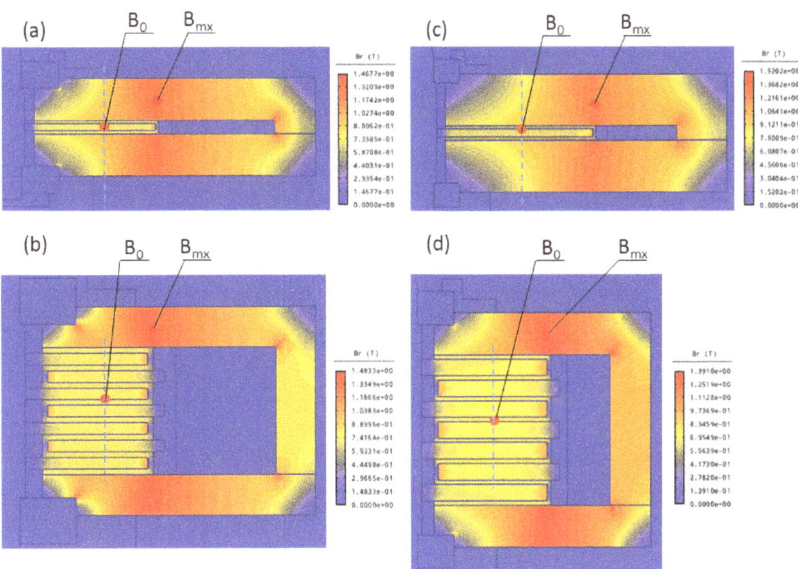

Figure 9. Magnetic field distribution for the selected variants: variant 1 (**a**), variant 2 (**b**), variant 3 (**c**), variant 4 (**d**).

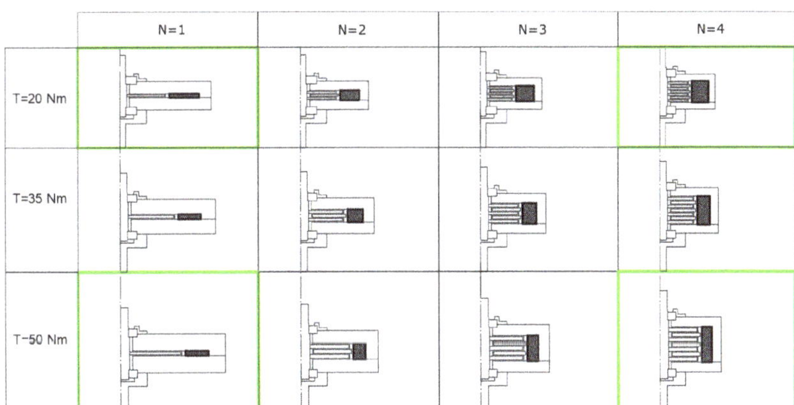

Figure 10. Graphical overview of cross-sections for the 12 designed variants: $T_c = 20$ Nm, N = 1,2,3,4, $T_c = 35$ Nm, N = 1,2,3,4 and $T_c = 50$ Nm, N = 1,2,3,4.

For a comparative study, which is the main goal and essence of the paper, we use our own analytical-field design method—the idea of which, was briefly described in this chapter. This method was verified experimentally on the example of the constructed 5-disc MR clutch presented in [21]. The detailed step-by-step description of this integrated method is presented in [22]. Someone who wishes to replicate the comparative study can also use his own design method (field, analytical or field-analytical) and his own method for choosing the most favourable starting point.

Table 1. The results of the design calculations for the selected variants.

		var. 1	var. 2	var. 3	var. 4
radius of shaft	r_s (mm)	5.9	6.1	8	8.1
thickness of discs	d (mm)	2	2	3	4
thickness of MR fluid gap	g (mm)	1	1	1	1
inner radius of secondary member discs	r_{2i} (mm)	11	12.5	13.5	13.5
external radius of primary member discs	r_{1e} (mm)	47	30.2	63.7	40.5
external radius of secondary member discs	r_{2e} (mm)	50	33.2	66.7	43.5
inner radius of coil	r_{ci} (mm)	52	35.2	68.7	45.5
external radius of coil	r_{ce} (mm)	83.5	52.9	93.1	56.1
external length of coil	L_c (mm)	4	22	5	36
thickness of cover and cylinder yoke	Y (mm)	12.7	7	17.5	10.2
number of coil turns	z (-)	364	1364	367	1363
yoke factor	k_Y (-)	1.259	1.176	1.270	1.176
total magnetomotive force	Θ(A)	173.6	347.3	520.9	694.5
external radius of magnetic circuit	r_e (mm)	96.2	59.9	110.6	66.3
external length of magnetic circuit	L_e (mm)	29.4	36	40	56.4

3. Results

3.1. Results of Comparative Study for the MR Clutches $T_c = 20, 35, 50$ Nm with Number of Discs $N = 1,2,3,4$

A comparative analysis of the external (overall) dimensions, masses and volumes, as well as the characteristic coefficients relating to the use of active materials and the space occupied, was carried out for MR clutches developing a clutching torque $T_c = 20, 35, 50$ Nm with the number of primary member discs varying from N = 1 to N = 4. A group of charts referring to total lengths L, external radius r_e, their masses m and volumes V [6], as well as the coefficients: T_c/m ratio and T_c/V ratio are presented in Figures 11a,b, 12a,b and 13a,b, respectively.

To select the correct number of discs, different aspects must be considered, the most important of which is the limitations imposed either on the outer dimensions or on the mass and volume specified by the user, based on the planned or expected applications, e.g., in automotive/motorcycle drive systems or disengagement auxiliaries in automotive/aviation/ship industries.

In the electromechanical industry, the values of torque per mass and torque per volume play an important role when comparing various possible solutions. The greater the developed torque, for the same mass or volume of an electric machine or electromagnetic device, the better the obtained solution is at utilising the full use of construction materials (or a better use of space occupied by the machine or device).

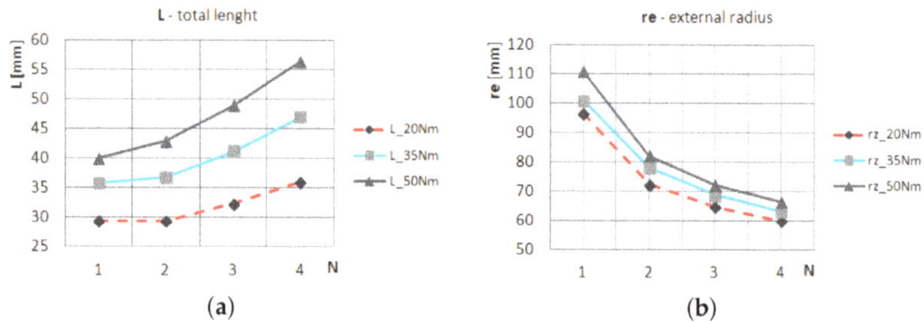

Figure 11. Total length vs. number of discs N (**a**) and external radius vs. number of discs N (**b**) for MR clutches $T_c = 20, 35, 50$ Nm.

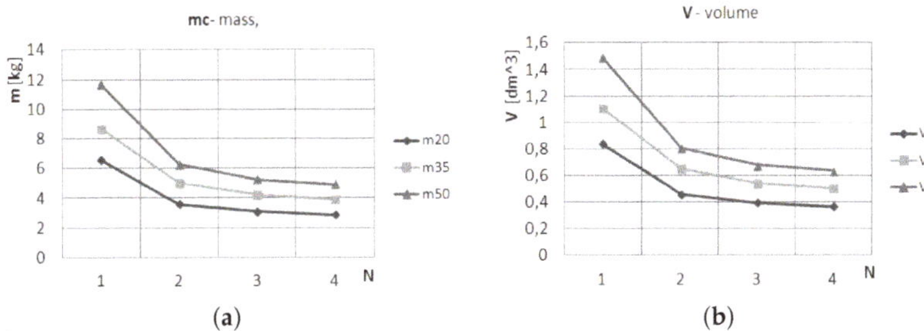

Figure 12. Mass vs. number of discs N (**a**) and volume vs. number of discs N (**b**) for MR clutches $T_c = 20, 35, 50$ Nm.

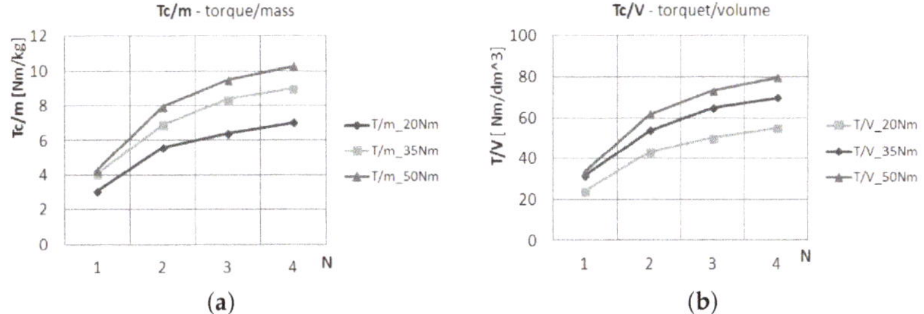

Figure 13. T_c/m ratio vs. number of discs N (**a**) and T_c/V ratio vs. number of discs N (**b**) for MR clutches $T_c = 20, 35, 50$ Nm.

As can be seen in Figure 11a, increasing the number of discs leads to an increase in the length of the clutch (an increase in length is more significant when the torque developed by the clutch is greater) and a decrease in the external radius of the clutch Figure 11b.

It should be noted that the clutches $T_c = 20$ Nm with N = 1 and N = 2 have the same total length L (Figure 11a), which results from the fact that the clutch with N = 2 has a much thinner cover. For the abovementioned clutches, the increase in length of the movement region is compensated for by a decrease in the thickness of the yoke covers: $Y_{N=1} = 12.7$ mm, $Y_{N=2} = 9.7$ mm. For the clutches $T_c = 20$ Nm with N = 3 and N = 4, the increase in total length L (in relation to the length of clutch N = 1) is 10% and 22.5%, respectively, while the decrease in the cover thickness is less pronounced $Y_{N=3} = 8.2$ mm, $Y_{N=4} = 7$ mm.

As seen in Figure 12a,b, the increase in the number of discs leads to a distinctly visible reduction in the masses and volumes of the MR clutches. The selected number of discs is typically a compromise between the mass or volume of the clutch, which decreases with the number of discs and the technological complexity or production costs that increase with the number of discs. As seen in Figure 12, an asymptotic trend of the charts occurs, which means that it does not make sense to consider variants with $N > 4$ discs, due to the rapidly increasing complexity of the design, as well as increasing manufacturing costs.

Regarding the ratios: T_c/m and T_c/V (Figure 13a,b), a clearly visible increase in their value is noticeable when changing the number of discs from N = 1 to N = 2. The course of the charts for clutches with N > 2 also becomes increasingly flat. From the point of view of the above ratios, the recommended number of primary discs is N = 2.

The graphs shown in Figures 11a,b or 12a,b are useful for determining the number of discs where total length L, external radius r_e, mass m or volume V are limited. Graphical overview of cross sections for the 12 designed variants: $T_c = 20$ Nm, N = 1,2,3,4, $T_c = 35$ Nm, N = 1,2,3,4 and $T_c = 50$ Nm, N = 1,2,3,4 are given in Figure 10.

3.2. Detailed Analysis of the Case: $T_c = 20$ Nm and Physical Explanation of the Obtained Results

This chapter is devoted to a detailed analysis of the impact of the number of primary member discs (N = 1, 2, 3, 4) on selected design data on the example of an MR clutch developing torque $T_c = 20$ Nm.

The consecutive Figures 14–16 show total lengths L (Figure 14) and radii r_e, masses m and volumes V (Figure 15), ratios T_c/m and T_c/V (Figure 16) together in a combined figure, which makes it possible to evaluate the rate of change of important constructional design parameters when increasing the number of discs.

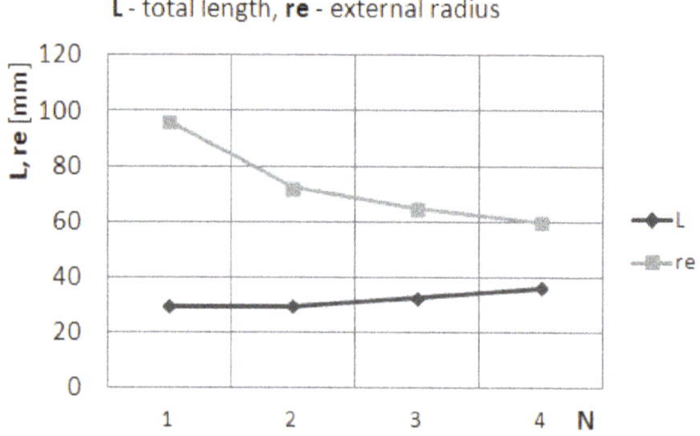

Figure 14. Total lengths L and external radius r_e vs. number of discs N.

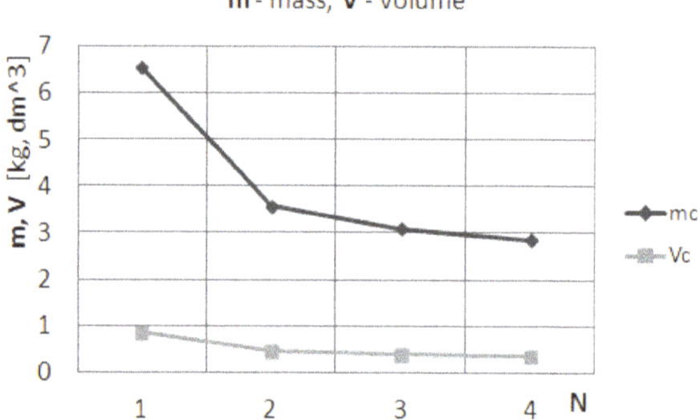

Figure 15. Mass m and volume V vs. number of discs N.

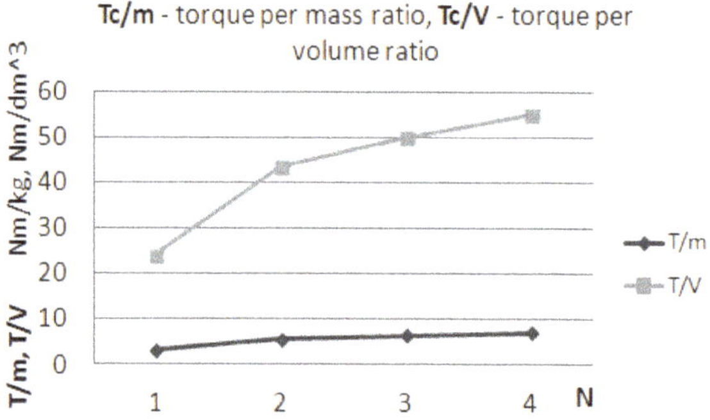

Figure 16. Torque per mass ratio T_c/m and torque per volume ratio T_c/V vs. number of discs N.

An important aspect that must be taken into account in the mass production of electrical machines and electromagnetic devices with magnetorheological (MR) fluid is the total mass of copper (Cu), the total mass of ferromagnetic steel (Fe) and the total mass of MR fluid, due to the constantly changing market prices of construction materials. The design that is accepted and approved for production is always a compromise between the best possible technical supplies (operating data) and the lowest production costs. The charts presented in Figure 17 will be important in reducing the total cost of the above-mentioned active materials (magnetic steel, copper and MR fluid) used in the mass production of clutches.

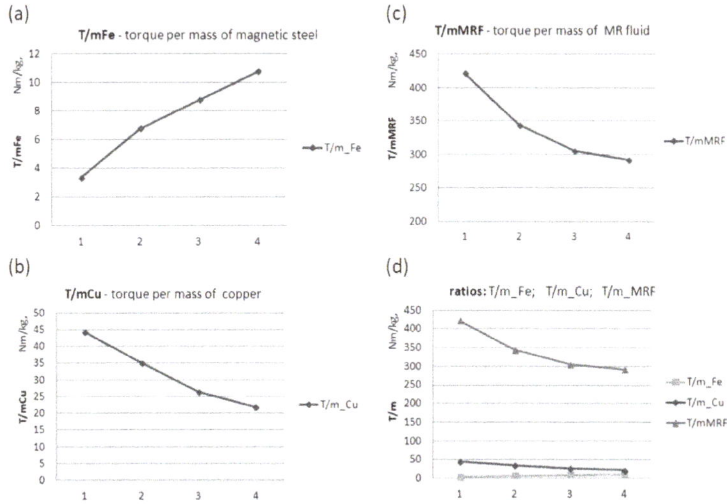

Figure 17. Torque per magnetic steel mass (**a**), torque per copper mass (**b**), torque per MR fluid mass (**c**) and the complete set of charts (**d**).

Looking at Figure 10, we can easily explain significant decreases in the clutch mass and significant increases in torque per magnetic steel mass (especially when considering the variant N = 2 in reference to variant N = 1). The reason for that is the remarkable drop in the thickness of cover and cylinder yokes (Figure 18).

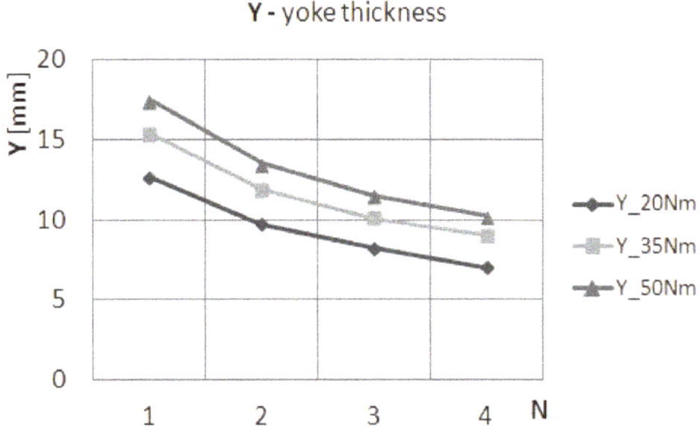

Figure 18. Yoke thickness vs. number of discs N for the clutch $T_c = 20$ Nm.

4. Discussion

This paper indicates the possibility of a different number of discs (from N = 1 to N = 4) in MR clutches and shows that each number of discs is associated with different properties, with respect to the overall dimensions, geometrical proportions, masses, volumes, clutching torque per mass ratio T_c/m and clutching torque per volume ratio T_c/V.

These numerous possibilities of varying the geometrical proportions of clutches and increasing the values of the abovementioned ratios by selecting different numbers of discs appears to be a particularly promising advantage. This is especially important in embedded drive systems applied in the automotive, aerospace, shipbuilding and robotics industries.

The results of this study are a set of charts and guidelines that allow designers to choose the appropriate number of discs from the viewpoint of various criteria and various constraints concerning the geometry, mass and costs of active materials.

Where there are no limitations regarding geometry and mass of a designed clutch or restrictions on the amount of active materials used in its manufacturing process, the optimal (recommended) number of discs is equal to N = 2.

In the case of devices with MR fluids, there is also the problem of possible degradation of the fluid properties during prolonged or improper use. Regarding the MR clutch, this may refer to excessively long running time of the clutch in case of slip. Determining the value of this unacceptable working time, and the value of the permissible slip, requires supplementing the developed clutch model with thermal calculations focused on the analysis of the temperature field distribution [23].

Author Contributions: Conceptualization, K.K. and Z.P.; methodology, K.K.; software, Z.P.; validation, Z.P., K.K.; formal analysis, Z.P., K.K.; writing—original draft preparation, Z.P.; writing—review and editing, Z.P.; visualization, Z.P.; supervision, K.K. All authors have read and agreed to the published version of the manuscript.

Funding: This research was conducted at the Faculty of Electrical and Computer Engineering, Cracow University of Technology.

Institutional Review Board Statement: Not applicable.

Informed Consent Statement: Not applicable.

Data Availability Statement: MDPI Research Data Policies.

Conflicts of Interest: The authors declare no conflict of interest. The funders had no role in the design of the study; in the collection, analyses, or interpretation of data; in the writing of the manuscript, or in the decision to publish the results.

Sample Availability: Samples of the compounds are available from the authors.

Abbreviations

The following abbreviations are used in this manuscript:

MR clutches   Magnetorheological clutches

References

1. Chen, J.Z.; Liao, W.H. Design, testing and control of a magnetorheological actuator for assistive knee braces. *Smart Mater. Struct.* **2010**, *19*, 035029. [CrossRef]
2. Kciuk, M.; Chwastek, K.; Kluszczyński, K.; Szczygłowski, J. A study on hysteresis behaviour of SMA linear actuators based on unipolar sigmoid and hyperbolic tangent functions. *Sens. Actuators A Phys.* **2016**, *243*, 52–58. [CrossRef]
3. Kim, K.J.; Lee, C.W.; Koo, J.H. Design and modelling of semi-active squeeze film dampers using magneto-rheological fluids. *Smart Mater. Struct.* **2008**, *17*, 035006. [CrossRef]
4. Martynowicz, P. Study of Vibration Control Using Laboratory Test Rig of Wind Turbine Tower-Nacelle System with MR Damper Based Tuned Vibration Absorber. *Bull. Pol. Acad. Sci. Tech. Sci.* **2016**, *64*, 347–359. [CrossRef]
5. Rossa, C.; Jaegy, A.; Lozada, J.; Micaelli, A. Design Considerations for Magnetorheological Brakes. *Trans. Mechatronics* **2014**, *19*, 1669–1680. [CrossRef]

6.   Nguyen, Q.H.; Choi, S.B. Selection of magnetorheological brake types via optimal design considering maximum torque and constrained volume. *Smart Mater. Struct.* **2011**, *21*, 015012. [CrossRef]

7.   Li, W.H.; Du, H. Design and Experimental Evaluation of a Magnetorheological Brake. *Int. J. Adv. Manuf. Technol.* **2003**, *21*, 508–515. [CrossRef]

8.   Rizzo, R.; Musolino, A.; Bucchi, F.; Forte, P.; Frendo, F. Magnetic FEM Design and Experimental Validation of an Innovative Fail-Safe Magnetorheological Clutch Excited by Permanent Magnets. *IEEE Trans. Energy Convers.* **2014**, *29*, 628–640 [CrossRef]

9.   Karakoc, K.; Park, E.J.; Suleman, A. Design considerations for an automotive magnetorheological brake. *Mechatronics* **2008**, *18*, 434–447. [CrossRef]

10.   East, W.; Turcotte, J.; Plante, J.-S.; Julio, G. Experimental assessment of a linear actuator driven by magnetorheological clutches for automotive active suspensions. *J. Intell. Mater. Syst. Struct.* **2021**, *32*, 955–970. [CrossRef] [PubMed]

11.   Park, E.J.; Falcao, L.; Suleman, A. Multidisciplinary design optimization of an automotive magnetorheological brake design. *Comput. Struct.* **2008**, *28*, 207–216. [CrossRef]

12.   Sohn, J.W.; Jeon, J.; Nguyen, Q.H.; Seung-Bok, C. Optimal design of disc-type magneto-rheological brake for mid-sized motorcycle: Experimental evaluation. *Smart Mater. Struct.* **2015**, *24*, 85–109. [CrossRef]

13.   Horváth, B.; Szalai, I. Nonlinear Magnetic Properties of Magnetic Fluids for Automotive Applications. *Hung. J. Ind. Chem.* **2020**. *48*, 61–65. [CrossRef]

14.   Chen, G.; Lou, Y.; Shang, T. Mathematic Modeling and Optimal Design of a Magneto-Rheological Clutch for the Compliant Actuator in Physical Robot Interactions. *IEEE Robot. Autom. Lett.* **2019**, *4*, 3625–3632. [CrossRef]

15.   Standard Specification for Standard Nominal Diameters and Cross-Sectional Areas of AWG Sizes of Solid Round Wires Used as Electrical Conductors, ASTM International, 2014. Available online: http://www.astm.org/Standards/B258.htm (accessed on 1 August 2021).

16.   Lord Corporation Product Bulletins. Available online: https://lordfulfillment.com/pdf/44/DS7012_MRF-140CGMRFluid.pdf (accessed on 1 August 2021).

17.   Nguyen, Q.H.; Choi, S.B. Optimal design of an automotive magnetorheological brake considering geometric dimensions and zero-field friction heat. *Smart Mater. Struct.* **2010**, *19*, 11. [CrossRef]

18.   Kluszczyński, K.; Pilch, Z. MR multi disc clutches—Construction, parameters and field model. In Proceedings of the 20th International Conference on Research and Education in Mechatronics (REM), Wels, Austria, 23–24 May 2019. [CrossRef]

19.   Kluszczyński, K.; Pilch, Z. Basic features of MR clutches-resulting from different number of discs. In Proceedings of the 2019 15th Selected Issues of Electrical Engineering and Electronics (WZEE), Zakopane, Poland, 8–10 December 2019. [CrossRef]

20.   Available online: http://www.agros2d.org/ (accessed on August 2021).

21.   Kowol, P.; Pilch, Z. Analysis of the magnetorheological clutch working at full slip state. *Electr. Rev.* **2015**, *91*, 108–111. [CrossRef]

22.   Kluszczyński, K.; Pilch, Z. Integrated analytical-field design method of multi-disc magnetorheological clutches for automotive applications. *Bull. Pol. Acad. Sci. Tech. Sci.* **2021**, in print.

23.   Pilch, Z. Analysis of Established Thermal Conditions for Magnetorheological Clutch for Different Loading Conditions. In *Analysis and Simulation of Electrical and Computer Systems*; Springer International Publishing: Berlin/Heidelberg, Germany, 2015; pp. 197–213. [CrossRef]

Article

Design and Optimization of a High-Speed Switched Reluctance Motor

Stefan Kocan [1,*], Pavol Rafajdus [1], Ronald Bastovansky [2], Richard Lenhard [3] and Michal Stano [1]

[1] Department of Power Systems and Electric Drives, Faculty of Electrical Engineering and Information Technology, University of Zilina, 010 26 Žilina, Slovakia; pavol.rafajdus@feit.uniza.sk (P.R.); michal.stano@feit.uniza.sk (M.S.)
[2] Department of Design and Machine Elements, Faculty of Mechanical Engineering, University of Zilina, 010 26 Žilina, Slovakia; bastovansky@uniza.sk
[3] Department of Power Engineering, Faculty of Mechanical Engineering, University of Zilina, 010 26 Žilina, Slovakia; lenhard@uniza.sk
* Correspondence: stefan.kocan@feit.uniza.sk

Abstract: Currently, one of the most used motor types for high-speed applications is the permanent-magnet synchronous motor. However, this type of machine has high costs and rare earth elements are needed for its production. For these reasons, permanent-magnet-free alternatives are being sought. An overview of high-speed electrical machines has shown that the switched reluctance motor is a possible alternative. This paper deals with design and optimization of this motor, which should achieve the same output power as the existing high-speed permanent-magnet synchronous motor while maintaining the same motor volume. The paper presents the initial design of the motor and the procedure for analyses performed using analytical and finite element methods. During the electromagnetic analysis, the influence of motor geometric parameters on parameters such as: maximum current, average torque, torque ripple, output power, and losses was analyzed. The analysis of windage losses was performed by analytical calculation. Based on the results, it was necessary to create a cylindrical rotor shape. The rotor modification method was chosen based on mechanical analysis. Using thermal analysis, the design was modified to meet thermal limits. The result of the work was a design that met all requirements and limits.

Keywords: high-speed motor; switched reluctance motor; finite element method; electromagnetic analysis; mechanical analysis; thermal analysis

Citation: Kocan, S.; Rafajdus, P.; Bastovansky, R.; Lenhard, R.; Stano, M. Design and Optimization of a High-Speed Switched Reluctance Motor. *Energies* **2021**, *14*, 6733. https://doi.org/10.3390/en14206733

Academic Editors: Sérgio Cruz and Ryszard Palka

Received: 18 August 2021
Accepted: 12 October 2021
Published: 16 October 2021

Publisher's Note: MDPI stays neutral with regard to jurisdictional claims in published maps and institutional affiliations.

1. Introduction

High-speed machines are used in an increasing number of applications, such as compressors, turbochargers, spindles, flywheel energy-storage systems, turbomolecular pumps, and microturbine generators. These machines are defined by rotational speed and frequency, but also the peripheral speed of the rotor, because this is dominant limitation in the design of these machines [1].

Higher rotational speed brings many advantages, such as reducing the volume of the machine while maintaining the same output power. The machine can be smaller and lighter, which is especially desirable in automotive applications in which any weight reduction leads to reduced fuel consumption and emissions. Another advantage in some applications is the elimination of the gearbox between the drive and the driven equipment, which leads to reduced costs and increased reliability [2].

The choice of the electrical machine type for high-speed application depends mainly on the specific application. In the past, universal motors were used. The advantage of these machines was especially their simple speed control, but the disadvantages were low efficiency and short commutator lifetime, which were associated with high maintenance costs [3]. Because of these disadvantages, they are currently being replaced by other types of electrical machines, including:

Energies **2021**, *14*, 6733. https://doi.org/10.3390/en14206733   https://www.mdpi.com/journal/energies

- Induction machines (IMs);
- Permanent-magnet synchronous machines (PMSMs); and
- Reluctance machines.

IMs with a laminated rotor are used in common industrial applications. For high-speed applications, their usage is significantly limited due to the high centrifugal forces acting on the squirrel cage. According to [4], it is a problem for them to work above 50,000 rpm. From a mechanical point of view, IMs with a solid rotor are more suitable for high-speed applications. According to the overview in [1], up to 180,000 rpm can be achieved with this type of IM. The disadvantages of these machines compared to PMSMs are a lower efficiency and higher rotor heating, which are caused by losses in the rotor winding [5].

Another type of electrical machine that has been the focus of many researchers is the PMSM. The advantages of this type of machine are especially high efficiency and smaller volume compared to IMs. The disadvantages are increased costs and the risk of PM demagnetization due to high temperatures and strong magnetic fields. In addition, these machines have high rotor inertia, which leads to higher bearing load and shortened bearing lifetime [6]. PMs can be located on the rotor surface (SPM) or inside (interior) the rotor (IPM). In the case of SPMs, PMs are fixed with a high-strength adhesive, and a sleeve is often required [7]. A comparison of these two topologies presented in [8] showed that from an electromagnetic point of view, higher iron losses occurred in the IPM topology. A disadvantage of IPM topology compared to the SPM is also higher susceptibility to mechanical damage. However, the cost of IPM topology is lower due to the smaller number of PMs needed to achieve the same output power. According to the overview in [1], the SPM rotor is used more frequently for high-speed machines. According to this article, PMSMs are preferred in applications with a higher speed and lower output power, and vice versa for IMs.

The rising price of rare earth elements has resulted in the search for an alternative to PMSMs without significantly affecting performance. Alternatively, reluctance machines can be used, including the switched reluctance motor (SRM) and reluctance synchronous motor (RSM). For high-speed applications, the SRM is mainly used. From the overview given in [9], the highest speed for which an RSM has been designed is 54,000 rpm.

An SRM has a simple and robust construction, formed by salient poles on the stator and rotor. The concentrated winding is wound around the stator poles. The individual winding coils can be prepared in advance and then slid onto the individual stator poles, which simplifies production and reduces costs. This winding also allows better heat dissipation, and since the end winding can be shorter compared to those in IMs, there is a decrease in winding losses [10]. The individual coils are connected in series or in parallel, and form phases that are gradually excited by current pulses. This winding increases fault tolerance, which puts this type of electrical machine above competing machines in terms of reliability, because this machine can operate even if one of the phases fails [11]. The rotor is without winding or PMs, which reduces losses, costs, and rotor heating. The machine has less strict thermal limits because there is no risk of PM demagnetization. Thermal limits are determined only by the winding insulation class and the thermal limit of sheets [12]. The rotor does not contain other components that would complicate mechanical or electromagnetic design, as in the case of PMSMs or IMs, which also reduces rotor inertia.

The absence of a rotor magnetic field reduces the power density, therefore an SRM needs a smaller air gap compared to a PMSM. Another disadvantage is high torque ripple and associated noise and vibration, which can be problematic in some applications; e.g., turbomolecular pumps or passenger vehicles. In addition, this machine has high windage losses, due to the salient poles and small air gap [13].

This paper deals with the design and optimization of a high-speed switched reluctance motor. The next section presents the state of design before this analysis. The third section presents the procedure of performed analysis. The fourth section presents the results of the performed analyses. The fifth section presents the final design of motor, and finally the results are discussed.

2. Basic Design of the SRM

The SRM design was based on the existing high-speed PMSM used in a turbocharger. This PMSM achieved an output power of 8 kW at speed of 100,000 rpm. The same application was considered during the design, and the main goal was to achieve output power of the PMSM while maintaining the same volume, rotation speed, and supply voltage. In addition, the motor had to be designed to meet mechanical and thermal requirements. The initial design of a high-speed SRM has already been presented in [14]. To obtain dynamic parameters, the model compiled in MATLAB—Simulink presented in [15] was used.

In addition to the initial design, several analyses of motor parameters were performed in Ansys. Based on the analysis given in [16], the number of turns was set to three, because this presented the highest average torque. A higher number of turns caused a high back EMF, which caused a decrease in the maximum phase current and average torque. The consequence of this change was a greater torque ripple and winding losses.

The analysis of the air gap size presented in [17] was performed. The analysis showed that by reducing the air gap below 0.22 mm, it was possible to increase average torque, but this increase was small compared to the increase in torque ripple. By increasing the air gap, the exact opposite could be achieved, but another disadvantage was the increase in phase current and winding losses. However, a larger air gap facilitates the manufacturing process. Based on this analysis, the size of the air gap was not selected because the calculation of windage losses was performed, which considered this parameter.

The analysis of rotor geometry given in [18] showed that average torque depends mainly on the length and width of rotor poles. The rotor diameter affects average torque only minimally, but with a larger diameter, it is possible to use longer poles. However, by reducing rotor diameter, it is possible to achieve a significant reduction in torque ripple, but there is a rise in iron losses. Hysteresis control has so far been used to control the maximum phase current. In this analysis, a sampling and switching frequency of 100,000 kHz was selected, based on which the control method was changed to single-pulse operation. The pole geometry was adjusted to achieve the highest average torque and limit the maximum phase current to the desired value. Different materials for the stator and rotor were also considered in the initial design: material NO10 for the stator, and M235-35A for the rotor. During this analysis, the sheet material was changed to 10JNEX900, reducing iron losses by 180%. These sheets were used for the stator and rotor.

The analysis resulted in several designs for different rotor diameters, and all designs had a maximum phase current limited to 200 A. A summary of the geometrical dimensions of the motor at which the highest average torque was reached is given in Table 1, and the electromagnetic parameters in Table 2. To better understand the meaning of the individual geometric dimensions in Table 1, they are additionally shown in Figure 1.

Table 1. Summary of the mechanical parameters of the basic high-speed SRM design.

Parameter	Symbol	Value
Number of stator poles	N_s	6
Number of rotor poles	N_r	4
Outer stator diameter	d_s	90 mm
Outer rotor diameter	d_r	38 mm
Length of stator poles	l_{ps}	20.18 mm
Length of rotor poles	l_{pr}	2.52 mm
Stator pole arc	β_s	30°
Rotor pole arc	β_r	34°
Yoke thickness of stator	h_{ys}	6.6 mm
Yoke thickness of rotor	h_{yr}	6.95 mm

Table 2. Summary of the electrical parameters of the basic high-speed SRM design.

Parameter	Symbol	Value
Number of phases	m	3
Number of turns	N	3
Maximum phase current	I_{max}	200 A
Supply voltage	V_{DC}	48 V
Phase resistance at 75 °C	R	0.85 mΩ
Average torque	T_{av}	0.538 Nm
Torque ripple	ΔT	235.57%
Winding losses	ΔP_j	21,47 W
Iron losses	ΔP_{Fe}	177.23 W
Windage losses	ΔP_w	2.17 kW
Bearing losses (50% Co)	ΔP_b	196.3 W
Addition losses	ΔP_{ad}	15.3 W
Output power	P_2	3.05 kW
Efficiency	η	54.2%

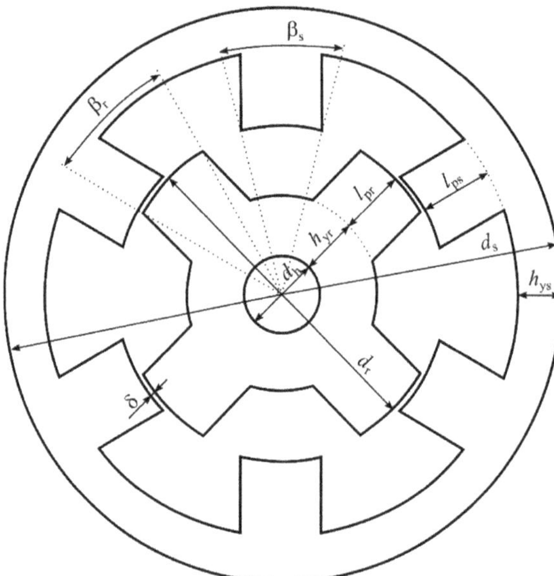

Figure 1. Definitions of the individual geometric dimensions. The displayed geometry is for illustration only; i.e., the values of individual dimensions do not correspond to the values in the table.

An important parameter not listed in Table 1 was the initial air gap size of 0.22 mm. Another important parameter was the active length of machine, with a value of 55 mm, which together with stator outer diameter, defined the motor volume.

Table 2 shows that this design did not achieve the required output power. Therefore, the analysis was performed and is presented in the next chapter. This table also shows the very low efficiency, which was mainly due to high windage losses. This calculation was performed using an analytical calculation that was not intended for high-speed motors. Therefore, a more appropriate calculation was performed, which also will be presented in the following sections.

3. Methods of Analysis

Most analyses were performed using software based on the finite element method (FEM). It is a numerical method used to calculate electromagnetic parameters of various physical models, but also to calculate mechanical parameters, heat flow, fluid flow, etc. Its principle is to discretize the model into a certain number of elements using a network of finite elements. Subsequently, a system of partial differential equations is solved at each point of this network. For this calculation, specialized programs are used that can solve the problem in a 2D or 3D environment. In this case, the program Ansys was used for the FEM calculation, and the model was solved in 2D. Several types of analyses were performed using this program.

The first was magnetostatic analysis, which was used to solve the steady state of the magnetic field. The finite element model, together with the network of elements, is shown in Figure 2. This figure shows three models at different stages of motor design. Figure 2a shows the state of the design shown in [14], on which no optimization had been performed. The geometric and electromagnetic parameters of this design are compared with the final design in Section 4.6. Figure 2b shows the state of the design after several optimizations, which were described in Section 2. The geometrical and electromagnetic parameters of this design are given in Tables 1 and 2. Finally, Figure 2c shows the design after optimization that met all the specified requirements. These models were generated using Ansys—RMxprt. The default mesh determined by Ansys—Maxwell after generating the model was used in the analyses. This mesh was sufficient in terms of accuracy.

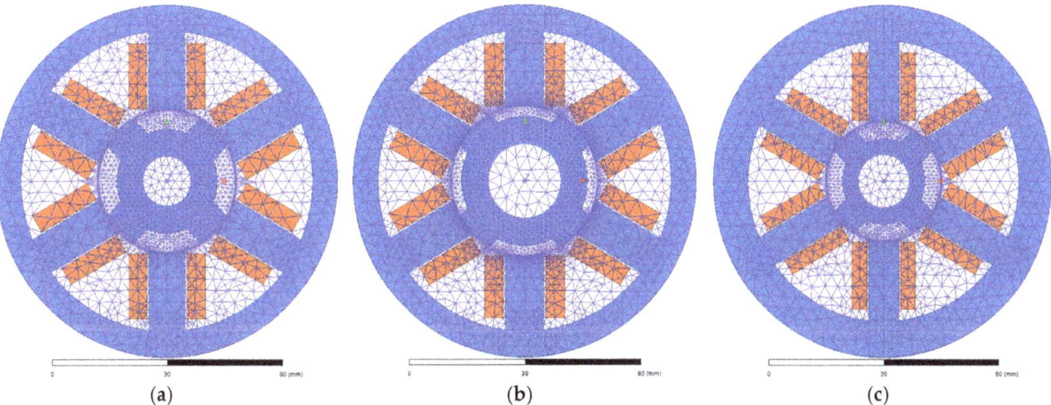

(a)   (b)   (c)

Figure 2. Finite element model from Ansys—Maxwell 2D: (**a**) initial design—before optimization; (**b**) basic design—after several optimizations; (**c**) final design—after optimization, and without a sleeve. The individual colors represent the materials used: blue—sheet material; orange—winding material; white—air.

In the magnetostatic analysis, only one phase was supplied with constant current. The aim of this analysis was to obtain static parameters of the motor, which included: phase inductance, magnetic flux, co-energy, electromagnetic torque, and phase resistance. In this analysis, the rotation of the motor at a certain speed was not considered, and only static parameters of given geometry and phase current were obtained. All these parameters except the phase resistance depended on the current and rotor position. For this reason, parameters were calculated for rotor positions from 0 to 45° and for phase currents of 5 to 400 A. The parameters obtained from Ansys—Maxwell were also verified using the program FEMM. The phase resistance was calculated using the program Ansys—RMxprt, which is used for fast analytical calculation of motors. The calculated phase resistance was verified by different analytical calculations. Using RMxprt, suitable switching angles were also found, which were chosen to achieve the maximum torque.

The values of these angles were needed for the transient analysis. These angles were inserted into the supply circuit required for this analysis. The supply circuit is presented in Figure 3.

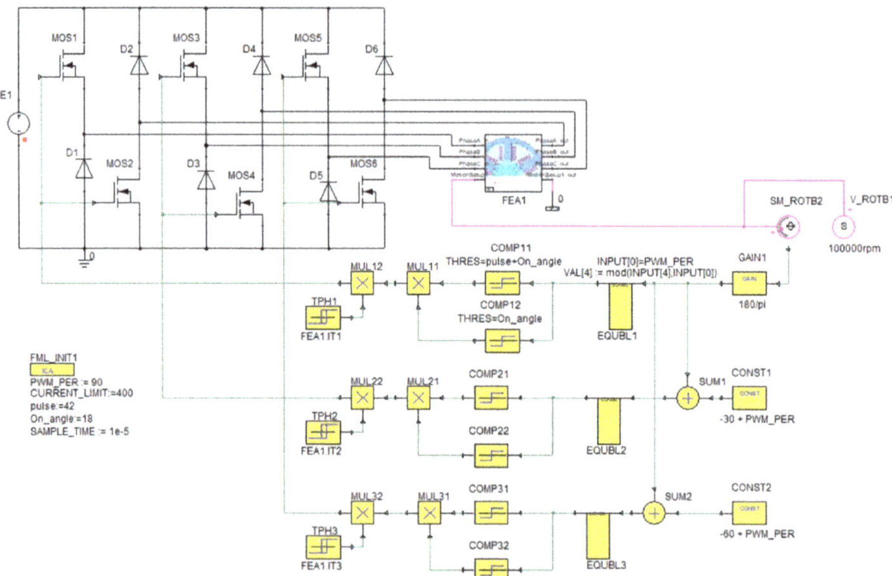

Figure 3. Supply circuit from Ansys—Twin Builder. Description of individual symbols: E—voltage source; D—diode; MOS—MOSFET transistor; FEA—FEM model of motor; SM_ROTB—rotational angle meter; V_ROTB—rotational angular velocity source; GAIN—proportional gain; EQUBL—equation block; CONST—constant value; COMP—comparator; MUL—multiplier; TPH—signal processing block—two-point element with hysteresis.

This circuit was created in Ansys—Twin Builder, in which it was connected to the finite element model in Figure 2. This circuit was based on the circuit presented in [19], but several adjustments were made to the circuit in Figure 3, such as adding another phase. This model did not solve motor start-up, only the steady state at the rated speed. At the beginning of this simulation, it was assumed that the motor rotated at the rated speed, and then this speed was maintained throughout the simulation. Using this analysis, it was possible to obtain dynamic parameters of the motor at the rated speed. These parameters included waveforms: voltage, magnetic flux, supply current, phase inductance, and electromagnetic torque. Dynamic parameters were verified using a combination of FEMM with MATLAB—Simulink. Using these parameters, other parameters were calculated, such as average torque, torque ripple, and individual types of losses. The torque ripple was calculated as:

$$\Delta T = \frac{T_{\max} - T_{\min}}{T_{av}},$$ (1)

where $T_{\max}$ is maximal torque and $T_{\min}$ is minimal torque. Winding losses were calculated as:

$$\Delta P_j = m \cdot R_{f75} \cdot I_{frms}^{2},$$ (2)

where R_{f75} is resistance at 75 °C and I_{frms} is the RMS phase current. Iron losses were calculated with the Bertotti equation as:

$$\Delta P_{Fe} = k_{hys} \cdot f \cdot B^2 + k_{ec} \cdot f^2 \cdot B^2 + k_{exc} \cdot f^{1.5} \cdot B^{1.5}$$ (3)

where k_{hys} is the coefficient of hysteresis losses, f is the frequency, B is the peak flux density, k_{ec} is the coefficient of eddy current losses, and k_{exc} is the coefficient of excess losses. The individual loss coefficients were calculated from the catalog data of the electromagnetic sheets using Ansys—Maxwell. Mechanical losses consisted of windage losses and bearing losses. Both types of losses were determined outside Ansys using analytical calculations presented in [14] and [16]. However, as mentioned in the previous chapter, windage losses must be calculated using a more appropriate method. The original calculation did not consider the air gap size, and losses were calculated for a certain range of loss coefficients. For these reasons, windage losses were calculated as in [20]:

$$\Delta P_w = C_f \cdot \pi \cdot \rho \cdot l_{Fe} \cdot r_r^4 \cdot \Omega^3, \tag{4}$$

where C_f is the friction coefficient, ρ is the air density, l_{Fe} is the length of the motor, r_r is the rotor radius, and Ω is the angular speed. Several methods have been published for calculating the friction coefficient as a function of the Reynolds number, Re. In this case, the Wendt method was chosen. The Wendt friction coefficient can be calculated for $400 < \mathrm{Re} < 10^4$ as [20]:

$$C_f = 0.46 \cdot (\delta + \delta^2)^{0.25} / \mathrm{Re}^{0.5}, \tag{5}$$

or for $10^4 < \mathrm{Re} < 10^5$ as:

$$C_f = 0.037 \cdot (\delta + \delta^2)^{0.25} / \mathrm{Re}^{0.3}, \tag{6}$$

where δ is the air gap size. The Reynolds number can be calculated as [21]:

$$\mathrm{Re} = \frac{r_r \cdot \delta \cdot \Omega \cdot \rho}{\nu}, \tag{7}$$

where ν is the air viscosity. However, the SRM had salient poles, and this must also be included in the calculation using the salient coefficient. This coefficient can be calculated as [21]:

$$C_{sp} = \left(8.5 \cdot \frac{l_{pr}}{r_r} + 2.2\right) - 1, \tag{8}$$

where l_{pr} is the length of the rotor pole. Windage losses were then calculated as [21]:

$$\Delta P_w = C_{sp} \cdot C_f \cdot \pi \cdot \rho \cdot l_{Fe} \cdot r_r^4 \cdot \Omega^3. \tag{9}$$

Additional losses were determined as 0.5% of output power. In addition to electromagnetic analysis, thermal analysis and mechanical analyses were also performed using FEM.

Thermal analysis was carried out by Ansys—Motor CAD, which was also used to verify the electromagnetic parameters obtained by transient analysis. In the case of the analysis in Motor CAD, it was also considered that the winding was divided into several parallel conductors, which was necessary due to the high phase current. The method of cooling is shown in Figure 4. The motor was closed, and around the stator was a frame with a diameter of 110 mm, in which axial water cooling was placed. Aluminum was chosen for the frame material and cast iron for the shaft material, which were available in the materials library. The ambient and cooling water temperature were set at 40 °C. The preliminary water flow value was set to 10 L/min. All types of losses calculated by electromagnetic analysis were entered into the program. The analysis considered the change in winding losses as a function of temperature, and the change in other types of losses as a function of rotational speed.

The mechanical analysis was performed in cooperation with the Faculties of Mechanical Engineering using Ansys—Workbench. The results of this analysis will be presented in the following sections.

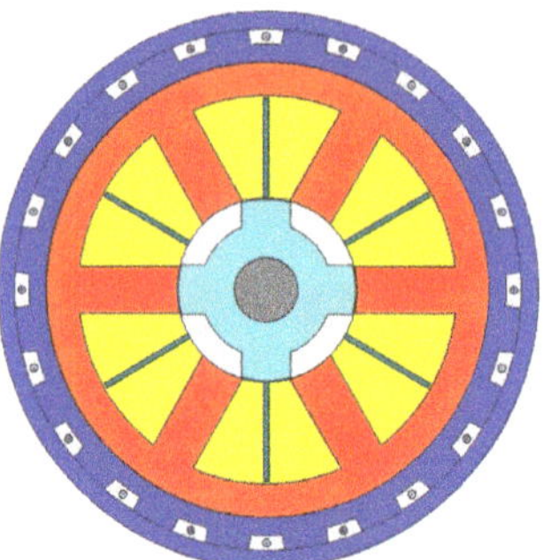

Figure 4. The method of cooling. The individual colors represent parts of the motor: dark blue—frame with cooling channels; red—stator sheets; yellow—stator winding; green—coil divider, insulation between coils; white—empty space, filled with water in the case of cooling channels or air in the case of the air gap between the rotor and stator; light blue—rotor sheets; grey—shaft.

4. Results

In this section, the results of electromagnetic, mechanical, and thermal analyses are presented, and at the end, the final construction of the motor is presented. In all analyses presented in this section, the following parameters had a constant value: length of the machine, number of poles, number of phases, number of turns, and supply voltage. These values are given in Section 2.

4.1. Analysis of Windage Losses

The analysis was performed for a rotor diameter of 20 to 38 mm and for an air gap size of 0.1 to 0.5 mm. The analysis was carried out for the rated speed, and in the case of the rotor with salient poles, a length of the rotor pole of 2.52 mm was considered. The results for the case of the rotor with salient poles are shown in Figure 5a. For a rotor diameter of 38 mm and an air-gap size of 0.22 mm, these losses had a value of 1922 W. Such high losses caused excessive heat that could not be reduced below the thermal limit of materials. For this reason, it was necessary to adjust the rotor to reduce these losses.

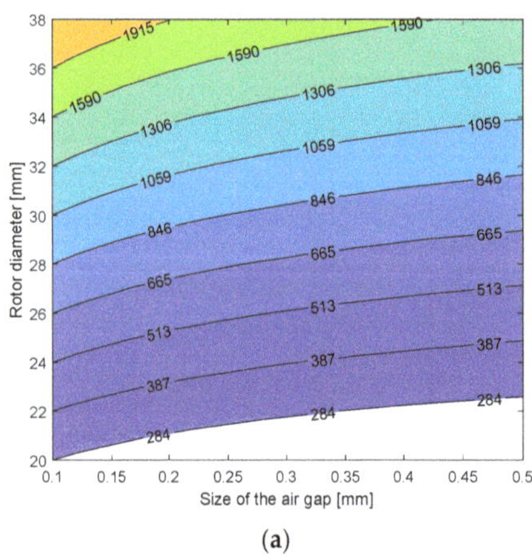

(a)

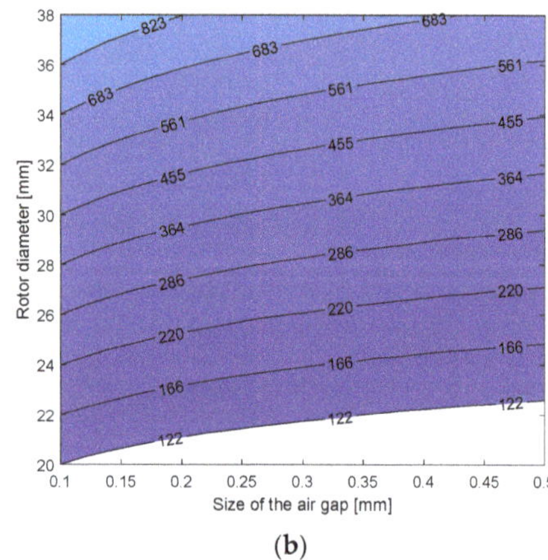

(b)

Figure 5. Results of the windage losses analysis: (**a**) salient poles; (**b**) smooth cylindrical rotor.

This analysis was also performed for the case of a smooth cylindrical rotor, and the results are shown in Figure 5b. Unlike the previous windage-losses analysis, this analysis assumed that the space between rotor poles was filled with material, thus changing the shape of the rotor into a cylinder. This modification did not increase the rotor diameter or reduce air gap size. In this case, there was a significant decrease in windage losses. For a rotor diameter of 38 mm and an air-gap size of 0.22 mm, these losses had a value of 806.1 W. Based on these results, it was decided to modify the rotor geometry to a smooth cylindrical rotor.

4.2. Mechanical Analysis of Various Rotor Modifications

Mechanical analysis was performed to determine a suitable method of forming a smooth cylindrical rotor. Von Mises stress and sheet deformation were monitored during the analysis. All mechanical calculations were performed for the motor design listed in Table 1.

4.2.1. Flux Bridges between Rotor Poles

In the first modification, the rotor poles were connected by bridges that were supported in the middle of the gap between the rotor poles. These bridges and supports were composed of the same material as the rotor. This rotor modification was used in the construction of the SRM for a speed of 50,000 rpm in [22]. From an electromagnetic point of view, this modification caused a decrease in torque; therefore, these bridges should be as thin as possible. However, with the thickness of bridges shown in Figure 6, this construction did not meet the mechanical aspect. At 100,000 rpm, mechanically stronger rotor material would be required for this modification.

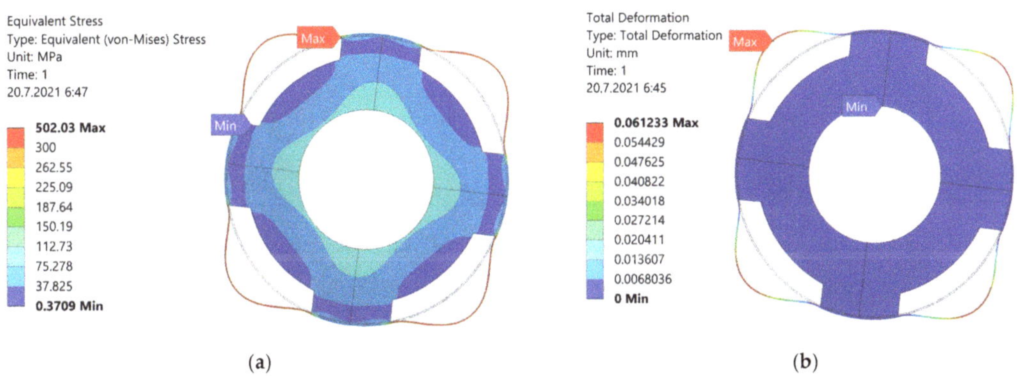

Figure 6. Mechanical analysis of the sleeve glued to the rotor: (**a**) Von Mises stress; (**b**) total deformation.

4.2.2. Sleeve around the Rotor

A better modification of the rotor at the rated speed was to use a sleeve, as in the case of PMSMs. In contrast to the previous modification, in this modification, the sleeve was not supported in the middle of the gap between the rotor poles, but nevertheless, there was a decrease in Von Mises stress of 76 MPa. This rotor modification was used in the construction of the SRM for speed of 750,000 rpm in [23]. The thickness of the sleeve was 0.2 mm, which was the same thickness as that of the bridges in the case of the previous analysis. However, when using this modification, it was necessary to increase the air-gap size due to higher deformation, but above all to provide sufficient space for the sleeve. It was necessary to use a material such as carbon fiber for the sleeve in which there were no losses caused by eddy currents, which would cause further heating of the rotor. Figure 6 shows the results of a mechanical analysis of a rotor with a sleeve. The sleeve was glued to the rotor. The glue was placed between upper part of the rotor pole and the sleeve itself. The contact between the sleeve and the rotor was defined in Ansys Workbench as Bonded. The rotor with the sleeve was loaded with a rotational velocity of 10,472 rad/s.

Much better results could be obtained if the sleeve was slid onto the rotor. In this case, the sleeve was heated during manufacturing to increase its diameter so that it could then be slid onto the rotor. After the sleeve had cooled, its diameter was reduced, and a force was created between the sleeve and the rotor to hold the sleeve in place. Figure 7 shows the results of the mechanical analysis of a sleeve that was only slid onto the rotor (without gluing). The sleeve is shown without a rotor. The sleeve was loaded at a rotational velocity of 10,472 rad/s.

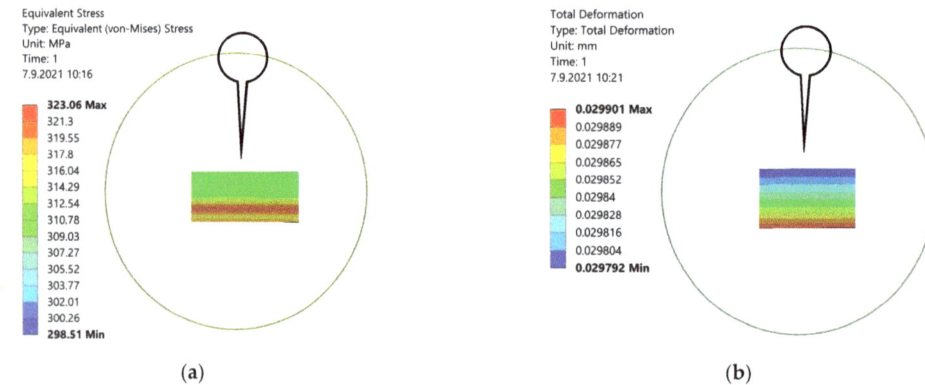

(**a**)   (**b**)

Figure 7. Mechanical analysis of the sleeve slid onto the rotor: (**a**) Von Mises stress; (**b**) total deformation.

In both cases, there was a slight increase in windage losses, as the air-gap size decreased by the width of the sleeve, and the rotor diameter increased by twice the width of the sleeve.

4.2.3. Material between Poles

Finally, the possibility of filling the space between the poles with nonconductive and nonmagnetic material was analyzed. Several modifications to the rotor shape were made to fix the material. These modifications included placing pole heads on the tip of the rotor poles, a cutout in the rotor yoke, and a cutout at the points where the rotor poles protruded from the rotor yoke. Out of these three variants, the third variant was the best, but nevertheless, the maximum Von Mises stress reached up to 925 MPa. From a mechanical point of view, none of these variants was suitable. At the end, the possibility of fixing the material with glue was analyzed. The analysis showed a significant decrease in Von Mises stress to 587 MPa and minimal deformation to 0.01 mm. However, the use of a sleeve was still superior.

4.3. Preliminary Thermal Analysis

Two thermal analyses were performed. In the first analysis, the losses values obtained at the rated speed were used. These losses were constant during the simulation. In the second analysis, a duty cycle was implemented in which the motor started from zero speed to the rated speed, and then the speed was reduced until the machine stopped. In this analysis, the losses varied with speed. The aim of the analyses was to determine whether the creation of a smooth cylindrical rotor would reduce losses so that the thermal limits of the motor were not exceeded. In the simulations, the motor temperature changed over time, so only the steady state of this thermal analysis, which occurred after the temperature had stabilized, is shown in Figure 8. Bearing losses at 50% load and windage losses calculated for the smooth cylindrical rotor were included in the simulation. All thermal calculations were performed for the motor design listed in Table 1.

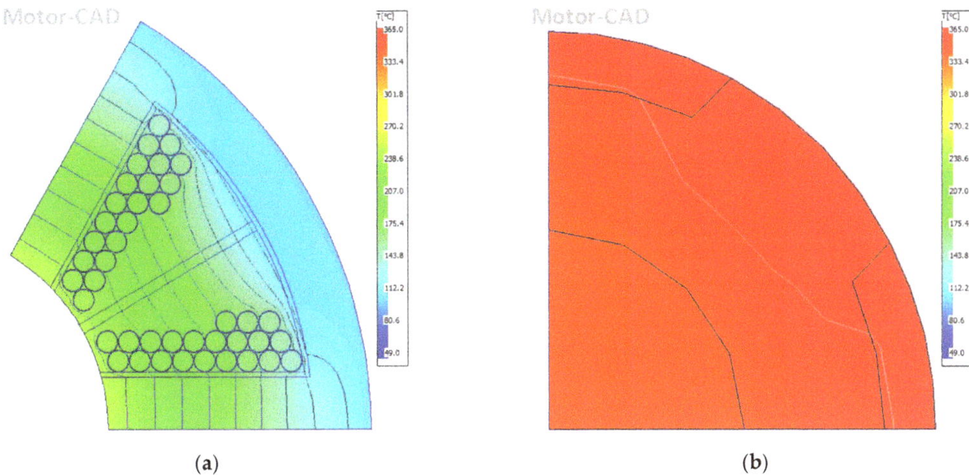

(a)   (b)

Figure 8. The distribution of temperature at steady state: (**a**) stator; (**b**) rotor.

High heat was generated near the air gap, which affected the entire motor. This heat could not be reduced by increasing the water flow or adjusting the cooling channels. This heating was generated in the machine during continuous operation.

Based on the results shown in Figure 9, it was shown that this warm-up occurred within approximately 10 min of machine operation. Figure 9 shows the temperature change of the most important parts of the motor from the beginning of the simulation until the temperature stabilization, as shown in Figure 8. However, due to its application, this motor will not run continuously, but in a defined duty cycle. Therefore, thermal analysis respecting this cycle was performed, and the results are shown in Figure 10.

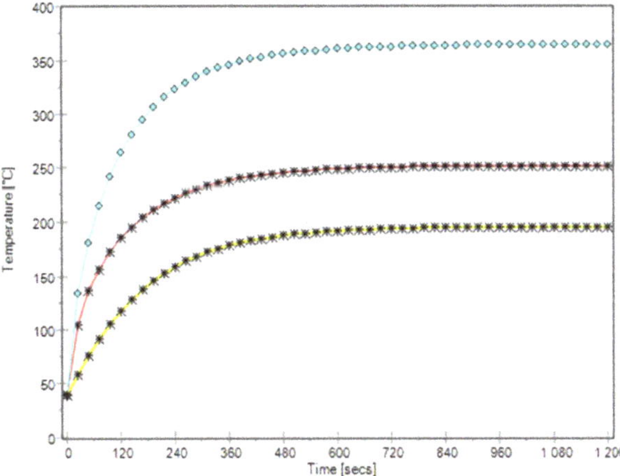

Figure 9. The temperatures of individual motor parts as a function of time: blue curve—rotor heating; red curve—stator heating; yellow—windage heating.

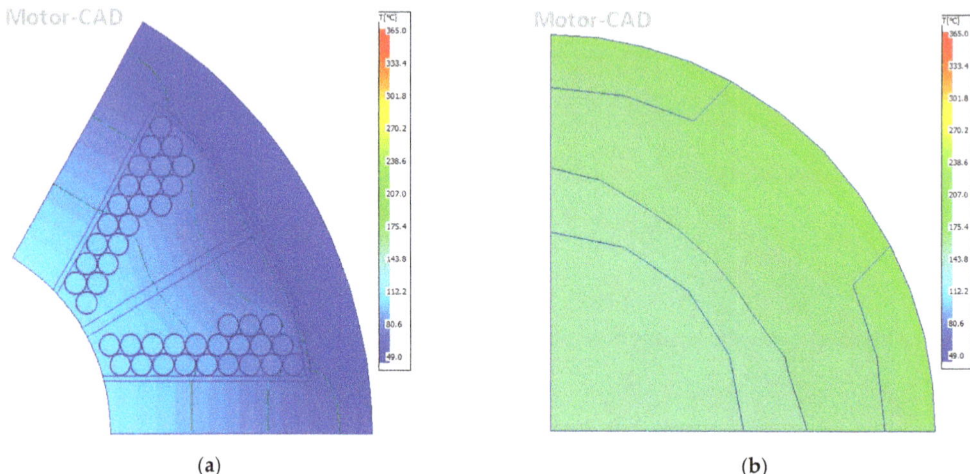

Figure 10. The distribution of temperature at steady state with duty cycle: (**a**) stator; (**b**) rotor.

The duty cycle consisted of a shorter interval during which the motor was connected to the supply voltage, and a longer interval during which the motor was disconnected from the supply voltage. In the shorter interval, the motor started from zero and increased to the rated speed, and worked at this speed for some time. In the longer interval, the electromagnetic losses decreased almost immediately to zero, but the mechanical losses decreased gradually depending on the speed. At the end of longer interval, a complete stopping of the motor was assumed.

The implementation of the motor duty cycle improved cooling and significantly reduced the temperature. However, the rotor was still heated to 172.1 °C, which was more than the limit of the sheet. The stator and rotor sheets had a temperature limit of 150 °C. Therefore, it was necessary to reduce the rotor diameter to reach this limit. This analysis is presented in the following section.

4.4. Rotor Optimization

In this section, the rotor was optimized to achieve required output power while maintaining thermal and mechanical limits. In these analyses, all stator parameters had constant values as shown in Table 1.

4.4.1. Increase of Phase Current

In a previous work [18], by optimizing the rotor geometry for a maximum phase current of 200 A, the average torque was increased to 0.538 Nm. However, it still failed to achieve the required output power of 8 kW. Therefore, the maximum phase current increased. During this analysis, the shaft diameter and the length and width of the rotor poles were variable parameters. The rotor diameter had a constant value of 38 mm and an air gap of 0.22 mm. Increasing length and decreasing width of rotor poles led to an increase in the maximum phase current. Therefore, an analysis was performed in which the geometric dimensions of the poles were changed regardless of the maximum phase current. The goal was to achieve the required output power. By changing the length of the rotor poles, the shaft diameter was changed in this analysis. However, the required output power could not be achieved, even with very long rotor poles, which would reduce the shaft diameter to 7 mm. According to [10], such a small shaft diameter is at the limit of manufacturability. In addition, this analysis was performed with a rotor diameter of 38 mm. However, according to the results of the preliminary thermal analysis, it was necessary to reduce rotor diameter, which made the situation even worse.

An increase in the maximum phase current can also be achieved by shifting the switching angles. Therefore, an analysis was performed in which the geometry of the rotor poles was adjusted to achieve the required output power at different values of the advance angle. The aim was to find a compromise between changing the advance angle and changing the rotor pole geometry. The SRM control used the parameters On angle ("On_angle" in Figure 3) and pulse length ("pulse" in Figure 3). An On angle equal to $0°$ usually represents the unaligned position of the rotor. In this position, the rotor pole was located exactly between the two stator poles. The advance angle represents the motor control parameter at which the phase of motor is switched on before unaligned position. This parameter is usually negative. However, since the supply circuit in Figure 3 had a problem operating with negative values, an On angle of $0°$ represented $-30°$. The pulse length was held constant at $42°$ throughout this analysis. The results of this analysis are shown in Figure 11.

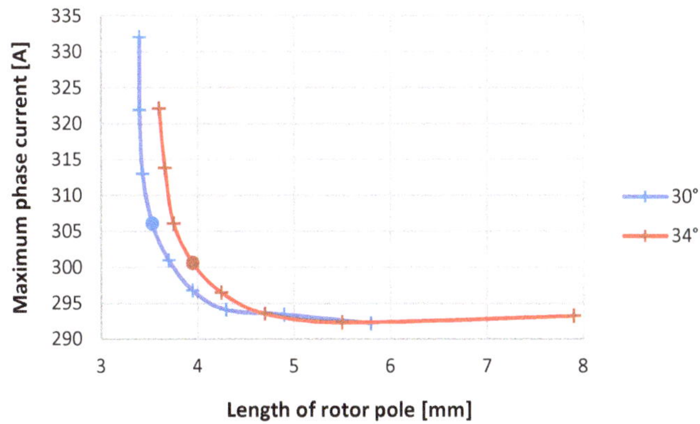

Figure 11. Possible designs with an output power of 8 kW. Each discrete point indicates a possible motor design, and the symbol • indicates the finally selected motor design.

The analysis was performed for an advance angle of -14 to $-22°$. It was not possible to achieve the required output power with a smaller advance angle. The individual points in Figure 11 represent designs for different advance angles, with the value of this angle varying from $-22°$ to $-15°$ from right to left. The analysis was performed for two values of the rotor pole arc. All points in Figure 11 represent possible motor designs. At lower values of the advance angle, the maximum phase current did not change significantly, but by increasing the advance angle, the rotor poles could be much shorter. By further increasing the advance angle, the maximum phase current began to increase, but was still possible to achieve a significant reduction in the length of the rotor poles. An advance angle greater than $-20°$ had no effect on shortening the rotor poles, and only caused an excessive increase in the maximum phase current. Based on these results, an advance angle of $-19°$ was chosen for further analysis. The results of torque ripple and losses are shown in Table 3.

Table 3. Torque ripple and losses for different advance angles for a rotor pole arc of $34°$.

Angle (°)	8 (−22)	9 (−21)	10 (−20)	11 (−19)	12 (−18)	13 (−17)	14 (−16)	15 (−15)
ΔT (%)	205	198	193	189	184	181	179	175
ΔP_{j} (W)	44.7	42.3	40.1	38.3	36.8	35.6	34.7	34.1
ΔP_{Fe} (W)	299	304	309	313	316	320	328	327

As more emphasis was placed on shortening the rotor poles so that the shaft diameter could be as large as possible, the rotor pole arc was also changed to $30°$. This allowed the rotor poles to be shortened even more.

4.4.2. Rotor Diameter

In this analysis, the rotor diameter, shaft diameter, and length of the rotor pole were variable parameters. The length of the rotor poles was adjusted for different rotor diameters to achieve the desired output power, as in the previous analysis, but the switching angles were constant. The air gap also had a constant value of 0.22 mm and a rotor pole arc of $30°$. Electromagnetic parameters and the highest rotor heating were monitored during the analysis. The results of this analysis are shown in Table 4.

Table 4. Results of the rotor-diameter analysis.

d_r (mm)	d_h (mm)	I_{max} (A)	T_{av} (Nm)	ΔT (%)	ΔP_j (W)	ΔP_{Fe} (W)	ΔP_w (W)	P_2 (kW)	ϑ_{max} [1] (°C)
38	19.5	302	0.897	184	37.4	295.8	806	8.04	175
36	18.3	295.3	0.879	178	37.5	295.4	676	8	164
34	17	294.7	0.871	167	37.9	296	562	8.02	156.7
32	15.7	294.4	0.861	160	38.6	294.8	461	8.02	147.1
30	14.2	295	0.851	151	39.8	295.3	374	8.02	142.7
28	12.4	298.1	0.847	143	42.3	293	299	8.04	134.6
26	10.4	303.4	0.837	134	45.8	283.4	235	8	130.4

[1] Maximum rotor temperature.

As shown in to Table 4, the required output power was achieved with all mentioned rotor diameters. The largest possible rotor diameter in terms of heating was 32 mm. However, the gaps between the poles were not covered in this analysis. This heating would therefore increase further due to poorer heat dissipation by convention. For this reason, a rotor diameter of 30 mm was chosen.

Reducing the rotor diameter reduced the shaft diameter, but it was necessary to reduce heating. Table 4 shows that reducing the rotor diameter also significantly reduced the torque ripple.

4.4.3. Air-Gap Size

By reducing rotor diameter, the thermal limit of the rotor sheets was reached, so a change in the air gap size was not necessary to reduce the heating. However, the air gap had to be enlarged to provide sufficient space for the sleeve to form the smooth cylindrical rotor. In this analysis, shaft diameter, length of the rotor pole, and air-gap size were variable parameters. The rotor pole arc had a constant value of $30°$, and the rotor diameter was 30 mm. The results of this analysis are shown in Table 5.

Table 5. Results of the air-gap-size analysis.

δ (mm)	l_{pr} (mm)	I_{max} (A)	T_{av} (Nm)	ΔT (%)	ΔP_j (W)	ΔP_{Fe} (W)	ΔP_w (W)	P_2 (kW)
0.22	3.1	295	0.851	151	39.8	295.3	374	8.02
0.3	3.4	315.2	0.847	145	43	287.6	346.3	8
0.4	3.9	331.7	0.847	133	48.9	281	322.6	8.02
0.5	4.5	346.3	0.844	122	54.9	271.1	305.3	8.04

As the air-gap size increased, the maximum phase current required to achieve the required output power was significantly increased. This also caused an increase in winding losses. In addition, longer rotor poles were required to achieve the required output power. Conversely, iron losses and windage losses were reduced, which reduced motor heating. Increasing the air-gap size also further reduced the torque ripple.

To provide sufficient space for the sleeve, which had to be at least 0.2 mm thick, and to simplify production, an air gap size of 0.5 mm was chosen. With this size of the air gap, both the rotor and the stator could be made of one piece of material, which also reduced costs. Smaller sizes required additional material for the rotor.

4.5. Stator Optimization

The entire optimization was focused on the rotor, because the possibilities of stator optimization were considerably limited to ensure sufficient space for winding. Nevertheless, an analysis was performed, but the change in geometric parameters was limited to provide sufficient space for the winding. The rotor parameters, air-gap size, and stator outer diameter were constant throughout the analysis. The rotor diameter was 30 mm, the length of rotor pole was 4.5 mm, the rotor pole arc was 30°, the shaft diameter was 13.24 mm, the air-gap size was 0.5 mm, and the stator outer diameter was 90 mm.

4.5.1. Length of the Stator Poles

Because most of the parameters were kept constant, changing this parameter also changed the thickness of the stator yoke. The analysis was performed for stator poles with lengths from 21.5 to 23.5 mm in 0.5 mm steps. The lower limit was limited by the space for winding, and the upper limit by the saturation of the stator yoke, which reached 1.1 T at this pole length. The results of this analysis are shown in Table 6.

Table 6. Results of the analysis of the length of the stator poles.

l_{ps} (mm)	h_s (mm)	I_{max} (A)	T_{av} (Nm)	ΔT (%)	ΔP_j (W)	ΔP_{Fe} (W)	ΔP_{mech} [1] (W)	P_2 (kW)
23.5	6	344.1	0.834	123	54.4	290.3		7.89
22.9	6.6	346.3	0.844	122	54.9	271.1	$305.3 +$	8.01
22.5	7	348.3	0.852	123	55.4	250.8	196.3 [2]	8.12
22	7.5	350.6	0.858	122	55.7	229.8		8.2
21.5	8	351.3	0.863	123	56	213.6		8.27

[1] Windage and bearing losses. [2] These values are the same for all rows.

When the length of stator poles was changed, it was possible to observe the opposite trend as the length of rotor poles was changed. The average torque increased with decreasing stator pole length. Winding losses increased with decreasing pole length, which was caused by an increase in the maximum phase current. However, this increase was not high, and could be reduced if the rotor poles were adjusted to achieve the required output power. There was a much larger difference in the case of iron losses, the decrease of which occurred mainly due to the increase in the rotor yoke thickness. The disadvantage of short poles was lower saturation of the stator yoke and higher material consumption.

4.5.2. Rounding at the Junction of Poles

Similar results to those of the previous analysis could be obtained by inserting rounding with radius r_{sp} at the point where the stator pole protruded from the yoke. The results are shown in Table 7. This modification also reduced the length of the stator poles, but the yoke thickness did not increase along the entire circumference of the motor, but only at the point of rounding.

Table 7. Results of the analysis of the rounding at the junction of poles.

r_{sp} (mm)	1	2	3	4	5	6	7	8
I_{max} (A)	346.3	346.6	346.9	347.7	348.1	350.1	350.4	351.1
T_{av} (Nm)	0.844	0.845	0.847	0.85	0.853	0.858	0.862	0.865
ΔP_j (W)	54.9	55	55.2	55.5	55.7	56.1	56.6	57
ΔP_{Fe} (W)	266.5	262.5	257.4	251.1	243.5	235.3	224.9	214
P_2 (kW)	8.02	8.03	8.06	8.1	8.13	8.19	8.25	8.29

With this modification, it was possible to achieve an unchanged saturation in the narrower part of stator yoke; and in rounding, it was possible to cut out excess material. However, this was also possible in the case of the previous analysis by reducing the outer stator diameter. According to [19], for motors with very high speeds, it is more appropriate to use a thicker stator yoke than is electromagnetically necessary to reduce the vibration of the stator. When using a large radius of rounding, there is also deformation of the winding space.

4.5.3. Stator Pole Arc

This analysis was performed for values from $30°$ to $38°$ in $1°$ steps. The upper limit was again limited by the space for winding. During this analysis, the length of stator poles was constant at 22.9 mm, and the rotor yoke thickness was 6.6 mm. The results are shown in Table 8.

Table 8. Results of the analysis of the stator pole arc.

β_s (°)	I_{max} (A)	T_{av} (Nm)	ΔT (%)	ΔP_j (W)	ΔP_{Fe} (W)	ΔP_{mech} (W)	P_2 (kW)
30	346.3	0.844	122.3	54.9	271.1		8.01
31	336.9	0.821	120.9	51.8	269.5		7.78
32	326.6	0.797	119.3	48.9	269.4		7.53
33	316.8	0.773	118.6	46.2	269	$305.3 +$	7.28
34	308.5	0.751	118.8	43.8	267.4	196.3 [1]	7.05
35	298.2	0.724	117.1	41.3	266		6.77
36	289.3	0.698	116.3	39	264.7		6.51
37	280.1	0.674	114.9	36.9	263.6		6.26
38	271.1	0.648	113.4	34.7	260.4		6

[1] These values are the same for all rows.

The analysis confirmed the theoretical assumption that a smaller torque ripple could be achieved with a larger pole stator arc. However, the average torque also decreased. It was possible to achieve the same reduction in torque ripple with a lower average torque reduction by changing the switching angles. Thus, changing the stator pole arc was not advantageous.

Based on these analyses, the stator poles were shortened to 21.5 mm. The increase in output power was reduced by changing the length of the rotor poles to 4.05 mm. The stator pole arc was left at $30°$. In addition, in the area where the stator pole protruded from the yoke, a rounding with a radius of 3 mm was inserted to reduce the sharp edge and adjust the saturation that occurred at this point.

4.5.4. Number of Parallel Conductors

The last stator optimization performed was the winding optimization. The winding of the designed motor consisted of six coils, which were wound on the stator poles and were connected to each other in series. The individual coils consisted of three turns; this number had already been fixed based on the analysis presented in [16]. However, the high values of phase currents required to achieve the required output power would require the use of larger diameter conductors, which would cause an increase in losses and production problems. For these reasons, it was considered to divide the individual turns into several thinner conductors, which were connected in parallel. Thus, an analysis was performed to determine the appropriate number of these conductors.

Parallel conductors with diameter of 2.12 mm were considered in the simulations. Their number was modified so that the current density did not exceed $J = 10$ A/mm^2. However, for the given maximum phase current and conductor cross section, this led to 11 parallel conductors, which could be problematic during winding manufacturing. According to [1], the current density can be higher when using water cooling $J = 23.3$–31 A/mm^2.

For these reasons, an analysis of the parallel conductors number was performed. In this analysis, only the winding was adjusted, and all geometric dimensions were constant. The results are shown in Figures 12 and 13.

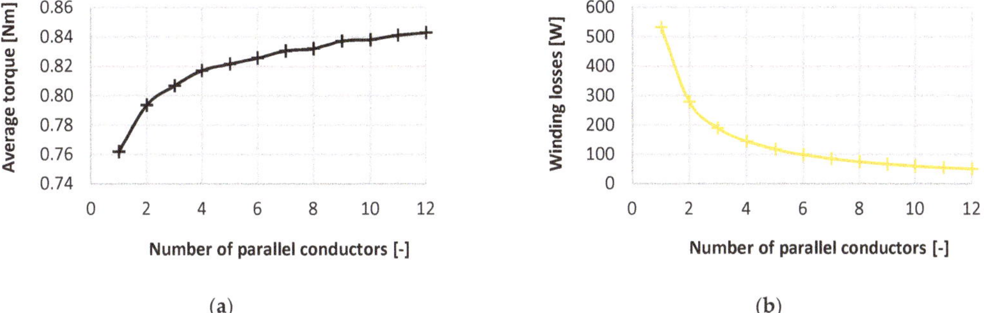

(a)

(b)

Figure 12. Results of the electromagnetic analysis for different numbers of parallel conductors: (**a**) average torque; (**b**) winding losses.

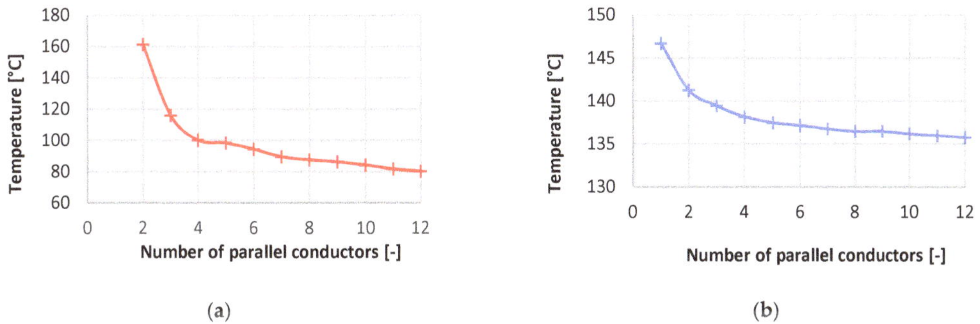

(a)

(b)

Figure 13. Results of the thermal analysis for different numbers of parallel conductors: (**a**) temperature of winding; (**b**) temperature of rotor surface.

As the number of parallel conductors decreased, average torque also decreased, which was caused by the decrease in the maximum phase current, which decreased with this change. Iron losses also decreased, but this decrease was small. However, this did not apply to winding losses. With a very small number of parallel conductors, these losses reached very high values.

From a thermal point of view, the thermal limit was met for each number of parallel conductors except $z_p = 1$. At this number, the winding reached temperatures of up to 364 °C. The permissible heating temperature of the winding depends on its insulation class, with the highest insulation class having a thermal limit of 180 °C. Based on Figure 13a, it should be appropriate to use at least four parallel conductors, due to the significant increase in winding heating when using a smaller number of parallel conductors.

The current density range specified at the beginning of this section was best matched by four parallel conductors ($J = 27$ A/mm^2) or five parallel conductors ($J = 21.7$ A/mm^2). Since, according to Figure 12a, reducing the number of parallel conductors reduced the average torque and thus the output power, it was necessary to adjust the stator geometry to increase the output power. As shown in Table 6, it was possible to increase the output power by shortening the stator poles. This modification was realized by increasing the yoke thickness, while maintaining a constant outer stator diameter. This modification would also cause increase in phase current and thus also increase in current density. For this reason, it was more appropriate to use a higher number of parallel conductors. However, six parallel conductors (J = 18.4 A/mm^2) were finally selected to ensure higher reliability in the event of a single-phase outage and a consequent increase in the maximum phase current.

4.6. Final Design

This section presents the final design of the high-speed SRM. Table 9 shows the geometrical dimensions of the final design and a comparison with the initial design. Both geometries had the same number of stator and rotor poles, outer stator diameter, and stator pole arc.

Table 9. Comparison of the geometric parameters.

Parameter	Initial Design	Final Design
Outer stator diameter	90 mm	90 mm
Outer rotor diameter	36 mm	30 mm
Length of stator poles	20.18 mm	20.32 mm
Length of rotor poles	4.72 mm	4.05 mm
Stator pole arc	30°	30°
Rotor pole arc	32°	30°
Yoke thickness of stator	6.6 mm	9.18 mm
Yoke thickness of rotor	6.95 mm	4.18 mm
Air-gap size	0.22 mm	0.5 mm
Shaft diameter	12.68 mm	13.54 mm

Compared to the initial design, the outer rotor diameter and rotor yoke thickness were reduced, the stator yoke thickness was increased, the stator and rotor poles were shortened, and the same pole arc was used for both parts of the motor. In addition, the stator and rotor material was changed to 10JNEX900, the control method was changed to single-pulse operation, and the number of parallel conductors was changed to six.

4.6.1. Electromagnetic Parameters

Dynamic parameters were calculated for this geometry, and are shown in Figure 14.

Using these parameters, other parameters such as average torque and losses were calculated. These parameters are listed and compared with the initial design in Table 10.

Table 10. Comparison of the electromagnetic parameters.

Parameter	Initial Design	Final Design
Supply voltage	48 V	48 V
Maximal phase current	203.4 A	347.1 A
Sampling frequency	-	100 kHz
Average torque	0.4846 Nm	0.844 Nm
Torque ripple	194.4%	122.7%
Winding losses	23.66 W	99.89 W
Iron losses	510.89 W	184.21 W
Windage losses	1840 W	305.3 W
Bearing losses	196.3 W	196.3 W
Additional losses	12.5 W	40 W
Output power	2.49 kW	8.01 kW
Efficiency	49.13%	-
Efficiency (cylindrical rotor)	72.08%	90.63%

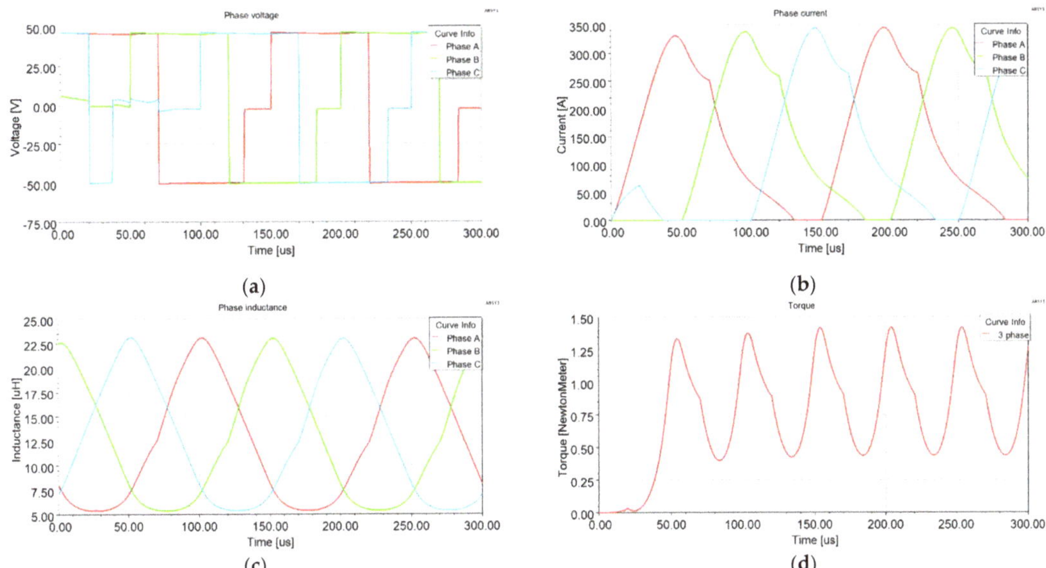

Figure 14. Dynamic parameters of the high-speed SRM at the rated speed: (**a**) phase voltage; (**b**) phase current; (**c**) phase inductance; (**d**) torque of all three phases.

The flux density distribution for position of the highest yoke saturation is shown in Figure 15. White areas in the figures indicate that the scale range had been exceeded.

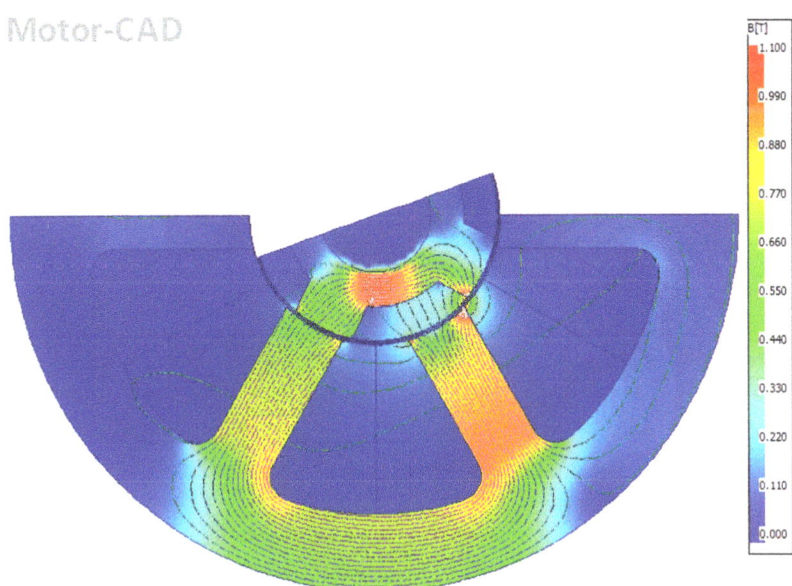

Figure 15. Flux density distribution for rotor position of the highest yoke saturation that occurred in the rotor.

The thickness of the rotor yoke was adjusted so that the flux density did not exceed 1.1 T. This value represented the saturation point of the material, which should not be exceeded to reduce iron losses. The flux density in the stator yoke was much lower because a thicker yoke was used than was necessary from an electromagnetic point of view. However, this thicker yoke was advantageous in terms of reducing vibration. At the edge of stator yoke, there were areas where the material was very weakly saturated. These places could be cut out and used to better secure the stator to the frame.

4.6.2. Air-Flow Analysis

This analysis was performed in the case of the salient poles and in the case of the cylindrical rotor in the program Fluent 2D. In Figure 16, the velocity field for the rated speed is represented.

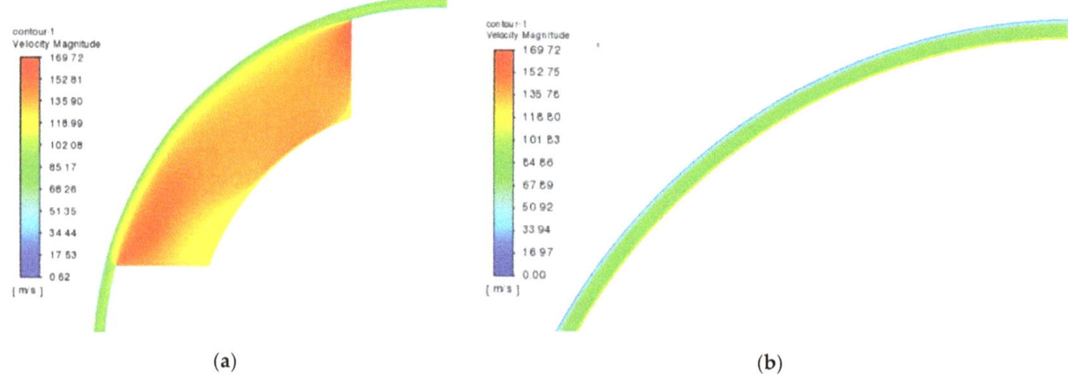

(a)   (b)

Figure 16. Velocity field contours inside the air gap at 100,000 rpm: (**a**) salient poles; (**b**) smooth cylindrical rotor.

In addition, the windage losses were calculated based on the force acting against the movement of the rotor. In the case of the rotor with salient poles, these losses reached a value of 1111.8 W, which was a difference of 4.2% compared to the analytical calculation; and 432 W in the case of the smooth cylindrical rotor, which was a difference of 41.5%. These values corresponded to the rated speed. A comparison of windage losses for different speeds is shown in Figure 17.

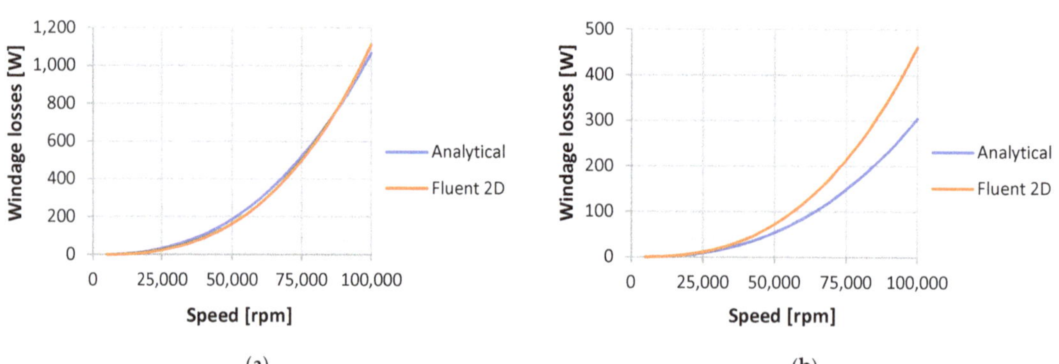

(a)   (b)

Figure 17. Comparison of windage losses for different speeds: (**a**) salient poles, (**b**) smooth cylindrical rotor.

4.6.3. Thermal Analysis

Finally, Figure 18 shows the thermal analysis of the final design of the SRM. The analysis confirmed that all thermal limits were met.

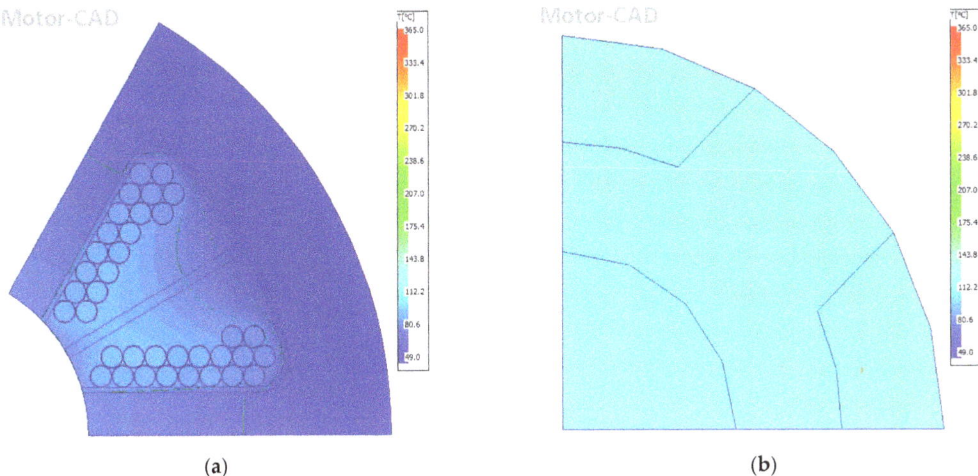

Figure 18. The distribution of temperature at steady state of the final design: (**a**) stator; (**b**) rotor.

5. Discussion

In the previous work, the initial design of a high-speed SRM was presented. However, this design did not achieve the required output power, and therefore optimization was required. Several analyses were performed, but it was not possible to determine the rotor diameter and air-gap size using electromagnetic parameters alone. Therefore, the analysis of windage losses, mechanical analysis, and preliminary thermal analysis were performed in this paper. Based on the analysis of windage losses, it was concluded that a smooth cylindrical rotor was needed to reduce these losses. Mechanical analysis has shown that the most suitable method for creating this rotor is to use a sleeve around the rotor. Preliminary thermal analysis has shown that the rotor diameter must be reduced to 30 mm to meet the thermal limits. To use a sleeve, it was necessary to increase the air-gap size to 0.5 mm.

In previous work, the highest possible average torque was achieved at a phase current of 200 A. The rotor was optimized from a thermal point of view, but the required output power still was not achieved. Thus, to achieve a higher average torque, it was necessary to increase the phase current. First, the increase in phase current was realized by increasing the length of the rotor poles. However, the required output power could not be achieved, even with very long rotor poles. The phase current could also be influenced by changing the advance angle. Therefore, optimization was performed to find a compromise between the phase current and the length of the rotor poles. Using this analysis, several possible designs were found that achieved the required output power of 8 kW, while the design with the shortest rotor poles was chosen so that a shaft with a larger diameter could be used.

The entire optimization was focused on the rotor, because modification of the rotor was limited due to the provision of sufficient space for winding. An analysis of the stator parameters was also performed for completeness. The analysis showed that it was possible to increase the average torque by shortening the length of the stator pole. The disadvantages of this change were the reduction of space for windings and a lower saturation of the stator yoke, because this shortening causes an increase in the thickness of the stator yoke. An increase in the average torque could also be achieved by inserting a rounding at the point where the stator pole protruded from the stator yoke. By increasing the pole width, it was possible to reduce the torque ripple, but the decrease in average torque was too high, so this change was not advantageous. The number of parallel conductors was reduced from 11 to 6 to facilitate winding manufacturing. This change caused a decrease in the average torque and an increase in the winding losses, which caused a decrease in output power. This decrease was replaced by shortening the stator poles.

Finally, the final geometry of motor was presented together with dynamic parameters, flux density distribution, and the results of the thermal analysis. The given geometry met the set requirements and limits. However, it was still possible to optimize the stator to better distribute the flux density or to reduce the maximum phase current. It was possible to reduce the stator's outer diameter or to cut out areas with low saturation to improve the flux density distribution. By further shortening the stator poles and subsequently reducing the advance angle, it was possible to achieve a reduction in the phase current. In addition, cooling could be optimized by reducing the water flow or the number of cooling channels. However, the simulation results for the motor prototype will be experimentally verified before further optimization.

6. Conclusions

The initial design of the high-speed SRM achieved high total losses and low efficiency, and did not achieve the required output power. Therefore, several analyses were performed to optimize this design. From an electromagnetic point of view, the geometry was adjusted to achieve the required output power and a good flux density distribution; and from a mechanical point of view, to use the largest possible shaft diameter.

Compared to the initial design, in addition to achieving the required output power, the total losses were also reduced, which led to an increase in efficiency of 41.5%, and thus to a reduction in energy consumption. This improvement was achieved by reducing windage losses by modifying the rotor design, since a smooth cylindrical rotor was not considered in the initial design. If a smooth cylindrical rotor was considered in the case of the initial design, this improvement would be lower. Nevertheless, efficiency was improved by 18.5%. This improvement was achieved by further reducing windage losses by reducing the rotor diameter, but also by reducing iron losses by selecting more suitable material.

Another improvement over the initial design was a 71.7% reduction in torque ripple. However, to achieve the required output power, it was necessary to increase the maximum phase current by 143.7 A, which led to more than an increase in winding losses of more than four times.

The main result of this paper was the optimized electromagnetic design high-speed SRM, which achieved an output power of 8 kW at a rotational speed of 100,000 rpm. In addition, windage losses were calculated using a procedure suitable for high-speed motors; the method for reducing them was determined based on mechanical analyses; and finally, thermal analysis was performed for the design. In all the analyses, modern and scientific methods and programs were used, and were based on the finite element method or analytical methods.

Author Contributions: Conceptualization, methodology, S.K. and P.R.; validation, P.R.; formal analysis, S.K., R.B., M.S., and R.L.; software, investigation, resources, data curation, writing—original draft preparation, visualization, S.K.; first review, P.R.; funding acquisition, supervision, P.R. All authors have read and agreed to the published version of the manuscript.

Funding: This work was supported by the Slovak Scientific Grant Agency VEGA (No. 1/0615/19). This publication was realized with the support of the Operational Program Integrated Infrastructure 2014–2020 of the project: Innovative Solutions for Propulsion, Power and Safety Components of Transport Vehicles, code ITMS 313011V334, co-financed by the European Regional Development Fund.

Institutional Review Board Statement: Not applicable.

Informed Consent Statement: Not applicable.

Data Availability Statement: Data sharing is not applicable.

Conflicts of Interest: The authors declare no conflict of interest.

References

1. Gerada, D.; Mebarki, A.; Brown, N.L.; Gerada, C.; Cavagnino, A.; Boglietti, A. high-speed electrical machines: Technologies, trends, and developments. *IEEE Trans. Ind. Electron.* **2014**, *61*, 2946–2959. [CrossRef]
2. Rahman, M.A.; Chiba, A.; Fukao, T. Super high speed electrical machines-summary. In Proceedings of the IEEE Power Engineering Society General Meeting, Denver, CO, USA, 6–10 June 2004; pp. 1–4. [CrossRef]
3. Hieu, P.T.; Lee, D.H.; Ahn, J.W. Design of a high speed 4/2 switched reluctance motor for blender application. In Proceedings of the IEEE Transportation Electrification Conference and Expo, Asia-Pacific (ITEC Asia-Pacific), Harbin, China, 7–10 August 2017; pp. 1–5.
4. McGuiness, D.T.; Gulbahce, M.O.; Kocabas, D.A. A performance comparison of different rotor types for high-speed induction motors. In Proceedings of the ELECO 2015 IEEE: 9th International Conference on Electrical and Electronics Engineering, Bursa, Turkey, 26–28 November 2015; pp. 584–589.
5. Uzhegov, N.; Barta, J.; Kurfurst, J.; Ondrusek, C.; Pyrhonen, J. Comparison of High-Speed Electrical Motors for a Turbo Cir-culator Application. *IEEE Trans. Ind. Appl.* **2017**, *53*, 4308–4317. [CrossRef]
6. Lim, M.-S.; Kim, J.-M.; Hwang, Y.S.; Hong, J.-P. Design of an ultra-high-speed permanent-magnet motor for an electric tur-bocharger considering speed response characteristics. *IEEE/ASME Trans. Mechatron.* **2017**, *22*, 774–784. [CrossRef]
7. Bianchi, N.; Bolognani, S.; Luise, F. Potentials and limits of high speed PM motors. In Proceedings of the 38th IAS Annual Meeting on Conference Record of the Industry Applications Conference, Salt Lake City, Salt Lake City, UT, USA, 12–16 October 2003; pp. 1570–1578. [CrossRef]
8. Dong, J.; Huang, Y.; Jin, L.; Lin, H. Comparative study of surface-mounted and interior permanent-magnet motors for high-speed applications. *IEEE Trans. Appl. Supercond.* **2016**, *26*, 1–4. [CrossRef]
9. Vag, L.; Laksar, J. Overview of technology, problems and comparison of high speed synchronous reluctance machines. In Proceedings of the 12th International Conference on ELEKTRO, Mikulov, Czech Republic, 1 March–23 May 2018; pp. 1–4.
10. Chayopitak, N.; Pupadubsin, R.; Karukanan, S.; Champa, P.; Somsiri, P.; Thinphowong, Y. Design of a 1 5 kW high speed switched reluctance motor for electric supercharger with optimal performance assessment. In Proceedings of the 15th International Conference on Electrical Machines and Systems (ICEMS), Sapporo, Japan, 21–24 October 2012; pp. 1–5.
11. Fairall, E.W.; Bilgin, B.; Emadi, A. State-of-the-art high-speed switched reluctance machines. In Proceedings of the 2015 IEEE International Electric Machines & Drives Conference (IEMDC), Coeur d'Alene, ID, USA, 10–13 May 2015; pp. 1621–1627.
12. Besharati, M.; Atkinson, G.; Widmer, J.; Pickert, V.; Pullen, K. Investigation of the mechanical constraints on the design of a super-high-speed switched reluctance motor for automotive traction. In Proceedings of the 7th IET International Conference on Power Electronics, Machines and Drives (PEMD 2014), Manchester, UK, 8–10 April 2014; Volume 1.
13. Kozuka, S.; Tanabe, N.; Asama, J.; Chiba, A. Basic characteristics of 150,000 r/min switched reluctance motor drive. In Proceedings of the 2008 IEEE Power and Energy Society General Meeting-Conversion and Delivery of Electrical Energy in the 21st Century, Pittsburgh, PA, USA, 20–24 July 2008; pp. 1–4.
14. Kocan, S.; Rafajdus, P.; Makys, P.; Bastovansky, R. Design of high speed switched reluctance motor. In Proceedings of the 2018 IEEE International Conference and Exposition on Electrical And Power Engineering (EPE), Iasi, Romania, 18–19 October 2018; pp. 421–426.
15. Kocan, S.; Rafajdus, P. Dynamic model of high speed switched reluctance motor for automotive applications. *Transp. Res. Procedia* **2019**, *40*, 302–309. [CrossRef]
16. Kocan, S.; Rafajdus, P.; Makys, P.; Bastovansky, R. Number of turns influence on the parameters of high speed switched reluctance motor. In Proceedings of the 2019 IEEE International Aegean Conference on Electrical Machines and Power Electronics (ACEMP) & 2019 International Conference on Optimization of Electrical and Electronic Equipment (OPTIM), Istanbul, Turkey, 2–4 September 2019; pp. 240–245.
17. Kocan, S.; Rafajdus, P.; Sumega, M. Effect of air gap size on parameters of high speed switched reluctance motor. In Proceedings of the 13th International Conference Elektro, Taormina, Italy, 25–28 May 2021; pp. 1–5.
18. Kocan, S.; Rafajdus, P.; Kovacik, M. Investigation of rotor parameters of high speed switched reluctance motor for auto-motive application. *Transp. Res. Procedia* **2021**, *55*, 1003–1010, in press. [CrossRef]
19. Gong, C.; Habetler, T. Electromagnetic design of an ultra-high speed switched reluctance machine over 1 million rpm. In Proceedings of the 2017 IEEE Energy Conversion Congress and Exposition (ECCE), Cincinnati, OH, USA, 1–5 October 2017; pp. 2368–2373. [CrossRef]
20. Sadr, S.; Abdelli, A.; Ben-NachouaneIntuitive, A.; Friedrich, G.; Vivier, S. Comprehension and estimation of windage losses in rotor slotted air gaps of electrical machines using CFD-LES methods. In Proceedings of the 2019 IEEE Energy Conversion Congress and Exposition (ECCE), Baltimore, MD, USA, 29 September–3 October 2019; pp. 6078–6083. [CrossRef]
21. Zhang, J.; Wang, H.; Chen, L.; Tan, C.; Wang, Y. Salient pole rotor with partially covered shrouds for the windage loss re-duction of bearingless switched reluctance motors. In Proceedings of the 13th IEEE Conference on Industrial Electronics and Applications (ICIEA), Wuhan, China, 31 May–2 June 2018; pp. 868–873.

22. Dang, J.; Haghbin, S.; Du, Y.; Bednar, C.; Liles, H.; Restrepo, J.; Mayor, J.R.; Harley, R.; Habetler, T. Electromagnetic design considerations for a 50,000 rpm 1 kW Switched Reluctance Machine using a flux bridge. In Proceedings of the 2013 IEEE International Electric Machines & Drives Conference, Chicago, IL, USA, 12–15 May 2013; pp. 325–331.
23. Kunz, J.; Cheng, S.; Duan, Y.; Mayor, J.R.; Harley, R.; Habetler, T. Design of a 750,000 rpm Switched Reluctance Motor for Micro Machining. In Proceedings of the IEEE Energy Conversion Congress and Exposition (ECCE), Atlanta, GA, USA, 12–16 September 2010; pp. 3986–3992.

 *energies*

Article

Experimental Investigation of a Double-Acting Vane Pump with Integrated Electric Drive

Marek Pawel Ciurys [1],* and Wieslaw Fiebig [2],*

1   Department of Electrical Machines, Drives and Measurements, Faculty of Electrical Engineering, Wroclaw University of Science and Technology, 50-370 Wroclaw, Poland
2   Faculty of Mechanical Engineering, Wroclaw University of Science and Technology, 50-371 Wroclaw, Poland
*   Correspondence: marek.ciurys@pwr.edu.pl (M.P.C.); wieslaw.fiebig@pwr.edu.pl (W.F.)

Abstract: The article presents an innovative design solution of a balanced vane pump integrated with an electric motor that has been developed by the authors. The designed and constructed bench, which enables testing of the system: power supply, converter, ntegrated motor—pump assembly and hydraulic load at different motor speeds and different pressures in the hydraulic system, is described. The electromagnetic and hydraulic processes in the motor-pump unit are investigated, and new, previously unpublished, results of experimental studies at steady and dynamic states are presented. The results of the study showed good dynamics of the integrated motor-pump assembly and proved its suitability to control the pump flow rate, and thus, the speed of the hydraulic cylinder or the speed of the hydraulic motor.

Keywords: vane pump; electric motor; integrated motor-pump assembly (IMPA); balanced vane pump; fluid power drives; brushless DC electric motor; permanent magnet machine; measurements

Citation: Ciurys, M.P.; Fiebig, W. Experimental Investigation of a Double-Acting Vane Pump with Integrated Electric Drive. *Energies* **2021**, *14*, 5949. https://doi.org/10.3390/en14185949

Academic Editor: Sérgio Cruz

Received: 11 August 2021
Accepted: 15 September 2021
Published: 18 September 2021

1. Introduction

Hydraulic drives are used wherever it is necessary to obtain high force values with a relatively low weight of the device. They have the highest of all known types of drives and power density [1] in a unit of volume. They are used in stationary machines (presses, machine tools, machines for processing, plastics, robotics, etc.), in mobile machines (construction, mining), and in vehicles (cars, trucks, etc.). Most applications require high reliability and good controllability. The possibility of controlling the flow of displacement pumps directly through the rotational speed has many advantages and is an important simplification of fluid power drives. The proposed solution uses a double vane pump driven by a permanent magnet electric motor with variable rotational speed. In such a solution, it is possible to adjust the power of the electric motor to the power of hydraulic actuators (hydraulic cylinders or hydraulic motors), which leads to increased energy efficiency.

Vane pumps belong to the group of displacement machines in which the displacement elements are vanes placed in the radial slots of the rotor. The inter-vane volume changes depending on the angle of rotation and the working fluid is transported from the suction to the delivery channels.

There are two types of vane pumps, i.e., single-acting and double-acting (balanced) pumps (Figure 1). In double-acting pumps, during one rotation of the rotor, the suction and discharge cycles occur twice during one revolution.

Double -acting pumps have advantages such as: low bearing loads due to the compensation of pressure forces [2,3] and the associated higher maximum pressure, as well as high durability, low noise emission, and relatively high energy efficiency. Vane pumps are widely used due to their simple design and relatively easy manufacturing technology.

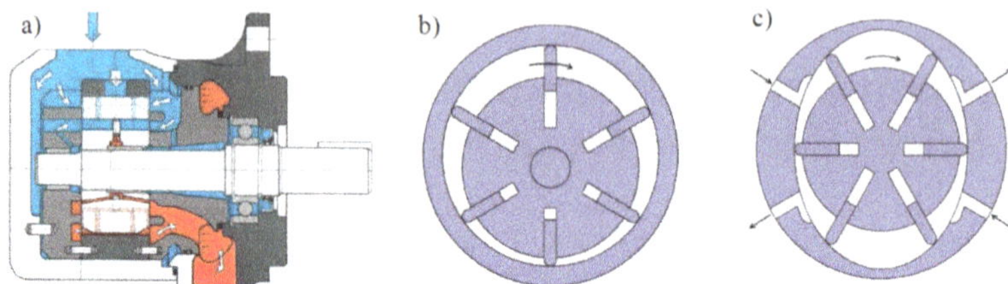

Figure 1. The vane pump: (**a**) Cross-over; (**b**) Single-acting; (**c**) Double-acting.

In conventional hydraulic drive solutions, the fluid flow is usually controlled by means of proportional valves and servo valves. Apart from the high price, electrohydraulic control systems have a number of disadvantages, including problems with stability at high loads and high sensitivity to contamination of the working fluid [4,5]. Due to the mechatronic pump drive, it would be possible to control its flow rate directly by varying the speed of the electric motor, which is a significant simplification of control systems in hydraulic drives.

The paper concerns a new design solution in Figure 5 which a displacement pump is built into the rotor of the electric motor [6,7]. The stator of the electric motor is a fixed element of the housing of the entire device. The rotor of the electric motor causes the displacement pump housing to rotate. The inner part of the pump (with the vanes) is stationary. Due to the lack of centrifugal forces, the vanes are pressed against the cam ring by springs [8].

In the case of currently used solutions, where the pump together with the casing and the motor are separate elements, a mounting console and a clutch are usually needed. These elements constitute an additional mass. In the integrated motor-pump assembly (IMPA) solution, these elements are not not present.

The authors are convinced that the advantages of this solution will include, above all, a higher power density in a volume unit compared to conventional solutions, the ability to control the pump flow rate by changing the rotational speed of the electric motor (hence better controllability), greater energy efficiency, better dynamics, and lower noise emissions due to the installation of positive displacement pump inside the electric motor, which is a kind of soundproofing housing. Such systems are more efficient due to the possibility of direct adjustment of the motor power consumption to the load of the actuators (actuator, hydraulic motor).

A similar solution has been used in Voith's company EPAI pump [9]. In this solution, the internal gear pump is built into the interior of a three-phase, four-pole, squirrel cage induction motor with internal cooling with hydraulic fluid.

Despite the many types of electric motors available on the market, three-phase squirrel cage induction motors are the most widely used source of electric drive [10,11]. This is due to their simple construction, high reliability, low price, and low maintenance requirements. However, their efficiency, especially for low-rated power, is much lower compared to permanent magnet motors [10–12].

Permanent magnet motors are used in high-performance drives with a wide range of rotational speed regulation. They have the highest energy efficiency of all electric motors, the highest rated and maximum torque values per unit of mass and volume, and high torque overload [12,13]. Moreover, they enable high dynamics [14] and precision of the drive control, i.e., fast response of the drive to load changes and changes in control settings. They are widely used in automation and robotics [12,13,15], computer equipment [12], aviation [12,16–18], and electric and hybrid vehicles [12,19–27]. They are also used in medical devices [28–31] as well as in aerospace devices [12,32,33].

Due to the advantages of machines with permanent magnets, as a motor in the developed solution, a brushless DC motor with a rotational speed regulated by a converter was used. While, for example, in the publication of [34], solutions in which the pump is driven by brushless DC motors are known, the developed construction in which the vane pump is built into the rotor of the BLDC motor (Figures 2 and 3) is a novel design solution.

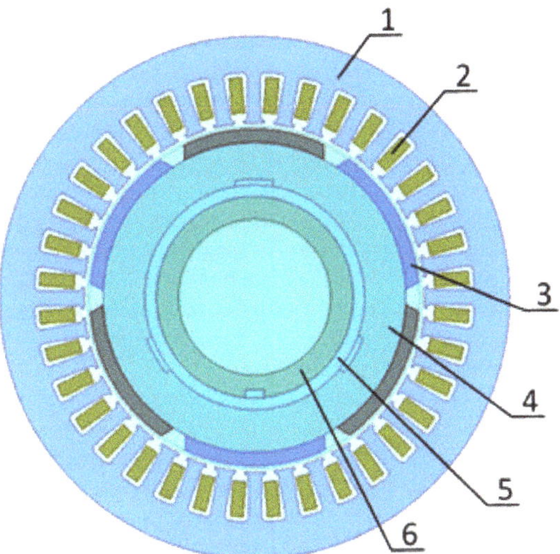

Figure 2. Cross-section of the brushless DC motor. 1—stator core, 2—winding, 3—permanent magnet, 4—rotor yoke, 5—non-magnetic sleeve, 6—pump.

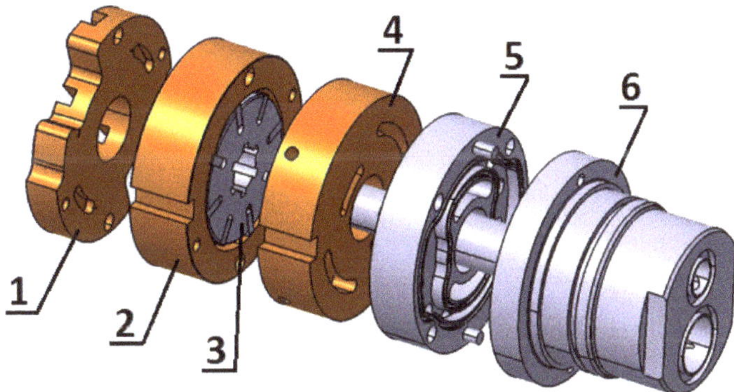

Figure 3. Vane pump module [6]. (Copyright by ASME, 2017) Number of vanes 8, unit displacement 3.45 cm^3/rev. 1—left pressure plate, 2—cam ring, 3—internal part of the pump with vanes, 4—right pressure plate/rotational part of the flow distributor, 5—stationary part of the flow distributor, 6—delivery channels.

In the further part of the article, the novel integrated motor-pump assembly is presented, as well as its experimental investigation on the designed and built test stand in the steady and dynamic states.

2. Design of the Novel Pump-Motor Assembly

The BLDC motor of the developed IMPA solution has been designed at the Department of Electrical Machines, Drives and Measurements of the Wrocław University of Science and Technology for the possibility of integrating a vane pump, and in order to obtain the required parameters: power of P_{max} = 2.5 kW at the maximum rotational speed n_{max} = 3000 rpm.

Preliminary design calculations for the motor have been performed using the proprietary BLDC motor design program. With its help, the winding and initial dimensions of the magnetic circuit of the motor were selected. To reduce motor dimensions, a three-pole pairs construction has been applied. Then, using the ANSYS Maxwell software, a finite element method computational analysis of the influence of the motor magnetic circuit dimensions on the value of the magnetic flux, the cogging torque, the electromagnetic torque, the magnetic flux penetration into the pump, and the resistance of the magnets to demagnetization have been performed. The results of these computations made it possible to select the final motor design solution. In the ANSYS Maxwell software, using the developed field-circuit model of the: supply network—converter—motor system [35], the motor parameters were determined and then verified by measurements [36,37].

The cross-section of the motor magnetic circuit is shown in Figure 2. The stator consists of a laminated core (1) with a three-phase winding (2). High-energy neodymium magnets (3) are mounted on the rotor yoke surface (4). Between the rotor yoke (4) and the pump (6), a nonmagnetic bushing (5) is used. Its purpose is to prevent the penetration of the magnetic flux into the pump interior.

In Figure 3, the developed vane pump module is shown. The pump elements 1, 2 and 4 are movable, while the elements 3, 5 and 6 are stationary. The connection to the suction and delivery line occurs with the pressure compensated plate 5 and the connecting part 6 with the suction and discharge channels. Several problems had to be overcome through the different stages of the vane pump design process. One of the important issues was to select the geometry of the cross-over areas between the suction and discharge ports and to provide sufficient support for the vanes in these areas.

In the proposed integrated motor-pump assembly solution (Figures 4 and 5), the pump cartridge 3 (Figure 4) is placed inside the rotor of the permanent magnet brushless DC motor (Figures 2, 4 and 5).

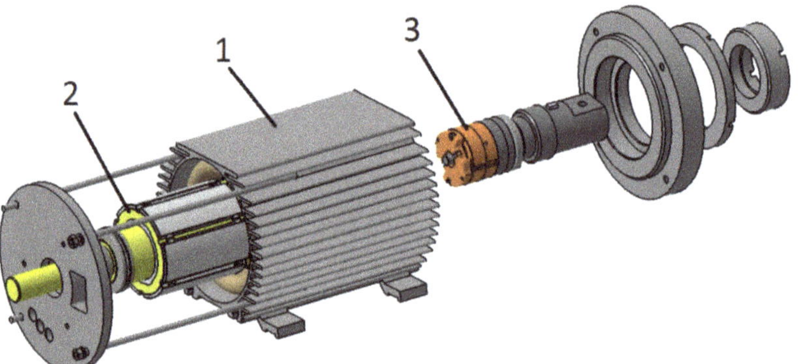

Figure 4. Vane pump integrated with BLDC motor. 1—stator, 2—rotor of electric motor, 3—pump cartridge.

The rotor 5 of the motor (Figure 5) and the pump cartridge 8 (cam ring and two pressure plates) are coupled together. In the new solution, internal part of the pump 4 with vanes is stationary, therefore, the additional support of vanes with springs has been introduced.

The rotor 5 of the brushless DC motor (Figure 5) drives the pump cartridge 8. Due to the fact that the inner part 4 of the balanced vane pump is stationary, the vanes are pressed to a cam ring with spring elements. Changing the motor speed using the pulse width modulation method provides the ability to control the pump flow.

The developed IMPA solution has been patented (displacement pump with the integrated electric drive P. 213898), and is predestined for applications where the drive must be compact (weight reduction) and have reduced noise emissions. One of the possible application examples can be an electric driven vehicle with hydraulic ramp for lifting loads. The electric motor would then be powered by the battery of the vehicle.

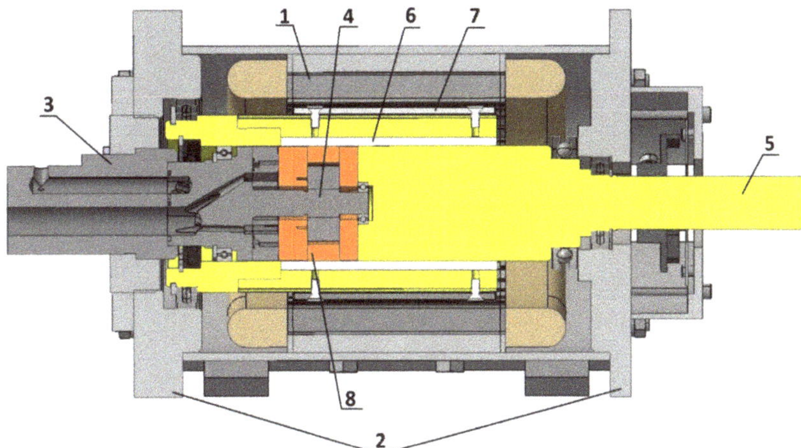

Figure 5. Integrated BLDC motor-vane pump assembly. 1—stator core, 2—covers, 3—suction and discharge channels, 4—internal part with vanes, 5—rotor of brushless DC motor, 6—nonmagnetic bushing, 7—neodymium magnets, 8—pump cartridge [6]. (Copyright by ASME, 2017).

3. Description of the Research Test Bench

The research bench (Figures 6–9) enables the measurement of electrical, mechanical, and hydraulic quantities in the system: power supply—converter—integrated motor pump assembly—hydraulic system, at different pressures in the hydraulic system, and at different speeds.

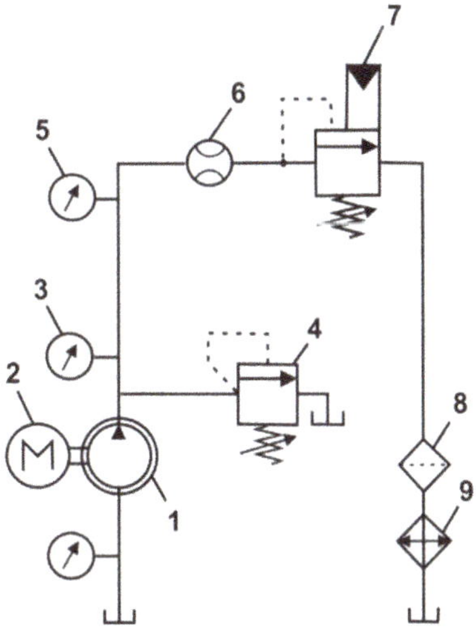

Figure 6. The schematic view of the hydraulic test rig. 1, 2—pump motor assembly, 3, 5—pressure gauge, 4—pressure relief valve, 6—flow meter, 7—pressure relief valve, 8—filter, 9–cooler.

Figure 7. View of the test bench.

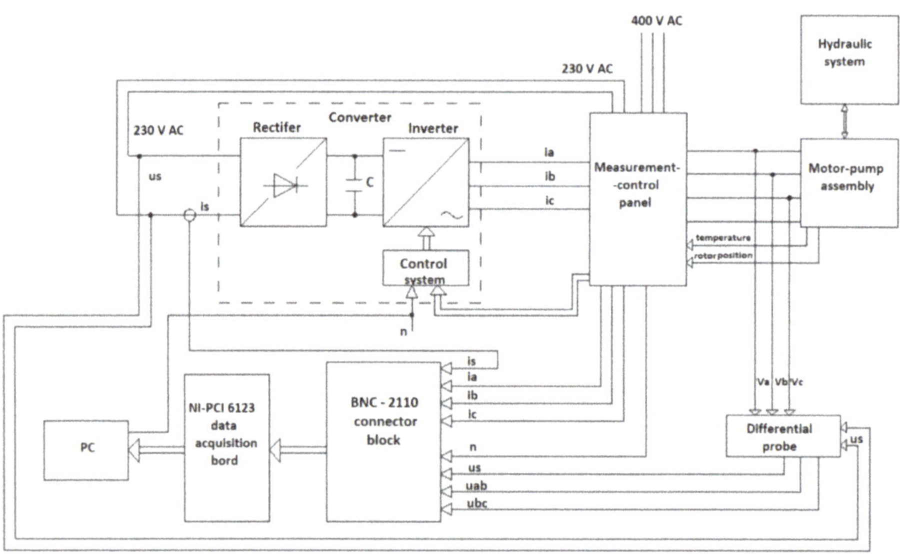

Figure 8. Diagram of the system for measuring electrical quantities of the motor-pump unit. is— converter current, us—supply voltage, ia, ib, ic—phase currents, va, vb, vc—potentials of the phases a, b, and c, uac, ubc—line-to-line voltages, and n—speed.

Absolute pressure (5), pressure pulsation (3), and flow rate (6) are measured on the delivery side of the pump (Figure 6). On the return line, the filter (8) and cooler (9) are located. The load of the pump is realized by the proportional pressure relief valve (7). The safety valve (4) secures the system maximal pressure.

The Parker SCP 400 pressure transducer (3) with the 0–400 bar measuring range and accuracy of 0.5% has been used. The SCFT-060 turbine flow rate meter (6) is from Parker (measuring range: 3–60 Lpm, measured value accuracy +/− 0.5%). The oil temperature is controlled using a temperature sensor.

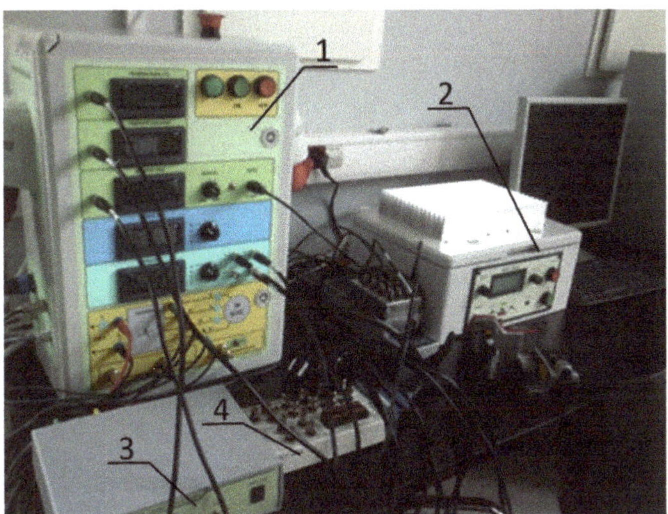

Figure 9. View of the system elements for measuring electrical quantities of the motor-pump unit. 1—measurement-control panel, 2—converter, 3—differential probe, 4—BNC-2110 connector block.

The test bench and the developed integrated motor-pump assembly unit are shown in Figure 7.

The BLDC motor is powered (Figure 8) by a converter (rectifier—capacitor—inverter) from a single-phase 230 VAC, 50 Hz mains. The measurement-control panel (Figures 8 and 9) contains rms and average value meters, current transducers as well as switching and signalling elements.

LA55-P transducers have been used to measure phase currents, and a differential probe has been applied for the purpose of measuring voltages.

The instantaneous values of motor speed, currents and voltages, as well as the converter supply voltage and current, are given to the PCI-6123 multifunction DAQ device and PC through the BNC-2110 terminal block (Figure 8). Information about the position of the rotor as well as the speed signal are obtained by means of optical sensors.

The rotational speed of the IMPA may be controlled manually by the control panel potentiometer or by setting the value or trajectory of the speed control signal by the LabVIEW program. The virtual instrument developed in the LabVIEW programming environment provides communication with the DAQ card, previewing, and registering of the measured hydraulic, electrical, and mechanical signals.

4. Steady-State Tests of the Motor-Pump Assembly

Measurements of the waveforms of electrical and hydraulic quantities in steady states have been performed at different values of speed stabilized by the converter. The sampling frequency of electrical quantity signals was set to the maximum value supported by the measurement card at 500,000 samples per second per channel. Such a high sampling frequency of electrical signals resulted from the pulse width modulation (PWM) of the converter supplying the motor pump unit with the 15,625 Hz frequency. The sampling rate of the hydraulic quantities was 100 samples/s.

The results for speeds of 1000 rpm and 1500 rpm at the pressure of 60 bar are presented in Figures 10–13.

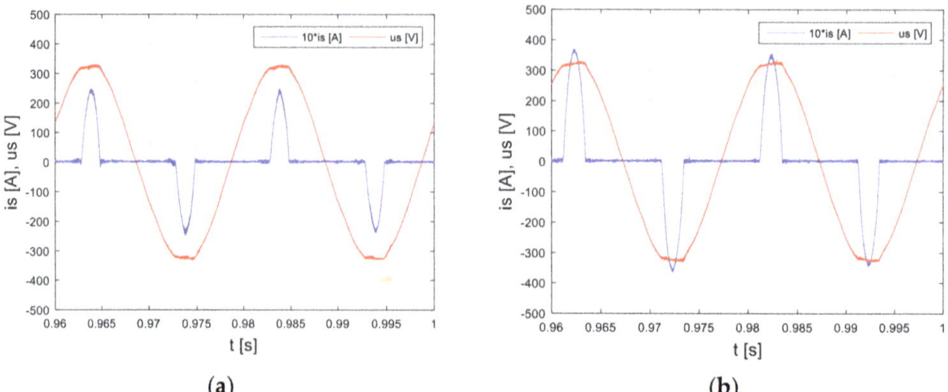

(a) **(b)**

Figure 10. Waveforms of the converter input current (is) and voltage (us) at the 60 bar pressure for the speed of: (**a**) 1000 rpm; (**b**) 1500 rpm.

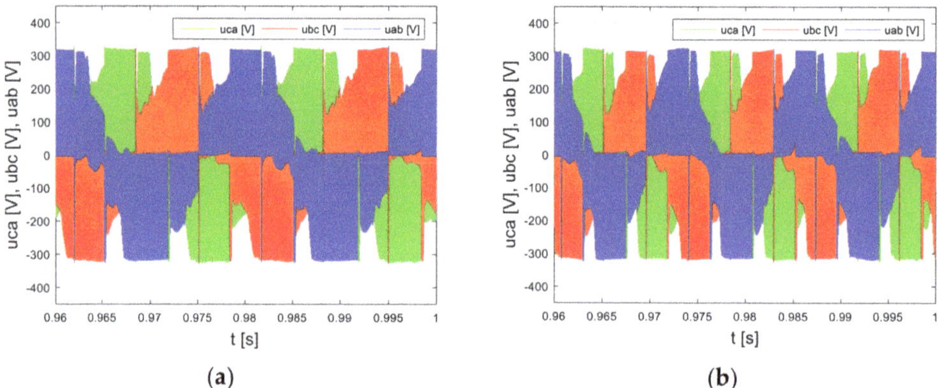

(a) **(b)**

Figure 11. Waveforms of the line-to-line voltages (uab, ubc, uca) at 60 bar pressure for the speed of: (**a**) 1000 rpm; (**b**) 1500 rpm.

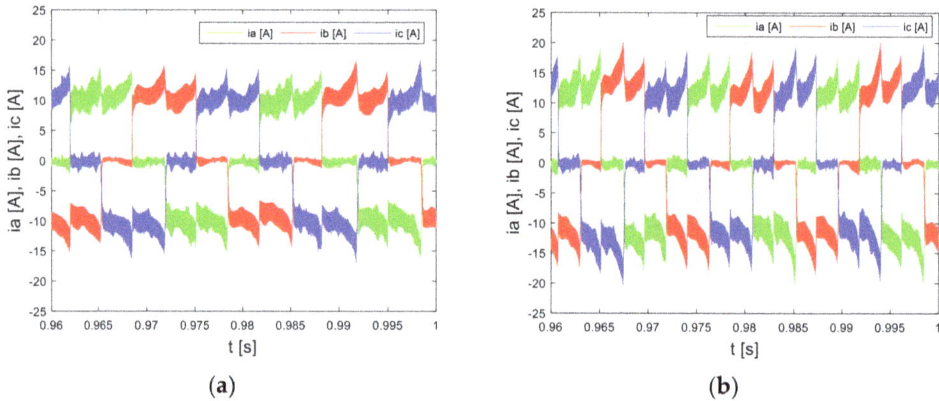

(a) **(b)**

Figure 12. Phase-current waveforms (ia, ib, ic) at the 60 bar pressure for the speed of: (**a**) 1000 rpm; (**b**) 1500 rpm.

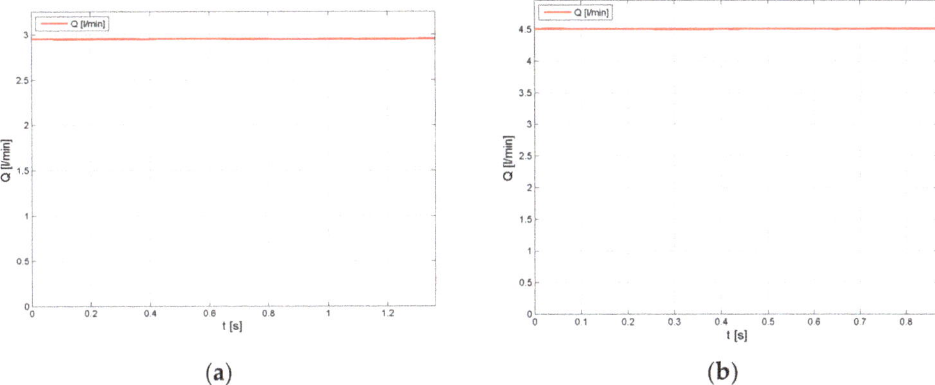

Figure 13. Waveforms of the pump flow rate (Q) at 60 bar pressure for the speed of: (**a**) 1000 rpm; (**b**) 1500 rpm.

The waveforms of the converter input current and voltage are presented in Figure 10.

The converter input current is not continuous. It is taken from the supply network in the time periods in which the instantaneous value of the voltage at the input of the converter is higher than the voltage of capacitor C (Figure 8). The frequency of the capacitor voltage pulsation, i.e., the voltage at the input of the inverter, is 100 Hz.

Pulsations of this voltage are transferred to the waveforms of the phase-to-phase voltages (Figure 11) and phase currents (Figure 12) of the motor. They are more noticeable, the higher the output power of the motor-pump unit.

Apart from pulsations with a frequency of 100 Hz, the waveforms of phase-to-phase voltages and phase currents also contain pulsations resulting from the pulse width modulation (PWM) operation of the inverter and from the commutation of the phase windings.

In the waveforms of the motor phase currents, the share of pulsations resulting from the commutation is the highest.

Similarly to the phase-to-phase voltages (Figure 11), differences can be noticed for different rotational speeds in the phase currents' (Figure 12) commutation frequency. This is due to the operation of the converter which supplies the winding phases on the basis of information about the rotor position.

Pulsations in the waveforms of electrical quantities (currents and voltages) and the resulting pulsations of the mechanical torque of the motor practically do not cause the pulsation of rotational speed. The rotational speed ripples do not exceed 2%, which is due to the high frequency of mechanical torque pulsations, the rotor moment of inertia, and the operation of the converter speed stabilization system.

The pump flow rate at 60 bar pressure for the speed of 1000 and 1500 rpm is shown in Figure 13.

5. Dynamic Testing of the Motor-Pump Assembly at Constant Rotational Speed and Pressure Changes

Measurements of electrical and hydraulic quantities' waveforms have been performed at different values of speed (1000, 1500 rpm) stabilized by the converter while forcing pressure changes with the use of the proportional solenoid valve.

Changes in the pressure waveforms were forced by the LabVIEW program by setting the voltage signal trajectory that controls the operation of the proportional solenoid valve. These waveforms corresponded to a typical "ramp" function, i.e., the valve control signal increased linearly in the initial time interval, then it was kept constant, and in the final phase, it decreased linearly.

The waveform of the given voltage signal controlling the solenoid valve is shown in Figure 14. The ramp time (signal rise and fall time) was one second.

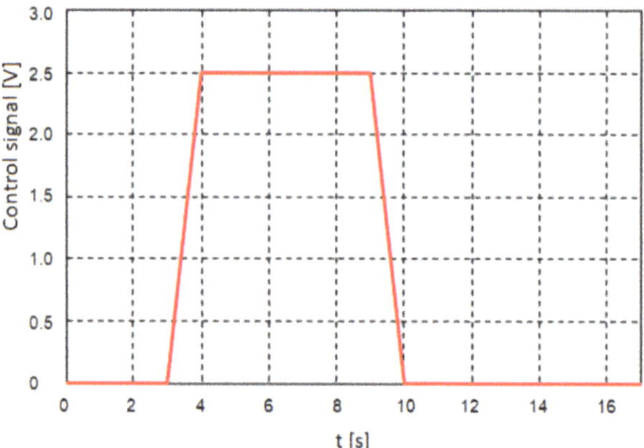

Figure 14. Waveform of the control signal.

The performed tests showed that the converter input voltage and the motor phase-to-phase voltages practically did not change with pressure changes, and therefore they were not included in the article. This proves that the converter and the motor can be loaded with more power than that required by the hydraulic system during the measurements.

The converter input currents for the speeds of 1000 and 1500 rpm are presented, respectively, in Figure 15, while the phase currents of the motor-pump assembly are shown in Figure 16.

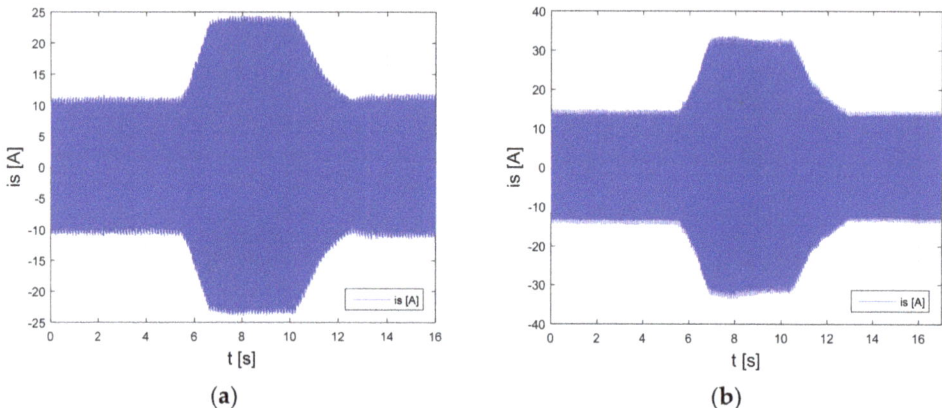

Figure 15. Waveforms of the input current of the converter for the speed of: (**a**) 1000 rpm; (**b**) 1500 rpm.

During the pressure changes in the hydraulic system (Figure 17), the rotational speed (Figure 18) practically did not change. At the speed of 1000 rpm, the speed changes did not exceed 1.4%, while at the speed of 1500 rpm, the changes were less than 1.1%. This proves a good quality stabilization of the rotational speed of the motor-pump drive system.

The drive system reacts without delay and overshoot to load changes caused by pressure changes in the hydraulic system. This results from a comparison of the waveforms of the converter input current (Figure 15), the phase currents of the motor-pump assembly (Figure 16), the waveforms of the actual pressure changes (Figure 17), and the pump flow rate (Figure 19). This proves the good dynamics of the pump drive system. However, the discharge pressure variation occurred with a significant delay from the LabVIEW voltage control signal (Figures 14 and 17).

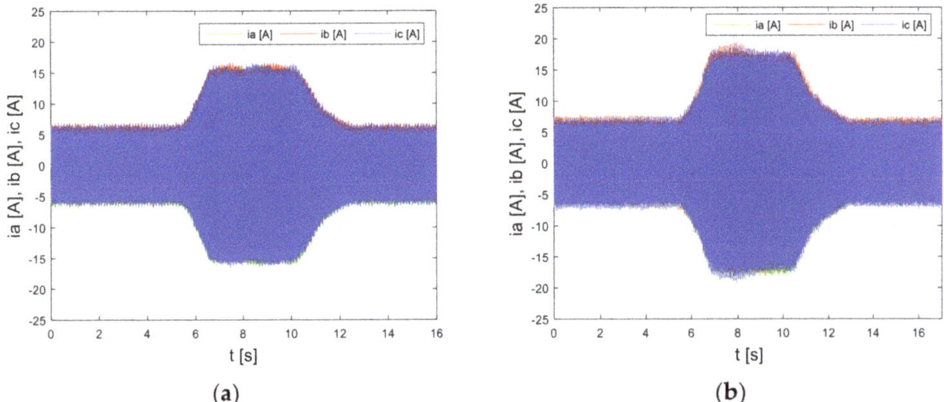

(a)

(b)

Figure 16. Phase-current waveforms (ia, ib, ic) for the speed of: (**a**) 1000 rpm; (**b**) 1500 rpm.

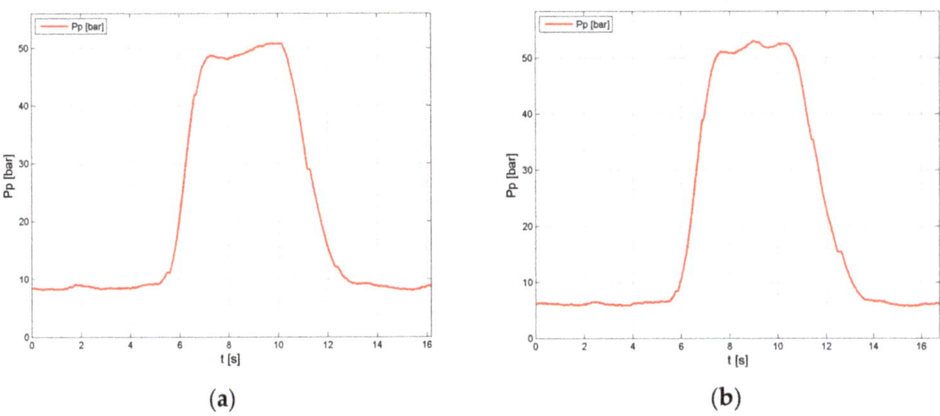

(a)

(b)

Figure 17. Waveforms of the pressure (Pp) for the speed of: (**a**) 1000 rpm; (**b**) 1500 rpm.

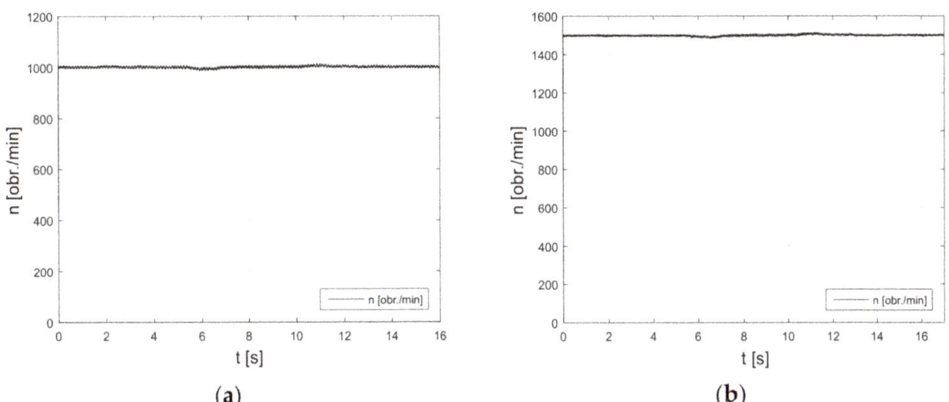

(a)

(b)

Figure 18. Waveforms of rotational speed (n) for the speed of: (**a**) 1000 rpm; (**b**) 1500 rpm.

The tests have confirmed the good quality of the pump drive system control. During pressure changes, there were no excesses, which usually occur with this type of change in the hydraulic system.

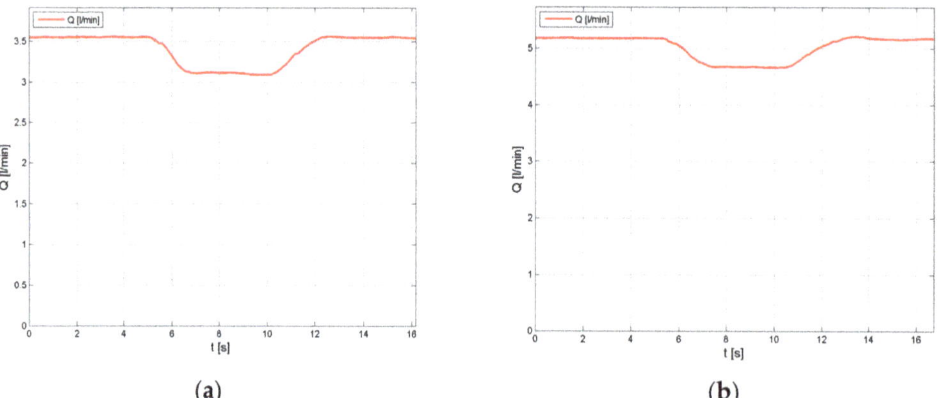

(a)

(b)

Figure 19. Waveforms of the pump flow rate (Q) for the speed of: (**a**) 1000 rpm; (**b**) 1500 rpm.

6. Dynamic Testing of the Motor-Pump Assembly at Constant Pressure and Changes in the Rotational Speed

Measurements of the waveforms of electrical and hydraulic quantities of the motor pump unit in dynamic states at different changes of rotational speed were performed. Measurements have been carried out without the voltage signal controlling the pressure in the hydraulic system. The sampling frequency of the electrical quantity signals was 500,000 samples/s and the sampling frequency of the hydraulic quantity signals was 100 samples/s. The changes in rotational speed were set manually by means of the converter.

The waveforms of rotational speed changes are presented in Figure 20. The measured waveforms of the electrical and hydraulic quantities of the motor-pump unit at various rotational speed changes are presented in Figures 21–23.

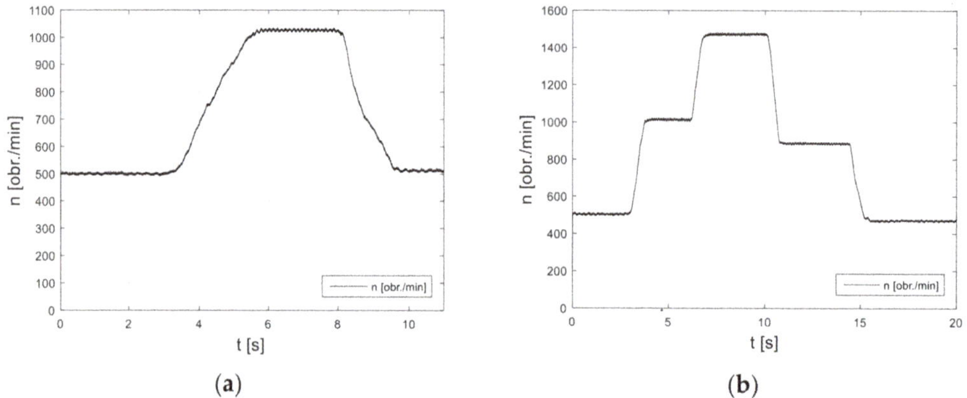

(a)

(b)

Figure 20. Waveforms of the forced rotational speed changes of the motor-pump assembly: (**a**) Single speed ramp; (**b**) Double speed ramp.

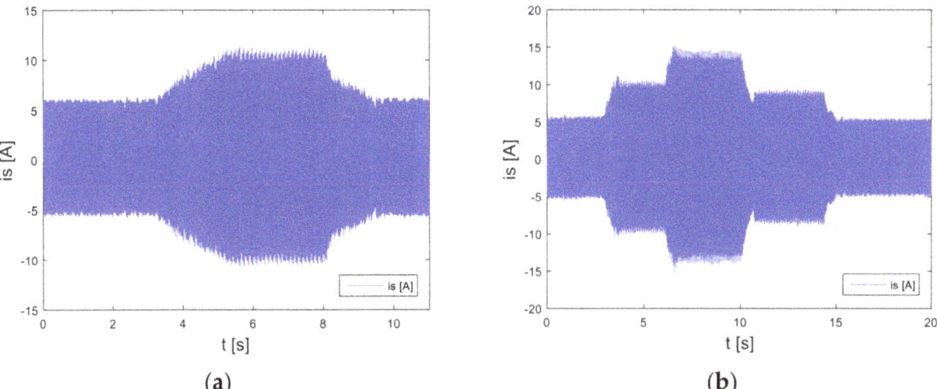

Figure 21. Waveforms of the converter input current: (**a**) Single speed ramp; (**b**) Double speed ramp.

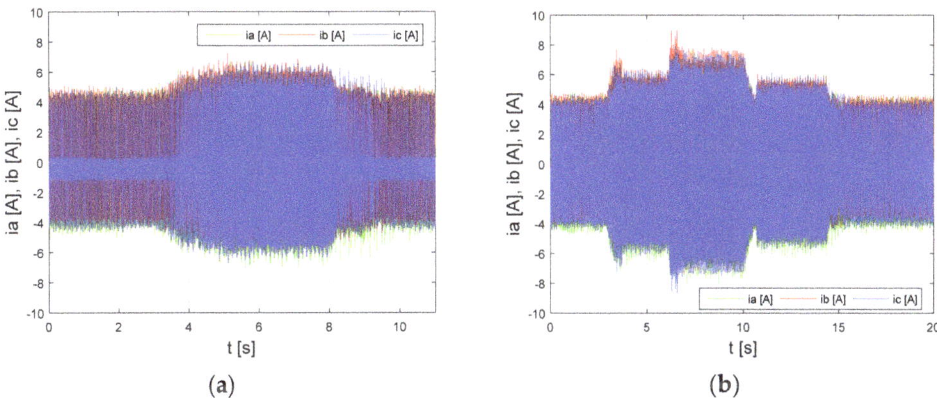

Figure 22. Motor-pump assembly phase-current waveforms: (**a**) Single speed ramp; (**b**) Double speed ramp.

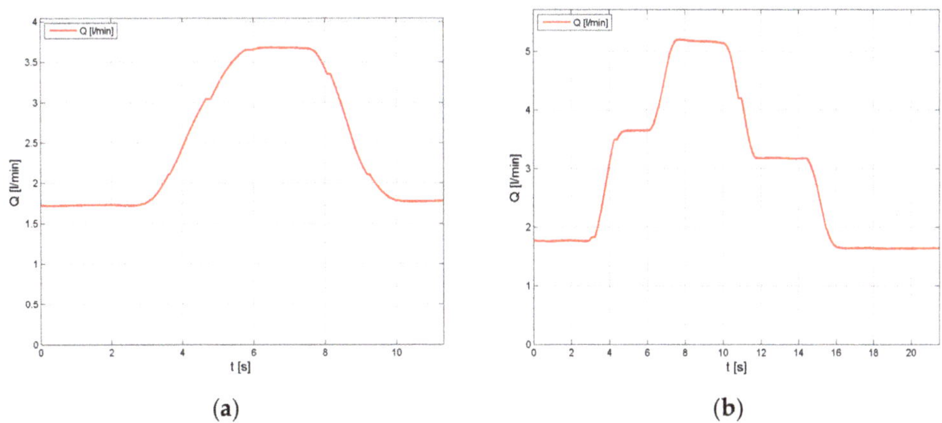

Figure 23. Flow rate: (**a**) Single speed ramp; (**b**) Double speed ramp.

Variations in the rotational speed of the motor (Figure 20) lead to the corresponding changes in the flow rate of the pump. By changing the rotational speed of the electric motor, it is possible to control the pump flow rate and thus the speed of the hydraulic cylinder or the speed of the hydraulic motor. The control of the speed of the electric motor has a high dynamic, comparable to that of control by means of hydraulic valves. During the changes in rotational speed, no overshoots, usually accompanying such changes in the hydraulic system, were observed.

7. Conclusions

The integrated motor-pump assembly has been tested at different speeds of the pump and different pressures in the hydraulic system. The measured waveforms of electrical quantities (current and voltage at the converter input, motor phase currents, and phase-to-phase voltages) are correct, i.e., characteristic for a drive system with a brushless DC motor and with the PWM speed control. The tests have shown good dynamics of the developed integrated motor-pump assembly and proved its suitability to control the pump flow rate and thus the speed of the hydraulic cylinder or the speed of the hydraulic motor.

8. Patents

Displacement pump with the integrated electric drive P. 213898.

Author Contributions: Conceptualization, M.P.C. and W.F.; methodology, M.P.C. and W.F.; software, M.P.C. and W.F.; validation, M.P.C. and W.F.; formal analysis, M.P.C. and W.F.; investigation, M.P.C. and W.F.; resources, M.P.C. and W.F.; data curation, M.P.C. and W.F.; writing—original draft preparation, M.P.C. and W.F.; writing—review and editing, M.P.C. and W.F.; visualization, M.P.C. and W.F.; supervision, M.P.C. and W.F. Both authors have read and agreed to the published version of the manuscript.

Funding: This research was funded by The Polish National Centre of Research and Development, grant number 208471.

Conflicts of Interest: The authors declare no conflict of interest.

References

1. Ivantysyn, J.; Ivantysynova, M. *Hydrostatic Pumps and Motors: Principles, Design, Performance, Modelling, Analysis, Control and Testing*; Tech Books International: New Delhi, India, 2003.
2. Brusiani, F.; Bianchi, G.; Costa, M.; Squarcini, R.; Gasperini, M. *Analysis of Air/Cavitation Interaction inside a Rotary Vane Pump for Application on Heavy Duty Engine*; SAE Technical Paper 2009-01-1943; SAE International: Warrendale, PA, USA, 2009.
3. Sullivan, P.; Sehmby, M. *Internal Force Analysis of a Variable Displacement Vane Pump*; SAE Technical Paper 2012-01-0409; SAE International: Warrendale, PA, USA, 2012. [CrossRef]
4. Frendo, F.; Novi, N.; Squarcini, R. Numerical and Experimental Analysis of Variable Displacement Vane Pumps. In Proceedings of the International Conference on Tribology, Parma, Italy, 20–22 September 2006.
5. Inaguma, Y. Friction Characteristics of Vane for a Balanced Vane Pump. *Trans. Jpn. Fluid Power Syst. Soc.* **2014**, *45*, 58–65. [CrossRef]
6. Fiebig, W.; Cependa, P.; Jędraszczyk, P.; Kuczwara, H. Innovative Solution of an Integrated Motor Pump Assembly. In *Fluid Power Systems Technology*; American Society of Mechanical Engineers: New York, NY, USA, 2017.
7. Fiebig, W.; Dudzikowski, I.; Ciurys, M.; Hubert, K. A Vane Pump Integrated with an Electric Motor. In Proceedings of the 9th International Fluid Power Conference, IFK, Aachen, Germany, 9 March 2014.
8. Fiebig, W.; Cependa, P.; Kuczwara, H.; Wang, F. Analysis of vane loads and motion in a hydraulic double vane pump with integrated electrical drive. *Arch. Civ. Mech. Eng.* **2021**, *21*, 1–15. [CrossRef]
9. *Motor/Pump Hybrid System EPAI for High and Medium- Pressure Application*; Voith Brochure: Rutesheim, Germany, 2012.
10. Almeida, A.; Ferreira, F.J.T.E.; Fong, J. Standards for Efficiency of Electric Motors. *IEEE Ind. Appl. Mag.* **2011**, *17*, 12–19. [CrossRef]
11. De Almeida, A.T.; Ferreira, F.J.T.E.; Baoming, G. Beyond Induction Motors—Technology Trends to Move Up Efficiency. *IEEE Trans. Ind. Appl.* **2014**, *50*, 2103–2114. [CrossRef]
12. Gieras, J.F. *Permanent Magnet Motor Technology: Design and Applications*, 3rd ed.; CRC Press: Boca Raton, FL, USA, 2009; ISBN 978-1-4200-6440-7.
13. Hendershot, J.R.; Miller, T.J.E. *Design of Brushless Permanent-Magnet Machines*, 2nd ed.; Motor Design Books LLC: Venice, FL, USA, 2010; ISBN 978-0-9840687-0-8.
14. Hughes, A. *Electric Motors and Drives: Fundamentals, Types and Applications*, 3rd ed.; Newnes: Amsterdam, NY, USA; Boston, MA, USA, 2006; ISBN 978-0-7506-4718-2.

15. Seo, J.-M.; Rhyu, S.-H.; Kim, J.-H.; Choi, J.-H.; Jung, I.-S. Design of Axial Flux Permanent Magnet Brushless DC Motor for Robot Joint Module. In Proceedings of the 2010 International Power Electronics Conference ECCE ASIA, Sapporo, Japan, 21–24 June 2010; pp. 1336–1340.
16. Yang, S.; Jung, Y.; Seo, J.; Lee, M.; Kim, J.-H. Numerical and Experimental Study on the Cooling Performance Affected by Ventilation Holes of a BLDC Motor for Multi-Copters. In Proceedings of the 2018 IEEE 18th International Power Electronics and Motion Control Conference, Budapest, Hungary, 26–30 August 2018; pp. 293–298. [CrossRef]
17. Cao, W.; Mecrow, B.C.; Atkinson, G.J.; Bennett, J.W.; Atkinson, D.J. Overview of Electric Motor Technologies Used for More Electric Aircraft (MEA). *IEEE Trans. Ind. Electron.* **2012**, *59*, 3523–3531. [CrossRef]
18. Carev, V.; Roháč, J.; Šipoš, M.; Schmirler, M. A Multilayer Brushless DC Motor for Heavy Lift Drones. *Energies* **2021**, *14*, 2504. [CrossRef]
19. Chau, K.T.; Chan, C.C.; Liu, C. Overview of Permanent-Magnet Brushless Drives for Electric and Hybrid Electric Vehicles. *IEEE Trans. Ind. Electron.* **2008**, *55*, 2246–2257. [CrossRef]
20. Liu, X.; Chen, H.; Zhao, J.; Belahcen, A. Research on the Performances and Parameters of Interior PMSM Used for Electric Vehicles. *IEEE Trans. Ind. Electron.* **2016**, *63*, 3533–3545. [CrossRef]
21. Pellegrino, G.; Vagati, A.; Guglielmi, P.; Boazzo, B. Performance Comparison between Surface-Mounted and Interior PM Motor Drives for Electric Vehicle Application. *IEEE Trans. Ind. Electron.* **2011**, *59*, 803–811. [CrossRef]
22. De Santiago, J.; Bernhoff, H.; Ekergård, B.; Eriksson, S.; Ferhatovic, S.; Waters, R.; Leijon, M. Electrical Motor Drivelines in Commercial All-Electric Vehicles: A Review. *IEEE Trans. Veh. Technol.* **2012**, *61*, 475–484. [CrossRef]
23. Wang, J.; Yuan, X.; Atallah, K. Design Optimization of a Surface-Mounted Permanent-Magnet Motor with Concentrated Windings for Electric Vehicle Applications. *IEEE Trans. Veh. Technol.* **2013**, *62*, 1053–1064. [CrossRef]
24. Yang, Z.; Shang, F.; Brown, I.P.; Krishnamurthy, M. Comparative Study of Interior Permanent Magnet, Induction, and Switched Reluctance Motor Drives for EV and HEV Applications. *IEEE Trans. Transp. Electrif.* **2015**, *1*, 245–254. [CrossRef]
25. Yang, Y.-P.; Luh, Y.-P.; Cheung, C.-H. Design and Control of Axial-Flux Brushless DC Wheel Motors for Electric Vehicles—Part I: Multiobjective Optimal Design and Analysis. *IEEE Trans. Magn.* **2004**, *40*, 1873–1882. [CrossRef]
26. Zeraoulia, M.; Benbouzid, M.E.H.; Diallo, D. Electric Motor Drive Selection Issues for HEV Propulsion Systems: A Comparative Study. *IEEE Trans. Veh. Technol.* **2006**, *55*, 1756–1764. [CrossRef]
27. Xu, W.; Zhu, J.; Guo, Y.; Wang, S.; Wang, Y.; Shi, Z. Survey on Electrical Machines in Electrical Vehicles. In Proceedings of the 2009 International Conference on Applied Superconductivity and Electromagnetic Devices, Chengdu, China, 25–27 September 2009; pp. 167–170.
28. Antaki, J.; Paden, B.; Piovoso, M.; Banda, S. Award-winning control applications. *IEEE Control. Syst. Mag.* **2002**, *22*, 8–19. [CrossRef]
29. Piccigallo, M.; Scarfogliero, U.; Quaglia, C.; Petroni, G.; Valdastri, P.; Menciassi, A.; Dario, P. Design of a Novel Bimanual Robotic System for Single-Port Laparoscopy. *IEEE/ASME Trans. Mechatron.* **2010**, *15*, 871–878. [CrossRef]
30. Lawson, B.E.; Mitchell, J.E.; Truex, D.; Shultz, A.H.; LeDoux, E.; Goldfarb, M. A Robotic Leg Prosthesis: Design, Control, and Implementation. *IEEE Robot. Autom. Mag.* **2014**, *21*, 70–81. [CrossRef]
31. Yamashita, H.; Kim, D.; Hata, N.; Dohi, T. Multi-Slider Linkage Mechanism for Endoscopic Forceps Manipulator. In Proceedings of the 2003 IEEE/RSJ International Conference on Intelligent Robots and Systems (IROS 2003) (Cat. No.03CH37453), Las Vegas, NV, USA, 27–31 October 2003; Volume 3, pp. 2577–2582.
32. Li, H.; Li, W.; Ren, H. Fault-Tolerant Inverter for High-Speed Low-Inductance BLDC Drives in Aerospace Applications. *IEEE Trans. Power Electron.* **2017**, *32*, 2452–2463. [CrossRef]
33. Huang, X.; Goodman, A.; Gerada, C.; Fang, Y.; Lu, Q. Design of a Five-Phase Brushless DC Motor for a Safety Critical Aerospace Application. *IEEE Trans. Ind. Electron.* **2012**, *59*, 3532–3541. [CrossRef]
34. Yoon, K.-Y.; Baek, S.-W. Robust Design Optimization with Penalty Function for Electric Oil Pumps with BLDC Motors. *Energies* **2019**, *12*, 153. [CrossRef]
35. Ciurys, M.P. Electromagnetic phenomena analysis in brushless DC motor with speed control using PWM method. *Open Phys.* **2017**, *15*, 907–912. [CrossRef]
36. Ciurys, M.P. Analysis of the Influence of Inverter PWM Speed Control Methods on the Operation of a BLDC Motor. *Arch. Electr. Eng.* **2018**, *67*, 939–953. [CrossRef]
37. Ciurys, M.P.; Dudzikowski, I.; Pawlak, M. Laboratory tests of a PM-BLDC motor drive. In Proceedings of the 2015 Selected Problems of Electrical Engineering and Electronics (WZEE), Kielce, Poland, 17–19 September 2015; pp. 1–6. [CrossRef]

 *energies*

Article

Prediction of Electromagnetic Characteristics in Stator End Parts of a Turbo-Generator Based on MLP and SVR

Likun Wang [1], Yutian Sun [1], Baoquan Kou [2], Xiaoshuai Bi [3,*], Hai Guo [4], Fabrizio Marignetti [5] and Huibo Zhang [6]

[1] Harbin Electric Machinery Company Limited, Harbin 150040, China; wlkhello@hrbust.edu.cn (L.W.); sunyutian@hec-china.com (Y.S.)

[2] School of Electrical Engineering and Automation, Harbin Institute of Technology, Harbin 150001, China; koubq@hit.edu.cn

[3] National Engineering Research Center of Large Electric Machines and Heat Transfer Technology, Harbin University of Science and Technology, Harbin 150080, China

[4] The College of Computer Science and Engineering, Dalian Minzu University, Dalian 116600, China; guohai@dlnu.edu.cn

[5] Department of Electrical and Information Engineering, The University of Cassino and South Lazio, 03043 Rome, Italy; marignetti@unicas.it

[6] Georgia Institute of Technology, College of Engineering, Atlanta, GA 30332, USA; kevinzhang@gatech.edu

* Correspondence: 1920300024@stu.hrbust.edu.cn

Abstract: In order to study the multiple restricted factors and parameters of the eddy current loss of generator end structures, both the multi-layer perceptron (MLP) and support vector regression (SVR) are used to study and predict the mechanism of the synergistic effect of metal shield conductivity, relative permeability of clamping plates and structural characteristics of eddy current losses. Based on the eddy current losses of generator end structures under different metal shielding thicknesses and electromagnetic properties, the calculation accuracy of the MLP and SVR is compared. The prediction method gives an effective means for the complex design of the end region of the generator, which reduces the effort of the designers. It also promotes the design efficiency of the electrical generator.

Keywords: turbo-generator; eddy current losses; data driven; support vector regression; multi-layer perceptron

Citation: Wang, L.; Sun, Y.; Kou, B.; Bi, X.; Guo, H.; Marignetti, F.; Zhang, H. Prediction of Electromagnetic Characteristics in Stator End Parts of a Turbo-Generator Based on MLP and SVR. *Energies* **2021**, *14*, 5908. https://doi.org/10.3390/en14185908

Academic Editor: Ryszard Palka

Received: 17 August 2021
Accepted: 14 September 2021
Published: 17 September 2021

Publisher's Note: MDPI stays neutral with regard to jurisdictional claims in published maps and institutional affiliations.

1. Introduction

Due to the special nature of the end structure of turbo-generators, the distribution of magnetic flux leakage is complex. The distribution of eddy current loss is affected by many factors, such as the size, shape and physical properties of the end structure. The existing literature studies the influence of single factors on the electromagnetic loss of generator end structures and draws some basic research conclusions. Since the eddy current loss of generator end structures is affected by the conductivity, permeability and thickness of the structure, the relationship between these factors and the eddy current loss is difficult to display and record. Even if this corresponding function relationship makes sense to a certain extent, it may be a hypothetical one as a result of the ignorance of certain factors. The deviation analysis may be large and with a limited generalization ability of the obtained results.

Recently, scholars and researchers have performed extensive investigations on magnetic fields and eddy current issues of electrical machines. S. Utegenova et al. analyzed the magnetic issue of a wound-rotor motor by using an equivalent circuit method and introducing the principle of the magnetic equivalent circuit model [1]. J. Nam et al. proposed a new closed-path magnetic system. A mapping method was proposed to utilize the FEM and polynomial regression in order to analyze the magnetic field [2]. In [3], the impact of the leading degree on the eddy losses was analyzed. J. J. Perez-Loya et al. calculated the generator loss with a parallel path of the stator. Considering the unbalanced magnetic pull, both the currents of the damper bar and the circulating currents of stator winding were researched [4]. S. Kahourzade et al. conducted an electromagnetic analysis for a tapered axial flux PM machine. A new procedure of loss breakdown and efficiency estimation was introduced by both the experiment and the FEM [5]. In [6], some laws were proposed for using analytical methods to analyze eddy current losses of AC generators. J. L. Risti'c-Djurovi' et al. introduced a new method to add extra stator windings to enlarge the length of the variation along a test volume direction [7]. In [8], the circulating currents of the double-stator Roebel bars were calculated by using several models. The calculation results and test data were compared, and the advantage of each model was discussed. J. Lee et al. proposed a calculation method to include the additional losses, which is not considered in many cases [9]. In [10], A. Tessarolo et al. conducted research on the eddy currents by the time-harmonic FEA. In some methods, they are alternative. In [11], the influence of the main deformation of the shape and size of the winding section of the circular solenoid on the magnetic field's distribution and uniformity was studied. Y. Kwon et al. used two simplified nonlinear magnetic equivalent circuit models to analyze the magnetic field capability caused by the change in design parameters of the new soft magnetic composite prototype as compared with the basic model prototype [12]. In [13], a multi-objective optimization design for a non-core PM synchronous motor was introduced. By solving two Laplace equations, both the 3D performance analysis and the magnetic field distribution were obtained under open circuit conditions. In [14], Vraisanen et al. proposed a time harmonic model, which can be used to deal with multi-layer cylindrical rotors. In order to consider the influence of the stator slot on eddy current loss, the calculation model was linked to a finite element solution by covering the stator. M. Z. Youssef et al. introduced a new electromagnetic analysis method by optimizing the cost of the electromagnetic system based on mathematical analysis [15]. In [16], a numerical model of the magnetic bearing was proposed. The 3D magnetic field's distribution between the stator and the rotor was calculated. In addition, the magnetic forces of the hybrid magnetic bearing system were studied under different stator currents. P. Hekmati et al. established the magnetic analytical models for different rotor structures of electrical machines. The electromagnetic parameters of both stator and rotor sides were obtained [17]. G. G. Sotelo et al. suggested a new design method for a motor. By changing the load condition, the proposed motor can operate both under a synchronous state and a hysteresis state [18]. S. G. Min et al. presented a novel analytical solution to obtain the best electromagnetic performances of concentrated windings of the stator and of the permanent magnet machines [19]. J. Lee et al. conducted an electromagnetic analysis for a PM sensor. Both the position and structure of the PM Hall sensor were considered. The proposed magnetic equivalent circuit model gained a fast calculation result [20]. P. R. Eckert et al. developed a model for obtaining the flux distribution and the stator voltage. The method was validated by an experiment of an actual prototype [21]. In [22], a support vector machine (SVM) was used to classify and evaluate induction motor faults. The calculation results showed that, compared with the other two machine learning algorithms, the SVM calculation results were more accurate. The fuzzy C-Means machine learning algorithm was used to analyze the influence of the flux sensor position on the automatic classification. The results proved the potential of the method for its future incorporation into autonomous condition monitoring systems that can be satisfactorily applied to determine the health of these machines [23]. In [24], the linear prediction coefficients and mel frequency cepstral coefficients were extracted from the machine sound to develop. Machine learning (ML) models were created to monitor and identify the malfunctioning machines based on the operating sound. The experimental results showed the performance of ML models for the machine sound recorded, with different signal-to-noise ratio levels for normal and abnormal operations.

In this paper, a mathematical model of a 3D magnetic field in the complex end domain of the generator end is established by a time-step FEM. A neural network and support vector regression are used to study and predict the mechanism of the synergistic effect of the metal shield conductivity, relative permeability of clamping plates and structural characteristics of the eddy current loss of end structures. The prediction accuracy of the MLP and SVR are compared. This research provides an effective means for the complex design of generator end regions, which reduces the effort of designers. In addition, it promotes the design efficiency of the electrical generator.

2. 3D Electromagnetic Field Analysis

The stator end windings have an involute structure. There are many different metal structures, such as a finger plate, copper screen and clamping plate. Table 1 gives the basic parameters of this 330 MW hydrogen turbo-generator, which is used to calculate the flux of the end domain.

Table 1. Basic parameters.

Parameters	Values
Power	330 MW
Stator voltage	20 kV
Stator current	11.2 kA
Speed	3000 rpm
Rated efficiency	98.8%
Cooling medium	Hydrogen

Figure 1 shows the end structure of a hydrogen-cooled turbo-generator prototype. Based on the actual size of the generator end domain, a 3D electromagnetic field model was established, which is shown in Figure 2. Because the generator pole number is small, both the end winding span and the total end domain space are relatively large.

In order to truly reflect the actual results, the end domain of the 330 MW generator was established based on the actual shape and dimensions of the prototype. The whole solution domain Ω contains the eddy current domain V_1 and the non-eddy current domain V_2.

Figure 1. End structure of hydrogen-cooled turbo-generator prototype.

Figure 3 shows the solution domain of the 330 million W water–hydrogen–hydrogen-cooling turbo-generator. The mathematical model contains vector potential T and the scalar potential Ψ. The solution formulas are shown in (1)–(5) [25].

$$\begin{cases} \nabla \times \rho \nabla \times T - \nabla \rho \nabla \cdot T + \frac{\partial \mu (T - \nabla \psi)}{\partial t} + \frac{\partial \mu H_s}{\partial t} = 0 \\ \nabla \cdot \mu (T - \nabla \psi) = -\nabla \cdot \mu H_s \end{cases} \tag{1}$$

$$\nabla \cdot \mu \nabla \psi = \nabla \cdot \mu H_s \text{ in } V_2 \tag{2}$$

where

$$H_s = \frac{1}{4\pi} \int_{\Omega_s} \frac{J_s \times r}{r^3} d\Omega \tag{3}$$

Boundary conditions:

$$\begin{cases} \left.\dfrac{\partial \psi}{\partial \mathbf{n}}\right|_{S_1,S_2} = 0 \\ \psi|_{S_3} = \psi_0 \end{cases} \tag{4}$$

The initial conditions:

$$\begin{cases} T|_{V_1} = T_0 \\ \psi|_{\Omega} = \psi_0 \end{cases} \tag{5}$$

where J_s is the source current density in the windings(in A/m^2), μ is the permeability(in H/m), ρ is the resistivity (in $\Omega \cdot$m), T_0 is the electric vector potential at the initial time, ψ_0 is the scalar magnetic potential at the initial time, t is the time(in s), and $\mathbf{n}$ is the normal vector of the surface.

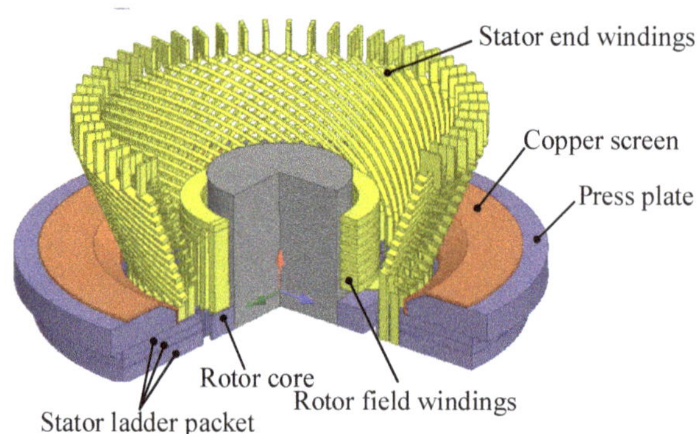

Figure 2. Physical model of the 330 million W water–hydrogen–hydrogen-cooling generator.

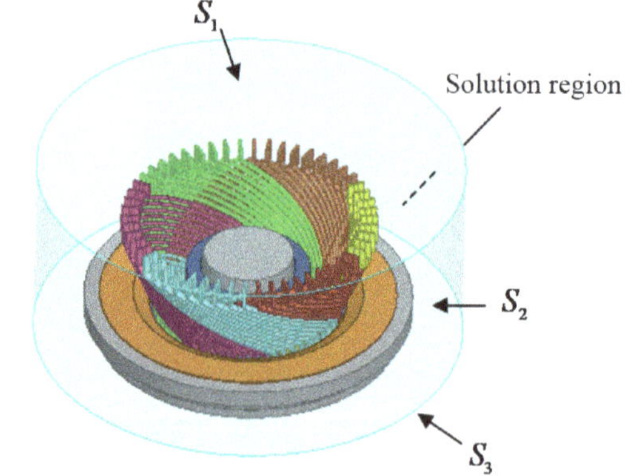

Figure 3. Solution region of the 3D transient electromagnetic field.

3. Electromagnetic Losses of Metal Parts

3.1. Electromagnetic Loss Calculation and Analysis

Figure 4 gives the distribution of the leakage flux field in the end domain of the turbogenerator. This shows that the magnetic flux leakage passes around the armature windings. The magnetic flux leakage is essentially parallel to the outer surface of the copper screen.

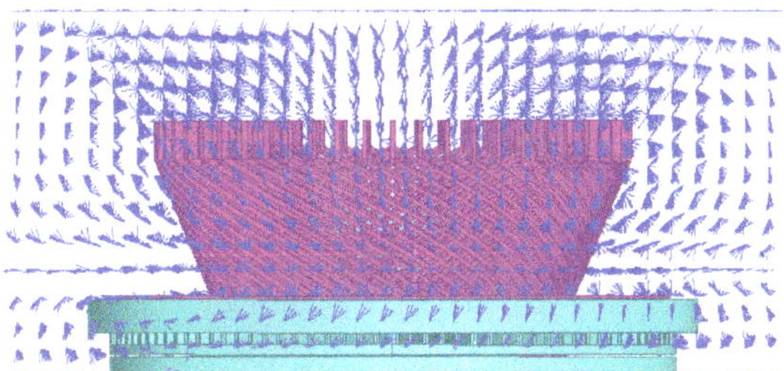

Figure 4. Leakage magnetic field.

Figure 5 shows the eddy current distribution in the copper screen. The yellow arrow shows the path of the eddy current in the copper screen. The distribution of the eddy current density indicates that the copper screen is essential for preventing the intrusion of the end leakage flux into the clamping plate.

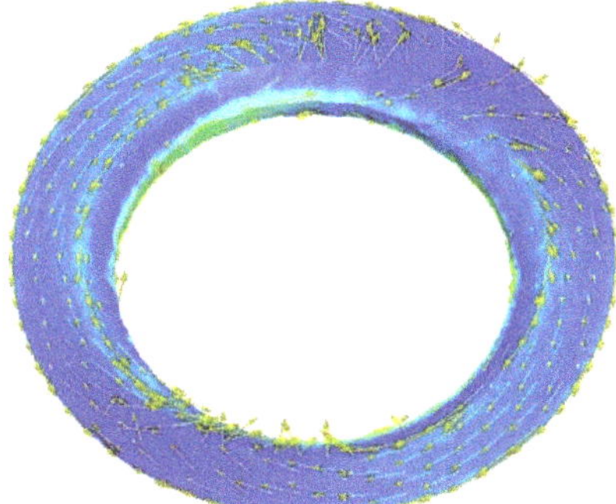

Figure 5. The path of eddy current.

Figure 6 gives the value of the magnetic flux density of the finger plates. The magnetic flux density is high at the front of the finger plates. The value reduces from the front parts to the back parts. In the end domain, the eddy current losses are not only impacted by the radial component of the total flux; due to the complex structure, the axial component of the flux also exists in this domain.

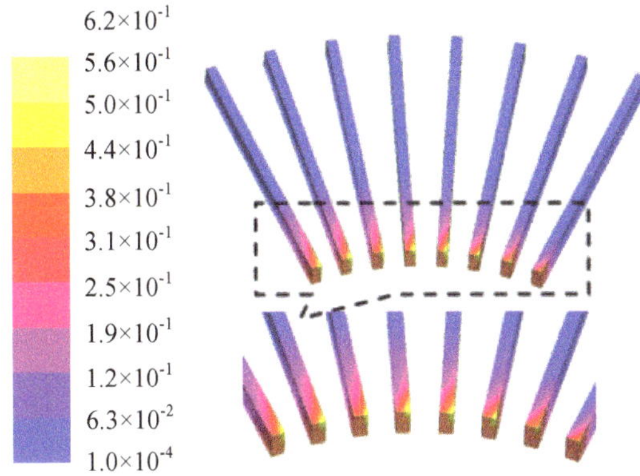

Figure 6. Flux density of the finger plate (T).

The eddy current loss of metal structures is calculate by (6)

$$P_e = \frac{1}{T_c} \int_{T_c} \sum_{i=1}^{k} J_e^2 \Delta_e \sigma_r^{-1} dt \tag{6}$$

where P_e is losses of the element (in W), J_e is the eddy current density (in A/m^2), Δ_e is the element volume (in m^3), σ_r is the metal component conductivity (in S/m), T_c is the period (in s), k is the total element number in volume, and e is the element number.

3.2. Verification for Electromagnetic Loss Calculation by Thermal Test

Using the results of the electromagnetic losses as heat sources, the temperature field of the end domain can be calculated [26]. The fluid–solid coupled model is given in Figure 7. Figure 8 gives the mesh results of the solution domain. The total number of mesh elements is 7,932,399.

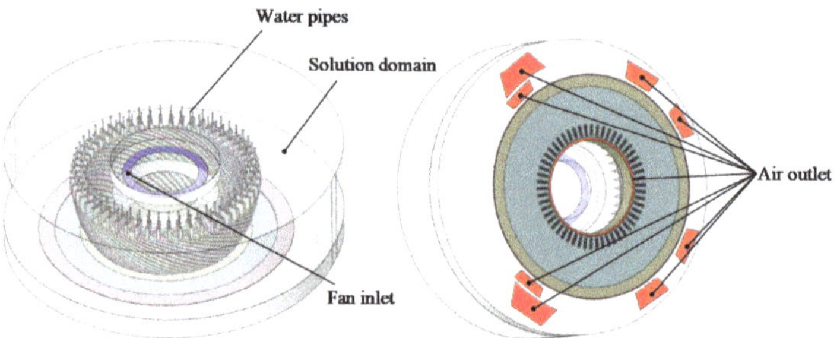

Figure 7. 3D fluid–solid coupled model of the end domain.

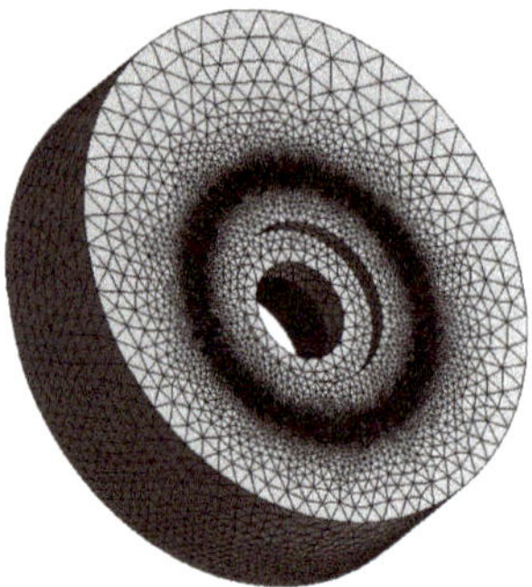

Figure 8. Mesh plot of end region for thermal–fluid solution model.

Figure 9 gives the temperature of the copper screen. Table 2 shows the measured values of the temperature. Figures 10 and 11 give the locations of the temperature sensors and the copper screen used for the test.

Table 2. Test values.

	Position M	Position N	Position P
Temperature (°C)	74.3	63.6	56.9

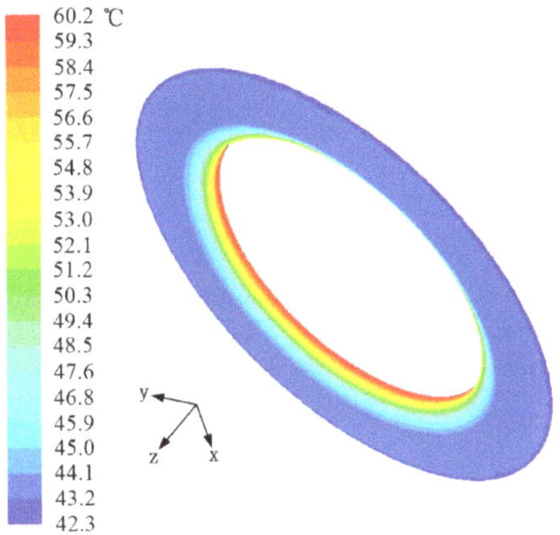

Figure 9. Temperature distribution of the copper screen.

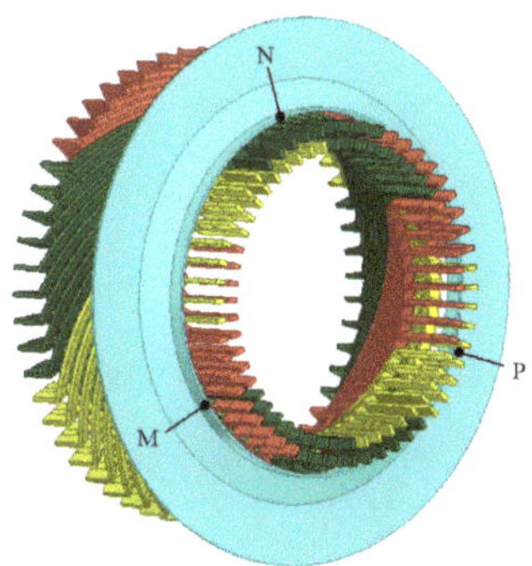

Figure 10. Locations of test sensors.

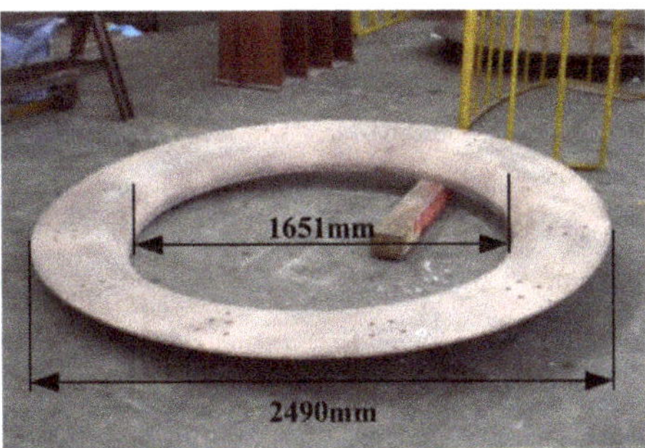

Figure 11. The practical copper screen.

The end structures may not be absolutely symmetric, such as the distance between the adjacent windings and the distance between the adjacent water pipes in the stator windings. These factors could cause the velocity distribution of the cooling medium to be asymmetric. The deformation of the end structures could also result in asymmetric distribution of loss. The measured results are different values at different positions, which may be caused by these asymmetric factors. For the simulation results, the highest temperature of the copper screen is 60.2 °C. The average temperature of the copper screen is 57.3 °C.

4. Prediction and Result Analysis Using Multi-Layer Perceptron

4.1. Prediction and Analysis Based on Multi-Layer Perceptron

In order to research the collaborative impact of multiple factors on the eddy current losses of end structures, the back propagate neural network (BPNN) model is established to study the known samples that are calculated by the FEM. The input nodes are the multi-factors, which have the thickness of a metal screen, the conduction characteristics of a metal screen and the permeability performance of a clamping plate. These factors are the elements of the input vector of the BPNN model. The output vector is the total eddy current of the end structures. The BPNN model is shown in Figure 12. To improve the prediction accuracy of forecasting samples and the generalization ability of the BPNN model, the middle hidden layer has multiple layers. In this paper, the hidden layer of the BPNN model has three layers, which are 5, 6, and 5.

The input vector A = (ur, thk, sim), the output vector Y = (loss).

where ur is the relative permeability of the clamping plate, thk is the thickness of the copper screen, sim is the conductivity of the metal screen, and loss is the total eddy current losses of end structures.

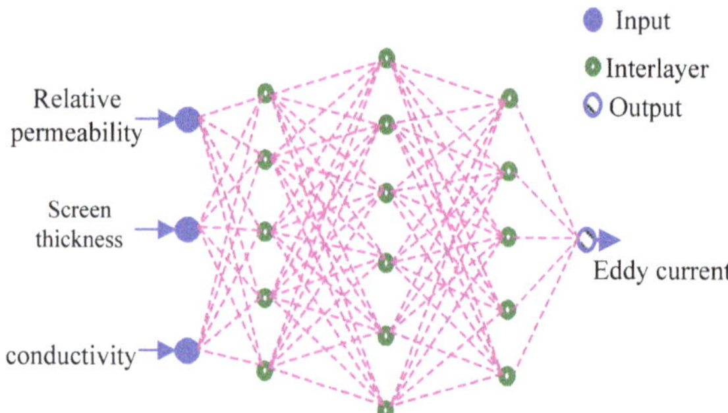

Figure 12. BPNN model with different hidden layers of a 5–6–5 structure.

(1) In the process of information forward propagation, if $a_j^{(1)} = x_i$ is the input value (activation value) of layer 1 neurons, the activation value of the next layer is [27]:

$$\begin{cases} a_j^{(1)} = x_i \\ a_j^{(l+1)} = f\left(z_j^{(l+1)}\right) \\ z_j^{(l+1)} = \sum_{i=1}^{n} W_{ji}^{(l)} a_i^{(l)} b_j^{(l)} \end{cases} \tag{7}$$

where x_i is the input value of neuron i node data of the first layer of sample data; $a_i^{(l)}$ represents the output value of the i-th node of the l-th layer; and $z_j^{(l)}$ represents the activation value of node j of layer 1. $W_{ji}^{(l)}$ is the connection weight parameter between the i-th node of layer l and the j-th node of layer $l + 1$; $b_j^{(l)}$ is the intercept term of node j on layer $l + 1$. f is the sigmoid activation function, and the expression is $\varphi(x) = \frac{1}{1+e^{-x}}$.

(2) Error back propagation process [28]

$$C(W, b) = \frac{1}{2} \sum_{i \in \text{outputs}} \|y_i - a_i^2\| \tag{8}$$

where y_i is the true value of node i traffic of the sample data output layer; a_i is the flow output value of node i in the sample data output layer.

(3)   Determination of optimization objectives

W (weight) and b (bias) minimize the loss function $C(W,b)$, and the flow prediction value output by the model will be closer to the real value. The iterative formula of W and b is as follows [29]:

$$\begin{cases} W_{ji}^{(l)} = W_{ji}^{(l)} - \alpha \dfrac{\partial C(W,b)}{\partial W_{ji}^{(l)}} \\ b_{j}^{(l)} = b_{j}^{(l)} - \alpha \dfrac{\partial C(W,b)}{\partial b_{j}^{(l)}} \end{cases} \tag{9}$$

where α is the learning rate, and the value range is $(0, 1)$.

4.2. Deviation Analysis and Generalization Ability Based on Multi-Layer Perceptron Prediction

Table 3 shows the training sample sets of the total eddy current losses of end structures. Before predicting the eddy current losses of test sample sets, data training and learning should be conducted for the training sample sets.

Table 3. The training sample sets of eddy current losses.

Sample	Relative Permeability	Thickness (mm)	Conductivity (S/m)	Eddy Current Loss (kW)
1	1	12	46,082,949	25.42
2	10	12	46,082,949	24.40
3	20	12	46,082,949	24.01
4	30	12	46,082,949	23.95
5	40	12	46,082,949	24.01
6	50	12	46,082,949	24.09
7	1	12	46,082,949	25.42
8	1	14	46,082,949	22.80
9	1	16	46,082,949	22.12
10	1	18	46,082,949	21.24
11	1	20	46,082,949	20.90
12	1	22	46,082,949	20.33
13	1	12	6,418,485	53.79
14	40	12	6,418,485	37.86
15	100	12	6,418,485	26.22

The highly precise learning results are gained through the training samples of the eddy current loss of each structure based on the BPNN with a middle layer of the 5–6–5 type, as shown above. Figure 13 gives the learning results of the eddy current losses of each of the structures based on the BPNN with a middle layer of the 5–6–5 type. From Figure 13, we learn that the variation trend of the FEM and MLP is the same, and the eddy current loss is the largest around the seventh sampling point.

Table 4 gives a comparison between the predicted results of the test sample and the calculated results by the finite element method. It is shown that even if the electrical conductivity of metal aluminum material is not provided in the training sample, the MLP predicts that the loss value of the end structure parts is close to the calculated values by the finite element method when the end of generator is shielded by metal aluminum.

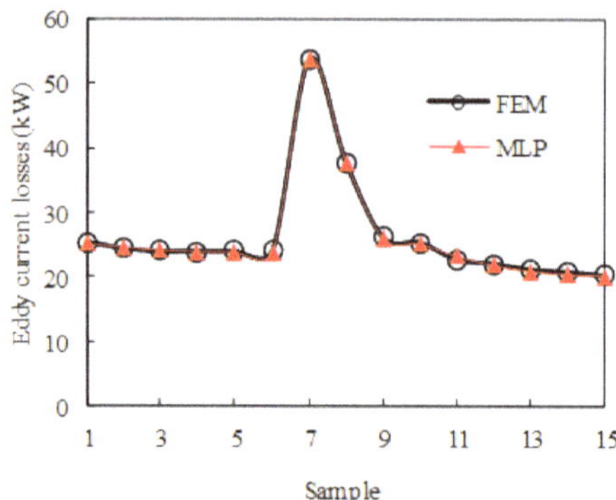

Figure 13. Learning results based on MLP.

Table 4. Predicted eddy current losses of the test sample set (Kw).

Relative Permeability	Thickness (mm)	Conductivity (S/m)	Losses	
			FEM	MLP
1	12	28,589,902	32.57	32.77
1	20	28,589,902	28.02	27.99
2	12	46,082,949	25.06	25.07
4	12	46,082,949	24.99	24.93
8	12	46,082,949	24.57	24.67

When the hidden layers are changed to two layers, the predicted result of (32.77, 27.99, 25.07, 24.93, 24.67) is changed to (33.81, 30.00, 25.22, 25.02, 24.67). It shows that, for the prediction loss results of the end structure parts of a turbo-generator by the BPNN, the deviation between the eddy current loss of end structure parts by the MLP and its calculated results by the finite element method decreases with the increase in hidden layers of the neural network.

5. Prediction and Results Analysis by SVR

5.1. Mathematical Principle of Support Vector Regression

Support vector regression (SVR) belongs to the category of statistics, and the idea is to use classification as the leading factor. SVR has the characteristics of low risk, which avoid the defects of blind training, over-learning and entering the minimum region of traditional prediction methods. SVR is suitable for the data mining of small sample sets, and its generalization ability is strong. SVR is often the first choice in order to study small sample data. This data-mining model maps high-dimensional space to low-dimensional space by selecting the kernel function, which makes the problem less complex. During this period, it does not increase the difficulty of calculation and effectively avoids the issue of dimension. Therefore, SVR is widely used in predicting engineering problems [30–35].

Let the sample set be $\{(x_i, y_i)|x_i \in R^n; y_i \in \{-1, +1\}, i = 1, \ldots, I\}$, and find an optimal hyperplane that has two types of points labeled +1 and −1 that are not only separated but also have the largest separation interval.

Linear separation can be achieved in n-dimensional Euclidean space; that is, there is a hyperplane that divides the sample set on both sides according to the labels −1 and +1. Since the mathematical expression of the hyperplane in n-dimensional Euclidean space is a linear equation $< w, x > +b = 0$, this means that among them, w is a coefficient vector, x is a n-dimensional variable, $< w, x >$ is an inner product, and b is a constant. The distance from point x_i to hyperplane L in space is denoted as $d(x_i, L) = \frac{|<w,x_i>+b|}{||w||}$. For maximization, $d(x_i, H)$ is equivalent to $\frac{1}{2}\left\|w\right\|^2$ minimum. Next, we obtain an extreme value problem under the following constraints.

$$\begin{cases} \min \frac{1}{2}\left\|w\right\|^2 \\ y_i(<w, x_i> +b) \geq 1, i = 1, 2, \ldots, I \end{cases} \tag{10}$$

By introducing Lagrange multiplier $\alpha = (\alpha_1, \alpha_2, \ldots, \alpha_I)$, we can solve the equation about the parameter by (11).

$$Q(\alpha) = \sum_{i=1}^{I} \alpha_i - \frac{1}{2} \sum_{i,j=1}^{I} \alpha_i \alpha_j y_i y_j < x_i, x_j > \tag{11}$$

The above formula is called the Lagrange dual function, and its constraint is expressed as (12).

$$\sum_{i,j=1}^{I} \alpha_i y_i = 0, \alpha_i \geq 0, i = 1, 2, \ldots, I \tag{12}$$

Under this constraint, if α makes $Q(\alpha)$ reach the maximum value, there are many α_i whose values are 0. However, the sample x_i corresponds to α_i, which is not 0 and is the support vector.

When linear separation cannot be achieved in the input space, it is assumed that non-linear mapping ϕ is found. It can map the sample set that is expressed as $\{(x_i, y_i) | x_i \in R^n; y_i \in \{-1, +1\}, i = 1, \ldots, I\}$ into the high-dimensional feature space H.

Presently, we consider the linear classification of the set $\{(\phi(x_i), y_i) | x_i \in R^n; y_i \in \{-1, +1\}, i = 1, \ldots, I\}$ in H by constructing a hyperplane in H. Its weight coefficient w satisfies similar extreme value problems. Since exceptions are allowed in some areas, slack terms can be introduced, that is, rewritten as:

$$\begin{cases} \min \frac{1}{2} ||w||^2 + C \sum_{i=1}^{L} \xi_i \\ y_i(<w, x_i> +b) \geq 1 - \xi_i, \xi_i \geq 0, i = 1, 2, \ldots, I \end{cases} \tag{13}$$

A classification problem is an extreme case, but it is very useful. Let $\{x_i | x_i \in R^n, i = 1, \ldots, I\}$ be a finite observation point in space R^n. Find the smallest sphere containing these points with a as the center and R as the radius. Therefore, a classification is the best method for finding the minimum envelope surface of a compound component. Exactly as above, let ϕ be the embedded mapping derived from a kernel function $K(x, s)$ from the input space to the feature space, and finally we understand the quadratic programming problem.

$$\begin{cases} \min_\alpha \frac{1}{2} \alpha' D\alpha + c' \alpha \\ y' \alpha = 0, 0 \leq \alpha = (\alpha_1, \ldots \alpha_I)^T \leq A = (C, \ldots, C)^T \end{cases} \tag{14}$$

where $y = (y_1, \ldots, y_I)^T$, $c = (-1, \ldots, -1)^T$, and $D = (K(x_i, x_j) y_i y_j)_{1 \leq i,j \leq I}$ are matrixes. $K(x, s)$ is a kernel function. Then,

$$f(x) = K(x, x) - 2\sum_{i=1}^{L} \alpha_i K(x, x_i) + \sum_{j=1}^{L} \sum_{i=1}^{L} \alpha_i \alpha_j K(x_i, x_j) \tag{15}$$

where all points satisfy the relationship with $f(x) \leq R^2$. The parameter C controls the number of points that fall outside the ball. The interval of change is $1/L < C < 1$.

5.2. Prediction of Eddy Current Loss Based on SVR

According to the prediction principle of SVR, the mathematical prediction model of the eddy current loss of generator end structures with multiple factors, such as metal shielding thickness, metal shielding conductivity and relative permeability of clamping plates, is constructed. The training sample set above is studied again, and the test sample set is predicted. Figure 14 shows the learning result of the eddy current losses of turbo-generators based on SVR. From Figure 14 displaying the learning results of eddy current loss based on SVR, it can be observed that there are deviations in individual points of the learning results, but the deviations in the overall learning results are small.

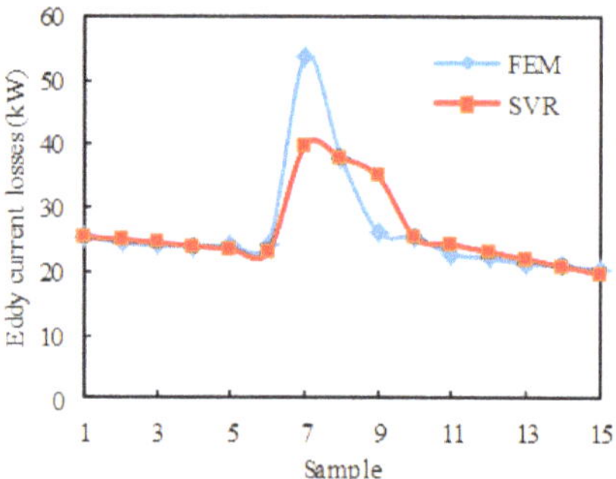

Figure 14. Learning results based on SVR.

Table 5 gives the prediction results of the loss of generator end structures in the test samples based on SVR. It is not difficult to see that the eddy current loss has a high prediction accuracy and strong generalization ability based on SVR. The deviation of learning results of individual elements in the training set does not affect the accurate prediction of eddy current loss of the test samples from SVR.

Table 5. Predictive eddy current losses in the test sample based on SVR.

Relative Permeability	Thickness (mm)	Conductivity (S/m)	Losses (kW)	
			FEM	SVR
1	12	28,589,902	32.57	31.72
1	20	28,589,902	28.02	27.20
2	12	46,082,949	25.06	25.32
4	12	46,082,949	24.99	25.22
8	12	46,082,949	24.57	25.03

6. Conclusions

In this paper, in order to study the multiple restricted factors of the eddy current loss of generator end structures, a mathematical model of the 3D electromagnetic field in the complex end domain is established by the time-step FEM. Both the neural network and the support vector regression are used to study and predict the mechanism of the synergistic effect of metal shield conductivity, relative permeability of clamping plates and structural characteristics on the eddy current loss of end structures. The different prediction types are compared, and the accuracy of the prediction of loss results is studied.

(1) The learning results and predicted eddy current loss of the test samples fit well with the numerical calculation from the FEM. This shows that even if the electrical conductivity of metal aluminum material is not provided in the training sample, the MLP can predict that the loss value of end structure parts is close to the calculated values by the finite element method when the end of the generator is shielded by metal aluminum. When the relative permeability is 1, the conductivity is 28,589,902 S/m,

and the thickness increases from 12 to 20 mm, the eddy current loss obtained by the FEM is reduced by 14%, and the eddy current loss obtained by the MLP is reduced by 14.6%. When the relative permeability increases from 2 to 4, the conductivity is 46,082,949 S/m and the thickness is 12 mm, the eddy current loss obtained by the FEM is reduced from 25.06 to 24.99 kW, and the eddy current loss obtained by the MLP is reduced from 25.07 to 24.93 kW. When the relative permeability increases from 4 to 8, the conductivity is 46,082,949 S/m and the thickness is 12 mm, the eddy current loss results obtained by the FEM and MLP are also reduced.

(2) For the prediction results of the eddy current loss of end structure parts of the turbogenerator by the BPNN, the deviation between the eddy current loss of end structure parts by the MLP and the eddy current loss gained by the FEM decreases with the increase in hidden layers of the neural network.

(3) From the results of the eddy current loss learning based on SVR, there are deviations in individual points of the learning results, but the deviations in the overall learning results are small. Eddy current loss has a high prediction accuracy and strong generalization ability based on SVR. The deviation of learning results of individual elements in the training sets does not affect the accurate prediction results of the eddy current loss of the test samples based on SVR.

This method gives an effective means for the complex design of the end region of the generator, which reduces the effort of designers. It also promotes the design efficiency of the electrical generator.

In future studies, a large data sample for a three-dimensional mathematical model of the end transient electromagnetic field of a turbine generator will be constructed, and the effect of the end magnetic leakage on the loss of the structural parts will be studied separately in combination with deep learning. In addition, big data samples with more influencing factors will be constructed, and models with more layers will be applied to further improve the accuracy of the prediction model.

Author Contributions: Conceptualization, L.W., Y.S. and X.B.; methodology, L.W., Y.S. and X.B.; formal analysis, L.W.; validation, L.W.; investigation, L.W.; data curation, L.W. and X.B.; writing—original draft preparation, L.W.; writing—review and editing, L.W. and X.B.; project administration, L.W.; funding acquisition, L.W., Y.S., B.K., X.B., H.G., F.M. and H.Z. All authors have read and agreed to the published version of the manuscript.

Funding: This research was funded by Natural Science Foundation for Excellent Young Scholars of Heilongjiang Province, grant number: YQ2021E038.

Institutional Review Board Statement: Not Applicable.

Informed Consent Statement: Not Applicable.

Data Availability Statement: Not Applicable.

Conflicts of Interest: There is no conflict of interest.

References

1. Utegenova, S.; Dubas, F.; Jamot, M.; Glises, R.; Truffart, B.; Mariotto, D.; Lagonotte, P.; Desevaux, P.; Utegenova, S.; Desevaux, P. An Investigation Into the Coupling of Magnetic and Thermal Analysis for a Wound-Rotor Synchronous Machine. *IEEE Trans. Ind. Electron.* **2017**, *65*, 3406–3416. [CrossRef]
2. Nam, J.; Lee, W.; Jung, E.; Jang, G. Magnetic Navigation System Utilizing a Closed Magnetic Circuit to Maximize Magnetic Field and a Mapping Method to Precisely Control Magnetic Field in Real Tim. *IEEE Trans. Ind. Electron.* **2018**, *65*, 5673–5681. [CrossRef]
3. Wang, L.; Li, W. Influence of underexcitation operation on electromagnetic loss in the end metal parts and stator step packets of a turbogenerator. *IEEE Trans. Energy Convers.* **2014**, *29*, 748–757. [CrossRef]
4. Perez-Loya, J.J.; Abrahamsson, C.J.D.; Lundin, U.; Abrahamsson, J. Electromagnetic Losses in Synchronous Machines During Active Compensation of Unbalanced Magnetic Pull. *IEEE Trans. Ind. Electron.* **2018**, *66*, 124–131. [CrossRef]
5. Kahourzade, S.; Ertugrul, N.; Soong, W.L. Loss Analysis and Efficiency Improvement of an Axial-Flux PM Amorphous Magnetic Material Machine. *IEEE Trans. Ind. Electron.* **2018**, *65*, 5376–5383. [CrossRef]
6. Drubel, O.; Stoll, R.L. Comparison between analytical and numerical methods of calculating tooth ripple losses in salient pole synchronous machines. *IEEE Trans. Energy Convers.* **2001**, *16*, 61–67. [CrossRef]

7. Ristić-Djurović, J.L.; Gajić, S.S.; Ilić, A.Ž. Design and Optimization of Electromagnets for Biomedical Experiments with Static Magnetic and ELF Electromagnetic Fields. *IEEE Trans. Ind. Electron.* **2018**, *65*, 4991–5000. [CrossRef]

8. Haldemann, J. Transpositions in Stator Bars of Large Turbogenerators. *IEEE Trans. Energy Convers.* **2004**, *19*, 553–560. [CrossRef]

9. Jun, H.-W.; Lee, J.-W.; Yoon, G.-H.; Lee, J. Optimal Design of the PMSM Retaining Plate with 3D Barrier Structure and Eddy-Current Loss-Reduction Effect. *IEEE Trans. Ind. Electron.* **2017**, *65*, 1808–1818. [CrossRef]

10. Tessarolo, A.; Agnolet, F.; Luise, F.; Mezzarobba, M. Use of Time-Harmonic Finite-Element Analysis to Compute Stator Winding Eddy-Current Losses Due to Rotor Motion in Surface Permanent-Magnet Machines. *IEEE Trans. Energy Convers.* **2012**, *27*, 670–679. [CrossRef]

11. Beiranvand, R. Effects of the Winding Cross-Section Shape on the Magnetic Field Uniformity of the High Field Circular Helmholtz Coil Systems. *IEEE Trans. Ind. Electron.* **2017**, *64*, 7120–7131. [CrossRef]

12. Kwon, Y.-S.; Kim, W.-J. Electromagnetic Analysis and Steady-State Performance of Double-Sided Flat Linear Motor Using Soft Magnetic Composite. *IEEE Trans. Ind. Electron.* **2016**, *64*, 2178–2187. [CrossRef]

13. Min, S.G.; Sarlioglu, B. 3D Performance Analysis and Multiobjective Optimization of Coreless-Type PM Linear Synchronous Motors. *IEEE Trans. Ind. Electron.* **2018**, *65*, 1855–1864. [CrossRef]

14. Raisanen, V.; Suuriniemi, S.; Kurz, S.; Kettunen, L. Rapid computation of harmonic eddy-current losses in high-speed solid-rotor induction machines. *IEEE Trans. Energy Convers.* **2013**, *28*, 782–790. [CrossRef]

15. Abdelrahman, A.S.; Sayeed, J.; Youssef, M.Z. Hyperloop Transportation System: Analysis, Design, Control, and Implementation. *IEEE Trans. Ind. Electron.* **2018**, *65*, 7427–7436. [CrossRef]

16. Zad, H.S.; Khan, T.I.; Lazoglu, I. Design and Adaptive Sliding-Mode Control of Hybrid Magnetic Bearings. *IEEE Trans. Ind. Electron.* **2017**, *65*, 2537–2547. [CrossRef]

17. Hekmati, P.; Yazdanpanah, R.; Mirsalim, M.; Ghaemi, E. Radial-Flux Permanent-Magnet Limited-Angle Torque Motors. *IEEE Trans. Ind. Electron.* **2017**, *64*, 1884–1892. [CrossRef]

18. Sotelo, G.G.; Sass, F.; Carrera, M.; Lopez-Lopez, J.; Granados, X. Proposal of a Novel Design for Linear Superconducting Motor Using 2G Tape Stacks. *IEEE Trans. Ind. Electron.* **2018**, *65*, 7477–7484. [CrossRef]

19. Min, S.G.; Bramerdorfer, G.; Sarlioglu, B. Analytical Modeling and Optimization for Electromagnetic Performances of Fractional Slot PM Brushless Machines. *IEEE Trans. Ind. Electron.* **2018**, *65*, 4017–4027. [CrossRef]

20. Seol, H.-S.; Lim, J.; Kang, D.-W.; Park, J.S.; Lee, J. Optimal Design Strategy for Improved Operation of IPM BLDC Motors with Low-Resolution Hall Sensors. *IEEE Trans. Ind. Electron.* **2017**, *64*, 9758–9766. [CrossRef]

21. Eckert, P.R.; Filho, A.F.F.; Perondi, E.A.; Dorrell, D.G. Dual Quasi-Halbach Linear Tubular Actuator with Coreless Moving-Coil for Semi-Active and Active Suspension. *IEEE Trans. Ind. Electron.* **2018**, *65*, 9873–9883. [CrossRef]

22. Choudhary, A.; Goyal, D.; Letha, S.S. Infrared Thermography-Based Fault Diagnosis of Induction Motor Bearings Using Machine Learning. *IEEE Sens. J.* **2021**, *21*, 1727–1734. [CrossRef]

23. Iglesias-Martinez, M.E.; Antonino-Daviu, J.; de Cordoba, P.F.; Conejero, J.A.; Dunai, L. Automatic Classification of Winding Asymmetries in Wound Rotor Induction Motors based on Bicoherence and Fuzzy C-Means Algorithms of Stray Flux Signals. *IEEE Trans. Ind. Appl.* **2021**. [CrossRef]

24. Natesha, B.V.; Guddeti, R.M.R. Fog-Based Intelligent Machine Malfunction Monitoring System for Industry 4.0. *IEEE Trans. Ind. Inform.* **2021**, *17*, 7923–7932. [CrossRef]

25. Bi, X.; Wang, L.; Marignetti, F.; Zhou, M. Research on Electromagnetic Field, Eddy Current Loss and Heat Transfer in the End Region of Synchronous Condenser with Different End Structures and Material Properties. *Energies* **2021**, *14*, 4636. [CrossRef]

26. Zhang, S.; Li, W.; Li, J.; Wang, L.; Zhang, X. Research on Flow Rule and Thermal Dissipation between the Rotor Poles of a Fully Air-cooled Hydro-generator. *IEEE Trans. Ind. Electron.* **2015**, *62*, 3430–3437. [CrossRef]

27. Hussain, I.; Agarwal, R.K.; Singh, B. MLP Control Algorithm for Adaptable Dual-Mode Single-Stage Solar PV System Tied to Three-Phase Voltage-Weak Distribution Grid. *IEEE Trans. Ind. Inform.* **2018**, *14*, 2530–2538. [CrossRef]

28. Sharifi, A.; Sharafian, A.; Ai, Q. Adaptive MLP neural network controller for consensus tracking of Multi-Agent systems with application to synchronous generators. *Expert Syst. Appl.* **2021**, *184*, 115460. [CrossRef]

29. Devi, R.M.; Murugesan, P.; Venkatesan, M.; Keerthika, P.; Sudha, K.; Kannan, J.; Suresh, P. Development of MLP-ANN model to predict the Nusselt number of plain swirl tapes fixed in a counter flow heat exchanger. *Mater. Today* **2021**, *46*, 8854.

30. Borrás, M.D.; Bravo, J.C.; Montaño, J.C. Disturbance Ratio for Optimal Multi-Event Classification in Power Distribution Networks. *IEEE Trans. Ind. Electron.* **2016**, *63*, 3117–3124. [CrossRef]

31. Laufer, S.; Rubinsky, B. Tissue Characterization with an Electrical Spectroscopy SVM Classifier. *IEEE Trans. Biomed. Eng.* **2009**, *56*, 525–528. [CrossRef] [PubMed]

32. Koda, S.; Zeggada, A.; Melgani, F.; Nishii, R. Spatial and Structured SVM for Multilabel Image Classification. *IEEE Trans. Geosci. Remote Sens.* **2018**, *56*, 2357–2369. [CrossRef]

33. Hosseini, Z.S.; Mahoor, M.; Khodaei, A. AMI-Enabled Distribution Network Line Outage Identification via Multi-Label SVM. *IEEE Trans. Smart Grid* **2018**, *9*, 5470–5472. [CrossRef]

34. Adankon, M.M.; Cheriet, M.; Biem, A. Semisupervised Learning Using Bayesian Interpretation: Application to LS-SVM. *IEEE Trans. Neural Netw.* **2011**, *22*, 513–524. [CrossRef] [PubMed]

35. Jindal, A.; Dua, A.; Kaur, K.; Singh, M.; Kumar, N.; Mishra, S. Decision Tree and SVM-Based Data Analytics for Theft Detection in Smart Grid. *IEEE Trans. Ind. Inform.* **2016**, *12*, 1005–1016. [CrossRef]

 *energies*

Article

Application of the Harmonic Balance Method for Spatial Harmonic Interactions Analysis in Axial Flux PM Generators

Natalia Radwan-Pragłowska, Tomasz Węgiel * and Dariusz Borkowski

Faculty of Electrical and Computer Engineering, Cracow University of Technology, Warszawska 24 St., 31-155 Cracow, Poland; natalia.radwan-praglowska@pk.edu.pl (N.R.-P.); dborkowski@pk.edu.pl (D.B.)
* Correspondence: pewegiel@cyfronet.pl or tomasz.wegiel@pk.edu.pl

Abstract: In this paper, an application of the Harmonic Balance Method (HBM) for analysis of Axial Flux Permanent Magnet Generator (AFPMG) is carried out. Particular attention was paid to development of mathematical model equations allowing to estimate the machine properties, without having to use quantitative solutions. The methodology used here allowed for precise determination of Fourier spectra with respect to winding currents and electromagnetic torque (both quantitatively and qualitatively) in steady state operation. Analyses of space harmonic interaction in steady states were presented for the three-phase AFPMG. Satisfactory convergence was between the results of calculations and measurements which confirmed the initial assumption that the developed circuit models of AFPMG are sufficiently accurate and can be useful in the diagnostic analyses, tests and the final stages of the design process.

Keywords: permanent magnet machines; axial flux generator; spatial harmonic interaction; harmonic balance method

Citation: Radwan-Pragłowska, N.; Węgiel, T.; Borkowski, D. Application of the Harmonic Balance Method for Spatial Harmonic Interactions Analysis in Axial Flux PM Generators. *Energies* **2021**, *14*, 5570. https://doi.org/10.3390/en14175570

Academic Editor: Islam Safak Bayram

Received: 28 July 2021
Accepted: 3 September 2021
Published: 6 September 2021

Publisher's Note: MDPI stays neutral with regard to jurisdictional claims in published maps and institutional affiliations.

1. Introduction

Interest in Permanent Magnet (PM) synchronous generators is associated with a global trend to support local energy supplied with renewable energy sources such as wind or water power, in which synchronous generators excited by permanent magnets are being increasingly applied [1–6]. Generators in this class of applications typically have a cylindrical rotor with surface-mounted magnets, but axial magnetic flux disc generator solutions are also very popular [7,8]. For this reason, both design and mathematical modelling techniques of these machines are being constantly developed and improved [9–25]. The Axial Flux PM (AFPM) generator deserves special attention due to their variety of designs and popularity in low-power home applications. This article is a continuation of the authors' papers [9–11] on AFPMG modeling, however it does not limit the possibility of applying the presented methodology to other constructions of PM generators.

Mathematical models of PM synchronous generators should provide the possibility of solving, in a relatively simple manner, various operational issues associated with the generation of electrical energy. For this purpose, the most suitable are so-called circuit models, commonly used in conventional machines [9–12,16,22] and the purpose of this paper is to extend the mathematical modeling methodology for this specific class of machines, i.e., axial flux PM synchronous generators operating in steady state.

Specific design features of PM machines make the analysis of effects occurring in these machines relatively difficult. Mapping the actual shape of the air gap and PM and the real distribution of the stator windings leads to complex analytical models [13–15,17,19,20,25]. Finding solutions using FEM analysis [18,21,24] does not always enable a qualitative analysis of electromagnetic phenomena in the machine.

The Harmonic Balance Method (HBM) [26–33] gives the possibility of analyzing solutions of mathematic model equations of electrical machines in case of periodic variation of

Energies **2021**, *14*, 5570. https://doi.org/10.3390/en14175570   https://www.mdpi.com/journal/energies

their coefficients; it is a simple extension of the symbolic method and leads to algebraization of the machine description in steady states. This approach is competitive with FEM analysis and allows to combine the electromagnetic phenomena occurring in the process of energy conversion in electrical machines. Such calculations are much simpler and faster.

In this paper, we focused on the methodology of HBM application for AFPMG modelling in which we also considered the influence of higher harmonics of the flux density distribution and the influence of space harmonics influence on the winding currents and the electromagnetic torque. The HBM is known and has a well-established place in the modeling of electrical machines, however, it was not widely used for AFPMG. Due to the specific design features of AFPM machines, the methodology of modeling with the use of HBM [26–30] must be adapted to them. In the presented approach, the model parameters are linear because there are no nonlinearity problems for the AFPMG structure.

The main task of this presented study is full use of the HBM in analysis of the steady-state of [30,31] AFPM generators. This approach has not been widely published for this class of machines. The issue was investigated to determine whether by means of mathematical modelling and numerical calculations it is possible to distinguish and quantify the interaction of spatial harmonics [26,29–31]. This paper presents developed HBM mathematical models that enable the execution of such analyses also for diagnostic purpose. Another aspect of the usefulness of mathematical models is, in addition to the analysis of these effects that are usually of parasitic nature, the possibility to synthesize harmonic interactions at the stage of designing the machine. In steady-state operation, the mathematical model of a PM machine is reduced to a system of linear differential equations with periodically varying coefficients [30,31]. A detailed analysis and solution of this system of equations using the HBM enables qualitative (frequency determination) and quantitative (amplitude determination) assessment of Fourier spectrums of currents and electromagnetic torque.

2. Harmonic Balance Method for Modeling PM Machines

2.1. General Assumptions

In permanent magnet excited machines, the most important energy processing elements are coils that form windings and permanent magnets. For co-energy to clearly describe winding states of their characteristics, relationships describing the linkage fluxes as functions of currents, and the rotor position must be unambiguous. It is a fundamental assumption, and it means that it is not possible to include the phenomenon of magnetic hysteresis. Changes in the energy state of the permanent magnet under the influence of coil currents occur in accordance with the inner hysteresis loop, called a return curve, which is very narrow for modern magnets, and is usually approximated by a straight line, as a result the hysteresis effect practically disappears. In modern permanent magnets of rare earth metals, the return curves additionally coincide with the demagnetization curve for its rectilinear section. Therefore, it can be assumed that for changes induced by winding currents, the permanent magnet operating point moves along unique demagnetization curve, which corresponds to uniqueness of changes in the co-energy of a permanent magnet [11,31]. Since iron cores of the stator and rotor have very high magnetic permeability compared to air and rare earth magnet materials, it can be assumed that the energy of the magnetic field concentrates mainly in the air gap and permanent magnets, which means that magnetic voltage drops in the machine's iron yokes are linear and can be included in permeance function, which describes machine magnetic geometry or these drops can be neglected. Assumption of linearity of the magnetic circuit is the key element for further analyses contained in the paper. This assumption in PM machines is usually acceptable and allows for determination of winding characteristics and the relationship describing co-energy, using self and mutual inductances. Windings will then be described by functions depending on the angle of rotor position and are additionally linear functions of currents. These functions also have additional elements representing flux linkage excited by PM, which is also a function of the rotation angle φ.

Using Lagrange's formalism based on the characteristics of windings and co-energy function with respect to electromechanical components of the entire system lead to mathematical equations of 3-phase PM machines. The mathematical model equations of PM machine can be written generally in a standard matrix form [11,31]

$$\frac{d}{dt}\{L(\varphi)\cdot i + \Psi_{PM}(\varphi)\} + R_s \cdot i = u$$
$$J\frac{d\omega}{dt} = T_{em}(\varphi, i_1, i_2, i_3) + T_{cog}(\varphi) + T_E - D\cdot\omega \tag{1}$$
$$\frac{d\varphi}{dt} = \omega$$

wherein electromagnetic torque

$$T_{em}(i_1, i_2, i_3, \varphi) = \frac{1}{2}i^T\cdot\frac{\partial}{\partial\varphi}L(\varphi)\cdot i + i^T\cdot\frac{\partial}{\partial\varphi}\Psi_{PM}(\varphi) \tag{2}$$

cogging torque

$$T_{cog}(\varphi) = \frac{\partial E_{0PM}(\varphi)}{\partial\varphi} \tag{3}$$

where:

$$L(\varphi) = L_{\sigma s} + L_s(\varphi) = \begin{bmatrix} L_{\sigma s} & & \\ & L_{\sigma s} & \\ & & L_{\sigma s} \end{bmatrix} + \begin{bmatrix} L_{11}(\varphi) & L_{12}(\varphi) & L_{13}(\varphi) \\ L_{21}(\varphi) & L_{22}(\varphi) & L_{23}(\varphi) \\ L_{31}(\varphi) & L_{32}(\varphi) & L_{33}(\varphi) \end{bmatrix} \tag{4}$$

$$i = \begin{bmatrix} i_1 \\ i_2 \\ i_3 \end{bmatrix} \quad u = \begin{bmatrix} u_1 \\ u_2 \\ u_3 \end{bmatrix} \quad \Psi_{PM}(\varphi) = \begin{bmatrix} \psi_{PM1}(\varphi) \\ \psi_{PM2}(\varphi) \\ \psi_{PM3}(\varphi) \end{bmatrix} \quad R_s = \begin{bmatrix} R_s & & \\ & R_s & \\ & & R_s \end{bmatrix}, \tag{5}$$

i_a and u_a—stator phase "a" current and voltage, $a = 1, 2, 3$; R_s—stator winding resistance; $\psi_{PMa}(\varphi)$—flux linkage of winding "a", produced by permanent magnets, $a = 1, 2, 3$; $L_{\sigma s}$, $L_{aa}(\varphi)$, $L_{ab}(\varphi)$—represents the inductance of the windings (leakage, self and mutual) $a, b = 1, 2, 3$; $E_{0PM}(\varphi)$—component of PM machine co-energy independent of winding currents; ω—rotor speed; T_E—external driving torque; D—damping coefficient.

The equation of electromechanical torque (2) contains the reluctance torque component and the main electromagnetic torque generated by the interaction between winding and permanent magnet fluxes. Torque produced in a zero-current state (3), is called cogging torque T_{cog} and is formed by the tangential forces acting on the slot walls and edges of the permanent magnet poles. The above equations are common, and their structure is very similar to Lagrange equations concerning conventional electric machines. Additional elements appear due to the presence of PM and the inductance must consider the existence of PM in the magnetic circuit of the machine.

Mathematical models of electric machines used in spatial harmonic interactions impact assessment, require correct identification of the qualitative characteristics of flux linkage, as a function of the angle of rotor position [26,29–31]. The geometry of the magnetic circuit of PM machines is very diverse and simple relationships, sufficiently accurate for classic machines cannot always be used for PM machines. In many cases, the calculation of equation parameters for PM machines will require a numerical determination of the field distribution (FEM analysis) in the machine, and, on this basis, approximation of the necessary coefficients.

Generally, for most of the designed symmetrical windings, MMF harmonics belong to set $\{\ldots - 5p, -3p, -p, p, 3p, 5p\ldots\}$, where "$p$" is number of machine pole pairs. The magnetic circuit may be characterized using unit permeance function. In this case, when we consider regular shape of the magnetic circuit the inductance matrix can be written in the following form [26,29,31]

$$L_s(\psi) = \sum_{n=0,\pm2p,\pm4p,\ldots} L_n\cdot e^{jn\varphi} \tag{6}$$

and the vector of PM flux linkages can be presented as

$$\Psi_{PM}(\varphi) = \sum_{\varsigma=\pm p,\pm 3p,\pm 5p\dots} \Psi_{\varsigma}^{PM} \cdot e^{j\varsigma\varphi} \tag{7}$$

2.2. Aplication of HBM for Modeling Spatial Harmonic Interaction in 3-Phase PM Generators

The main problem here is to show how to define the parameters of the model, which should highlight the impact of all relevant harmonics of a spatial field distribution in the machine, and the conversion of the mathematical model to be able to track the interactions of these harmonics.

Very useful in determining this is the symmetrical components transform, which describes the machine in orthogonal bases, and puts in order inductance matrices and vectors of PM flux linkages, so that based on equations, we can analyze how spatial harmonics affect currents and electromagnetic torque [26,29–31].

If mathematical model Equation (1) is transformed into symmetrical components, using a matrix transformation

$$T_3 = \frac{1}{\sqrt{3}} \begin{bmatrix} 1 & 1 & 1 \\ 1 & a & a^2 \\ 1 & a^2 & a \end{bmatrix}_{(3x3)} \quad \text{where } a = e^{j\frac{2\pi}{3}} \tag{8}$$

then, the machine voltage equations take the form

$$\frac{d}{dt}\{[L_{\sigma s} + L_s^s(\varphi)] \cdot i^s\} + \frac{d}{dt}\Psi_{PM}^s(\varphi) + R_s \cdot i^s = u^s \tag{9}$$

and the equation for the electromagnetic torque can be defined as

$$T_{em}(i^{s0}\ i^{s1}\ i^{s2}, \varphi) = \frac{1}{2}(\overset{\vee}{i^s})^T \cdot \frac{\partial}{\partial\varphi} L_s^s(\varphi) \cdot i^s + (\overset{\vee}{i^s})^T \cdot \frac{\partial}{\partial\varphi}\Psi_{PM}^s(\varphi) \tag{10}$$

where:

$$u^s = T_3 \cdot u \ ; \ u^s = [u^{s0}\ u^{s1}\ u^{s2}]^T ; \ i^s = T_3 \cdot i \ \ ; \ i^s = [i^{s0}\ i^{s1}\ i^{s2}]^T \tag{11}$$

$$L_s^s(\varphi) = T_3 \cdot L_s(\varphi) \cdot T_3^{-1} = \sum_n L_n^s \cdot e^{jn\varphi} \tag{12}$$

$$\Psi_{PM}^s(\varphi) = T_3 \cdot \Psi_{PM}(\varphi) = \sum_{\varsigma}\Psi_{\varsigma}^{PMs} \cdot e^{j\varsigma\varphi} = [\psi^{s0}(\varphi)\ \psi^{s1}(\varphi)\ \psi^{s2}(\varphi)]^T \tag{13}$$

If we consider a synchronous PM machine operating in generator mode, then winding current arrows must indicate proper energy flow direction. In the considerations, receiver current arrows according to Figure 1 were adopted. For such adopted arrows, it is understood that the external torque $T_E > 0$ is for generator operation and then the values of currents obtained from the solutions of the machine model equations will have a negative sign.

Circuit Equation (9) for generator mode shown in Figure 1 can be defined as

$$\frac{d}{dt}\{(L_{\sigma s} + L_s^s(\varphi) + L_L^s) \cdot i^s\} + (R_s + R_L^s) \cdot i^s = e_L^s - \frac{d}{dt}\Psi_{PM}^s(\varphi) \tag{14}$$

where:

$$R_L^s = T_3 \cdot \begin{bmatrix} R_{L1} & & \\ & R_{L2} & \\ & & R_{L3} \end{bmatrix} \cdot T_3^{-1} + \begin{bmatrix} 3R_N & & \\ & 0 & \\ & & 0 \end{bmatrix} ; \ L_L^s = T_3 \cdot \begin{bmatrix} L_{L1} & & \\ & L_{L2} & \\ & & L_{L3} \end{bmatrix} \cdot T_3^{-1} \tag{15}$$

If we assume the EMFs on load side to be a balanced three-phase voltage system, although this assumption is not mandatory, then it will correspond to the generator coupled with the grid. These voltages in symmetrical components are of the form

$$
e_L^s = T_3 \cdot \sqrt{2}\, E_{Sph}
\begin{bmatrix}
\cos(\omega_0 t + \beta_0) \\
\cos(\omega_0 t + \beta_0 - \frac{2\pi}{3}) \\
\cos(\omega_0 t + \beta_0 - \frac{4\pi}{3})
\end{bmatrix}
= \sum_{\eta=\pm 1} E_\eta^s \cdot e^{j\eta\,\omega_0 t} = E_1^s \cdot e^{j\omega_0 t} + E_{-1}^s \cdot e^{-j\omega_0 t} =
\begin{bmatrix} 0 \\ E \\ 0 \end{bmatrix} \cdot e^{j\omega_0 t} +
\begin{bmatrix} 0 \\ 0 \\ \overset{\vee}{E} \end{bmatrix} \cdot e^{-j\omega_0 t}
\tag{16}
$$

where: $E = E_S \cdot e^{j\beta_0} = \sqrt{\frac{3}{2}}\, E_{Sph} \cdot e^{j\beta_0}$, E_{ph} is the RMS value of grid phase voltage (line-neutral).

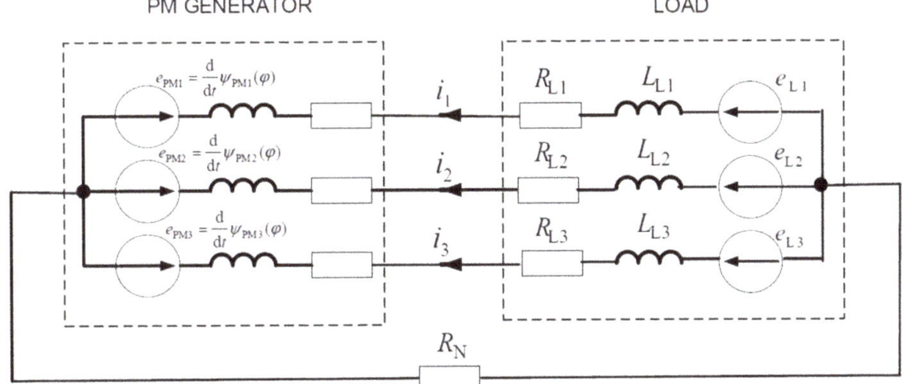

Figure 1. The PM generator under example load.

The steady state is considered when angular velocity of the rotor is constant $\omega = \Omega$ then $\varphi = \Omega \cdot t + \varphi_0$. The angle value φ_0 is related to the generator load. This angle will be important only in case of the generator's cooperation with the power grid (for standard monoharmonic models $p\varphi_0 - \beta_0 = \vartheta + \frac{3}{2}\pi$, where ϑ is a generator power angle). If synchronous steady-state dependency $\omega_0 = p\Omega$ is fulfilled, then the inductance matrix (12) and vector of PM flux linkages (13) become periodic, and we can assume solutions for the set of Equation (14), as

$$
i^s = \sum_\nu I_\nu^s \cdot e^{j\nu\Omega t};\ \ I_\nu^s = \begin{bmatrix} I_\nu^{s0} & I_\nu^{s1} & I_\nu^{s2} \end{bmatrix}^T
\tag{17}
$$

Solutions (17) fulfill, according to the HBM [26,29–31], an infinite dimensional system of algebraic equations

$$
\begin{aligned}
&\mathrm{diag}
\begin{bmatrix} \vdots \\ j3\,p\Omega\, E_{(3x3)} \\ j\,p\Omega\, E_{(3x3)} \\ -j\,p\Omega\, E_{(3x3)} \\ -j3\,p\Omega\, E_{(3x3)} \\ \vdots \end{bmatrix} \cdot
\begin{bmatrix}
\ddots & & & & \\
\cdots & L_0^{ss}+L_L^s & L_{2p}^{ss} & L_{4p}^{ss} & L_{6p}^{ss} & \cdots \\
\cdots & L_{-2p}^{ss} & L_0^{ss}+L_L^s & L_{2p}^{ss} & L_{4p}^{ss} & \cdots \\
\cdots & L_{-4p}^{ss} & L_{-2p}^{ss} & L_0^{ss}+L_L^s & L_{2p}^{ss} & \cdots \\
\cdots & L_{-6p}^{ss} & L_{-4p}^{ss} & L_{-2p}^{ss} & L_0^{ss}+L_L^s & \\
& & & & & \ddots
\end{bmatrix} \cdot
\begin{bmatrix} \vdots \\ I_{3p}^s \\ I_p^s \\ I_{-p}^s \\ I_{-3p}^s \\ \vdots \end{bmatrix} + \\[1em]
&+\mathrm{diag}
\begin{bmatrix} \vdots \\ R_s+R_L^s \\ R_s+R_L^s \\ R_s+R_L^s \\ R_s+R_L^s \\ \vdots \end{bmatrix} \cdot
\begin{bmatrix} \vdots \\ I_{3p}^s \\ I_p^s \\ I_{-p}^s \\ I_{-3p}^s \\ \vdots \end{bmatrix} =
\begin{bmatrix} \vdots \\ 0 \\ E_1^s \\ E_{-1}^s \\ 0 \\ \vdots \end{bmatrix} -
\mathrm{diag}
\begin{bmatrix} \vdots \\ j3\,p\Omega\, E_{(3x3)} \\ j\,p\Omega\, E_{(3x3)} \\ -\,j\,p\Omega\, E_{(3x3)} \\ -j3\,p\Omega\, E_{(3x3)} \\ \vdots \end{bmatrix} \cdot
\begin{bmatrix} \vdots \\ \Psi_{3p}^{ss} \\ \Psi_p^{ss} \\ \Psi_{-p}^{ss} \\ \Psi_{-3p}^{ss} \\ \vdots \end{bmatrix}
\end{aligned}
\tag{18}
$$

where:

$$E_{(3x3)} = \begin{bmatrix} 1 & & \\ & 1 & \\ & & 1 \end{bmatrix} \quad L_n^{ss} = \begin{cases} L_{\sigma s} + L_0^s & \text{for } n = 0 \\ L_n^s \cdot e^{jn\varphi_0} & \text{for } n \neq 0 \end{cases} \quad \text{and} \quad \Psi_\varsigma^{ss} = \Psi_\varsigma^{PMs} \cdot e^{j\varsigma\varphi_0} \tag{19}$$

The infinite dimensional system of linear Equation (18) with complex coefficients (19) determines the currents in the steady state. In order to make a practical use of this equation system it is necessary to limit the number of considered equations. From technical point of view, the limitations of the number equations depend on the content of the set of significant elements of PM flux linkages vector (7).

After formal mathematical transformations, similarly as it was described in [31] for analyses using HBM, we can derive the following general equation of the electromagnetic torque

$$T_{em} = -\tfrac{1}{2} Im \Big\{ \sum_{k_1=0,\pm1,\pm2....} \begin{bmatrix} \cdots & \overset{\vee}{I_{3p}^s} & \overset{\vee}{I_p^s} & \overset{\vee}{I_{-p}^s} & \cdots \end{bmatrix}$$

$$\cdot \begin{bmatrix} & \vdots & \vdots & \vdots & \\ \cdots & 2pk\ L_{0+2pk}^{ss} & 2p(k+1)\ L_{2p+2pk}^{ss} & 2p(k+2)\ L_{4p+2pk}^{ss} & \cdots \\ \cdots & 2p(k-1)\ L_{-2p+2pk}^{ss} & 2pk\ L_{0+2pk}^{ss} & 2p(k+1)\ L_{2p+2pk}^{ss} & \cdots \\ \cdots & 2p(k-2)\ L_{-4p+2pk}^{ss} & 2p(k-1)\ L_{-2p+2pk}^{ss} & 2pk\ L_{0+2pk}^{ss} & \cdots \\ & \vdots & \vdots & \vdots & \end{bmatrix} \cdot \begin{bmatrix} \vdots \\ I_{3p}^s \\ I_p^s \\ I_{-p}^s \\ \vdots \end{bmatrix} \cdot e^{j2pk\Omega t} \Big\} -$$

$$- Im \Bigg\{ \sum_{k=0,\pm1,\pm2....} \begin{bmatrix} \cdots & \overset{\vee}{I_{3p}^s} & \overset{\vee}{I_p^s} & \overset{\vee}{I_{-p}^s} & \cdots \end{bmatrix} \cdot \begin{bmatrix} \vdots \\ (2pk+3p)\cdot\Psi_{2pk+3p}^{ss} \\ (2pk+p)\cdot\Psi_{2pk+p}^{ss} \\ (2pk-p)\cdot\Psi_{2pk-p}^{ss} \\ \vdots \end{bmatrix} \cdot e^{j2pk\Omega t} \Bigg\} \tag{20}$$

Equation (20) contains the reluctance torque component and the main electromagnetic torque generated by the interaction between winding currents and PM fluxes. This approach allows for the use of the developed methodology to track interactions of spatial harmonics for every topology of generators excited by permanent magnets for different levels of external asymmetry. In the case of system analysis without a neutral wire, a very large R_N value should be assumed (for example 1 MΩ). Analyzing the Equations (18) and (20), it can be concluded that generally in PM machine operating in steady state, winding currents have the pulsation resulting from the interaction of spatial harmonics of orders $p\Omega, 3\,p\Omega, 5\,p\Omega, 7\,p\Omega\ldots$ while electromagnetic torque contains components of the pulsation of orders $2\,p\Omega, 4\,p\Omega, 6\,p\Omega\ldots$.

3. Model of AFPMG in Steady State Operation

3.1. Parameters of AFPMG Mathematical Model

The considerations were carried out for the three-phase AFPM generator. It can be said that this is a selected case of application of the spatial harmonic interactions modeling methodology for the PM machine presented in Section 2. The topology and structure of AFPMG are presented at Figure 2.

For AFPMG [11], due to its design features, the mathematical model is simpler and matrices of inductance $L_s(\varphi)$ (6) and after transformation $L_s^s(\varphi)$ (12) consist of elements independent of φ

$$L_s(\varphi) = L_0 = \begin{bmatrix} L_{ss} & M_{ss} & M_{ss} \\ M_{ss} & L_{ss} & M_{ss} \\ M_{ss} & M_{ss} & L_{ss} \end{bmatrix} \quad L_s^s(\varphi) = L_0^s = \begin{bmatrix} L_0^{ss0} & 0 & 0 \\ 0 & L_0^{ss1} & 0 \\ 0 & 0 & L_0^{ss2} \end{bmatrix} \tag{21}$$

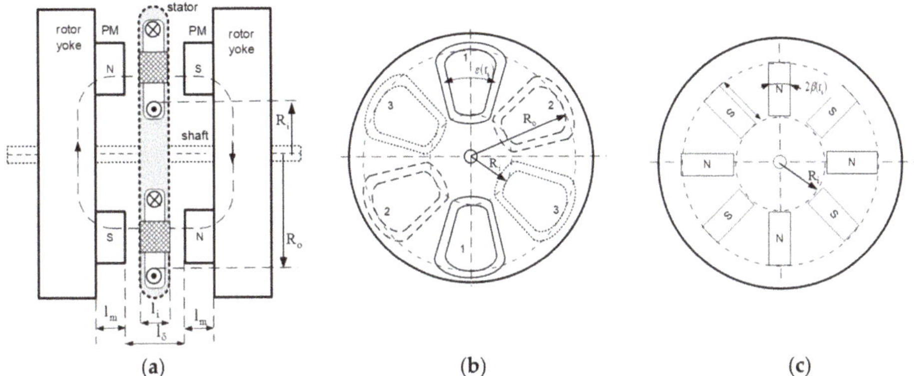

Figure 2. Construction of an AFPMG: (**a**) cross section, (**b**) stator, (**c**) rotor.

where:

$$L_0^{ss0} = L_{ss} + 2M_{ss} \quad L_0^{ss1} = L_0^{ss2} = L_{ss} - M_{ss} \tag{22}$$

Vector of PM flux linkages $\psi_{PM}(\varphi)$ (7), after the symmetrical component transformation $\psi_{PM}^s(\varphi)$ (13) contains elements which can be written as follows

$$\Psi_{\varsigma}^{PMs} = \begin{bmatrix} \psi_{\varsigma}^{s0} \\ 0 \\ 0 \end{bmatrix} \quad \text{for } \varsigma = \pm 3p, \pm 9p \ldots \qquad \text{where } \psi_{\varsigma}^{s0} = \sqrt{3}\, \psi_{\varsigma}^{PMs} \tag{23}$$

$$\Psi_{\varsigma}^{PMs} = \begin{bmatrix} 0 \\ \psi_{\varsigma}^{s1} \\ 0 \end{bmatrix} \quad \text{for } \varsigma = \ldots - 5p, p, 7p \ldots \quad \text{where } \psi_{\varsigma}^{s1} = \sqrt{3}\, \psi_{\varsigma}^{PMs} \tag{24}$$

$$\Psi_{\varsigma}^{PMs} = \begin{bmatrix} 0 \\ 0 \\ \psi_{\varsigma}^{s2} \end{bmatrix} \quad \text{for } \varsigma = \ldots - 7p, -p, 5p \ldots \quad \text{where } \psi_{\varsigma}^{s2} = \sqrt{3}\, \psi_{\varsigma}^{PMs} \tag{25}$$

The method of determining the above parameters L_{ss}, M_{ss}, ψ_{ς}^{PMs} ... for AFPMG was described in detail, for example, in the earlier papers of authors [9–11].

3.2. Spatial Harmonic Interaction Model for AFPMG

According to the previously presented considerations, the HBM Equations (18) and (20) for AFPM generator in the steady state can take the following form

$$\text{diag}\begin{bmatrix} \vdots \\ j3\,p\Omega\,E_{(3x3)} \\ j\,p\Omega\,E_{(3x3)} \\ -j\,p\Omega\,E_{(3x3)} \\ -j3\,p\Omega\,E_{(3x3)} \\ \vdots \end{bmatrix} \cdot \begin{bmatrix} \ddots \\ & L_0^{ss} + L_L^s \\ & & L_0^{ss} + L_L^s \\ & & & L_0^{ss} + L_L^s \\ & & & & L_0^{ss} + L_L^s \\ & & & & & \ddots \end{bmatrix} \cdot \begin{bmatrix} \vdots \\ I_{3p}^s \\ I_p^s \\ I_{-p}^s \\ I_{-3p}^s \\ \vdots \end{bmatrix} +$$

$$+\,\text{diag}\begin{bmatrix} \vdots \\ R_s + R_L^s \\ R_s + R_L^s \\ R_s + R_L^s \\ R_s + R_L^s \\ \vdots \end{bmatrix} \cdot \begin{bmatrix} \vdots \\ I_{3p}^s \\ I_p^s \\ I_{-p}^s \\ I_{-3p}^s \\ \vdots \end{bmatrix} = \begin{bmatrix} \vdots \\ 0 \\ E_1^s \\ E_{-1}^s \\ 0 \\ \vdots \end{bmatrix} - \text{diag}\begin{bmatrix} \vdots \\ j3\,p\Omega\,E_{(3x3)} \\ j\,p\Omega\,E_{(3x3)} \\ -j\,p\Omega\,E_{(3x3)} \\ -j3\,p\Omega\,E_{(3x3)} \\ \vdots \end{bmatrix} \cdot \begin{bmatrix} \vdots \\ \Psi_{3p}^{ss} \\ \Psi_p^{ss} \\ \Psi_{-p}^{ss} \\ \Psi_{-3p}^{ss} \\ \vdots \end{bmatrix} \tag{26}$$

where: $L_0^{ss} = L_{\sigma s} + L_0^s$ $\Psi_\varsigma^{ss} = \Psi_\varsigma^{PMs} \cdot e^{j\varsigma\varphi_0}$

$$T_{em} = -\mathrm{Im}\left\{ \sum_{k=0,\pm1,\pm2\,\ldots} \begin{bmatrix} \cdots & \overset{\vee}{I_{3p}^s} & \overset{\vee}{I_p^s} & \overset{\vee}{I_{-p}^s} & \cdots \end{bmatrix} \cdot \begin{bmatrix} \vdots \\ (2pk+3p)\cdot\Psi_{2pk+3p}^{ss} \\ (2pk+p)\cdot\Psi_{2pk+p}^{ss} \\ (2pk-p)\cdot\Psi_{2pk-p}^{ss} \\ \vdots \end{bmatrix} \cdot e^{j2pk\Omega t} \right\} \tag{27}$$

and matrices R_L^s, L_L^s (15) look as follows

$$R_L^s = \begin{bmatrix} \frac{1}{3}(R_{L1}+R_{L2}+R_{L3})+3R_N & \frac{1}{3}(R_{L1}+\underline{a}^2 R_{L2}+\underline{a}\,R_{L3}) & \frac{1}{3}(R_{L1}+\underline{a}\,R_{L2}+\underline{a}^2 R_{L3}) \\ \frac{1}{3}(R_{L1}+\underline{a}\,R_{L2}+\underline{a}^2 R_{L3}) & \frac{1}{3}(R_{L1}+R_{L2}+R_{L3}) & \frac{1}{3}(R_{L1}+\underline{a}^2 R_{L2}+\underline{a}\,R_{L3}) \\ \frac{1}{3}(R_{L1}+\underline{a}^2 R_{L2}+\underline{a}\,R_{L3}) & \frac{1}{3}(R_{L1}+\underline{a}\,R_{L2}+\underline{a}^2 R_{L3}) & \frac{1}{3}(R_{L1}+R_{L2}+R_{L3}) \end{bmatrix} \tag{28}$$

$$L_L^s = \begin{bmatrix} \frac{1}{3}(L_{L1}+L_{L2}+L_{L3}) & \frac{1}{3}(L_{L1}+\underline{a}^2 L_{L2}+\underline{a}\,L_{L3}) & \frac{1}{3}(L_{L1}+\underline{a}\,L_{L2}+\underline{a}^2 L_{L3}) \\ \frac{1}{3}(L_{L1}+\underline{a}\,L_{L2}+\underline{a}^2 L_{L3}) & \frac{1}{3}(L_{L1}+L_{L2}+L_{L3}) & \frac{1}{3}(L_{L1}+\underline{a}^2 L_{L2}+\underline{a}\,L_{L3}) \\ \frac{1}{3}(L_{L1}+\underline{a}^2 L_{L2}+\underline{a}\,L_{L3}) & \frac{1}{3}(L_{L1}+\underline{a}\,L_{L2}+\underline{a}^2 L_{L3}) & \frac{1}{3}(L_{L1}+L_{L2}+L_{L3}) \end{bmatrix} \tag{29}$$

Assuming a symmetrical load and introducing the following notations:

$$R_L = R_{L1} = R_{L2} = R_{L3};\ L_L = L_{L1} = L_{L2} = L_{L3};\ \underline{\psi}_\varsigma^{s(0,1,2)} = \psi_\varsigma^{s(0,1,2)} \cdot e^{j\varsigma\varphi_0} = \sqrt{3}\,\psi_\varsigma^{PMs} \cdot e^{j\varsigma\varphi_0} \tag{30}$$

it is possible to perform transformations of the system of Equation (26) rejecting the rows in which there are no voltage excitations and cutting out the corresponding columns of the system of equations. The system of Equation (26) is then significantly simplified to the following form:

$$\mathrm{diag}\begin{bmatrix} \ddots \\ j3p\Omega \\ jp\Omega \\ -jp\Omega \\ -j3p\Omega \\ \vdots \end{bmatrix} \cdot \begin{bmatrix} \ddots \\ L^{s0}+L_L & & & & \cdots \\ \cdots & L^{s1}+L_L & & & \\ \cdots & & L^{s2}+L_L & & \\ \cdots & & & L^{s0}+L_L & \\ & & & & \ddots \end{bmatrix} \cdot \begin{bmatrix} \vdots \\ I_{3p}^{s0} \\ I_p^{s1} \\ I_{-p}^{s2} \\ I_{-3p}^{s0} \\ \vdots \end{bmatrix} +$$

$$+\mathrm{diag}\begin{bmatrix} \vdots \\ R_s+R_L+3R_N \\ R_s+R_L \\ R_s+R_L \\ R_s+R_L+3R_N \\ \vdots \end{bmatrix} \cdot \begin{bmatrix} \vdots \\ I_{3p}^{s0} \\ I_p^{s1} \\ I_{-p}^{s2} \\ I_{-3p}^{s0} \\ \vdots \end{bmatrix} = \begin{bmatrix} \vdots \\ 0 \\ \overset{\vee}{E} \\ \overset{\vee}{E} \\ 0 \\ \vdots \end{bmatrix} - \mathrm{diag}\begin{bmatrix} \vdots \\ j3p\Omega \\ jp\Omega \\ -jp\Omega \\ -jp\Omega \\ \vdots \end{bmatrix} \cdot \begin{bmatrix} \vdots \\ \underline{\psi}_{3p}^{s0} \\ \underline{\psi}_p^{s1} \\ \underline{\psi}_{-p}^{s2} \\ \underline{\psi}_{-3p}^{s0} \\ \vdots \end{bmatrix} \tag{31}$$

After the mathematical transformations of the general the electromagnetic torque Equation (27), we also obtain a simplified equation defining the electromagnetic torque for the case of a symmetrical load in the following form:

$$T_{em} = -\mathrm{Imag}\left\{ \sum_{k=0,\pm1,\pm2,\ldots} \begin{bmatrix} \cdots & \overset{\vee}{I_{3p}^{s0}} & \overset{\vee}{I_p^{s1}} & \overset{\vee}{I_{-p}^{s2}} & \cdots \end{bmatrix} \cdot \begin{bmatrix} \vdots \\ (3p+k6p)\,\psi_{3p+k6p}^{s0} \\ (p+k6p)\,\psi_{p+k6p}^{s1} \\ (-p+k6p)\,\psi_{-p+k6p}^{s2} \\ \vdots \end{bmatrix} \cdot e^{jk6p\Omega t} \right\} \tag{32}$$

In this case, according to HBM, the vector of stator currents relates to a specific harmonic and should result in only one symmetrical non-zero component. These components can generally be related to other symmetrical components of the stator currents, but for AFPM generators there is some simplification and the components do not interact with each other. Moreover, from the symmetry properties of the system of Equation (31), it can be noticed that:

$$I^{s0}_{\nu} = \overset{\vee}{I^{s0}_{-\nu}} \quad \text{and} \quad I^{s1}_{\nu} = \overset{\vee}{I^{s2}_{-\nu}} \tag{33}$$

After transforming to the phase coordinates, an example of the dependency determining the time form of the current of the first phase is as follows:

$$i_1(t) = \frac{2}{\sqrt{3}} Re \left\{ \sum_{n=0}^{\infty} \frac{I^{s\,(2n+1)\,\mathrm{mod}\,3}}{(2n+1)p} \cdot e^{j(2n+1)\,p\,\Omega\,t} \right\} \tag{34}$$

During the analysis of Equation (32), it can be seen that in the case of full internal and external symmetry of the generator, components with pulsations of the order $6\,p\Omega$ are present in the electromagnetic torque signal, while in the general case, according to Equation (20), additional pulsations were equal to $2\,p\Omega$.

For a 3-wire system (without neutral) $I^{s0}_{3p+k6p} = 0$ for $k = 0, \pm1, \pm2, \ldots$ it means that the zero-sequence current does not flow and there should be no harmonics of this order $3\,p\Omega$ in the current spectrum. However, in actual measurements, it is noticed that these harmonics are present. The reason for their occurrence is not the external asymmetry of the generator, but the internal one, resulting in the generator's EMF voltages being slightly asymmetrical. For the real construction of the generator, where asymmetries may occur, caused, e.g., by the non-uniformity of the air gap, which is often the case with disk generators, the flux linkage vectors (23)–(25) are complete and contain not only

one element: $\Psi^{PMs}_{\varsigma} = \begin{bmatrix} \psi^{s0}_{\varsigma} \\ 0 \\ 0 \end{bmatrix}$ but also other elements for the positive and negative

components $\Psi^{PMs}_{\varsigma} = \begin{bmatrix} \psi^{s0}_{\varsigma} \\ \psi^{s1}_{\varsigma} \\ \psi^{s2}_{\varsigma} \end{bmatrix}$ for harmonics $\varsigma = \pm3p, \pm9p\ldots$. Then, there will also

appear currents with harmonics of the order of $3\,p\Omega$, but the symmetrical components of this current will have the order of one (positive) and two (negative). Solutions in cases of internal asymmetry of the generator, as well as in case of external asymmetry, will have to be searched for using the full model according to the Equations (26) and (27).

4. Laboratory Tests and Model Verifications

4.1. Description of Tested Generators

The verification of created models was carried out for generators with the main elements in a form of: two rotor discs (each disc with a diameter of 650 mm and 56-PM, 28-PM on one disc, $p = 14$), a stator with a diameter of 780 mm (with 21 coils, with non-overlapping windings). The models' verifications were extended in addition to the measurements by FEM analysis, performed in the ANSYS Maxwell environment. The generator model (Figure 3) was divided into ten regions, with a separate mesh defined for each. The total number of tetrahedral elements was 857,643.

Further analyses were carried out for two generator's stator topologies [11]. The authors considered the following stator topologies: coreless (a) and stator with iron cores (b) placed inside the coils. The main dimensions and parameters of AFPM generators are summarized according to Table 1.

The laboratory equipment included a tested generator coupled with a torque measuring shaft DATAFLEX with a DC drive machine (Figure 4). All data acquisition were performed using a National Instruments measurement card. Measurements were recorded during the measurement time of 10 s with the sampling frequency of 100 kHz.

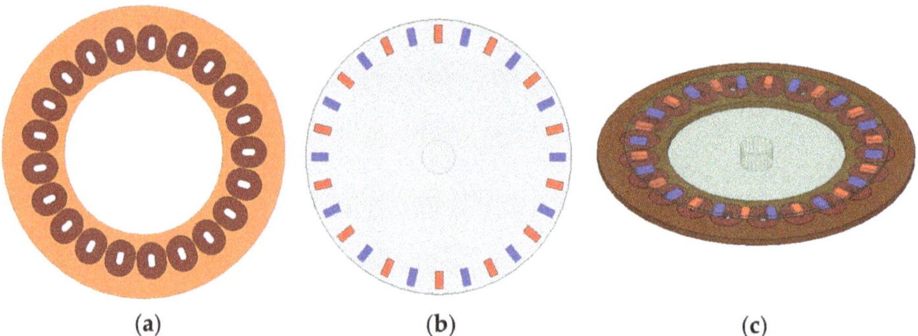

(a)   (b)   (c)

Figure 3. AFPM generator model with the ANSYS Maxwell program: (**a**) stator disc; (**b**) a rotor disc with PM arrangement; (**c**) assembling.

Table 1. Design AFPMG data according to Figure 2.

Parameters and Dimensions of the Permanent Magnets of AFPM Generators
• Magnets type: N40; Dimensions of a single magnet: $10 \times 18 \times 40$ mm; $l_c = 40$ mm   • $B_r = 1.2$ T; $H_c = 899$ kA/m; $\mu_{rm} = 1.07$; $\beta(r_s) = 0.0290$ rad; $l_m = 10$ mm;
Construction of the Stators in AFPM Generators
• $R_i = 270$ mm; $R_o = 310$ mm; $r_s = 290$ mm; $l_\delta = 26$ mm; $\varepsilon(r_s) = 0.1517$ rad;   • $l_i = 15$ mm—for structure with cores; $w_s = 980$—number of turns; $R_s = 2\ \Omega$

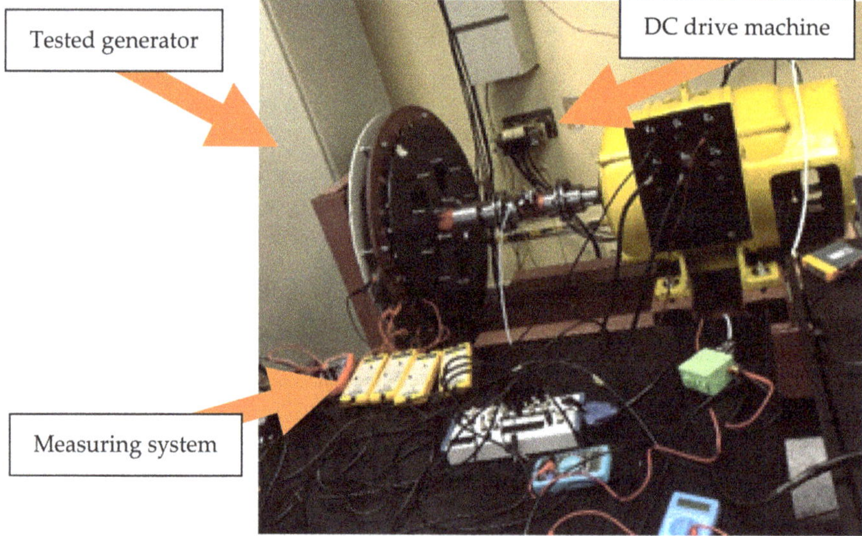

Figure 4. Laboratory test bench.

Figure 5 shows photos of the coreless (a) and with cores (b) stators. A photo of one rotor disc with the PM attached is shown in Figure 6.

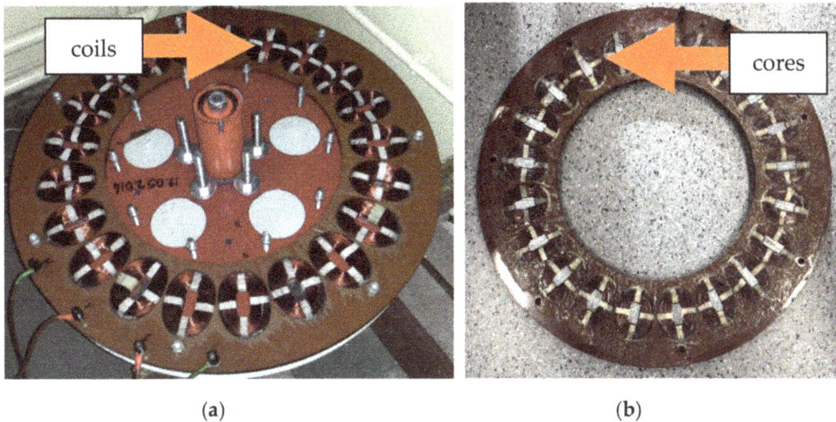

Figure 5. Stator discs; (**a**) coreless; (**b**) with cores.

Figure 6. Rotor disc.

4.2. Verification of Spatial Harmonic Interactions Models

The laboratory tests were carried out for modelled generators at a rotational speed of 206 rpm (48 Hz), with a three-phase (without neutral) symmetrical resistance load of the generator with the resistance value $R_L = 40\ \Omega$. The tests were performed for the AFPMG structure with a coreless stator and a stator with cores. The presented spectra (in dB) are for the adopted reference levels: 1 mV for voltages; 0.1 mA for currents and 1 mNm for torque. The parameters of the analytical models (inductances, PM flux linkage harmonics) were determined from the analytical equations presented in the authors' base article [11]. These parameters are summarized in Table 2. (inductance $M_{ss} \approx 0$).

Table 2. Main parameters of the analytical models.

	AFPMG	Inductances		PM Flux Linkage Harmonics $\psi_\varsigma^{PMs} = \psi_{-\varsigma}^{PMs}$; $\varsigma = p,3p,5p,7p,9p$				
		L_{ss}	$L_{\sigma s}$	ψ_p^{PMs}	ψ_{3p}^{PMs}	ψ_{5p}^{PMs}	ψ_{7p}^{PMs}	ψ_{9p}^{PMs}
(a)	coreless stator	4.7 mH	6.2 mH	0.897 Wb	18.2 mWb	0.30 mWb	0.03 mWb	0.007 mWb
(b)	stator with cores	6.0 mH	6.2 mH	1.398 Wb	57.2 mWb	4.9 mWb	1.2 mWb	0.7 mWb

First, the results for the induced EMF as a comparative measure of the PM flux linkage are presented in order to compare and indicate the main source of the generation of spatial harmonic interactions in AFPMG. The drawings below show the waveform of the induced EMF (voltage of phase one in the zero current state e_{PM1}) for the tested generators (Figure 7) and FFT analyses (Figure 8). Results were obtained from analytical models, measurements and FEM analyses.

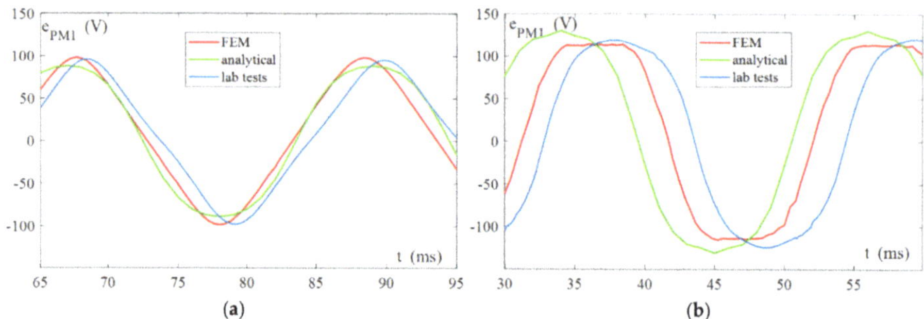

Figure 7. Waveform of EMF (phase 1): (**a**) AFPMG—coreless stator; (**b**) AFPMG—core stator.

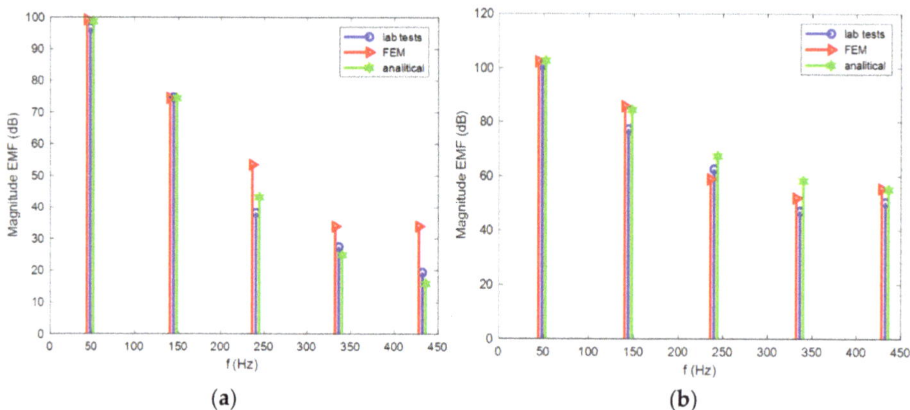

Figure 8. FFT spectrum of EMF (**a**) AFPMG—coreless stator (**b**) for AFPMG—core stator.

Table 3 contains comparisons of the values of the THD_{EMF} coefficient and the RMS values for the induced EMF of the generators, showing the relative error of calculations in relation to the measurements.

Table 3. Comparison of the results obtained from analytical models and laboratory tests for EMF.

	AFPMG	THD_{EMF}		$EMF_{(RMS)}$		
		Analytical Calculations	Measure	Analytical Calculations	Measure	$\lvert \Delta EMF_{(\%)} \rvert$
(a)	coreless stator	6.1%	6.5%	61.3 V	62.6 V	2.1%
(b)	stator with cores	6.0%	7.3%	101.3 V	95.8 V	5.7%

The results from the above table indicate EMF waveform distortions from the sinusoid, and thus also the winding flux linkages, although the proper verification of the presented mathematical models of spatial harmonic interaction in AFPMG required more detailed analyses of the winding currents and electromagnetic torque.

4.2.1. Verification of the Stator Currents

The drawings below (Figure 9) show the waveforms of the currents for the tested generators. The current waveforms can be considered to be similar, however, to assess them more precisely and analyse the effects of spatial harmonics interactions, FFT analyses were performed (Figure 10).

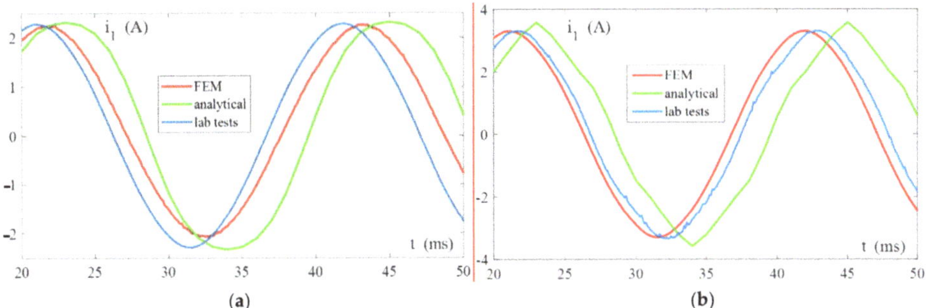

Figure 9. Waveform of current (phase 1): (**a**) AFPMG—coreless stator; (**b**) AFPMG—core stator.

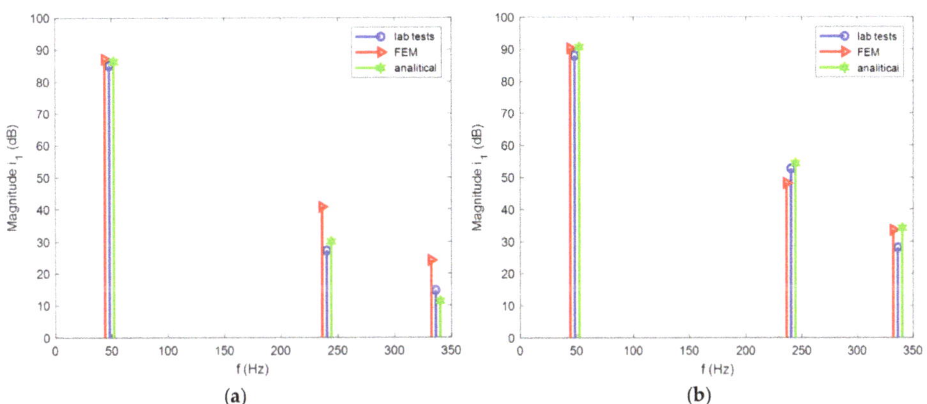

Figure 10. FFT spectrum of stator current (**a**) AFPMG—coreless stator (**b**) AFPMG—core stator.

When we analyse the spectrums from Figure 10, an ideal qualitative convergence can be noticed, while quantitatively, the measurement results and the analytical calculations differ by no more than 5 dB from each other. The values of the THD_I coefficient as an indicator of the content of higher harmonics, the RMS value of currents, including the relative error of calculations with respect to the measurements ($\Delta I_{(\%)}$), are summarized in Table 4. The obtained convergences can be considered to be satisfactory.

Table 4. Comparison of the results obtained from analytical models and laboratory tests for current i_1.

		THD_I		$I_{(RMS)}$		
	AFPMG	**Analytical Calculations**	**Measure**	**Analytical Calculations**	**Measure**	$\mid \Delta I_{(\%)} \mid$
(a)	coreless stator	0.16%	0.23%	1.65 A	1.69 A	2.4%
(b)	stator with cores	1.67%	1.95%	2.29 A	2.23 A	2.6%

4.2.2. Verification of the Electromagnetic Torque

The verification of the results obtained for the electromagnetic torque was not fully possible due to the relatively low level of harmonics generated in the waveforms of the electromagnetic torque. The values of the analytically obtained electromagnetic torque components of the order 6 $p\Omega$, 12 $p\Omega$ were practically at the noise level, and it was impossible to measure them. Therefore, analysis was limited to verifying the average value of the electromagnetic torque based on analytical calculations and measurements (Table 5) as well as comparisons of idealized cases based on numerical FEM analyses and analytical calculations (Figure 11).

Table 5. Comparison of the results obtained from analytical models and laboratory tests for average value of the electromagnetic torque.

AFPMG		$T_{em(AV)}$		
		Analytical Calculations	**Measure**	$|T_{em(\%)}|$
(a)	coreless stator	11.9 Nm	12.3 Nm	3.3%
(b)	stator with cores	31.1 Nm	29.3 Nm	6.1%

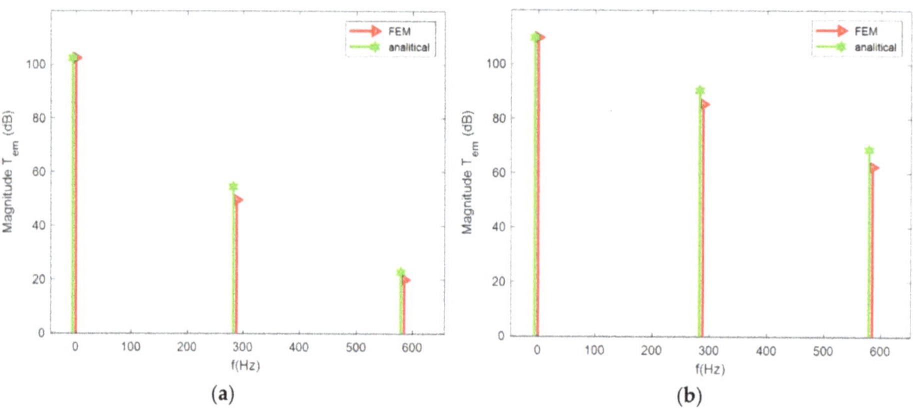

Figure 11. FFT spectrum of electromagnetic torque (**a**) AFPMG—coreless stator (**b**) AFPMG—core stator.

Electromagnetic torque mean value convergence up to 6%, as presented in Table 5 can be considered acceptable. The electromagnetic torque spectra presented in Figure 11 show the perfect qualitative consistency as well as the satisfactory quantitative convergence. The differences between the individual bands of the spectrum do not exceed the value of 5 dB.

5. Conclusions

The methodology presented in the article, concerning modelling of spatial harmonic interactions in AFPMG, has been confirmed for three-phase generators with a symmetrical structure which, nonetheless, do not limit the possibility of creating models using the presented methodology for the cases of machines with internal asymmetry of windings and electromagnetic circuit.

The parameters of the created models have integral form, so the accuracy of the results obtained from the presented AFPMG circuit models are limited. Certain discrepancies are present because many phenomena occurring in real models are not and cannot be represented in the mathematical models. The main reason for some discrepancies in the results are also the inaccuracy and imprecision in the assembly of the real AFPMG models.

It should be noted that for cases where there were differences in the results of calculations obtained from analytical models compared to the measurements, the results were not consistent with the use of FEM analysis either. In the presented results, in most cases (for the following quantities: EMF, currents, electromagnetic torque) the obtained results were consistent with the measurement results at a level below 10%, which can be considered as satisfactory. These results confirm the preliminary assumption that the use of HBM for AFPMG circuit models allows for sufficiently accurate analyses and the presented models can be useful for the analysis of various operational and design issues.

Author Contributions: Conceptualization, N.R.-P. and T.W.; methodology N.R.-P. and T.W.; validation, N.R.-P., T.W. and D.B.; investigation, N.R.-P., T.W. and D.B.; data curation, N.R.-P. and T.W.; writing—original draft preparation, N.R.-P. and T.W.; writing—review and editing, N.R.-P. and T.W.; supervision, N.R.-P. and T.W. All authors have read and agreed to the published version of the manuscript.

Funding: This work presented in this paper was funded by subsidies on science granted by Polish Ministry of Science and Higher Education for Cracow University of Technology.

Institutional Review Board Statement: Not applicable.

Informed Consent Statement: Not applicable.

Data Availability Statement: Not applicable.

Conflicts of Interest: The authors declare no conflict of interest.

References

1. Chalmers, B.; Spooner, E. An axial-flux permanent-magnet generator for a gearless wind energy system. *IEEE Trans. Energy Convers.* **1999**, *1z4*, 251–257. [CrossRef]
2. Caricchi, F.; Crescimbini, F. Modular Axial-Flux Permanent-Magnet Motor for Ship Propulsion Drives. *IEEE Trans. Energy Convers.* **1999**, *14*, 673–679. [CrossRef]
3. Chen, Y.; Pillay, P.; Khan, A. PM Wind Generator Topologies. *IEEE Trans. Ind. Appl.* **2005**, *41*, 1619–1626. [CrossRef]
4. Chan, T.F.; Lai, L.L. An Axial-Flux Permanent-Magnet Synchronous Generator for a Direct-Coupled Wind-Turbine System. *IEEE Trans. Energy Convers.* **2007**, *22*, 86–94. [CrossRef]
5. Park, Y.-S.; Jang, S.-M.; Choi, J.-H.; Choi, J.-Y.; You, D.-J. Characteristic Analysis on Axial Flux Permanent Magnet Synchronous Generator Considering Wind Turbine Characteristics According to Wind Speed for Small-Scale Power Application. *IEEE Trans. Magn.* **2012**, *48*, 2937–2940. [CrossRef]
6. Kirtley, J.L. *Permanent Magnets in Electric Machines. Electric Power Principle*; John Wiley & Sons Ltd.: Hoboken, NJ, USA, 2019.
7. Gieras, J.; Wang, R.; Kamper, M. *Axial Flux Permanent Magnet Brushless Machines*; Springer: Berlin/Heidelberg, Germany, 2004.
8. Aydin, M.; Huang, S.; Lipo, T.A. *Axial Flux Permanent Magnet Disc Machines: A Review*; Wisconsin Electric Machines & Power Electronics Consortium: Madison, WI, USA, 2004.
9. Radwan-Pragłowska, N.; Borkowski, D.; Węgiel, T. Model of Coreless Axial Flux Permanent Magnet Generator. In Proceedings of the 2017 International Symposium on Electrical Machines (SME), Naleczow, Poland, 18–21 June 2017. [CrossRef]
10. Radwan-Pragłowska, N.; Węgiel, T.; Borkowski, D. Parameters identification of coreless axial flux permanent magnet generator. *Arch. Electr. Eng.* **2020**, *67*, 391–402.
11. Radwan-Pragłowska, N.; Węgiel, T.; Borkowski, D. Modeling of Axial Flux Permanent Magnet Generators. *Energies* **2020**, *13*, 5741. [CrossRef]
12. Kamper, M.J.; Wang, R.-J.; Rossouw, F.G. Analysis and performance of axial flux permanent-magnet machine with air-cored nonoverlapping concentrated stator windings. *IEEE Trans. Ind. Appl.* **2008**, *44*, 1495–1504. [CrossRef]
13. Zhilichev, Y.N. Three-dimensional analytic model of permanent magnet axial flux machine. *IEEE Trans. Magn.* **1998**, *34*, 3897–3901. [CrossRef]
14. Parviainen, A.; Niemelä, M.; Pyrhönen, J. Modeling of axial flux permanent-magnet machines. *IEEE Trans. Ind. Appl.* **2004**, *40*, 1333–1340. [CrossRef]
15. Azzouzi, J.; Barakat, G.; Dayko, B. Quasi-3D analytical modeling of the magnetic field of an axial flux permanent-magnet synchronous machine. *IEEE Trans. Energy Conv.* **2005**, *20*, 746–752. [CrossRef]
16. Wang, R.J.; Kamper, M.J.; Van der Westhuizen, K.; Gieras, J.F. Optimal design of a coreless stator axial flux permanent-magnet generator. *IEEE Trans. Magn.* **2005**, *41*, 55–64. [CrossRef]
17. Virtic, P.; Pisek, P.; Marcic, T.; Hadziselimovic, M.; Bojan, S. Analytical Analysis of Magnetic Field and Back Electromotive Force Calculation of an Axial-Flux Permanent Magnet Synchronous Generator with Coreless Stator. *IEEE Trans. Magn.* **2008**, *44*, 4333–4336. [CrossRef]
18. Hosseini, S.M.; Agha-Mirsalim, M.; Mirzaei, M. Design, prototyping, and analysis of a low cost axial-flux coreless permanent-magnet generator. *IEEE Trans. Magn.* **2008**, *44*, 75–80. [CrossRef]

19. Choi, J.-Y.; Lee, S.-H.; Ko, K.-J.; Jang, S.-M. Improved Analytical Model for Electromagnetic Analysis of Axial Flux Machines with Double-Sided Permanent Magnet Rotor and Coreless Stator Windings. *IEEE Trans. Magn.* **2011**, *47*, 2760–2763. [CrossRef]
20. Sung, S.-Y.; Jeong, J.-H.; Park, Y.-S.; Choi, J.-Y.; Jang, S.-M. Improved Analytical Modeling of Axial Flux Machine with a Double-Sided Permanent Magnet Rotor and Slotless Stator Based on an Analytical Method. *IEEE Trans. Magn.* **2012**, *48*, 2945–2948. [CrossRef]
21. Stamenkovic, I.; Milivojevic, N.; Schofield, N.; Krishnamurthy, M.; Emadi, A. Design, Analysis, and Optimization of Ironless Stator Permanent Magnet Machines. *IEEE Trans. Power Electron.* **2013**, *28*, 2527–2538. [CrossRef]
22. Jin, P.; Yuan, Y.; Minyi, J.; Shuhua, F.; Heyun, L.; Yang, H.; Ho, S.L. 3-D Analytical Magnetic Field Analysis of Axial Flux Permanent-Magnet Machine. *IEEE Trans. Magn.* **2014**, *50*, 11. [CrossRef]
23. Huang, Y.K.; Zhou, T.; Dong, J.N.; Lin, H.Y.; Yang, H.; Cheng, M. Magnetic equivalent circuit modeling of yokeless axial flux permanent magnet machine with segmented armature. *IEEE Trans. Magn.* **2014**, *50*, 1–4. [CrossRef]
24. Maryam, S.; Naghi, R.; Vahid, B.; Juha, P.; Majid, R. Comparison of Performance Characteristics of Axial-Flux Permanent-Magnet Synchronous Machine with Different Magnet Shapes. *IEEE Trans. Magn.* **2015**, *51*, 1–6.
25. Węgiel, T. Cogging torque analysis based on energy approach in surface-mounted PM machines. In Proceedings of the 2017 International Symposium on Electrical Machines (SME), Naleczow, Poland, 18–21 June 2017. [CrossRef]
26. Sobczyk, T. A reinterpretation of the Floquet solution of the ordinary differential equation system with periodic coefficients as a problem of infinite matrix. *Compel* **1986**, *5*, 1–22. [CrossRef]
27. Gilmore, R.J.; Steer, M.B. Nonlinear Circuit Analysis Using the Method of Harmonic Balance—A Review of the Art. Part, I. Introductory Concepts. *Int. J. Microw. Millim.-Wave Comput.-Aided Eng.* **1991**, *1*, 22–37. [CrossRef]
28. Gilmore, R.J.; Steer, M.B. Nonlinear Circuit Analysis Using the Method of Harmonic Balance—A Review of the Art. Part II. Advanced Concepts. *Int. J. Microw. Millim.-Wave Comput.-Aided Eng.* **1991**, *1*, 159–180. [CrossRef]
29. Sobczyk, T. Direct determination of two-periodic solution for nonlinear dynamic systems. *Compel* **1994**, *13*, 509–529. [CrossRef]
30. Rusek, J. Category, slot harmonics and the torque of induction machines. *Compel* **2003**, *22*, 388–409. [CrossRef]
31. Wegiel, T. *Space Harmonic Interactions in Permanent Magnet Generator*; Cracow University of Technology: Cracow, Poland, 2013.
32. Esparza, M.; Segundo-Ramirez, J.; Kwon, J.B.; Wang, X.; Blaabjerg, F. Modeling of VSC-based power systems in the extended harmonic domain. *IEEE Trans. Power Electron.* **2017**, *32*, 5907–5916. [CrossRef]
33. Ludowicz, W.; Wojciechowski, R.M. Analysis of the Distributions of Displacement and Eddy Currents in the Ferrite Core of an Electromagnetic Transducer Using the 2D Approach of the Edge Element Method and the Harmonic Balance Method. *Energies* **2021**, *14*, 3980. [CrossRef]

Article

Lumped Parameter Model and Electromagnetic Performance Analysis of a Single-Sided Variable Flux Permanent Magnet Linear Machine

Basharat Ullah [1,*], Faisal Khan [1], Muhammad Qasim [1], Bakhtiar Khan [1], Ahmad H. Milyani [2], Khalid Mehmood Cheema [3] and Zakiud Din [3]

1 Department of Electrical and Computer Engineering, COMSATS University Islamabad, Abbottabad 22060, Pakistan; faisalkhan@cuiatd.edu.pk (F.K.); qasimrajput217@gmail.com (M.Q.); engr.bakhtiarkhan@gmail.com (B.K.)
2 Department of Electrical and Computer Engineering, King Abdulaziz University, Jeddah 21589, Saudi Arabia; ahmilyani@kau.edu.sa
3 School of Electrical Engineering, Southeast University, Nanjing 210096, China; kmcheema@seu.edu.cn (K.M.C.); zaki@seu.edu.cn (Z.D.)
* Correspondence: basharat.bigb@gmail.com

Citation: Ullah, B.; Khan, F.; Qasim, M.; Khan, B.; Milyani, A.H.; Cheema, K.M.; Din, Z. Lumped Parameter Model and Electromagnetic Performance Analysis of a Single-Sided Variable Flux Permanent Magnet Linear Machine. *Energies* **2021**, *14*, 5494. https://doi.org/10.3390/en14175494

Academic Editors: Ryszard Palka and Marcin Wardach

Received: 6 August 2021
Accepted: 30 August 2021
Published: 3 September 2021

Abstract: A new Single-sided Variable Flux Permanent Magnet Linear Machine with flux bridge in mover core is proposed in this paper. The flux bridge prevents the leakage flux from the mover and converts it into flux linkage, which greatly influences the performance of the machine. First, a lumped parameter model is used to find the suitable coil combination and no-load flux linkage of the proposed machine, which greatly reduces the computational time and drive storage. Secondly, the proposed machine replaces the expensive rare earth permanent magnets with ferrite magnets and provides improved flux controlling capability under variable excitation currents. Multivariable geometric optimization is utilized to optimize the leading design parameters like split ratio, stator pole width, width and height of permanent magnet, flux bridge width, the width of mover's tooth, and stator slot depth at constant electric and magnetic loading. The optimized design increases the flux linkage by 44.11%, average thrust force by 35%, thrust force density by 35.02%, minimizes ripples in thrust force by 23%, and detent force by 87.5%. Furthermore, the results obtained by 2D analysis are verified by 3D analysis. Thermal analysis is done to set the operating limit of the proposed machine.

Keywords: finite element analysis; flux switching machine; flux bridge; magnetic flux leakage; variable flux machine

1. Introduction

Due to simple and robust structure, linear flux switching permanent magnet machines (LFSPMMs) have been widely researched over the past few years [1–4]. In LFSPMM, the short mover carries all the excitation sources leaving the secondary completely robust, which makes it a prominent candidate for long stroke applications [1,5]. Since LFSPMMs possess high power density, faster dynamic response [6], high thrust force density [7], bipolar flux linkage [8] and good overload capability [9], they have been widely used in many applications from household appliances to transportation [2,10,11]. However, LFSPMM has some related disadvantages, especially the difficulty of adjusting the magnetic field, which is undesirable for applications that require flexible magnetic flux control or having a wide speed range [12].

Unfortunately, the prices of rare earth permanent magnet (PM) materials (such as dysprosium, neodymium, and terbium) have been rising for the past decade, which in turn increases the overall cost of the machine. In order to reduce the manufacturing cost of the machine, rare earth PMs are replaced by ferrite magnets in the design proposed in this

111

paper. To further lower the cost of the machine, overcome the aforementioned problems associated with LFSPMM, and achieve the flux regulation capability, another solution is Variable Flux Switching Machines i.e., Field Excited LFSM (FELFSM) and Hybrid Excited LFSM (HELFSM). LFSPMMs have only armature excitation (AE) and PMs. FELFSMs have both AE and field excitation (FE) [13]. Since FELFSM avoids the use of PM, it has the advantage of better controllability of flux and is cost effective [14], but it possesses an extensive downside of low thrust force density with weak flux linkage [15]. While HELFSMs have all the three excitation sources i.e., AE, PMs, and FE. PM along with FE becomes the main source of flux in HELFSM [16]. Compared to LFSPMM machines HELFSM has higher efficiency, higher thrust force density, and better controllability of flux [17]. Based on the field flux traveling path, HELFSMs can be further classified into series or parallel hybrid-excited ones. According to the placement of PMs and FE various HELFSM have been proposed [18–20]. In [18] a new type of HELFSM is proposed and studied based on the rotary hybrid excited flux switching machine proposed in [21]. In [18] the authors replaced the rare earth PMs with ferrite magnets and additional FE windings were used to obtain better flux regulation. This proposed design is then converted into double-sided HELFSM [16]. In [22], three different models of a hybrid excited FSM with FE coils located in stator slots are proposed and investigated. In [23] three HELFSM are proposed based on PM placement at the bottom, middle, and top, respectively. The analysis revealed that the bottom PM machine shows the best magnet utilization and flux adjustment capability. All the excitation sources are placed on a short mover so the area of field slot, armature slot, and PM will compete with each other, thereby, affecting the electromagnetic performance of the machine [24–26]. However, this issue is not that significant if appropriate effective parameters, proper electric and magnetic loadings are selected, to meet the desired performance.

In this paper, a new Single-sided Variable Flux Permanent Magnet Linear Machine (VFPMLM) with flux bridge in mover core is proposed. The proposed design replaces the expensive rare-earth PMs with ferrite magnets, uses FE for better field weakening capability. VFPMLM offers high thrust force density, high thrust force per PM volume, with much lower copper losses than the conventional design in the literature, and is preferred for long stroke applications.

The rest of the paper is organized as, Section 2 presents the machine topology and its operating principle, in Section 3 no load flux linkage is obtained through lumped parameter model, in Section 4 various performance analysis parameters are discussed, in Section 5 geometric optimization is utilized to analyze the influence of different design parameters on the performance of the machine, the electromagnetic performance of the machine is analyzed, and compared in Sections 6 and 7 concludes the paper.

2. Machine Topology and Operating Principle

2.1. Machine Topology

The topology of the single-sided VFPMLM is shown in Figure 1, and Figure 2 shows the corresponding design variables. Table 1 enlists the values of leading design parameters. All the excitation sources are installed on the mover, while the stator is completely passive. The proposed single-sided VFPMLM has concentrated three-phase armature windings, each coil set is composed of two sets of AE coils. The field winding is overlapping the armature winding. The magnetic flux bridge denoted by w_{fb} in Figure 3 eliminates the flux leakage through the mover, so the magnetic flux passes through the mover core and converts the leakage into a flux linkage. The yoke of the mover and the flux bridge help in flux linkage and distribution through the air gap. This magnetic flux arrangement improves the flux modulation effect of PM and FE, which superimpose each other through the flux bridge and generate higher magnetic flux density, hence improving thrust force generating ability, reducing detent force, force ripples and the PM slot effect. In order to select the number of stator poles Equation (1) is utilized.

$$S_p = P_s \left(2 \pm \frac{n}{2q} \right) \tag{1}$$

In (1), S_p represents the number stator poles, P_s represents the number of mover slots, q denotes the number of phases and n is any natural number. Various values of natural numbers are considered and the performance of the respective pole/slot combination is analyzed. Table 2 enlists all the possible combinations that the machine has a bipolar flux linkage and unipolar thrust force. Thrust force for the 7/6 combination is higher, therefore it is considered for further analysis.

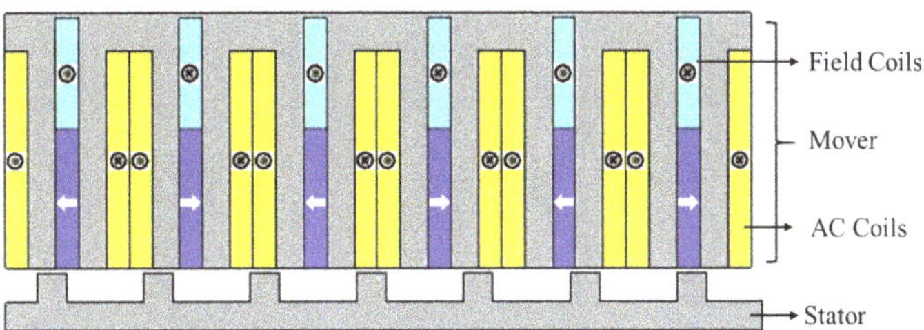

Figure 1. Design of proposed single-sided VFPMLM.

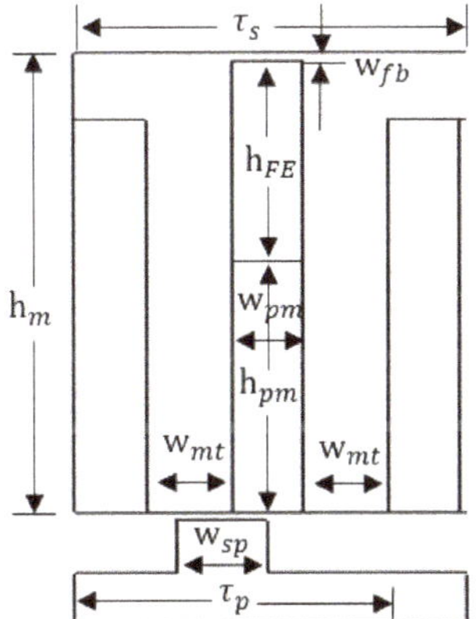

Figure 2. Design parameters of proposed single-sided VFPMLM.

2.2. Operating Principle

The operating principle of single-sided VFPMLM is based on no-load magnetic flux (excitation due to PM and FE) linkage achieved by 2D Finite Element Analysis (FEA). The proposed design has a bipolar flux linkage. The magnetic flux generated by the excitation sources at the mover has to flow through the mover's tooth, crosses the air gap, enters stator body through the stator tooth, then exits through the next stator tooth, crosses the air gap again, and enters into the second-mover tooth to link with the armature coil to complete its circuit.

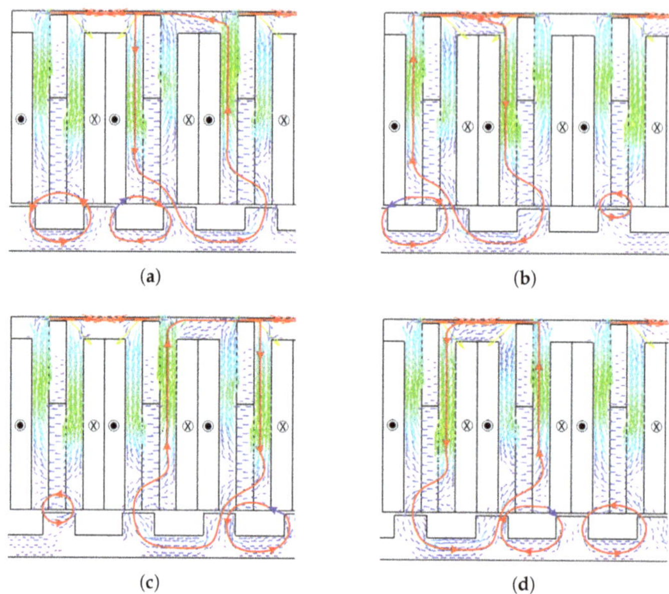

Figure 3. Principle of operation (**a**) $\theta = 0^o$ (**b**) $\theta = 90^o$ (**c**) $\theta = 180^o$ (**d**) $\theta = 270^o$.

Table 1. Design parameters of Single-sided VFPMLM.

Symbol	Description	Unit	Dimension
h_m	mover height	mm	43.5
L_T	whole mover length	mm	131
τ_s	mover pole pitch	mm	21.84
h_{AE}	AC slot height	mm	33.4
h_{FE}	DC slot height	mm	19.5
h_{PM}	PM height	mm	24
w_{fb}	flux bridge width	mm	1
N_{AE}	AC turns	–	220
N_{FE}	DC turns	–	50
v	velocity	m/sec	4
A_g	air gap	mm	0.8
L	stack length	mm	90
τ_p	stator pole pitch	mm	18.72
w_s	Stator pole width	mm	8.42
k_w	winding filling factor	–	0.5
I_{FE}	FE current	Amps	6

Table 2. Different stator poles combination.

Pole/Slot Combination	Thrust Force	Detent Force
5/6	112.714 N	76.59 N
7/6	165.212 N	36.05 N
8/6	Bipolar Force	–
10/6	Bipolar Force	–
11/6	139.653 N	49.01 N
13/6	65.42 N	9.53 N

Considering the movement of the mover along the x-axis as shown in Figure 3, when the relative position of the PM is in the center of the stator slot, the electrical angle (θ) is assumed to be zero degrees. The magnetic flux linkage in coil A is zero at the specified position is shown in Figure 3a, because the flux does not pass through the mover core. At 1/4th movement of electrical angle ($\theta = 90°$) the flux linkage in coil A is positive maximum because the direction of flux linkage and AE coil direction is same shown in Figure 3b. After half cycle ($\theta = 180°$) movement of the mover, the magnitude of flux linkage in coil A becomes zero again because the flux does not pass through the mover core as shown in Figure 3c. At 3/4th movement ($\theta = 270°$), the flux linkage in coil A attains a negative maximum because the direction of flux linkage and AE coil direction becomes opposite shown in Figure 3d. This all can be evident from Figure 4. The operating principle explained is repetitive, and has been verified for all three armature coils.

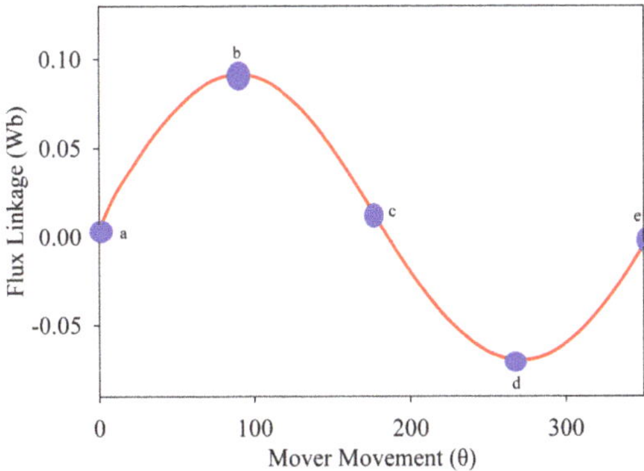

Figure 4. No-load flux linkage.

3. Lumped Parameter Model

Lumped parameter model (LPM) is widely used to model all the parts of machine (mover, airgap, stator) [27,28]. In this paper LPM is developed for 1/3rd machine as shown in Figure 5 to reduce the simulation time, drive storage and computational complexity. LPM is based on the following equations.

$$\phi_s = PF \tag{2}$$

$$F_{pm} = \frac{B_r l_{pm}}{\mu_0 \mu_r} \tag{3}$$

$$F_{FE} = N_{FE} I_{FE} \tag{4}$$

$$P_{pm} = \frac{\mu_0 \mu_r l_{stk} w_{pm}}{l_{pm}} \tag{5}$$

The permeance of iron parts can be calculated as:

$$P_{my/fb/mt/st/rc} = \int \frac{\mu_0 \mu_r}{l_i} dA_i \tag{6}$$

After calculating the permeance of all parts, and sources at nodes, nodal analysis is employed to calculate flux flowing out and flowing in to the corresponding nodes as:

$$\begin{bmatrix} \phi_s(1) \\ \vdots \\ \phi_s(N) \end{bmatrix} = \begin{bmatrix} P(1,1) & \cdots & P(1,N) \\ \vdots & \ddots & \vdots \\ P(M,1) & \cdots & P(M,N) \end{bmatrix} \begin{bmatrix} F(1) \\ \vdots \\ F(N) \end{bmatrix} \tag{7}$$

$P(M, N)$, the permeance matrix can be written as follows:

$$P(M, N) = \begin{cases} 1 & branch\ N\ begins\ from\ node\ M \\ 0 & no\ connection\ between\ branch\ N\ and\ node\ M \\ -1 & branch\ N\ ends\ at\ node\ M \end{cases} \tag{8}$$

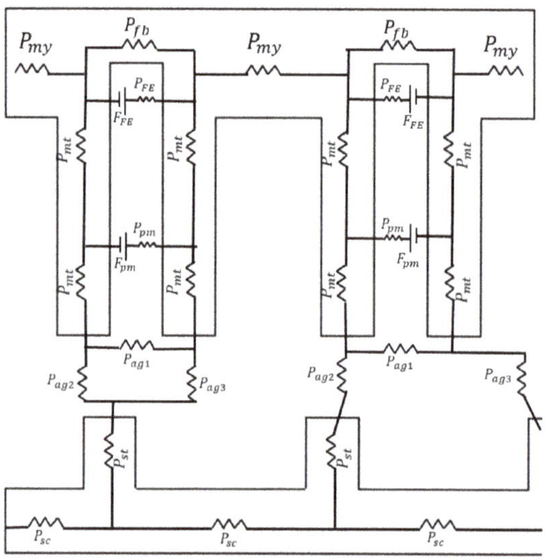

Figure 5. LPM of proposed Single-sided VFPMLM.

Equation (7) is solved in three steps [29], through iterative process to calculate the no-load flux linkage with considering the initial relative permeability of 4000. Firstly, H_i in the mover and stator part is calculated using Equation (9). Once H_i is calculated, relative permeability is updated using Equation (10). Finally, magnetic flux density (Equation (11)) is calculated using the same iterative process. Figure 6 shows the no-flux linkage obtained by FEA analysis and calculated through LPM. There is a small error between the two but considering computational time, drive storage it is not that significant.

$$H_i^{k-1} = \frac{\Delta F_i^{k-1}}{l_i} \tag{9}$$

$$\mu_r^k = \frac{\left[H_i^{k-1} + M_s \left(coth\left(\frac{H_i^{k-1}}{\beta} \right) - \frac{\beta}{H_i^{k-1}} \right) \right]}{H_i^{k-1}} \tag{10}$$

$$B_i^k = \frac{\Delta F_i^k P}{A_i} \tag{11}$$

The LPM is validated by JMAG v19.1 and detailed comparison based on computation time and drive storage is given in Table 3. The LPM greatly reduces the drive storage and computational time as compared to FEA. Both the LPM and FEA were performed on a Lenovo 64-bit operating system with 8 GB RAM and an Intel(R) Core(TM) i5-8500 CPU running at 3.00 GHz.

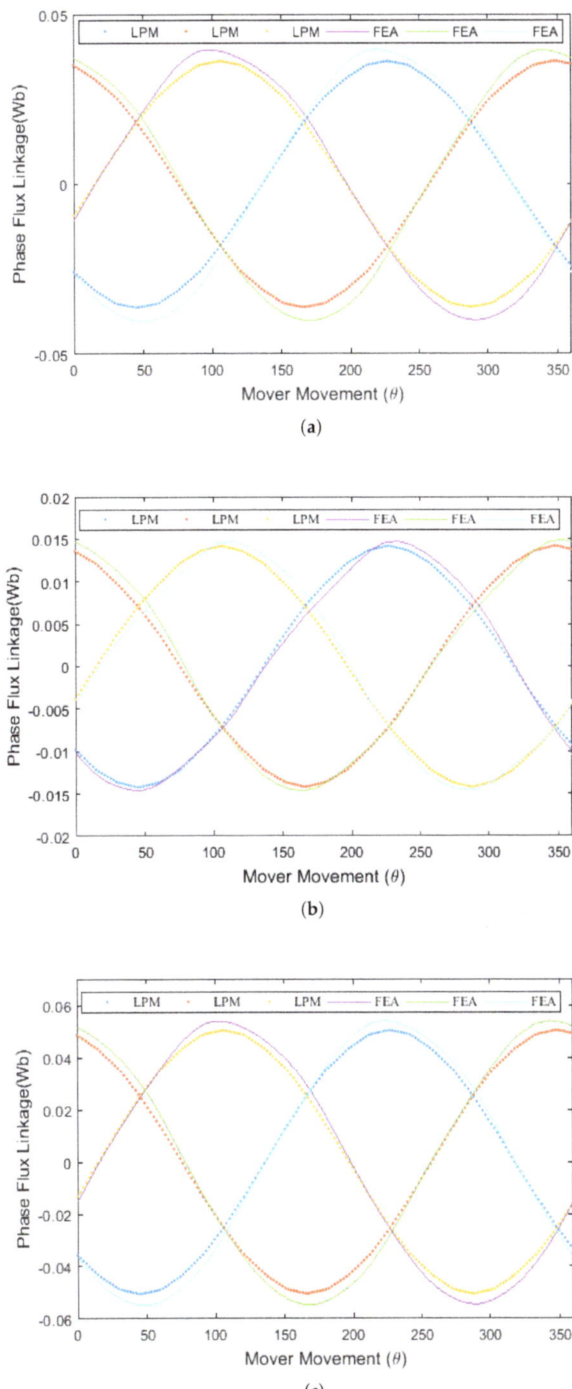

Figure 6. (**a**) PM flux linkage FEA vs. LPM (**b**) FE flux linkage FEA vs. LPM (**c**) PM+FE flux linkage FEA vs. LPM.

Table 3. Computational time and drive storage comparison.

Analysis Method	Time	Drive Storage
LPM	4.67 s	126 KBs
2D FEA	7 min and 48 s	217 MBs

4. Analysis of Single-Sided VFPMLM

After evaluating the initial performance of the proposed machine through LPM, different performance parameters like three-phase no-load flux linkage (FE and PM both), Total Harmonic Distortion (THD) of middle phase and its corresponding harmonic frequency spectrum, detent force, average thrust force (TF_{avg}), and thrust force density (TFD) are analyzed. Parameters like no-load flux linkage, detent, and thrust force are taken directly from 2D FEA solver while to find THD, TFD and TF_{avg} per PM volume, the following equations are utilized.

$$THD = \frac{\sqrt{\sum_{i=2}^{m} \Phi_i^2}}{\Phi_1} \tag{12}$$

$$TFD = \frac{TF_{avg}}{V_m} \tag{13}$$

$$TF_{avg} \text{ per PM volume} = \frac{TF_{avg}}{V_{PM}} \tag{14}$$

Φ_1 represents the flux linkage of fundamental component while Φ_2 up to Φ_m represent the harmonic components. V_m and V_{PM} is the volume of mover and PM respectively.

5. Multivariable Geometric Optimization

Multivariable geometric optimization (MGO) is employed to investigate the influence of different design parameters on the improvement of the initial electromagnetic performance of the proposed single-sided VFPMLM. MGO is a sequential optimization technique and does not depends on the previous values of the variables. Firstly, a set of variables are defined that has a combined influence on the objective function, and their respective ranges are defined. MGO is started to ensure the achievement of the global maximum. If the set target is not achieved, then the set of variables are refined. During MGO implementation, electric and magnetic loadings, slot area, air gap length, and stack length are kept constant to ensure that it is only applied to design parameters and their configuration. The flow chart for MGO is shown in Figure 7. MGO is used to improve the key parameters like TF_{avg}, thrust force ripples (TF_{rip}) and TFD. For the implementation of MGO, the objective function, constraints, set of variables to be optimized, and their respective ranges are defined as:

$$
\begin{aligned}
ObjectiveFunction : \quad & \max(TF_{avg}, TFD) \quad and \\
& \min(F_d, TF_{rip}, THD) \\
Constraints : \quad & TF_{avg} \geq 165.21N, \; TFD \geq 322kN/m^3 \\
& F_d \leq 36.1N \quad and \quad TF_{rip} \leq 60N
\end{aligned} \tag{15}
$$

$$
\begin{aligned}
Split\ Ratio \quad & 0.1436 \leq S.R \leq 0.2173 \\
Stator\ pole\ width \quad & 4.31 \leq w_{sp} \leq 9.36 \\
PM\ tooth\ width \quad & 3.5 \leq w_{pm} \leq 5 \\
PM\ tooth\ height \quad & 19.2 \leq h_{pm} \leq 27.43 \\
Flux\ bridge\ width \quad & 0.5 \leq w_{fb} \leq 1.5 \\
Mover\ tooth\ width \quad & 3.5 \leq w_{mt} \leq 5 \\
Secondary\ slot\ depth \quad & 3.5 \leq h_{st} \leq 7.0
\end{aligned} \tag{16}
$$

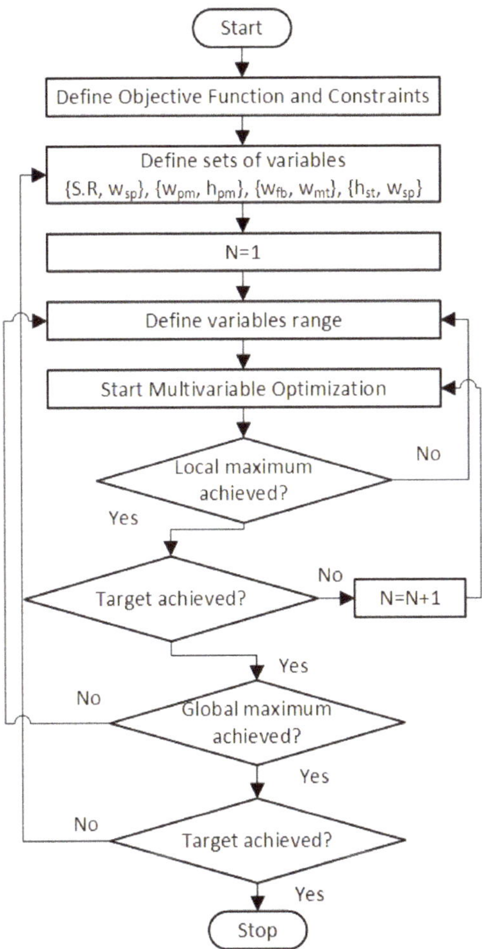

Figure 7. Flow chart for MGO.

5.1. Influence of S.R and w_{sp}

The split ratio defined by Equation (17) is the main parameter, its value is wisely chosen to avoid saturation of back iron, while the width of the stator pole has a great influence on no-load flux linkage. The value of these variables are varied in the specified range and the results are shown in Figure 8. From Figure 8 it is clear that at S.R = 0.1989 and w_{sp} = 7.5 mm, TF_{avg} increased from 165.21 N to 179.08 N, TFD is increased from 322 kN/m^3 to 349.1 kN/m^3. Ripples in the thrust force are reduced from 60 N to 57 N. THD is decreased from 3.47% to 2.85%.

$$S.R = \frac{h_s + A_g}{h_s + A_g + h_m} \tag{17}$$

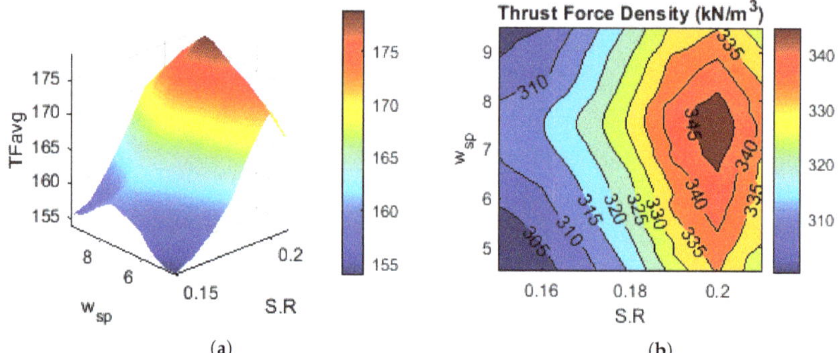

Figure 8. Influence of S.R and w_{sp} on (**a**) TF_{avg} (**b**) TFD.

5.2. Influence of w_{pm} and h_{pm}

Here the permanent magnet height and width are optimized in their specified ranges keeping the volume fixed, and their influence is observed in Figure 9. PM width variation is also dependent on mover tooth width so its value is carefully optimised. Analysis reveals that w_{pm} and h_{pm} of the PM improves TF_{avg} from 179.08 N to 184.95 N, TFD is increased from 349.1 kN/m^3 to 360.5 kN/m^3. TF_{rip} are decreased from 57 N to 55 N. Initially, the w_{pm} was 4 mm and h_{pm} was 24 mm are optimized to w_{pm} = 4.05 mm and h_{pm} = 23.7 mm. THD is reduced from 2.85% to 2.22%.

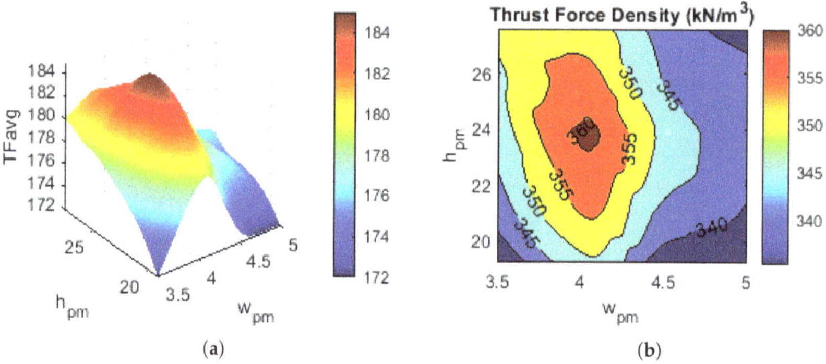

Figure 9. Influence of w_{pm} and h_{pm} on (**a**) TF_{avg} (**b**) TFD.

5.3. Influence of w_{mt} and w_{fb}

Single-sided VFPMLM is designed such that the flux bridge eliminates the flux leakage via mover, so the magnetic flux passes through the mover core and converts the leakage into a flux linkage. The width of the flux bridge was initially set to 1 mm but the improvement was observed in thrust force by minimizing its width. So its value is optimized along with the width of the mover's tooth (w_{mt}). The optimised w_{fb} is kept at 0.8 mm. Initially the w_{mt} was taken 4 mm which is optimized to 4.85 mm, TF_{avg} increased from 184.95 N to 208.12 N, TFD is increased from 360.5 kN/m^3 to 405.68 kN/m^3. TF_{rip} are decreased from 55 N to 51.5 N. Figure 10 shows the detailed analyses. THD is reduced from 2.22% to 1.89%.

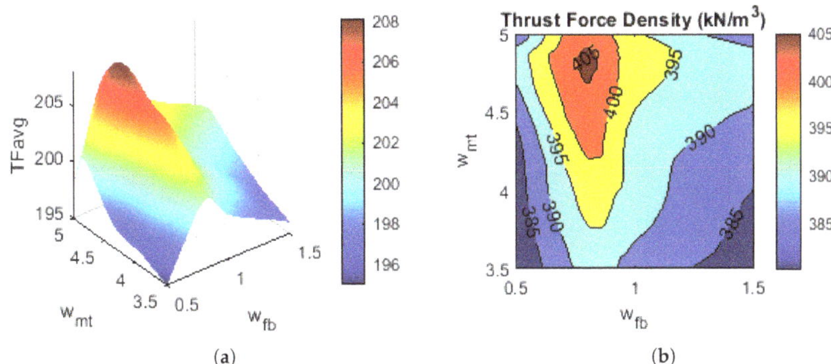

Figure 10. Influence of w_{fb} and w_{mt} on (**a**) TF_{avg} (**b**) TFD.

5.4. Influence of h_{st} and w_{sp}

The depth of stator slot (h_{st}) is optimized along stator pole width (w_{sp}) and its effect on the performance is shown in Figure 11. Initially h_{st} was taken 4.5 mm which is optimized to 5 mm, and w_{sp} is once again optimized with h_{st}, so the revised optimized value of w_{sp} is 5.2 mm that improves TF_{avg} from 208.12 N to 223.04 N, TFD is increased from 405.68 kN/m^3 to 434.76 kN/m^3. TF_{rip} and THD are decreased from 51.5 N to 46 N and from 1.89% to 1.27% respectively.

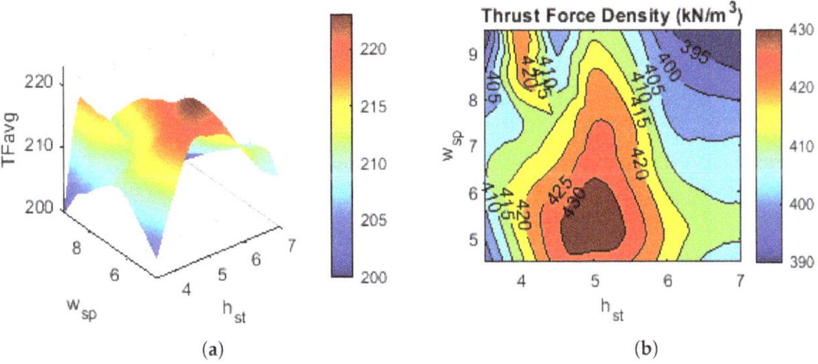

Figure 11. Influence of h_{st} and w_{sp} on (**a**) TF_{avg} (**b**) TFD.

After all optimization stages are completed, Table 4 is obtained which compares the performance of initial and optimized design.

Table 4. Performance based comparison of initial design and optimized design.

Parameter	Initial Values	Optimized Values	Improvements
Flux Linkage (Wb)	0.12112	0.17455	44.11%
THD (%)	3.47	1.27	63.40%
Detent Force (N)	36.05	4.51	87.49%
TF_{avg} (N)	165.21	223.04	35.0%
TFD (kN/m^3)	322	434.76	35.02%
TF_{rip} (N)	60	46	23.33%

6. FEA Based Performance Analysis

6.1. Flux Strength and Flux Regulation

When the machine is in the initial position, the flux line distribution of the optimized design is shown in Figure 12. Figure 13a shows the no-load flux linkage of the initial design and optimized design, while the corresponding harmonics are shown in Figure 13b,c respectively. Figure 14a shows the influence of field excitation on the flux linkage. Phase A is the middle phase and is not affected by the end effect, which is the problem in linear machines. Figure 14b shows the flux weakening and enhancing capability of the proposed machine under various FE current densities i.e., $J_e = 0, \pm 4, \pm 8, \pm 12$ (A/mm^2). When the current density is positive, the flux density in the air gap increases which results in an enhanced flux linkage whereas when the current density is negative, the flux density in the air gap is reduced causing a decrease in the flux linkage. To conclude, the flux regulation capability of the proposed single-sided VFPMLM mainly depends upon the FE current and the flux linkage is symmetrical and sinusoidal.

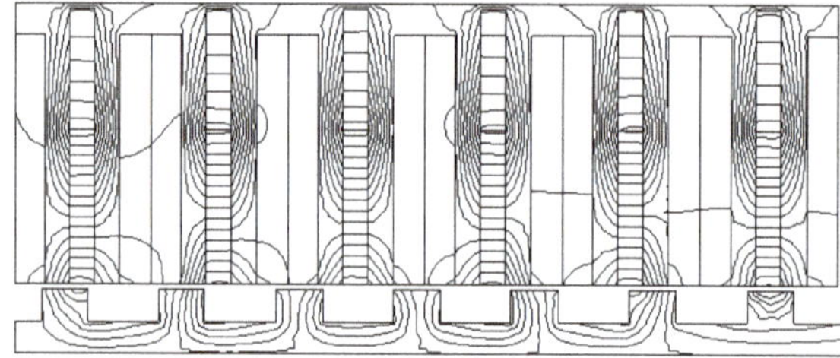

Figure 12. Flux line distribution of proposed design at initial position.

6.2. Detent Force and Average Thrust Force

Detent force is also one of the main factors in evaluating the performance of the machine to analyze the noise and vibration. It arises at no load due to the attractive force between the secondary core and the mover PM. The stronger force of attraction leads to higher vibration and noise and, hence, lower output TF_{avg}. Figure 15 depicts the detent force of the initial and optimized design. The average thrust force of the proposed single-sided VFPMLM is improved after optimization as shown in Figure 16. Compared with the initial design, the optimized model has an average thrust increase of 35%, which is an increase of 45.35% compared with the HEFSM proposed in literature [18] with the same stack length and area of the two machine movers. In the initial design, the TF_{avg} was 165.21 N, which is increased to 223.04 N after optimization under the same electric and magnetic loading. In the Figure 17, the armature current density (J_{AC}) is varied from 0 to 18 A/mm^2 and the variation in TF_{avg} is observed. The TF_{avg} grows linearly when the current density increases up to 14 A/mm^2, as can be observed. The thrust force grows slowly when the current density surpasses 14 A/mm^2 due to saturation of the iron components.

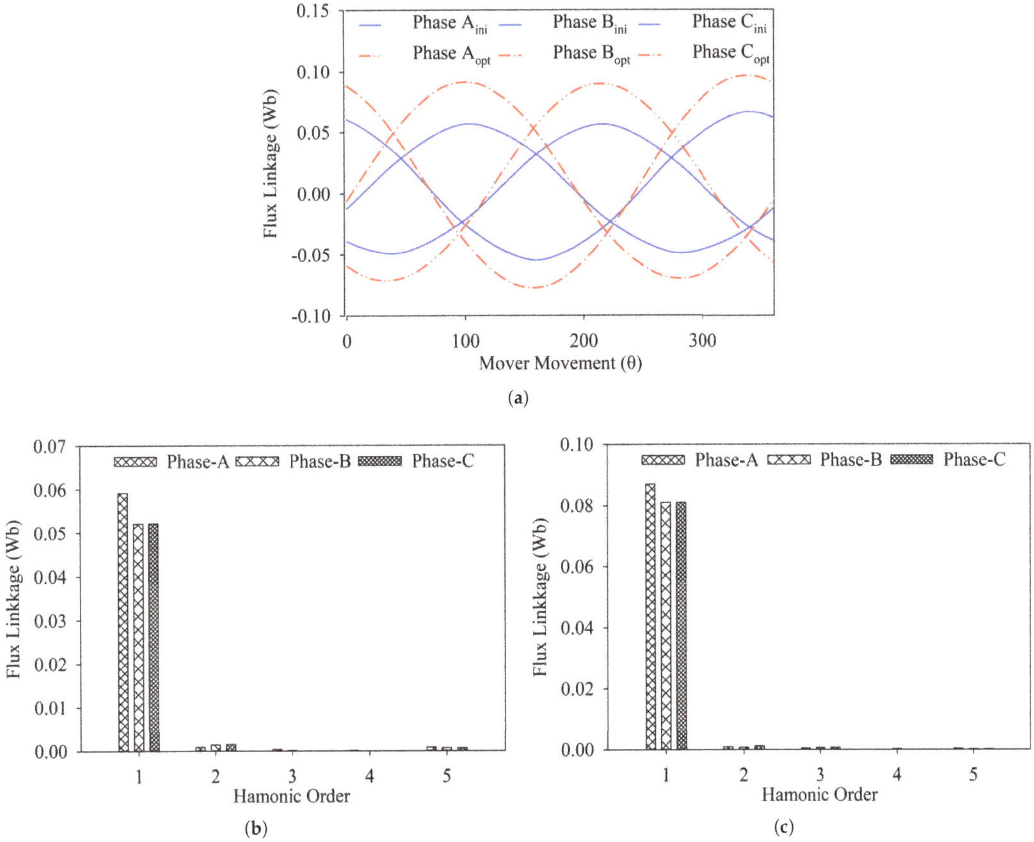

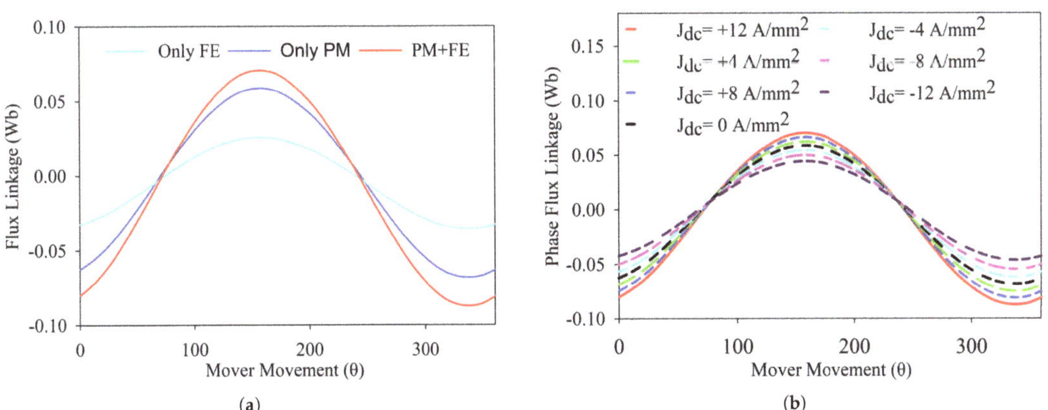

Figure 13. (**a**) Flux linkage of the proposed single-sided VFPMLM (**b**) Harmonics in initial design's flux linkage (**c**) Harmonics in optimized design's flux linkage.

Figure 14. (**a**) Influence of FE on flux linkage (**b**) Flux regulation capability of the proposed single-sided VFPMLM with field current density.

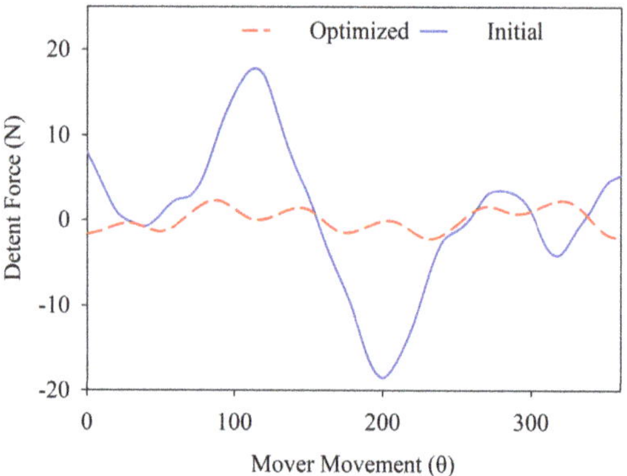

Figure 15. Detent force of the proposed single-sided VFPMLM.

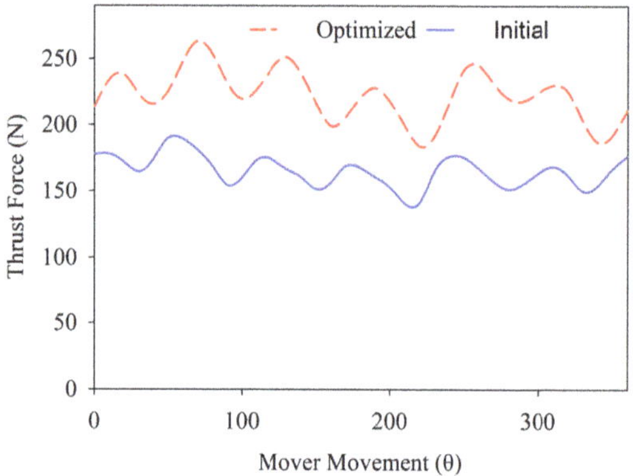

Figure 16. Thrust force of the proposed single-sided VFPMLM.

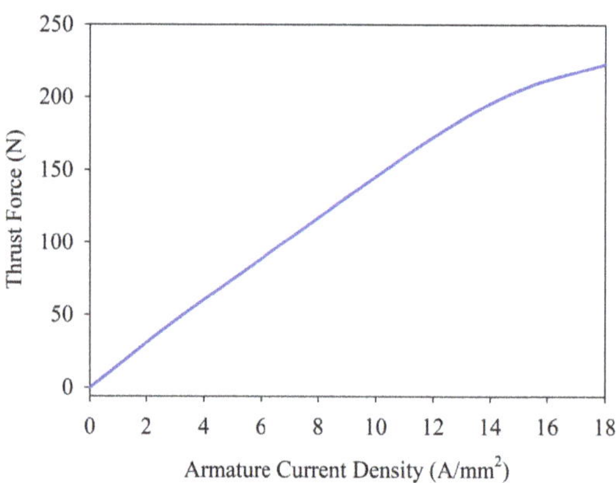

Figure 17. Variation of TF_{avg} with J_{AC}.

6.3. 3D Analysis

3D analysis of the proposed design is done to verify the results obtained by 2D analysis. Figure 18 compares the flux linkage of 2D and 3D designs. The flux linkage of 3D design shows a small decrement as compared to 2D design due to the end effects. Figure 19 shows the detent force and thrust force is depicted in Figure 20. Due to the longitudinal end effect, the thrust force of 3D is less in magnitude and has more ripples as compared to 2D design.

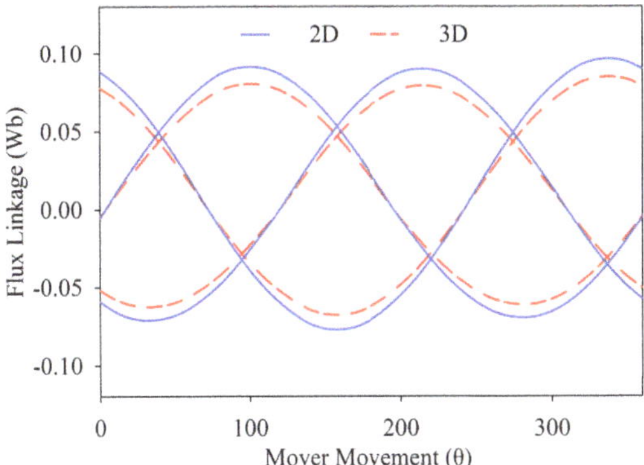

Figure 18. Flux linkage of 3D and 2D single-sided VFPMLM.

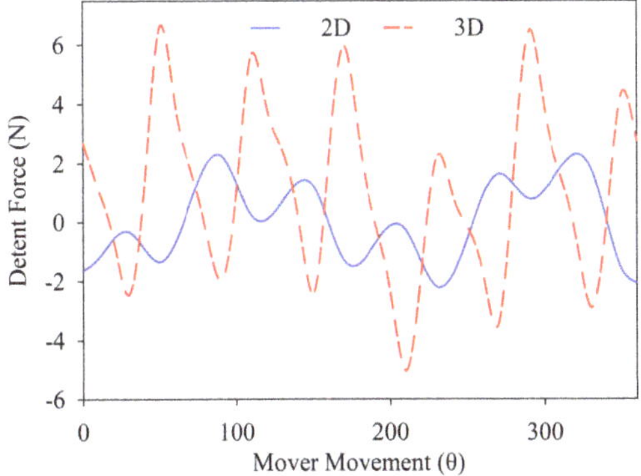

Figure 19. Detent force of 3D and 2D single-sided VFPMLM.

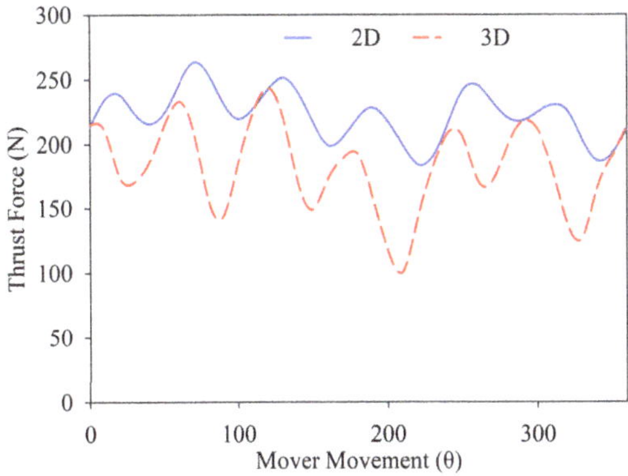

Figure 20. Thrust force of 3D and 2D single-sided VFPMLM.

6.4. Force-Velocity and Power-Velocity Characteristic Curve

Using the approach presented in [30,31], force–velocity and power–velocity curves are calculated for the proposed machine. The force–velocity curve is shown in Figures 21 and 22 shows the power velocity curve. The analysis reveals that the proposed single-sided VFPMLM achieved a maximum thrust force of 223.04 N at the velocity of 6.26 m/s while the power reaches 1393 W. Since force has an inverse relation with velocity, so at higher speeds, thrust force decreases maintaining a constant output power. It can be seen from Figure 22 that the proposed single-sided VFPMLM shows a better constant power operation capability.

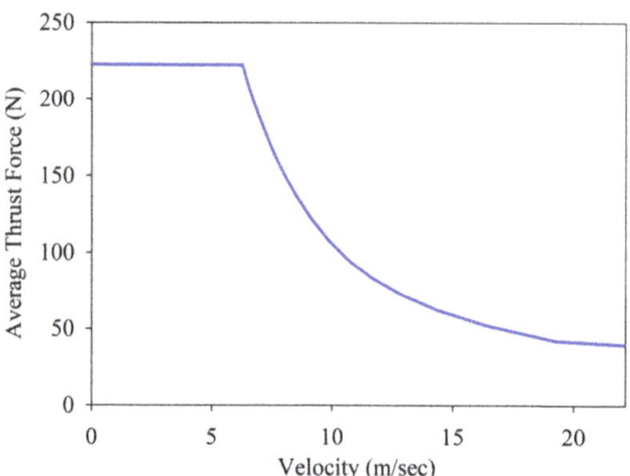

Figure 21. Average thrust force at different velocities.

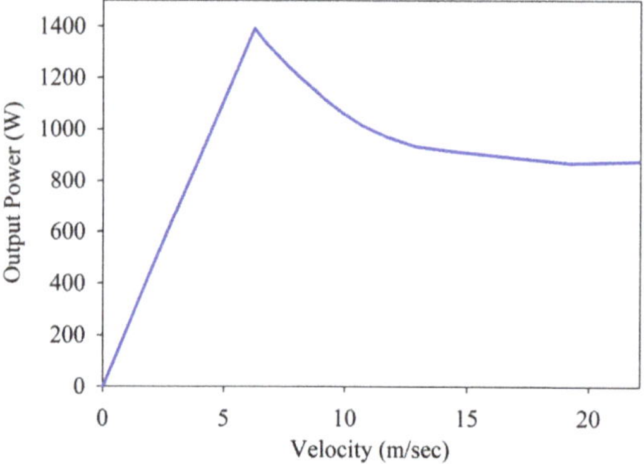

Figure 22. Output power at different velocities.

6.5. Efficiency Analysis

The output power in linear machines is the product of TF_{avg} and corresponding velocity while input power is calculated by the combination of output power and total losses (iron and copper losses). The losses are calculated at different points with the variable velocity and electric loading under the force–velocity graph. The points taken are comprised of current and current control angles shown in Figure 23. The copper losses can be calculated using Equation (20) while iron losses are calculated from 2D FEA using JMAG at all the selected points. For the efficiency calculation at each point, FE simulations are executed while iron and copper losses are considered. At point 1, the copper and iron losses of 150 W and 212 W are noted while at point 3, the iron losses were at a maximum (approximately 260 W), and hence efficiency is lower. The average efficiency of the proposed machine at different points is 72% as shown in Figure 24.

$$P_{cu} = P_{cu}(AE) + P_{cu}(FE) \tag{18}$$

Also from [32],

$$P_{cu} = I\rho JL(NQ)(1000) \tag{19}$$

(18) becomes

$$P_{cu} = 2I\rho JL(NQ)(1000) \tag{20}$$

where I, ρ, J, L, N, Q represents current (in rms if armature), resistivity $(\Omega - m)$, current density (A/mm^2), length of wire (mm), number of turns and number of slot pairs respectively.

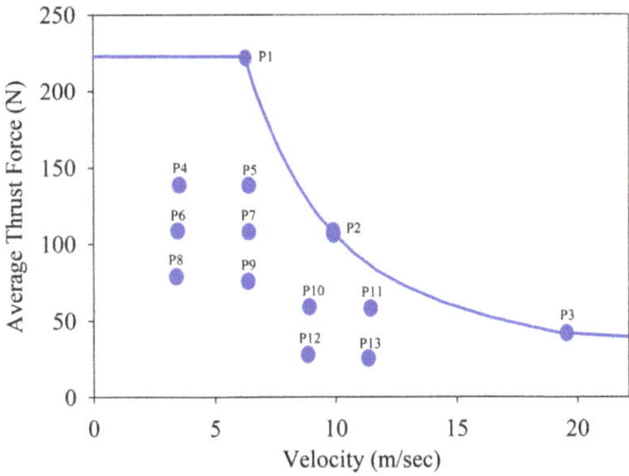

Figure 23. Points at which efficiency is calculated.

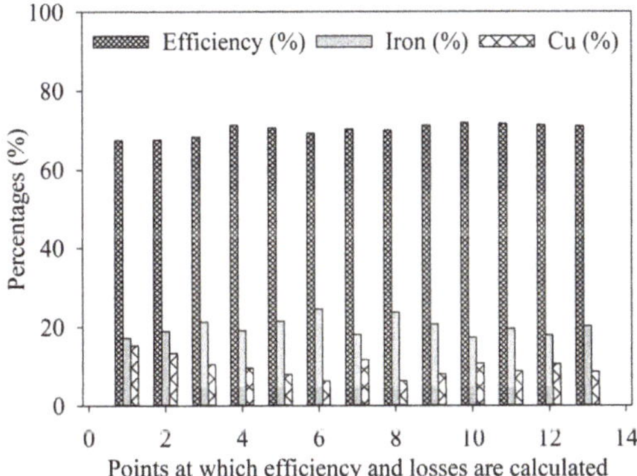

Figure 24. Efficiency at different points taken on Force-Velocity curve.

6.6. Thermal Analysis

In the electrical machine design process, thermal analysis plays an important role in setting operating limits and selecting insulating classes. The long-time operation of single-sided VFPMLM is limited by heat dissipation. If the temperature surpasses a particular allowable range, it will cause the electromagnetic performance to decrease and leads to inter turns short circuit fault [33]. The flow of current through the conductor causes heat dissipation and loss of power. This generation of heat in the winding will increase the heat of the entire machine. Apart from the copper losses associated with the windings, the machine also posses core losses. The core loss is the eddy current loss and hysteresis loss that occurs in the core of the machine. All these losses act as heat sources and cause a temperature rise. Therefore, the losses can be determined through the magnetic loss study, and thermal analysis can be performed to analyze the temperature distribution.

Firstly, losses are calculated in 3D FEA analysis of the proposed machine and then thermal analysis in 3D thermal studies is performed, as 3D analysis has more accuracy as compared to the 2D analysis. To acquire the temperature distribution of the complete machine, the 3D loss study is combined with the 3D heat study. The 3D thermal study reveals that because of the presence of all sources of excitation on the mover, the mover temperature rises very sharply, while the temperature of the secondary rises slightly, as the secondary of the proposed machine is made of iron only and is completely robust. The contour plot of the temperature distribution is shown in Figure 25. The temperature distribution in the mover attains a maximum value of 54.44 °C noticed in the windings. The maximum temperature at the secondary poles is 52.98 °C. Even at high temperatures, the proposed machine shows better thrust force. The temperature of the proposed machine can be lowered by using several cooling methods.

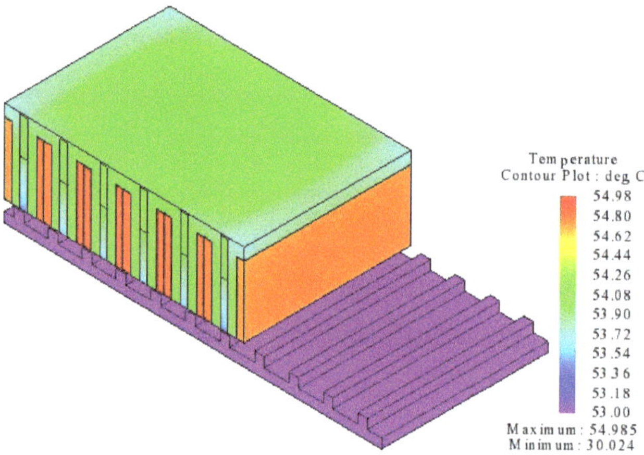

Figure 25. 3D thermal analysis of the proposed machine.

6.7. Comparison with Conventional Model

Finally, the single-sided VFPMLM presented is compared to the conventional design [18]. A detailed comparison is drawn in Table 5. The proposed design offers high average thrust force, high thrust force density, and high average thrust force per PM volume with the reduced number of armature excitation turns, field excitation turns while keeping the volume of PM the same as conventional design. Further, the proposed design offers constant power operation at high speeds.

Table 5. Detailed comparison of the proposed and conventional design.

Parameter	Proposed	Conventional
Mover length	131 mm	131 mm
Machine total height	54.3 mm	54.3 mm
Stack Length	90 mm	
Air gap	0.8 mm	
No. of armature slots	6	
No. of DC slots	3	7
Field Current	6 A	
Armature Turns	226	276
DC Turns	51	81
Thrust Force	223.04 N	153.45 N
TF_{avg} per PM Volume	4.302 N/cm^3	2.77 N/cm^3
TFD	434.76 kN/m^3	265.325 kN/m^3

7. Conclusions

This work proposed and investigated a single-sided VFPMLM which combines the advantages of permanent magnet machines (high thrust density) and wound field machines (flux adjustment capability). The flux bridge prevents the leakage flux from the mover and converts it into flux linkage, which greatly influences the performance of the machine. LPM is used to evaluate initial performance, which greatly reduces the computational time and drive storage. The proposed machine replaces the expensive rare earth PMs with ferrite magnets and improves flux controlling capability at various excitation currents. MGO is utilized to optimize the leading design parameters. The optimized design has improved flux linkage up to 44.11%, TF_{avg} up to 35%, TFD up to 35.02%, reduced thrust force ripples (TF_{rip}) by 23% and detent force (F_d) by 87.5%. The results obtained by 2D analysis are verified by 3D analysis. Finally, thermal analysis in 3D is done to set the operating limit and to select a proper insulating class for the proposed machine. Overall, the presented single-sided VFPMLM outperforms the conventional HEFSM that has been previously proposed in the literature.

Author Contributions: Conceptualization, B.U. and F.K.; methodology, B.U.; software, B.U., M.Q.; validation, B.U., B.K. and Z.D.; formal analysis, A.H.M., K.M.C.; investigation, A.H.M., Z.D.; resources, F.K., K.M.C.; data curation, Z.D.; original draft preparation, B.U.; review and editing, F.K.; supervision, F.K.; project administration, F.K., A.H. Milyani, K.M.C.; funding acquisition, A.H.M., K.M.C. All authors have read and agreed to the published version of the manuscript.

Funding: There is no external funding.

Acknowledgments: This work was supported by COMSATS University Islamabad, Abbottabad Campus and Higher Education Commission of Pakistan (No. TDF-03-067/R&D/HEC/2019).

Conflicts of Interest: The authors declare no conflict of interest.

References

1. Jin, M.J.; Wang, C.F.; Shen, J.X.; Xia, B. A modular permanent-magnet flux-switching linear machine with fault-tolerant capability. *IEEE Trans. Magn.* **2009**, *45*, 3179–3186. [CrossRef]
2. Huang, L.; Yu, H.; Hu, M.; Liu, H. Study on a long primary flux-switching permanent magnet linear motor for electromagnetic launch systems. *IEEE Trans. Plasma Sci.* **2013**, *41*, 1138–1144. [CrossRef]
3. Zhou, G.; Huang, X.; Jiang, H.; Tan, L.; Dong, J. Analysis method to a Halbach PM ironless linear motor with trapezoid windings. *IEEE Trans. Magn.* **2011**, *47*, 4167–4170. [CrossRef]
4. Mirzaei, M.; Abdollahi, S.E.; Lesani, H. A large linear interior permanent magnet motor for electromagnetic launcher. *IEEE Trans. Plasma Sci.* **2011**, *39*, 1566–1570. [CrossRef]
5. Zhu, Z.; Chen, J.; Pang, Y.; Howe, D.; Iwasaki, S.; Deodhar, R. Analysis of a novel multi-tooth flux-switching PM brushless AC machine for high torque direct-drive applications. *IEEE Trans. Magn.* **2008**, *44*, 4313–4316. [CrossRef]
6. Hellinger, R.; Mnich, P. Linear motor-powered transportation: History, present status, and future outlook. *Proc. IEEE* **2009**, *97*, 1892–1900. [CrossRef]
7. Cao, R.; Lu, M.; Jiang, N.; Cheng, M. Comparison between linear induction motor and linear flux-switching permanent-magnet motor for railway transportation. *IEEE Trans. Ind. Electron.* **2019**, *66*, 9394–9405. [CrossRef]
8. Azer, P.; Bilgin, B.; Emadi, A. Mutually Coupled Switched Reluctance Motor: Fundamentals, Control, Modeling, State of the Art Review and Future Trends. *IEEE Access* **2019**, *7*, 100099–100112. [CrossRef]
9. Wang, H.; Li, J.; Qu, R.; Lai, J.; Huang, H.; Liu, H. Study on high efficiency permanent magnet linear synchronous motor for maglev. *IEEE Trans. Appl. Supercond.* **2018**, *28*, 1–5. [CrossRef]
10. Zhao, W.; Cheng, M.; Ji, J.; Cao, R.; Du, Y.; Li, F. Design and analysis of a new fault-tolerant linear permanent-magnet motor for maglev transportation applications. *IEEE Trans. Appl. Supercond.* **2012**, *22*, 5200204. [CrossRef]
11. Cao, R.; Cheng, M.; Mi, C.; Hua, W.; Wang, X.; Zhao, W. Modeling of a complementary and modular linear flux-switching permanent magnet motor for urban rail transit applications. *IEEE Trans. Energy Convers.* **2012**, *27*, 489–497. [CrossRef]
12. Cao, R.; Cheng, M.; Hua, W. Investigation and general design principle of a new series of complementary and modular linear FSPM motors. *IEEE Trans. Ind. Electron.* **2012**, *60*, 5436–5446. [CrossRef]
13. Zhou, Y.; Zhu, Z. Comparison of wound-field switched-flux machines. *IEEE Trans. Ind. Appl.* **2014**, *50*, 3314–3324. [CrossRef]
14. Abdollahi, S.E.; Vaez-Zadeh, S. Back EMF analysis of a novel linear flux switching motor with segmented secondary. *IEEE Trans. Magn.* **2013**, *50*, 1–9. [CrossRef]

15. Ullah, N.; Basit, A.; Khan, F.; Shah, Y.A.; Khan, A.; Waheed, O.; Usman, A. Design and Optimization of Complementary Field Excited Linear Flux Switching Machine With Unequal Primary Tooth Width and Segmented Secondary. *IEEE Access* **2019**, *7*, 106359–106371. [CrossRef]
16. Liu, C.T.; Hwang, C.C.; Li, P.L.; Hung, S.S.; Wendling, P. Design optimization of a double-sided hybrid excited linear flux switching PM motor with low force ripple. *IEEE Trans. Magn.* **2014**, *50*. [CrossRef]
17. Gandhi, A.; Parsa, L. Hybrid flux-switching linear machine with fault-tolerant capability. In Proceedings of the 2015 IEEE International Electric Machines & Drives Conference (IEMDC), Coeur d'Alene, ID, USA, 10–13 May 2015; pp. 715–720.
18. Hwang, C.C.; Li, P.L.; Liu, C.T. Design and analysis of a novel hybrid excited linear flux switching permanent magnet motor. *IEEE Trans. Magn.* **2012**, *48*, 2969–2972. [CrossRef]
19. Xu, L.; Zhao, W.; Ji, J.; Liu, G.; Du, Y.; Fang, Z.; Mo, L. Design and analysis of a new linear hybrid excited flux reversal motor with inset permanent magnets. *IEEE Trans. Magn.* **2014**, *50*. [CrossRef]
20. Zeng, Z.; Lu, Q.; Ye, Y. Optimization design of E-core hybrid-excitation linear switched-flux permanent magnet machine. *Int. J. Appl. Electromagn. Mech.* **2017**, *54*, 351–366. [CrossRef]
21. Sulaiman, E.; Kosaka, T.; Matsui, N. Design and performance of 6-slot 5-pole PMFSM with hybrid excitation for hybrid electric vehicle applications. In Proceedings of the 2010 International Power Electronics Conference-ECCE ASIA, Sapporo, Japan, 21–24 June 2010; pp. 1962–1968.
22. Gaussens, B.; Hoang, E.; Lécrivain, M.; Manfe, P.; Gabsi, M. A hybrid-excited flux-switching machine for high-speed DC-alternator applications. *IEEE Trans. Ind. Electron.* **2013**, *61*, 2976–2989. [CrossRef]
23. Hua, W.; Zhang, G.; Cheng, M. Flux-regulation theories and principles of hybrid-excited flux-switching machines. *IEEE Trans. Ind. Electron.* **2015**, *62*, 5359–5369. [CrossRef]
24. Hua, H.; Zhu, Z. Novel partitioned stator hybrid excited switched flux machines. *IEEE Trans. Energy Convers.* **2017**, *32*, 495–504. [CrossRef]
25. Zhang, L.; Fan, Y.; Lorenz, R.D.; Cui, R.; Li, C.; Cheng, M. Design and analysis of a new five-phase brushless hybrid-excitation fault-tolerant motor for electric vehicles. *IEEE Trans. Ind. Appl.* **2017**, *53*, 3428–3437. [CrossRef]
26. Gao, Y.; Li, D.; Qu, R.; Fan, X.; Li, J.; Ding, H. A novel hybrid excitation flux reversal machine for electric vehicle propulsion. *IEEE Trans. Veh. Technol.* **2017**, *67*, 171–182. [CrossRef]
27. Zhu, Z.; Pang, Y.; Howe, D.; Iwasaki, S.; Deodhar, R.; Pride, A. Analysis of electromagnetic performance of flux-switching permanent-magnet machines by nonlinear adaptive lumped parameter magnetic circuit model. *IEEE Trans. Magn.* **2005**, *41*, 4277–4287. [CrossRef]
28. Ullah, W.; Khan, F.; Umair, M. Lumped parameter magnetic equivalent circuit model for design of segmented PM consequent pole flux switching machine. *Eng. Comput.* **2020**, *38*, 572–585. [CrossRef]
29. Chen, A.; Nilssen, R.; Nysveen, A. Analytical design of a high-torque flux-switching permanent magnet machine by a simplified lumped parameter magnetic circuit model. In Proceedings of the XIX International Conference on Electrical Machines-ICEM 2010, Rome, Italy, 6–8 September 2010; pp. 1–6.
30. Qi, G.; Chen, J.; Zhu, Z.; Howe, D.; Zhou, L.; Gu, C. Influence of skew and cross-coupling on flux-weakening performance of permanent-magnet brushless AC machines. *IEEE Trans. Magn.* **2009**, *45*, 2110–2117. [CrossRef]
31. Zhu, Z.; Shuraiji, A.L.; Lu, Q. Comparative study of tubular partitioned stator permanent magnet machines. In Proceedings of the 2015 Tenth International Conference on Ecological Vehicles and Renewable Energies (EVER), Monte Carlo, Monaco, 31 March–2 April 2015; pp. 1–7.
32. Hua, H.; Zhu, Z.; Zhan, H. Novel consequent-pole hybrid excited machine with separated excitation stator. *IEEE Trans. Ind. Electron.* **2016**, *63*, 4718–4728. [CrossRef]
33. Zhao, J.; Guan, X.; Li, C.; Mou, Q.; Chen, Z. Comprehensive evaluation of inter-turn short circuit faults in PMSM used for electric vehicles. *IEEE Trans. Intell. Transp. Syst.* **2020**, *22*, 611–621. [CrossRef]

Article

Analysis and Experimental Verification of the Demagnetization Vulnerability in Various PM Synchronous Machine Configurations for an EV Application

Gilsu Choi

Department of Electrical Engineering, Inha University, Incheon 22212, Korea; gchoi@inha.ac.kr

Abstract: Safety is a critical feature for all passenger vehicles, making fail–safe operation of the traction drive system highly important. Increasing demands for traction drives that can operate in challenging environments over wide constant power speed ranges expose permanent magnet (PM) machines to conditions that can cause irreversible demagnetization of rotor magnets. In this paper, a comprehensive analysis of the demagnetization vulnerability in PM machines for an electric vehicle (EV) application is presented. The first half of the paper presents rotor demagnetization characteristics of several different PM machines to investigate the impact of different design configurations on demagnetization and to identify promising machine geometries that have higher demagnetization resistance. Experimental verification results of rotor demagnetization in an interior PM (IPM) machine are presented in the latter half of the paper. The experimental tests were carried out on a specially designed locked-rotor test setup combined with closed-loop magnet temperature control. Experimental results confirm that both local and global demagnetization damage can be accurately predicted by time-stepped finite element (FE) analysis.

Keywords: permanent magnet machines; demagnetization; finite element analysis (FEA); interior PM synchronous motor; surface PM synchronous motor; traction applications

Citation: Choi, G. Analysis and Experimental Verification of the Demagnetization Vulnerability in Various PM Synchronous Machine Configurations for an EV Application. *Energies* **2021**, *14*, 5447. https://doi.org/10.3390/en14175447

Academic Editor: Ryszard Palka

Received: 10 August 2021
Accepted: 27 August 2021
Published: 1 September 2021

Publisher's Note: MDPI stays neutral with regard to jurisdictional claims in published maps and institutional affiliations.

1. Introduction

Due to their superior torque density and efficiency, permanent magnet synchronous machines (PMSMs) are frequently used in many high-performance applications. Despite recent price fluctuations, neodymium–iron–boron (NdFeB) rare-earth magnets have remained the popular choice for advanced applications due to their high coercive force and magnet remanence. However, most PM machines are naturally prone to irreversible demagnetization, which is typically caused by large demagnetizing fields from the stator windings and excessive temperature elevation. For high-performance PMSMs, reduction in magnet strength may result in significant performance degradation [1].

The properties of demagnetization characteristics of PMSMs have traditionally been examined based on the principles of Gauss's law and Ampere's law, commonly utilizing equivalent magnetic circuits as presented in many previous studies in the literature [2–4]. However, because of the secondary effects that are often omitted to simplify the study, the analytical models utilized in various previous research are insufficient to provide accurate answers for many real scenarios. Common examples include irreversible demagnetization, various leakage paths, and neglecting magnetic saturation, as well as an oversimplification of machine geometry.

The demagnetization characteristics of ferrite magnet material have been received more attention in recent years due to the fluctuations in the price of rare-earth magnet materials. Although the same fundamentals for demagnetization apply to both the magnet types, ferrite magnets have a significantly lower remanent flux density and coercivity. Therefore, machines using ferrite magnets must pay much more attention to demagnetization. In [5], the ratio of demagnetized magnet elements to total magnet materials for

131

a PM-assisted synchronous reluctance machine was calculated using finite element (FE) analysis. A more recent study uses a magnetic equivalent circuit model to investigate design requirements for ferrite-assisted synchronous reluctance machines [6]. Despite the use of an infinitely permeable core, the match between the analytical and FE results was excellent because of the lower level of core saturation induced by ferrite magnets.

Several methods for evaluating the demagnetization state of PMSMs have been proposed in the literature. Both hybrid [7] and FE-based methods [1,5–16] have been used to investigate the demagnetization characteristics of PM machines. Calculating the minimum flux density within the magnets under the effect of demagnetizing magnetomotive force (MMF) is one of the most extensively used ways for indicating the demagnetization state [8–12]. The reduction in back-emf voltage magnitude after exposure to a demagnetizing MMF has been employed by some authors to describe the demagnetization state [1,13]. This back-emf voltage-based method is useful for investigating experimental demagnetization because the quantities to be measured are directly available at the terminals of the machine. The reduction ratio of the magnetic flux linkage and the ratio of demagnetized magnet volume to the total magnet volume have also been employed [5,16]. In addition, a three-dimensional (3D) demagnetization analysis was performed in [16]. The process of localized demagnetization inside the magnets was studied using vector plots of the magnetic flux density [14,15]. Unfortunately, past experience has shown that there is no single representation that can include all of the detailed information to represent the demagnetization state of a PM machine.

More recent research trends focus on improving simulation accuracy by including the end effects into two-dimensional (2D) FE analysis [17], improved *B–H* curve modeling [18], and the complex spatial distribution of the demagnetizing MMF inside rotor magnets [19]. Novel rotor flux barrier designs that can mitigate demagnetization risks have been proposed in [20]. Design optimization considering demagnetization has been receiving increasing attention in recent studies [21,22].

Despite the fact that a typical demagnetization test requires a significant amount of time and expense, there have been some interesting findings published on the experimental verification of magnet demagnetization. The demagnetization characteristics of a dovetail PM machine was investigated under locked-rotor condition [13]. Reduction of the no-load back-emf voltage was used to evaluate the demagnetization damage. In another study, the demagnetization test results of a PM brushless dc machine under an inter-turn short-circuit fault condition were reported [23]. The demagnetization characteristics of permanent magnets outside the machine were investigated in [24] using a specially designed fixture. Another demagnetization test result presented in [25] was performed with the rotor magnets physically damaged. Unfortunately, most of the reported work to date may not be applicable to more generalized cases due to the specialized and constrained conditions.

A significant amount of demagnetization analysis results has been provided in [26] for several different PMSMs under the effect of demagnetizing MMF, highlighting the impact of the rotor geometry and winding type on demagnetization at an application level (i.e., EV). To date, the literature on this subject has received relatively little attention. As it is challenging to remagnetize the damaged magnets, identifying PM machine configurations with improved demagnetization resistance is critical. In most cases, the only way to recover the loss of magnet strength is to remove the rotor and magnets from the machine and remagnetize the rotor magnets with a huge external magnetizing MMF, which is a very time-consuming and expensive process. In this paper, the results presented in [26] are significantly expanded by investigating an interior PM (IPM) machine design with a new type of rotor configuration, which is called flux-intensifying (FI) IPM machine, and providing valuable experimental results of a fractional-power IPM machine as well as discussions on modeling accuracy issues to validate the FE results that are provided in the paper.

The rest of the paper is organized as follows: Section 2 describes the basics of magnet demagnetization and seven baseline PM machines to be studied. Section 2 also presents two different types of IPM machines—flux-weakening (FW) and flux-intensifying (FI) IPM machines—to highlight the differences in each of the machine configurations. Rotor demagnetization characteristics for eight different PM machine configurations under the influence of demagnetization MMF are investigated in Section 3. In Section 4, a collection of measured pre- and post-demagnetization test data is presented to verify the rotor demagnetization characteristics predicted by FE analysis. FE modeling accuracy issues are discussed in Section 5, and the conclusion is given in Section 6.

2. Demagnetization Principles and Baseline PM Machines

2.1. Demagnetization Principles

Irreversible magnet demagnetization is generally caused by the combined effect of demagnetizing field and temperature rise. In this paper, it was assumed that the irreversible demagnetization is caused by the application of an external demagnetizing field under constant magnet temperature. The variation of magnet operating point on the *B–H* curve was calculated based on predefined data arrays for each temperature level.

Figure 1a shows the magnet demagnetization characteristics on a *B–H* curve, taking into account the effect of demagnetizing MMF and temperature rise, respectively. With a large demagnetizing stator current exceeding a threshold value, the magnet operating point changes irreversibly beyond the knee point, from point (1) to (2) as in Figure 1a, and follows a recoil curve with reduced magnetization capability after removing the demagnetization current. Figure 1a also shows that the loss in magnet remanence does not recover once the magnet operating point changes from (2) to (3) due to temperature rise.

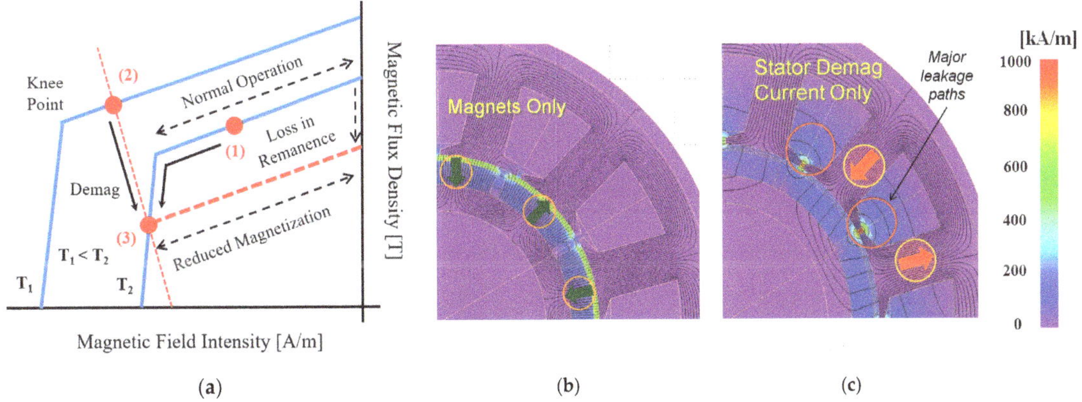

Figure 1. Magnet *B–H* curve characteristics and magnetic field strength H contours and flux lines for an SPM machine: (**a**) variation of magnet operating point on a *B–H* curve; (**b**) magnet flux alone; (**c**) demagnetizing MMF alone.

Figure 1b,c shows contour plots of the magnetic field strength H throughout a fractional-slot concentrated-winding (FSCW) surface PM (SPM) machine for excitation by magnets alone, and for 1 per-unit (pu) of negative *d*-axis current alone, respectively. The contour plots were calculated using a frozen permeability technique [27] to separate the combined loaded field components generated by both the negative *d*-axis current and rotor magnets. Figure 1c shows that most leakage flux in the SPM machine moves from a stator tooth to a neighboring stator tooth through the stator teeth-tips. In contrast to the SPM machine, the air cavity areas in the IPM rotor also provide an additional leakage path for demagnetizing MMFs, indicating that its demagnetization risk is lower than that of the SPM machine.

2.2. Baseline PM Machines

The demagnetization characteristics of several design configurations were investigated using seven different baseline PM machines that were initially developed in [26] for an EV application. The baseline designs were optimized by using a commercial optimization tool in order to meet the key performance characteristics of an 80 kW (pk) traction machine. This machine is rated for a 145 Apk stator current, which is defined to be 1 pu, for continuous operation and 565 Apk at peak power. As stated in [26], a constant magnet thickness of 8 mm was imposed on the optimization process for all of the designs to ensure a fair comparison between the seven machine designs.

Cross-sectional views of the baseline PM machines are shown in Figure 2. In order to highlight the impact of different stator MMF harmonics, a comparison between distributed winding (DW) and FSCW topologies is provided. The effect of different rotor structures was investigated by choosing IPM and SPM machines that are equipped with fractional-slot, concentrated windings. The variation of magnet positions in the rotor was considered in the analysis to show its impact on demagnetization—spoke type, flat type (Flat 1), the other flat type that is located closer to the airgap (Flat 2), single-layer V shape, and two-layer VU shape. Key performance metrics and dimensions for the seven baseline machines are summarized in Table 1.

2.3. FW-IPM vs. FI-IPM

Recently, there is a growing interest in FI-IPM machines as an alternative to the conventional FW-IPM machines due to the improved self-sensing capability as well as their potential for variable-flux capability [28]. In order to highlight the differences in demagnetization characteristics between the conventional FW-IPM and FI-IPM machines, an FI-IPM machine was designed based on the same performance specifications applied to the seven baseline PM machines.

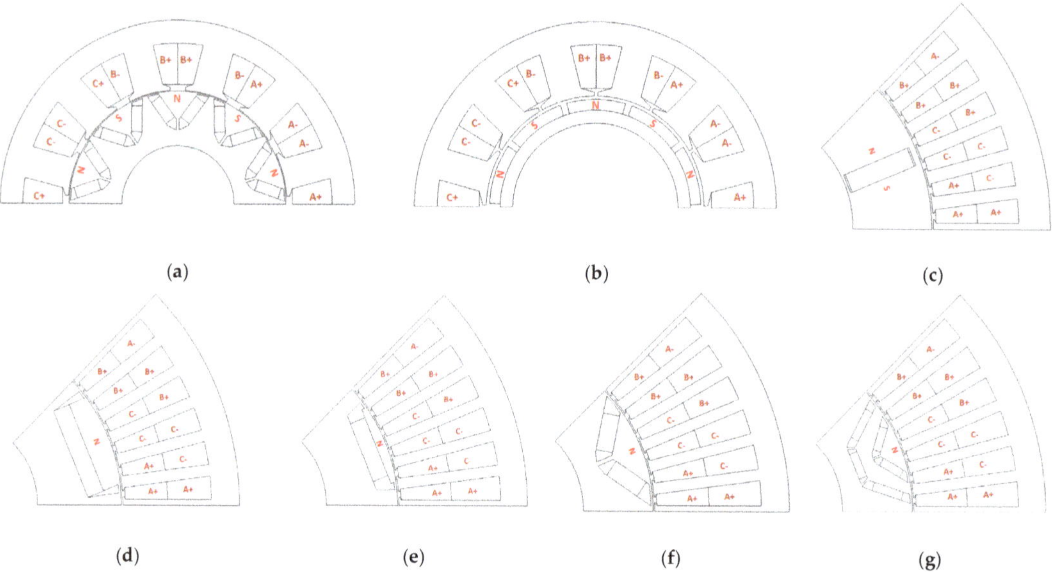

Figure 2. Cross-sectional views of seven baseline PM machines: (**a**) 12-slot/10-pole FSCW-IPM; (**b**) 12-slot/10-pole FSCW-SPM; (**c**) 48-slot/8-pole DW-IPM (spoke type); (**d**) 48-slot/8-pole DW-IPM (Flat 1); (**e**) 48-slot/8-pole DW-IPM (Flat 2); (**f**) 48-slot/8-pole DW-IPM (V shape); (**g**) 48-slot/8-pole DW-IPM (VU shape).

Table 1. Summary of key information of the seven baseline PM machines.

Dimensions/Metrics	Design 1	Design 2	Design 3	Design 4	Design 5	Design 6	Design 7
Winding type	FSCW	FSCW	ISDW	ISDW	ISDW	ISDW	ISDW
Rotor type	SPM	V shape	Flat 1	Flat 2	Spoke	V shape	VU shape
Slot/pole	12/10	12/10	48/8	48/8	48/8	48/8	48/8
Stator diameter [mm]	275	261	273	291	260	291	291
Rotor diameter [mm]	160	160	160	160	160	160	160
Stack length [mm]	96	98	113	88	148	88	85
Estimated volume [pu]	1.09	1	1.26	1.13	1.50	1.13	1.09
Magnet remanence [T] at 180 degC	1.01	1.01	1.01	1.01	1.01	1.01	1.01
Unsaturated saliency	≈1	1.51	2.18	2.21	1.58	2.39	2.29
Magnet mass [kg]	2.59	2.23	2.40	1.71	2.11	1.55	1.41

Contrary to the conventional FW-IPM machines, rotor magnets were embedded in a deeper position inside the rotor, while the flux barriers along the q-axis were placed next to the rotor surface. This helps to maximize the magnet torque as well as the difference between the dq inductances, L_d, and L_q, thereby achieving a saliency relationship of $L_d > L_q$ in FI-IPM machines [27], as shown in Figure 3a. In order to provide meaningful comparisons and highlight the differences in rotor design, both FI-IPM and FW-IPM machines had the same stator structure (i.e., same stator demagnetizing MMF), including the same number of turns that were designed based on the Flat 2 DW-IPM design shown in Figure 3b. However, the rotor structure of the FI-IPM machine is significantly different, as shown in Figure 3a, due to the machine characteristics explained above.

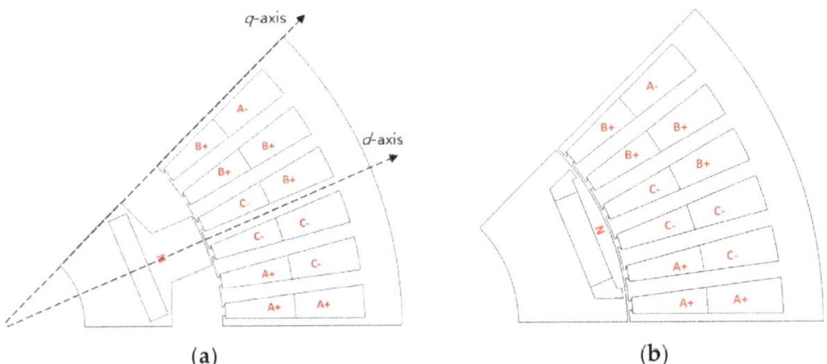

Figure 3. Cross-sectional views of the two IPM machines: (**a**) FI-IPM machine; (**b**) FW-IPM (Flat 2) machine.

A summary of the predicted performance metrics for the two PM machines is provided in Table 2. Although both FI- and FW-IPM machines were designed for identical design requirements with the same airgap length and rotor diameter constraints, the FI-IPM machine exhibits significant increases in volume and mass, compared to the FW-IPM machine.

Table 2 shows that the power density of the FI-IPM machine is degraded, compared to the conventional FW-IPM machine, primarily due to two reasons: (1) decreased magnet flux leakage due to the deeper embedded position of magnets and (2) degraded rotor permeance waveform due to the large flux barriers in the rotor q-axis. As a result, the peak power density of the FI-IPM machine is approx. 20% lower, compared to the FW-IPM machine. Both IPM machine designs exhibit relatively large torque ripple at rated operating conditions. On the other hand, the efficiency of the two IPM machines is as high as 97% at rated conditions.

Table 2. Key design parameters and performance metrics of the FI-IPM and FW-IPM machines.

Parameter	FI-IPM	FW-IPM
Slot/pole	48/8	48/8
Peak power/torque at base speed	80.5 kW/281 Nm	80.5 kW/281 Nm
Peak speed	10,000 r/min	10,000 r/min
Stator/rotor diameter	290.8 mm/160 mm	290.8 mm/160 mm
Airgap length	0.73 mm	0.73 mm
Magnet remanence	1.01 T at 180 degC	1.01 T at 180 degC
Active stack length	120 mm	88 mm
Magnet mass	2.10 kg	1.71 kg
Total active mass	47.02 kg	37.61 kg

3. Simulation Results

3.1. Baseline PM Machines

Since stator current aligned with the negative d-axis is considered as a major source of demagnetizing MMF, 5 pu negative d-axis stator current was selected as the worst-case scenario. Fixed-speed, time-stepping FE analysis was used to investigate the detailed demagnetization characteristics under the influence of demagnetizing MMFs. Each individual element inside the magnets was scheduled to follow a major B–H curve and a set of recoil curves based on the element's operating point to take into account irreversible demagnetization in the FE analysis. A commercial FE software was used for this study.

The number of demagnetized magnet elements to total magnet elements is defined as the demagnetization ratio. Figure 4 shows the demagnetization rate (in %) in terms of the applied demagnetizing MMF per rotor pole for the eight baseline PM machines including the FI-IPM machine. The stator MMF per rotor pole introduced in [2] is written as follows:

$$F_{ph} = \frac{4}{\pi} \frac{N_t I_m}{CP} \frac{N_{ph}}{2} \frac{k_{wh}}{h} \tag{1}$$

where I_m is the magnitude of stator current, N_t is series turn numbers, C is the number of parallel circuits, N_{ph} is the number of phases, P is the number of poles, and k_{wh} is the winding factor for the h-th order winding spatial harmonic component. The winding factor can be written as a product of winding distribution factor k_{dh}, slot opening factor k_{xh}, winding pitch factor k_{ph}, and skew factor k_{sh}, as shown in (2).

$$k_{wh} = k_{dh} \cdot k_{xh} \cdot k_{sh} \cdot k_{ph} \tag{2}$$

The skew factor and stator slot opening factors, k_{sh} and k_{xh}, are assumed to be equal to 1.

Simulation results in Figure 4 show that the magnets placed radially inner position (e.g., Flat 1 and V shape) are more resistant to demagnetization. However, when magnets are buried deeper, the weight and volume of the design generally become greater. The results in Figure 4 also show that the SPM machine is the most sensitive design to magnet demagnetization, followed by the spoke-type and FSCW-IPM machines. The relatively low resistance to demagnetization of both IPM designs can be explained by the greater d-axis inductance, which, in turn, reduces the number of series turns and lowers the fundamental MMF per rotor pole for a given current. The VU-shape machine, on the other hand, exhibits good demagnetization characteristics for the given demagnetizing MMF range since the lower d-axis inductance for this machine leads to higher fundamental MMF per pole for a given current. The V-shape, Flat 1, and FI-IPM designs have excellent demagnetization resistance. Notice that only fundamental demagnetizing MMF is considered in Figure 4 without including the impact of the spatial harmonics in FSCW-PM machines. If the harmonic MMFs were included, the overall demagnetizing MMF per pole for these machines would increase.

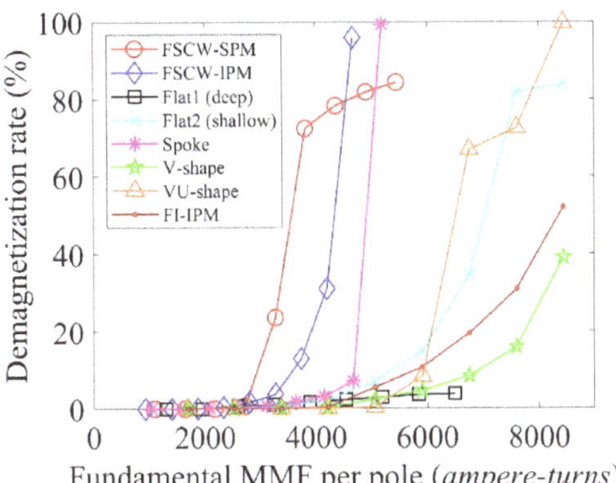

Figure 4. Demagnetization rate vs. fundamental MMF per pole for the eight baseline PM machines.

3.2. Remanence Ratio Contour Plots

The color contour plots of the remanence ratio within the entire rotor magnets give additional insight into the baseline machines' demagnetization characteristics. Figure 5 shows a rotor pole unit after applying the 5 pu negative d-axis current for one electrical cycle. The remanence ratio (in %) represents the reduction in magnet remanence and is expressed as follows:

$$\left(1 - \frac{B_r}{B_{ro}}\right) * 100(\%) \tag{3}$$

where B_{ro} is the initial magnet remanence and B_r is the post-demagnetization magnet remanence.

Consider the FSCW-SPM machine's results first. The results of the remanence ratio in Figure 5a show the extensive demagnetization in the rotor magnets, with most of the magnet material's remanence values decreasing by approx. 45% of its initial value. The SPM machine has the worst demagnetization among the eight baseline PM machines because its magnets are directly exposed in the air gap area without having additional leakage flux paths.

Figure 5b,d,g indicates that the demagnetization rates of the FSCW-IPM, Flat 2, and spoke machines are as high as the SPM machine values. Except for the areas near the leakage flux paths, the reduction in magnet strength for these three machines is rather modest, compared to the SPM machine, with remanence ratio values typically ranging from 5 to 25%. The rotor bridge and post provide magnetic shunts for the demagnetizing MMF, leading to the reduction of overall demagnetization risks. It should be noted that the flux passing through these bypasses demagnetizes the neighboring magnet material significantly.

Flat 1, V-shape, and FI-IPM machines are the three most favorable designs with regard to demagnetization resistance. Figure 5c,e,f shows very little decrease in flux density strength in most magnet materials for these three machines, except for significant losses near the magnet edge areas. Despite having a larger number of poles than the rest of the baseline machines, the two FSCW machines have generally higher demagnetization rates when compared with the DW-PM machines. This is primarily due to the harmonic demagnetizing MMFs that exist in these types of machines, as reported in [29].

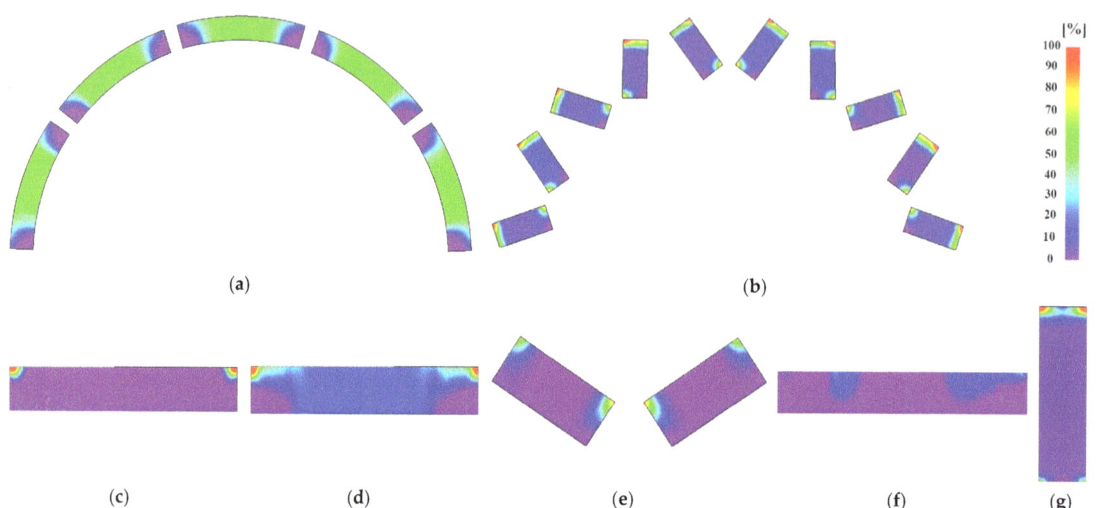

Figure 5. Contour plots of magnet remanence ratio values for the eight baseline PM machines ($i_d = -5$ p.u., $i_q = 0$ p.u.): (**a**) FSCW-SPM; (**b**) FSCW-IPM; (**c**) DW-IPM (Flat 1); (**d**) DW-IPM (Flat 2); (**e**) DW-IPM (V shape); (**f**) FI-IPM; (**g**) DW-IPM (spoke).

3.3. Demagnetization Index

Thus far, two different magnetization quantification methods are used to provide numerical and graphical representations of the demagnetization state for the baseline machines: demagnetization rate and remanence ratio. Unfortunately, these approaches do not take into account the impact of each machine's magnet mass on its demagnetization resistance. A new demagnetization index metric was introduced to reflect the impact of the total magnet mass on the demagnetization-withstand capability. The demagnetization index is expressed as follows:

$$\text{Demagnetization index} = \text{Normalized Demagnetization MMF/Magnet Mass [kg]} \quad (4)$$

The demagnetization MMF indicated in (4) is the stator MMF necessary to produce a 3% threshold demagnetization rate in each baseline machine employing negative d-axis current. This demagnetization MMF is then normalized by the demagnetizing MMF value necessary to produce the 3% demagnetization ratio value for the SPM machine (=1090 A-turns) and then divided by total magnet mass. The resultant demagnetization index value serves as a measure of the maximum demagnetizing MMF that a machine can withstand per unit PM mass.

The demagnetization index values for each of the eight baseline machines are shown in Figure 6. When the impact of magnet mass is considered, the Flat 1 machine and the FI-IPM machine, which scored high with regard to demagnetization resistance capability without considering the magnet mass, drop in ranks. Overall, the VU-shape machine, which is followed by the V-shape machine, has the greatest demagnetization index value.

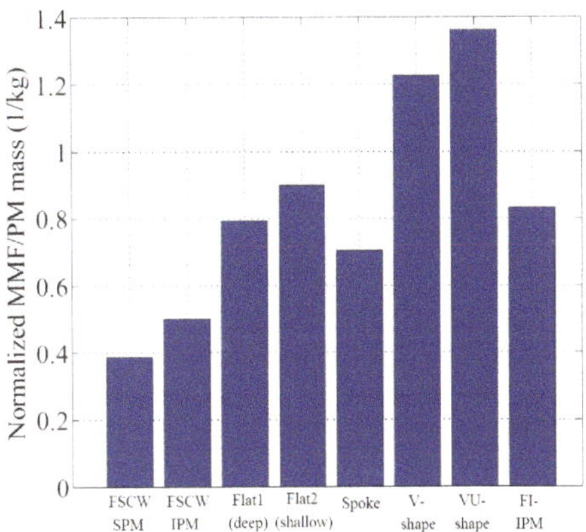

Figure 6. Demagnetization index representing normalized demagnetizing MMF per unit magnet mass.

3.4. Impact of q-Axis Current

It was shown in [26] that demagnetization can be mitigated by adding q-axis current. Although the impact of increasing q-axis current varies depending on the degree of demagnetization state and detailed rotor structure, the rate of demagnetization reduces as the magnitude of q-axis current rises. In order to provide a better explanation of why the presence of q-axis current tends to reduce demagnetization risks, Figure 7 shows vector plots of the magnetic flux density inside the VU-shape machine with two different values of q-axis current, $i_q = 0$ and 2 pu. It clearly shows that demagnetizing flux crosses the magnets along the d-axis and flows out the adjacent d-axis when there is negative d-axis current only. However, the addition of q-axis current shifts the direction of the demagnetizing flux towards the q-axis, parallel to magnet faces.

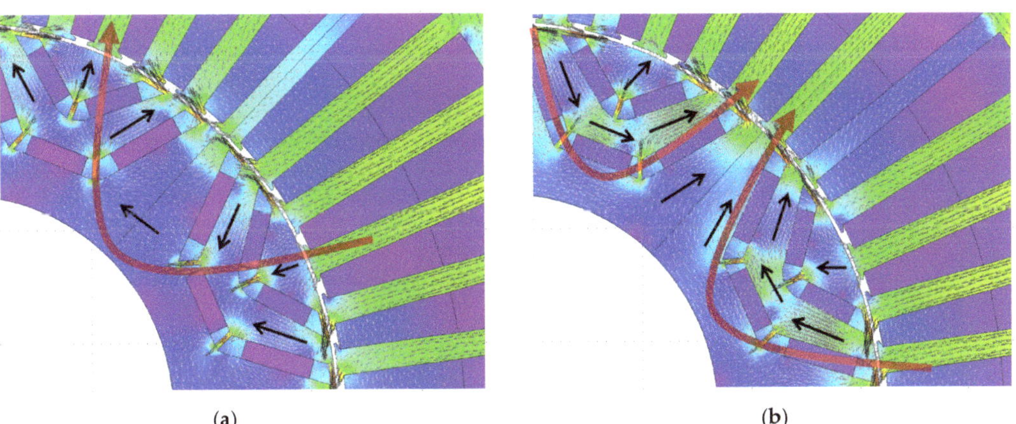

(a)   **(b)**

Figure 7. Magnetic flux density contour and vector plots for the VU-shape machine with 2 different q-axis current values in addition to -5 pu d-axis current: (**a**) $i_q = 0$ pu; (**b**) $i_q = 2$ pu.

4. Experimental Verification

Tested FSCW-IPM Machine

This section presents experimental results of a demagnetization test to collect measured data for comparison with FE analysis results. Due to the limited budget on this study, the PM machines shown in Figure 2 were not able to be manufactured for experimental verification. Instead, a commercial IPM machine shown in Figure 8a that was used in [30] was chosen for testing in this study. Table 3 summarizes the key parameters and dimensions of the tested machine. Custom-sized, low coercivity NdFeB magnets were used for testing to make the magnets more vulnerable to demagnetization during tests. Figure 8b shows measured *B–H* curve characteristics of the tested magnets at room temperature, indicating that the magnet properties are midway between the characteristics of N48 and N52 grades. Figure 8b also shows extrapolated *B–H* curves of the custom magnets at a higher temperature of 100 degC using an estimated thermal coefficient based on the measured properties at room temperature.

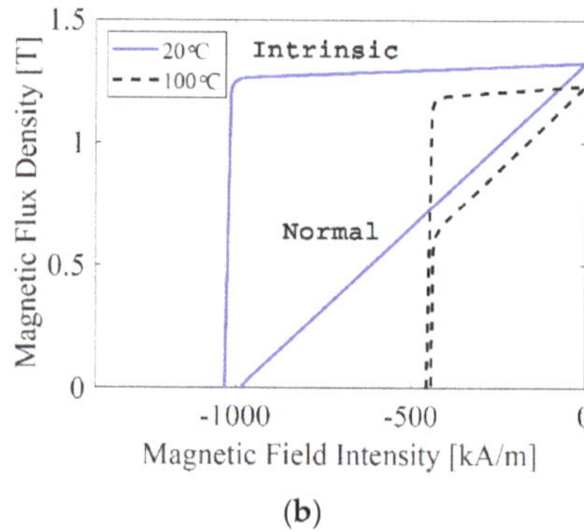

(a)

(b)

Figure 8. (**a**) Tested IPM machine; (**b**) *B–H* characteristics of the tested NdFeB magnets at 20 degC and 100 degC.

Table 3. Key parameters for the tested IPM machine.

Parameter	Value	Parameter	Value
Number of slots	9	Number of poles	6
Series turns	301.5	Number of coils/phase	3
Stator diameter	92.1 [mm]	Rotor diameter	45 [mm]
Active length	19.5 [mm]	Airgap length	1 [mm]

Figure 9a shows the test configuration that can perform demagnetization tests under a locked-rotor condition. Figure 10a,b shows that phase-*a* winding is excited by an externally applied 20A current pulse, generating a total of 2010 ampere-turns, at a rotor magnet temperature of 100 degC, while the terminals of the other two phases short-circuit each other. Figure 10c shows that the north poles are aligned with the Phase-*a* stator windings, while the south poles are located midway between the axes of phase *b* and *c* windings.

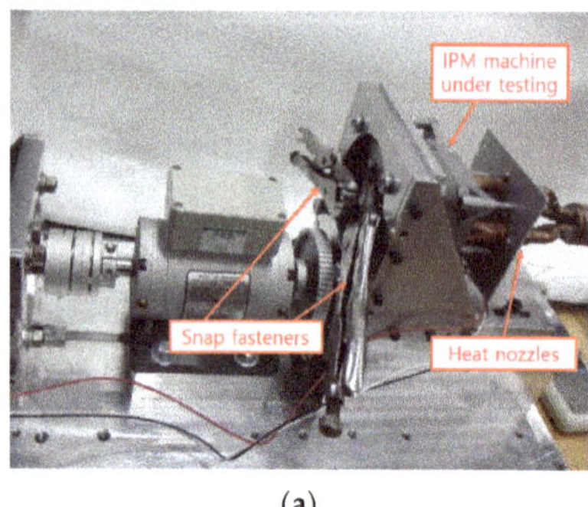

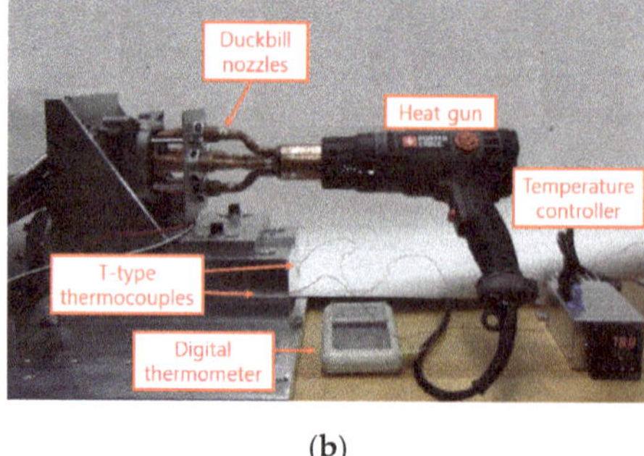

(a)

(b)

Figure 9. Dynamometer test configuration: (**a**) locked-rotor test setup; (**b**) heat gun with duckbill nozzles for temperature control.

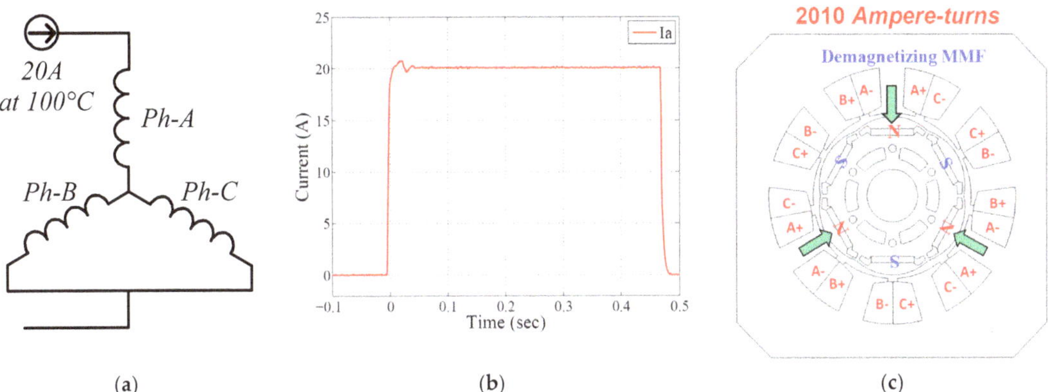

Figure 10. (**a**) Winding schematic; (**b**) current pulse waveform; (**c**) rotor position during locked-rotor test.

Time-stepped, 3D FE analysis was performed to simulate the rotor demagnetization at 100 degC under a locked-rotor condition. Figure 11 shows contour plots of the predicted remanence ratio throughout the magnets in two adjacent poles after applying the demagnetizing current pulse. The FE-calculated remanence ratio results shown in Figure 11 predict approx. 50% loss of remanence for most of the north pole, with losses reaching 100% in a very localized region at the north pole magnet corners. In contrast, relatively mild demagnetization is predicted in the south pole. More detailed studies revealed that the resulting demagnetization level in the magnets is very sensitive to the magnet properties used in FE analysis, indicating that neither of the thermal coefficients using N48- and N52-grade matches well with test results. As explained previously, magnet properties in Figure 8b that are midway between the characteristics of N48 and N52 grades were used for comparison between simulation and experimental results.

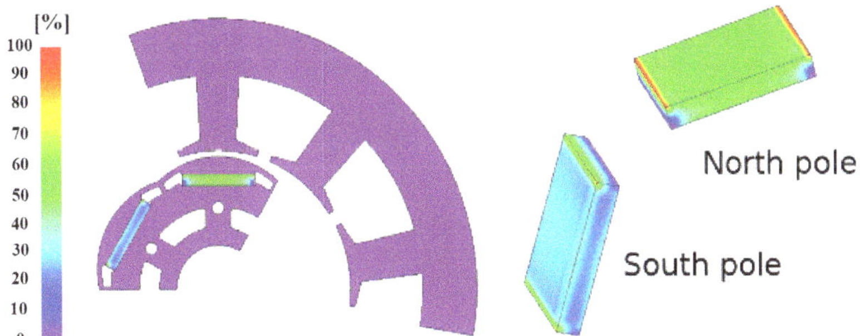

Figure 11. FE-predicted magnet remanence ratio in the experimental IPM after locked-rotor test using the estimated thermal coefficients (20 A at 100 degC).

Figure 12 shows a comparison between the measured surface flux density of the same North pole magnet in air measured before and after the locked rotor test with a 20A current pulse in phase-*a* at 100 degC, together with a % difference. For magnet flux density measurement, an AN_180KIT Hall probe with a very thin (<0.65 mm) sensor probe and a high flux density (±1 T) range was used. Key specifications and dimensions of this Hall probe are provided in Table 4. A commercial BW Bell Hall gaussmeter was used to calibrate the Hall probe. A commercial Zaber three-axis linear motion stage was used for precise measurement of the surface flux density of the individual magnets in the air.

Table 4. Key specifications for the AN_180KIT Hall probe.

Parameter	Value	Parameter	Value
Max. sensor tip thickness	0.65 mm	Linear magnetic field range	± 1 T
Frequency range	0 to 20 kHz	Operating temperature	0 to 50 °C
Sensitivity	1.069 mV/mT	Offset	2.54 V

The measured reduction of the magnet's surface flux density around the lower y-position area is approx. 44%, slightly lower than the FE-predicted reduction of 50% when using the B–H curve characteristics from Figure 8b. Figure 12 shows that the demagnetization damage is also more severe at the lower y-axis position, which is adjacent to the heat gun nozzles. This is due to the nonuniform temperature distribution of the magnets in the axial direction. Figure 12 shows that the measured demagnetization damage from the locked-rotor test in a magnet south pole is much less severe, as expected.

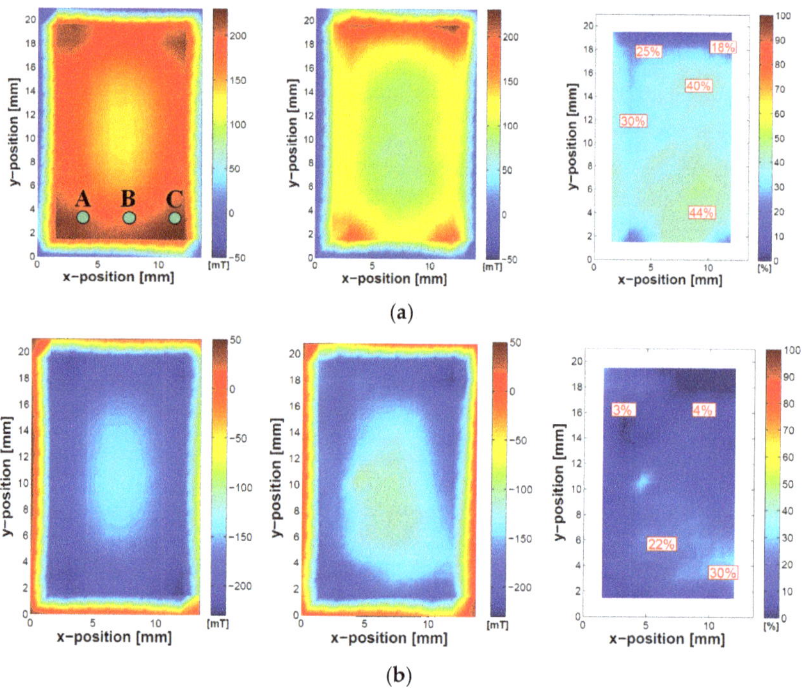

Figure 12. Comparison of measured surface flux density of the same magnets in air before (left) and after (middle) the locked-rotor test with 20 A current pulse at 100 degC and the reduction (in %) of surface flux density after the locked-rotor test (right): (**a**) north pole; (**b**) south pole. Three measurement points are marked in the north pole magnet (left).

In order to provide a more meaningful comparison between the FE predictions and measured results, the % difference of FE-predicted and measured demagnetization values at three different points on the north pole magnet after the locked-rotor test were compared; the results are shown in Table 5. The three measurement points marked on the north pole magnet are shown in Figure 12a. These particular measurement points are selected because the experimental magnet temperature from the three points is approx. 100 degC, which is assumed as magnet operating temperature in FE analysis. Results in Table 5 show that the FE results match well with the measured results with errors ranging from 2 % to 8 %.

The direct flux density measurement on the magnet surface, as shown in Figure 12, is a very efficient method to quantify any localized demagnetization in the magnets. However, it is very time consuming to measure the flux density directly from each of the tested magnets. It is also noted that the data obtained from this measurement cannot be represented as aggregated information of the test machine. This information can be measured through the air gap or stator windings when the magnets are mounted in the machine after the demagnetization.

Table 5. Differences between FE results and measured demagnetization values.

Measurement Point	FE Results	Measured
Point A	33.8%	32.3%
Point B	51.3%	43.3%
Point C	33.6%	35.7%

Figure 13 shows a comparison between the back-emf voltage waveforms with healthy and demagnetized magnets following the locked-rotor test using both FE-predicted and measured results. The terminal voltages were measured while the rotor was spinning at 3000 rpm under a no-load condition. Comparing the amplitude of the post-demagnetization back-emf voltages, the FE results underestimate by approx. 13%. Possible sources of discrepancy include the error in estimating the magnet thermal coefficients and the complex machine geometry. The waveshape is consistent with the previous results in Figure 12, showing different amplitudes for the positive and negative peak back-emf voltages due to the severe demagnetization concentrated on the north pole magnets. It is convenient to use the back-emf voltage measurement for evaluating the average reduction in the magnet flux linkage and thus the average torque production capability. However, the localized information available from the direct measurement is lost because the induced back-emf voltage is equal to the integration of the position-dependent flux density waveform over one electric cycle.

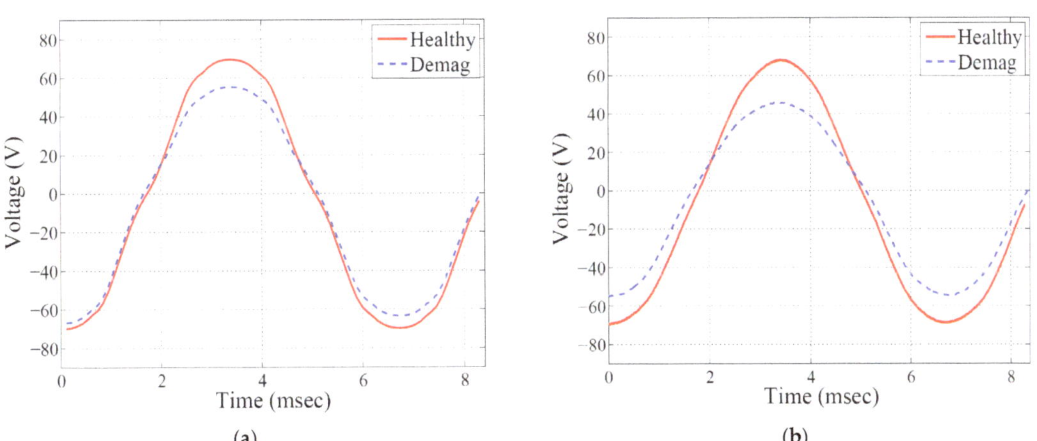

Figure 13. Comparison of the line-to-line back-emf waveforms for the IPM test machine. Both pre-test healthy magnets (solid lines) and demagnetized magnets (dashed line) after the locked-rotor test are shown: (**a**) FE-predicted; (**b**) measured.

5. Discussion

The accuracy of FE analysis can be compromised due to several different factors such as FE modeling accuracy in material property, machine geometry, and *B–H* curve characteristics, as well as manufacturing tolerances. Previous comparison results in Sections 3 and 4 show that the resulting demagnetization level in the magnets is very sensitive to the magnet thermal coefficients. Accurate thermal coefficient values can only be obtained from multiple experimental measurements over the wide temperature range in a temperature-controlled condition, which is often not possible in academic lab environments. In addition, any errors in modeling the *B–H* curve of rotor magnets can affect the accuracy of simulation results. Techniques based on piecewise-linear modeling, exponential modeling, and data-array-based methods are most popular among various modeling methods. It is very difficult to make machine parts as precise as initially designed due to the inevitable inaccuracy of the manufacturing process. For this reason, the impact of manufacturing tolerances such as eccentricity, errors in the properties of permanent magnets, and machine dimensions can also be a source of discrepancy. Any errors in estimating the saturation level of the machine lamination core (i.e., *B–H* curve data sheets provided by lamination manufacturers) can also be a source of discrepancy.

In order to improve the accuracy of FE predictions for comparison with experimental data, 3D FE analysis was carried out in Section 4. When compared with the results calculated by 2D analysis, 3D results show much better modeling accuracy because the end effects that are not included in the 2D analysis have a significant impact. Despite the complex nature of machine geometry and saturation effect, developing a hybrid method that combines 2D analysis and analytically calculated end leakages, such as in [17], has received continued attention. However, the exact calculation of the end leakages is still a challenging task.

As explained in Section 1, several demagnetization test results have been reported over the past decade. Unfortunately, the majority of past work was conducted in a special environment and thus failed at providing practical data. This paper presents valuable test results on a commercially constructed FSCW-IPM machine, including both local and global magnet demagnetization data for comparison with FE results.

6. Conclusions

This paper presents a comprehensive investigation of rotor demagnetization characteristics for several different PM machine configurations under the influence of demagnetization MMF by performing time-stepped FE simulations. New quantitative metrics that provide both aggregate and localized measurements of the demagnetization severity resulting from demagnetizing MMF were defined in order to evaluate the resulting magnet demagnetization in PM machines. The increased risks of demagnetization posed by moving magnets from deep inside the rotor poles to external mounting on the rotor core were clearly shown in the simulation results. The machines equipped with the distributed winding configurations show less vulnerability to demagnetization than the concentrated winding machines. In addition, the ability of the *q*-axis current to redirect rotor magnetic flux into more benign rotor paths that reduce the risk of demagnetization for IPM machines was graphically demonstrated.

A test configuration was built and utilized to experimentally verify the predicted rotor demagnetization characteristics for a fractional-power FSCW-IPM machine. Overall, the tests conducted using the FSCW-IPM test machine succeeded in providing a meaningful collection of measured pre- and post-demagnetization test data, which builds confidence in the FE predictions and key results developed in the paper regarding rotor demagnetization characteristics.

The work presented here can be used as design recommendations for machine designers to identify promising machine configurations when magnet demagnetization is of critical importance. Future work will be about identifying key machine parameters that are effective to mitigate the demagnetization risks and evaluating its engineering trade-offs.

Funding: This work was supported in part by Inha University Research Grant and by the BK21 Four Program funded by the Ministry of Education (MOE, Korea) and the National Research Foundation of Korea (NRF).

Institutional Review Board Statement: Not applicable.

Informed Consent Statement: Not applicable.

Acknowledgments: The author would like to thank the Wisconsin Electric Machines and Power Electronics Consortium (WEMPEC) for making its laboratory facilities available for the experimental work in this study.

Conflicts of Interest: The author declares no conflict of interest.

Abbreviations

2D	Two-dimensional
3D	Three-dimensional
DW	Distributed windings
EV	Electric vehicle
FE	Finite element
FI	Flux intensifying
FSCW	Fractional-slot concentrated winding
FW	Distributed winding
IPM	Interior permanent magnet
MMF	Magnetomotive force
NdFeB	Neodymium–iron–boron
PM	Permanent magnet
PMSM	Permanent magnet synchronous machine
pu	Per unit
SPM	Surface permanent magnet

References

1. Ruoho, R.; Kolehmainen, J.; Ikaheimo, J.; Arkkio, A. Interdependence of Demagnetization, Loading, and Temperature Rise in a PM Synchronous Motor. *IEEE Trans. Magn.* **2010**, *46*, 949–953. [CrossRef]
2. Lipo, T.A. *Introduction to AC Machine Design*; University of Wisconsin: Madison, WI, USA, 2007.
3. Hendershot, J.R.; Miller, T.J.E. *Design of Brushless Permanent Magnet Machines*; Clarendon: Oxford, UK, 2010.
4. Bianchi, N.; Jahns, T.M. (Eds.) Design, Analysis, and Control of Interior PM Synchronous Machines, Tutorial Course Notes. In Proceedings of the IEEE Industry Appl. Society Annual Meeting, Seattle, WA, USA, 3–7 October 2004.
5. Sanada, M.; Inoue, Y.; Morimoto, S. Rotor Structure for Reducing Demagnetization of Magnet in a PMASynRM with Ferrite Permanent Magnet and its Characteristics. In Proceedings of the IEEE Energy Conversion Congress and Exposition (ECCE), Phoenix, AZ, USA, 17–22 September 2011; pp. 4189–4194.
6. Vagati, A.; Boazzo, B.; Guglielmi, P.; Pellegrino, G. Design of Ferrite Assisted Synchronous Reluctance Machines Robust towards Demagne-tization. *IEEE Trans. Ind. Appl.* **2013**, *50*, 1768–1779. [CrossRef]
7. Farooq, J.; Srairi, S.; Djerdir, A.; Miraoui, A. Use of Permeance Network Method in the Demagnetization Phenomenon Modeling in a Permanent Magnet Motor. *IEEE Trans. Magn.* **2006**, *42*, 1295–1298. [CrossRef]
8. Kang, G.-H.; Hur, J.; Sung, H.-G.; Hong, J.-P. Optimal Design of Spoke Type BLDC Motor Considering Irreversible Demagnetiza-tion of Permanent Magnet. In Proceedings of the International Conference Electrical Machines and Systems (ICEMS), Beijing, China, 9–11 November 2003; Volume 1, pp. 234–237.
9. Bianchi, N.; Bolognani, S.; Luisem, F. Potentials and Limits of High-Speed PM Motors. *IEEE Trans. Ind. Appl.* **2004**, *40*, 1570–1578. [CrossRef]
10. Lee, B.-K.; Kang, G.-H.; Hur, J.; You, D.-W. Design of Spoke Type BLDC Motors with High Power Density for Traction Applications. In Proceedings of the IEEE Ind. Applicat. Conf. 39th IAS Annu. Meeting, Seattle, WA, USA, 3–7 October 2004; Volume 2, pp. 1068–1074.
11. Hosoi, T.; Watanabe, H.; Shima, K.; Fukami, T.; Hanaoka, R.; Takata, S. Demagnetization Analysis of Additional Permanent Magnets in Salient-Pole Synchronous Machines With Damper Bars Under Sudden Short Circuits. *IEEE Trans. Ind. Electron.* **2012**, *59*, 2448–2456. [CrossRef]
12. Kim, K.-C.; Kim, K.; Kim, H.J.; Lee, J. Demagnetization Analysis of Permanent Magnets According to Rotor Types of Interior Permanent Magnet Synchronous Motor. *IEEE Trans. Magn.* **2009**, *45*, 2799–2802.
13. Ruoho, S.; Kolehmainen, J.; Ikaheimo, J.; Arkkio, A. Demagnetization Testing for a Mixed-Grade Dovetail Permanent-Magnet Machine. *IEEE Trans. Magn.* **2009**, *45*, 3284–3289. [CrossRef]
14. McFarland, J.D.; Jahns, T.M. Investigation of the rotor demagnetization characteristics of interior PM synchronous machines during fault conditions. In Proceedings of the IEEE Energy Conversion Congress and Exposition (ECCE), Raleigh, NC, USA, 15–20 September 2012; pp. 4021–4028.
15. McFarland, J.D.; Jahns, T.M. Influence of d- and q-Axis Currents on Demagnetization in PM Synchronous Machines. In Proceedings of the Energy Conversion Congress & Exposition (ECCE), Denver, CO, USA, 15–19 September 2013; pp. 4380–4387.

16. Kakihara, W.; Takemoto, M.; Ogasawara, S. Rotor Structure in 50 kW Spoke-Type Interior Permanent Magnet Synchronous Motor with Ferrite Permanent Magnets for Automotive Applications. In Proceedings of the Energy Conversion Congress and Exposition (ECCE), Denver, CO, USA, 15–19 September 2013; pp. 606–613.
17. Lee, S.G.; Kim, K.-S.; Lee, J.; Kim, W.H. A Novel Methodology for the Demagnetization Analysis of Surface Permanent Magnet Synchronous Motors. *IEEE Trans. Magn.* **2016**, *52*, 1–4. [CrossRef]
18. Bekir, W.; Messal, O.; Benabou, A. Permanent Magnet Non-linear Demagnetization Model for FEM Simulation Environment. *IEEE Trans. Magn.* **2021**. [CrossRef]
19. Fukisaki, J.; Furuya, A.; Shitara, H.; Uehara, Y.; Kobayashi, K.; Hayashi, Y.; Ozaki, K. Demagnetization Correction Method by Using Inverse Analysis Considering Demagnetizing Field Distribution. *IEEE Trans. Magn.* **2020**, *56*, 1–4. [CrossRef]
20. Kong, Y.; Lin, M.; Yin, M.; Hao, L. Rotor Structure on Reducing Demagnetization of Magnet and Torque Ripple in a PMa-SynRM with Ferrite Permanent Magnet. *IEEE Trans. Magn.* **2018**, *54*, 1–5. [CrossRef]
21. Yoon, K.-Y.; Hwang, K.-Y. Optimal Design of Spoke-Type IPM Motor Allowing Irreversible Demagnetization to Minimize PM Weight. *IEEE Access* **2021**, *9*, 65721–65729. [CrossRef]
22. Wang, M.; Zhu, H.; Zhou, C.; Zheng, P.; Tong, C. Analysis and Optimization of a V-shape Combined Pole Interior Permanent-Magnet Synchronous Machine with Temperature Rise and Demagnetization Considered. *IEEE Access* **2021**, *9*, 64761–64775. [CrossRef]
23. Kim, K.; Lee, Y.; Hur, J.; Member, S. Transient Analysis of Irreversible Demagnetization of Permanent-Magnet Brushless DC Motor with Interturn Fault Under the Operating State. *IEEE Trans. Ind.* **2014**, *50*, 3357–3364. [CrossRef]
24. le Roux, W.; Harley, R.G.; Habetler, T.G. Detecting Rotor Faults in Low Power Permanent Magnet Synchronous Machines. *IEEE Trans. Power Electron.* **2007**, *22*, 322–328. [CrossRef]
25. Xiong, H.; Zhang, J.; Degner, M.W.; Rong, C.; Liang, F.; Li, W. Permanent Magnet Demagnetization Test Fixture Design and Validation. *IEEE Trans. Ind. Appl.* **2016**, *52*, 2961–2970. [CrossRef]
26. Choi, G.; Jahns, T.M. Demagnetization Characteristics of Permanent Magnet Synchronous Machines. In Proceedings of the 40th Annual Conference of the IEEE Industrial Electronics Society (IECON), Dallas, TX, USA, 29 October–1 November 2014; pp. 469–475.
27. Chu, W.Q.; Zhu, Z.Q. Average Torque Separation in Permanent Magnet Synchronous Machines using Frozen Permeability. *IEEE Trans. Magn.* **2013**, *49*, 1202–1210. [CrossRef]
28. Limsuwan, N.; Kato, T.; Akatsu, K.; Lorenz, R.D. Design and Evaluation of a Variable-Flux Flux-Intensifying Interior Permanent-Magnet Machine. *IEEE Trans. Ind. Appl.* **2014**, *50*, 1015–1024. [CrossRef]
29. Choi, G.; Jahns, T.M. Analysis and Design Recommendations to Mitigate Demagnetization Vulnerability in Surface PM Synchronous Machines. *IEEE Trans. Ind. Appl.* **2018**, *54*, 1292–1301. [CrossRef]
30. Choi, G.; Zhang, Y.; Jahns, T.M. Experimental Verification of Rotor Demagnetization in a Fractional-Slot Concentrated-Winding PM Synchronous Machine Under Drive Fault Conditions. *IEEE Trans. Ind. Appl.* **2017**, *53*, 3467–3475. [CrossRef]

Article

Research on Electromagnetic Field, Eddy Current Loss and Heat Transfer in the End Region of Synchronous Condenser with Different End Structures and Material Properties

Xiaoshuai Bi [1], Likun Wang [1,*], Fabrizio Marignetti [2] and Minghao Zhou [1]

1 National Engineering Research Center of Large Electric Machines and Heat Transfer Technology, Harbin University of Science and Technology, Harbin 150080, China; 1920300024@stu.hrbust.edu.cn (X.B.); zhouminghao@aliyun.com (M.Z.)
2 DIEI, Department of Electrical and Information Engineering, The University of Cassino and South Lazio, 03043 Cassino, Italy; marignetti@unicas.it
* Correspondence: wlkhello@hrbust.edu.cn

Abstract: Aiming at the problem of end structure heating caused by the excessive eddy current loss of large synchronous condensers used in ultra-high voltage (UHV) power transmission, combined with the actual operation characteristics of the synchronous condenser, a three-dimensional transient electromagnetic field physical model is established, and three schemes for adjusting the end structure of the condenser under rated condition are researched. The original structure has a copper shield and a steel clamping plate. Scheme 1 has no copper shield but has a steel clamping plate. Scheme 2 has no copper shield but has an aluminum clamping plate. By constructing a three-dimensional fluid–solid coupling heat transfer model in the end of the synchronous condenser, and giving the basic assumptions and boundary conditions, the eddy current loss of the structure calculated by the three schemes is applied to the end region of the synchronous condenser as the heat source, and the velocity distribution of the cooling medium and the temperature distribution of each structure under the three different schemes are obtained. In order to verify the rationality of the numerical analysis model and the effectiveness of the calculation method, the temperature of the inner edge of the copper shield in the end of the synchronous condenser is measured, and the temperature calculation results are consistent with the temperature measurement results, which provides a theoretical basis for the electromagnetic design, structural optimization, ventilation and cooling of the synchronous condenser.

Keywords: synchronous condenser; different materials; magnetic flux leakage in the end; eddy current loss; fluid–solid coupling

Citation: Bi, X.; Wang, L.; Marignetti, F.; Zhou, M. Research on Electromagnetic Field, Eddy Current Loss and Heat Transfer in the End Region of Synchronous Condenser with Different End Structures and Material Properties. *Energies* **2021**, *14*, 4636. https://doi.org/10.3390/en14154636

Academic Editors: Ryszard Palka and Marcin Wardach

Received: 12 June 2021
Accepted: 26 July 2021
Published: 30 July 2021

Publisher's Note: MDPI stays neutral with regard to jurisdictional claims in published maps and institutional affiliations.

1. Introduction

The 300 Mvar synchronous condenser can not only generate reactive power for transmission systems, improve the power factor of power grids and generate reactive power, but will also not decrease with the voltage drop of the power grid. This condenser is constructed with more attention paid to instantaneous performance, and is capable of fast voltage support, short-term over-current and over-voltage protection, which can provide more powerful guarantees of safe and stable operation of UHV power grids, Therefore, it has become an essential piece of equipment to improve the power quality and effective reactive power compensation of UHV transmission systems.

Due to the large capacity and large stator armature current of synchronous condensers, the stator current generates a strong magnetic leakage field in the end under rated operating conditions, which further induces eddy current loss in the copper shield, clamping plate and finger plate, causing local overheating of the end structure. When the end structure is heated significantly, it may cause abnormal shutdown of the condenser, or even threaten

147

the safety of the power system. Therefore, the reasonable design of the end structure of the condenser is very important in the improvement of the end structure and the reduction in its temperature.

In recent years, researchers have made many achievements in the research of the electromagnetic and heat transfer mechanisms of electromechanical energy conversion equipment [1–7]. In terms of the research methods of electromagnetic fields, a new idea for calculating the air gap magnetic circuit of a large turbo generator is proposed in [8]. Article [9] presents an analytical algorithm for calculating the leakage reactance of the strand slot in the transposition bar. In relation to the research methods of motor heat transfer performance, in [10], a lumped parameter network is proposed, which can better predict the thermal behavior of the machine. In [11], by employing the lumped-parameter model in the closed circuit of the flow network, the cooling medium flowrate and pressure drop in each element is calculated. In terms of the motor structure improvement and different material properties on the motor performance, a new ring structure is proposed, which can be better decrease the eddy current loss of the ring in [12]. Paper [13] described a new design scheme for a double three-phase asymmetric stator winding permanent magnet multiphase motor. Two kinds of permanent magnet synchronous motor, with the same capacity but different materials, were tested in [14], and the results show that compared with the traditional permanent magnet synchronous motor, the amorphous metal permanent magnet synchronous motor has very low no-load iron loss. In [15], the temperature distribution of the internal structural parts of a large generator with different shielding materials was researched. In paper [16], the traditional copper shield structure of a large turbo generator is transformed into a hollow structure, so that the temperature of the copper shield is reduced. Paper [17] analyzes the electromagnetic loss, stator temperature distribution and performance parameters of a high-speed induction motor with stator cores of different materials. Paper [17] analyzes the multi physical field distribution of a high-speed asynchronous motor with different stator core materials. However, compared with conventional large-scale motor, the end structure of the synchronous condenser for UHV transmission is complex, and its operation specification and assessment index requirements are more stringent than those of conventional motors, and thus, few researchers have researched the influence of the end structure change of the condenser and different material properties on the multiple internal physical fields of the motor.

In this paper, a 300 Mvar large synchronous condenser is taken as the research object, and three schemes of the end structure of the condenser are compared. By establishing a three-dimensional nonlinear transient electromagnetic field model in the end of the condenser, the distributions of magnetic flux leakage and eddy current loss of each structural member are researched. Based on the calculation results of electromagnetic eddy current loss in the end of the condenser, a three-dimensional fluid–solid coupling analysis model is established. The eddy current loss of each structural member in the end of the condenser calculated under the three schemes is applied to the end region of the condenser as a heat source. The velocity distribution of cooling medium and temperature distribution of each structure member under different schemes are obtained, which have significance as points of reference for improving the end structure of the condenser and reducing the temperature of the end structure of the condenser in the future.

2. Solution Domain Model of Electromagnetic Field in the End of Condenser

2.1. Physical Model of the End of Condenser

The 300 Mvar synchronous condenser researched in this paper is the first high-capacity condenser to be produced. According to the actual structural characteristics of the end of the condenser, it consists of a copper shield, clamping plate, far stator core side finger plate, magnetic shield, long finger plate, short finger plate, stator core, retaining ring and rotor. The specific structure of the model is shown in Figure 1. The rated parameters and some basic structural parameters of the synchronous condenser are given in Table 1.

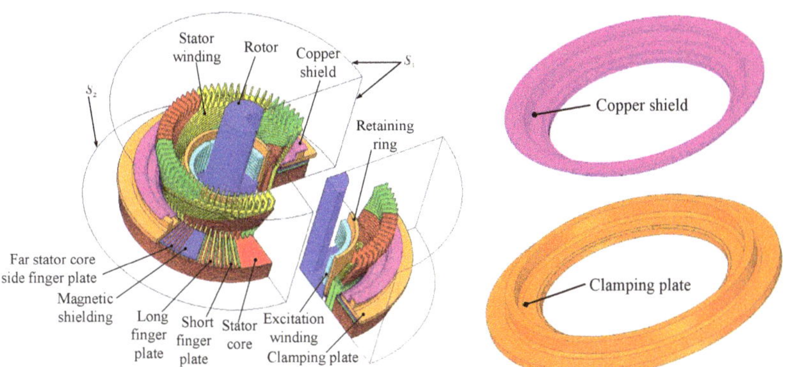

Figure 1. Physical model of the end region of the condenser.

Table 1. The rated parameters and some basic structural parameters of the condenser.

Parameters	Value
Rated Capacity/Mvar	300
Rated voltage/kV	20
Rated current of stator/A	8660
Rated rotate speed/r · min^{-1}	3000
Rated excitation current/A	2381
Stator external diameter/mm	2950
Number of stator slots	72
Number of poles	2

2.2. Three Dimensional Electromagnetic Field Mathematical Model of the End Region of the Condenser

The three-dimensional transient electromagnetic field solution domain of the end of the condenser is divided into the eddy current region and the non-eddy current region, in which the clamping finger, clamping plate and copper shield comprise the eddy current region, and the stator winding, rotor excitation winding and air region comprise the non-eddy current region. Taking current vector potential T and scalar magnetic potential ψ as unknown functions, the following mathematical model of the 3D transient electromagnetic field in the end of the large synchronous condenser is established [18–20]:

In the eddy current region:

$$\begin{cases} \nabla \times (\rho \nabla \times T) - \nabla(\rho \nabla \cdot T) + \frac{\partial \mu (T - \nabla \psi)}{\partial t} = \frac{\partial \mu H_s}{\partial t} \\ \nabla \cdot (\mu T - \mu \nabla \psi) = -\nabla \cdot (\mu H_s) \end{cases} \tag{1}$$

In the non-eddy region:

$$\nabla \cdot (\mu \nabla \psi) = \nabla \cdot (\mu H_s) \tag{2}$$

where H_s is the magnetic field intensity produced by the joint action of the armature current and the excitation current; ρ is resistivity; μ is permeability; t is time.

The boundary conditions of the three-dimensional transient electromagnetic field mathematical model are as follows:

$$\begin{cases} \frac{\partial \psi}{\partial n}|_{S_1} = 0 \\ \psi|_{S_2} = \psi_0 \end{cases} \tag{3}$$

where ψ_0 is the scalar magnetic potential at the initial time; n is the normal vector of the boundary.

The initial conditions are as follows:

$$\begin{cases} T|t = 0 = T_0(x, y, z) \\ \psi|t = 0 = \psi_0(x, y, z) \end{cases} \tag{4}$$

where T_0 is the current vector potential at the initial time.

According to the results of the transient electromagnetic field calculation, the instantaneous eddy current density $Je(t)$ can be obtained, and the instantaneous eddy current loss $P^{(e)}(t)$ of each structural part in the end of the condenser can be determined as:

$$P^{(e)}(t) = \int_{V_e} \frac{|Je(t)|^2}{\sigma} dV \tag{5}$$

where V_e is the volume of the structure and σ is the conductivity.

Let k be the number of finite element partitions in the calculation area, then according to Formula (5), the eddy current loss in a period T can be obtained as:

$$P_e = \sum_{e=1}^{k} \frac{1}{T} \int_0^T P^{(e)}(t) dt \tag{6}$$

3. Three-Dimensional Fluid–Solid Coupling Heat Transfer Model of the End Domain of the Condenser

The stator of the large synchronous condenser researched in this paper is cooled by air. An axial fan is used in the end of the condenser. Under the action of the fan, the cooling air enters the end domain of condenser from the fan, and a part of the cooling medium enters the stator side of the end to cool the structural parts in the end of the stator; a part of the cooling medium directly enters the rotor position in the end of the condenser, which directly cools the rotor. The cooling medium entering the rotor side in the end and the cooling medium entering the stator side in the end do not converge further, but flow into the straight section from the stator air chamber and the air outlet in the end of the rotor, respectively. Furthermore, the influence of rotor heating on the temperature increase in the end structure of the stator is ignored. The fluid–solid coupling heat transfer model is shown in Figure 2.

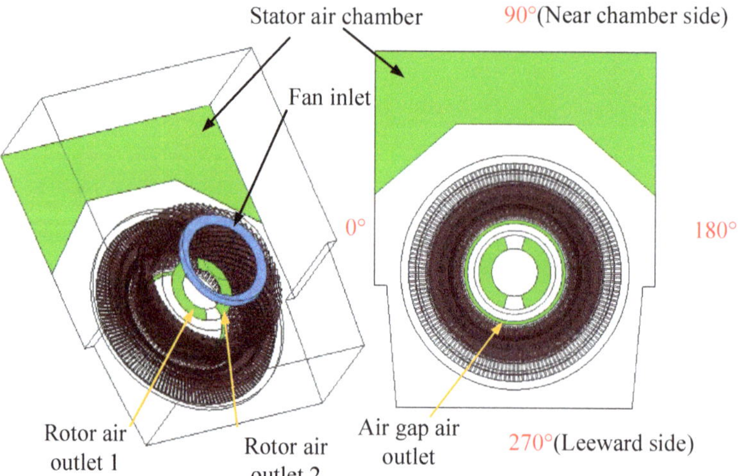

Figure 2. Physical model of the fluid–solid coupling field in the fluid domain.

Based on the tetrahedron and hexahedron hybrid subdivision, the end structures of the condenser are meshed. In the process of meshing, the stator winding, insulation layer and pressure finger are meshed by hexahedron, and the local size of these structures is controlled. The fluid domain is meshed by tetrahedron. The copper shield and clamping plate are meshed by tetrahedron and hexahedron.

Mathematical Model of Fluid–Solid Coupling Heat Transfer in the End of the Condenser

In view of the theory of fluid mechanics and heat transfer, the fluid flow and heat transfer in the condenser should meet the following three physical conservation laws. The specific conservation equations of mass, momentum and energy are shown below [21].

(1)   The conservation equation of mass is:

$$\frac{\partial \rho_0}{\partial t} + \mathrm{div}(\rho \mathbf{u}) = 0 \tag{7}$$

where ρ_0 is the fluid density (kg/m^3), t is time (s), $\mathbf{u}$ is the velocity vector (m/s).

(2)   The momentum conservation equation is:

$$\begin{cases} \frac{\partial(\rho_0 u)}{\partial t} + \mathrm{div}(\rho \mathbf{u} u) = \mathrm{div}(\eta \; grad \; u) - \frac{\partial P_0}{\partial x} + S_u \\ \frac{\partial(\rho_0 v)}{\partial t} + \mathrm{div}(\rho \mathbf{u} v) = \mathrm{div}(\eta \; grad \; v) - \frac{\partial P_0}{\partial y} + S_v \\ \frac{\partial(\rho_0 w)}{\partial t} + \mathrm{div}(\rho \mathbf{u} w) = \mathrm{div}(\eta \; grad \; w) - \frac{\partial P_0}{\partial z} + S_w \end{cases} \tag{8}$$

where u, v and w are, respectively, the components of $\mathbf{u}$ in the direction of x, y and z, m/s, η is the turbulent viscosity coefficient, (kg/(m·s)), P_0 is the fluid pressure, Pa, S_u, S_v and S_w are general source terms.

(3)   The energy conservation equation is:

$$\frac{\partial(\rho_0 T)}{\partial t} + \mathrm{div}(\rho \mathbf{u} T) = \mathrm{div}(\frac{\lambda}{c} \; grad \; T) + S_T \tag{9}$$

where T is the temperature, K; λ is the thermal conductivity, W/(m·K); c is the specific heat capacity, J/(kg·K); S_T is the power density calculated by FEA, W/m^3.

(4)   The k-epsilon equation is:

$$\begin{cases} \frac{\partial(\rho_0 k)}{\partial t} + \mathrm{div}(\rho_0 k u) = \mathrm{div}\left[\left(\mu + \frac{\mu}{\sigma_k}\right) grad \; k\right] + G_k - \rho_0 \varepsilon \\ \frac{\partial(\rho_0 \varepsilon)}{\partial t} + \mathrm{div}(\rho_0 \varepsilon u) = \mathrm{div}\left[\left(\mu + \frac{\mu}{\sigma_\varepsilon}\right) grad \; \varepsilon\right] + G_{1\varepsilon} \frac{\varepsilon}{k} G_k - G_{2\varepsilon} \rho_0 \frac{\varepsilon^2}{k} \end{cases} \tag{10}$$

where ε is the diffusion factor, (m^2/s^3), k is the kinetic energy of turbulence, (m^2/s^2); σ_k and σ_ε are the Prandtl numbers. $G_{1\varepsilon}$ and $G_{2\varepsilon}$ are the constants. G_k is the turbulence generation rate.

In order to simplify the solution process, the basic assumptions are as follows:

(1)   In this paper, the fluid flow state of the condenser researched is stable, and it is of a steady flow type.

(2)   The Reynolds number of the fluid in the condenser is much larger than 2300, and the flow is turbulent. The standard k-epsilon model is used to solve the fluid.

(3)   The velocity of fluid in the calculation domain of the condenser fluid field is far less than that of sound, so the compressibility of fluid is not considered.

Based on the above assumptions, all the equations are solved using the ANSYS Workbench finite element software. The standard k-ε model is used to deal with the turbulence. The solution parameters of the k-ε model are shown in Table 2.

Table 2. Parameters of the k-ε model.

Parameters	Value
σ_k	1.30
σ_ε	1.30
$G_{1\varepsilon}$	1.44
$G_{2\varepsilon}$	1.92
Energy Prandtl number	0.85
Wall Prandtl number	0.85

4. Numerical Analysis of Multiple Physical Fields in the End of the Condenser

4.1. Electromagnetic Field Analysis in the End of the Condenser

In order to better analyze the magnetic flux density distribution of the clamping plate, the circumference of the radius R1–R6 on the upper surface of the lower side of the clamping plate and the position of the inner circle and outer circle on the lower side of the clamping plate are shown in Figure 3. By solving the three-dimensional electromagnetic field in the end of condenser under original structure, the circumferential magnetic flux density distribution with a radius of R1–R6 on the upper surface of the lower side of the clamping plate under the original structure is obtained, as shown in Figure 4. The distributions of the magnetic flux density, radial magnetic flux density, tangential magnetic flux density and axial magnetic flux density of the circumference with radius of R1–R6 are, respectively, given in Figure 4a–d. It can be seen from Figure 4a that the magnetic flux density of the circumference of the clamping plate presents a sinusoidal regular change. The magnetic flux density decreases from R1 to R3 along the radial direction, and increases from R4 to R6 along the radial direction. At the circumference angle of 90° and 270°, the magnetic density reaches the peak value nearby. In Figure 4b, the radial magnetic flux density also presents a sinusoidal regular change. The radial magnetic flux density from R1 to R3 gradually decreases, and the radial magnetic flux density from R4 to R6 gradually increases. It can be seen from Figure 4c that the tangential magnetic flux density reaches a peak near 180° along the circumference. Figure 4d shows that the axial magnetic flux density of the circumference with radius R1–R5 shows a sinusoidal variation with smaller amplitude of magnetic flux density, while the axial magnetic flux density of the circumference with radius R6 shows a larger amplitude. Because there is copper shield above the circumference of radius R1–R5 and no copper shield above the circumference of radius R6, the axial magnetic flux density of the circumference of radius R6 is larger.

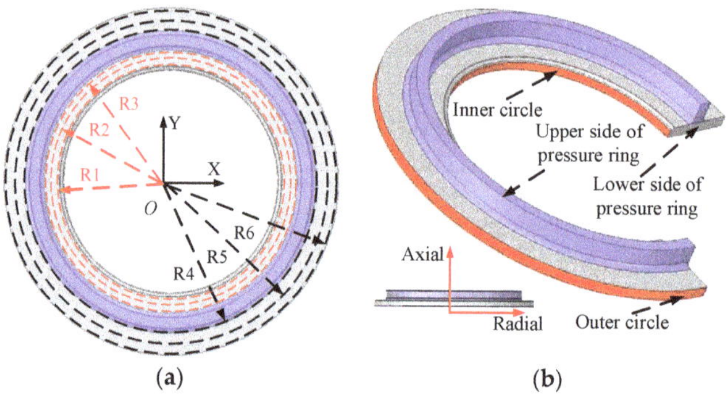

Figure 3. Clamping plate sampling position: (**a**) location of inner and outer surface of clamping plate; (**b**) inner circle and outer circle of clamping plate.

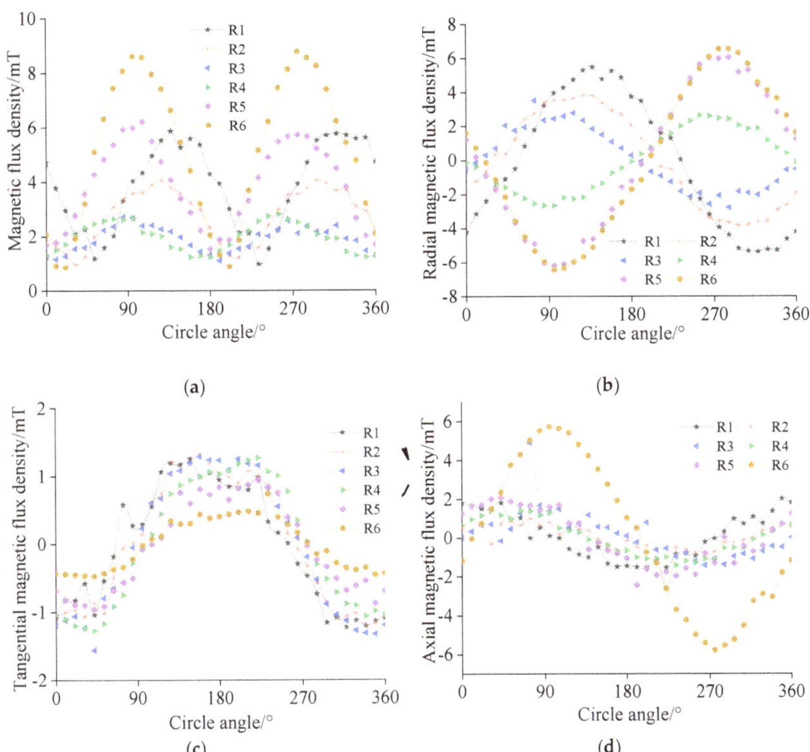

Figure 4. Magnetic flux density in different directions on the surface of the clamping plate of the original structure:
(**a**) magnetic flux density of the outer surface of the clamping plate; (**b**) radial magnetic flux density of the outer surface of
the clamping plate; (**c**) tangential magnetic flux density of the outer surface of the clamping plate; (**d**) axial magnetic flux
density of the outer surface of the clamping plate.

In order to further analyze the influence of the three different schemes on the magnetic
flux density distribution of the clamping plate, the magnetic flux density distribution of
the inner circle and the outer circle of the clamping plate under the three different schemes
are shown in Figure 5. It can be seen from Figure 5 that the magnetic flux density of the
inner circle and outer circle of clamping plate under the three schemes are sinusoidal
in the circumferential direction, and the magnetic flux density of the inner circle side is
higher than that of the outer circle side. In Figure 5a, the magnetic flux density of the inner
circle side of the original structure gradually decreases along the positive axial direction,
and the decreasing trend is larger. The maximum magnetic flux density of the inner
circle of the clamping plate is 30.75 mT, which is due to the copper shield in the original
structure that can effectively prevent the leakage flux from entering the clamping plate.
The maximum magnetic flux density of the outer circle is 7.25 mT. It can be seen from
Figure 5b,c that in Scheme 1 and Scheme 2, the maximum magnetic flux density of the
inner circle of the clamping plate is 83.04 mT and 112.93 mT, respectively, the maximum
magnetic flux density of the outer circle of the clamping plate is 9.73 mT and 7.11 mT,
respectively, and the magnetic flux density of the outer circle side gradually increases along
the positive axial direction. In conclusion, the change of the magnetic flux density of the
outer circle of the clamping plate along the axial direction is small under the three schemes.
However, under the three schemes, the magnetic density of the outer circle of the clamping
plate is the largest in Scheme 2, the second largest in Scheme 1, and the smallest in the
original structure.

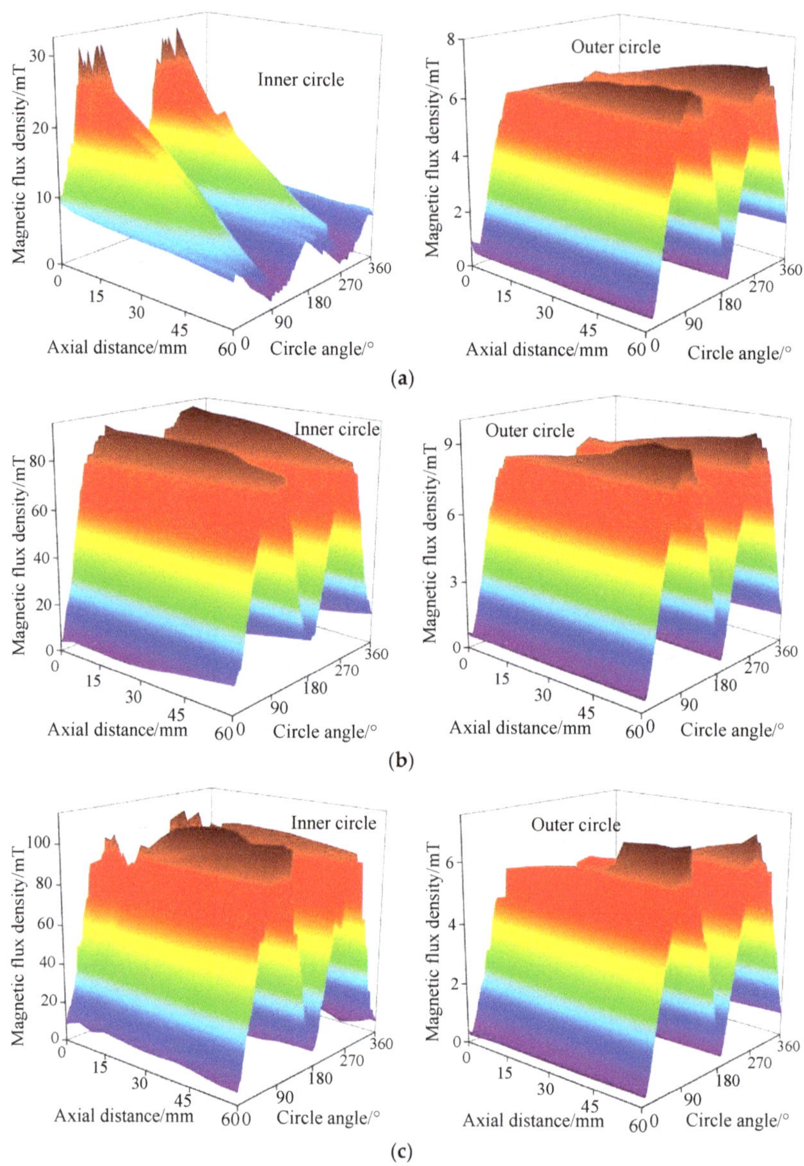

Figure 5. Magnetic density distribution in the inner and outer circle of the clamping plate: (**a**) original structure; (**b**) Scheme 1; (**c**) Scheme 2.

According to Formula (6), the eddy current loss of each structural part in the end of the condenser, under three different schemes, is shown in Figure 6. From the calculation results, compared with the original structure, in Scheme 1, the eddy current loss of the upper side of the clamping plate is increased by 5930 W, the eddy current loss of the lower side of the clamping plate is increased by 17,122 W, the eddy current loss of the retaining ring is reduced by 1.14%, and the eddy current loss of the finger plate is increased by 33.76%. Compared with the original structure, under Scheme 2, the eddy current loss increases by 665 W at the upper side of the clamping plate, 2368 W at the lower side of the clamping plate, 0.13% at the retaining ring and 18.37% at the finger plate.

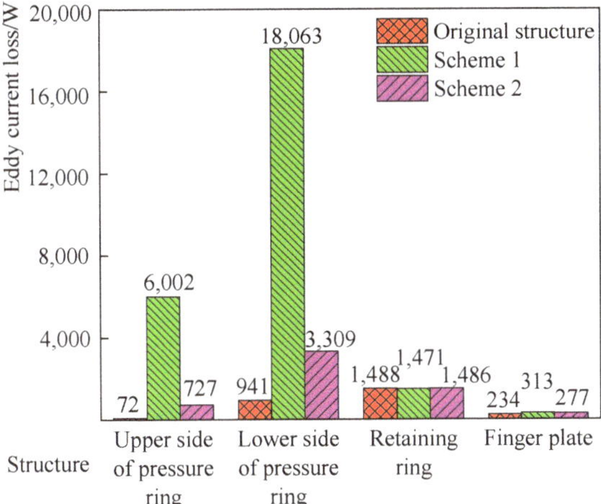

Figure 6. Distribution of eddy current loss in the end domain of the synchronous condenser.

4.2. Analysis of Calculation Results of Fluid–Solid Coupling Field in the End of the Synchronous Condenser

The eddy current loss of the structure is applied to the end domain as the heat source, and then the fluid–solid coupling field is solved. The velocity distribution of cooling medium and the temperature distribution of the structure near the end of synchronous condenser are obtained under three different schemes.

In order to describe the fluid distribution in the fluid domain, the calculated fluid trace distribution in the end region of the condenser is shown in Figure 7. According to the fluid trace distribution, most of the cooling medium flows through the fluid domain from the condenser fan inlet, and then flows out through the stator air cavity after cooling the end stator winding, copper shield, clamping plate and other components.

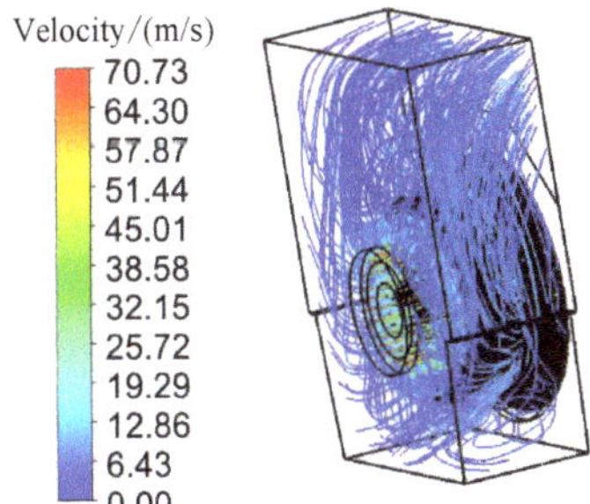

Figure 7. Trace distribution in the end of the synchronous condenser.

In order to research the fluid change law near the clamping plate under the three schemes, Figure 8a shows the velocity distribution and fluid temperature distribution of the clamping plate's leeward side (270°). Figure 8b shows the velocity distribution and fluid temperature distribution of the clamping plate near the chamber side (90°). As can be seen from Figure 8a, on the leeward side, under the original structure, the velocity in the inner circle of the clamping plate increases first and then decreases along the positive axial direction. Under Scheme 1 and Scheme 2, the velocity of cooling medium increases gradually along the positive direction of the inner circle axial direction, and they have the same flow velocity. The fluid temperature of the inner circle of the clamping plate under the original structure is higher than that of Scheme 1 and Scheme 2. This is because the gap between copper shield and clamping plate is small in the original structure, less cooling medium enters into the gap and the circulation is poor. Therefore, the flow rate of the cooling medium under the original structure is smaller and the fluid temperature is higher. Scheme 1 and Scheme 2 have no copper shield, so the cooling medium can flow to the inner circle of the clamping plate better and then cool the inner circle side of the clamping plate. Therefore, the fluid temperature in the inner circle of the clamping plate under the two schemes is lower. In the three schemes, the velocity of the outer circle of the clamping plate decreases slowly along the positive axial direction, but the temperature of the outer circle of the clamping plate in Scheme 1 is the highest, followed by Scheme 2, and that of the original structure is the lowest. This is because the velocity of the cooling medium on the outer circle side of the clamping plate is small, and it is not fully cooled to the outer circle side of the clamping plate. At this time, the fluid temperature on the outer circle side of the clamping plate has a great influence on the heat source of the clamping plate itself. In Figure 8b, on the inner circle side of the clamping plate, the fluid distribution on the near chamber side of the clamping plate is the same as that on the leeward side of the clamping plate. However, compared with the leeward side of the clamping plate, the fluid temperature on the near chamber side of the clamping plate lower than that of the leeward side.

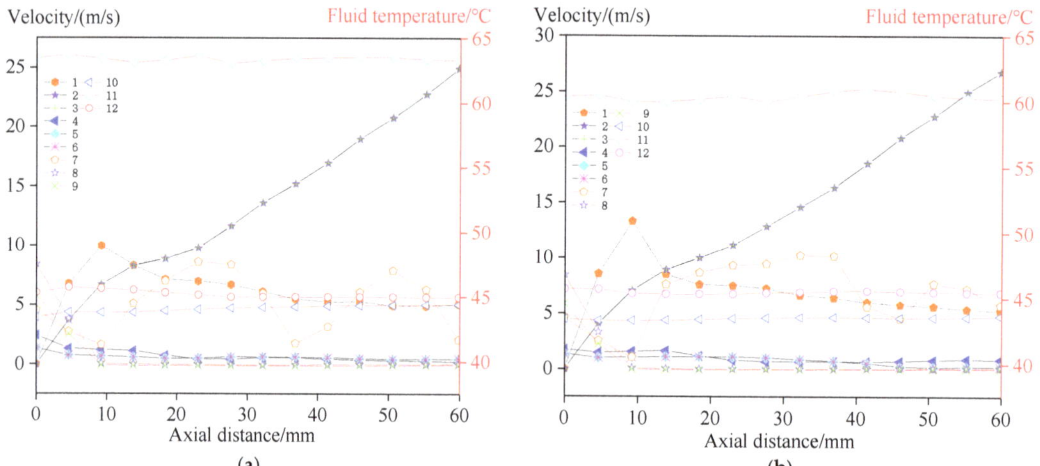

Figure 8. The fluid distribution of clamping plate on the leeward side and the near chamber side: (**a**) leeward side (270°); (**b**) near chamber side (90°).

1—Velocity of the inner circle side of the clamping plate of the original structure; 2—velocity of the inner circle side of the clamping plate of Scheme 1; 3—velocity of the inner circle side of the clamping plate of Scheme 2; 4—velocity of the outer circle side of the clamping plate of the original structure; 5—velocity of the outer circle side of the clamping plate of Scheme 1; 6—fluid temperature of the outer circle side of the clamping plate of Scheme 2; 7—fluid temperature of the inner circle side of the clamping plate of the original structure; 8—fluid temperature of the inner circle side of the clamping plate of Scheme 1; 9—fluid temperature of the inner circle side of the clamping plate of Scheme 2; 10—fluid temperature of the outer circle side of the clamping plate of the original structure; 11—fluid temperature of the outer circle side of the clamping plate of Scheme 1; 12—fluid temperature of the outer circle side of the clamping plate of Scheme 2;

In order to explore the influence of the three schemes on the temperature distribution of the clamping plate in detail, Figure 9 shows the temperature contour distribution of the clamping plate on the leeward side and the near chamber side under different schemes. It can be seen from Figure 9 that the temperature of the clamping plate on the leeward side is higher than that on the near chamber side. As shown in Figure 9a, the maximum temperature of the clamping plate is 47.95 °C under the original structure, and the temperature at A0 and D0 in the middle of the clamping plate is higher. From A0 along the radial direction to B0, A0 along the radial direction to C0, D0 along the radial direction to E0, and D0 along the radial direction to F0, the temperature of the clamping plate gradually decreases, and the temperature of the inner circle side of the clamping plate is less than that of the outer circle. It can be seen from Figure 9b that in Scheme 1, the temperature of the clamping plate decreases gradually from A1 to B1 and from D1 to C1. However, the temperature contour of the inner circle side of the clamping plate is dense, and the temperature contour of the outer circle side of the clamping plate is sparse. Therefore, the cooling effect of the inner circle side of the clamping plate is better. On the one hand, the cooling medium velocity is small on the outer circle side of clamping plate. Moreover, in Scheme 1, the heat source of the clamping plate is larger, which leads to the higher temperature outside the clamping plate and the same temperature in a large area. The maximum temperature of the clamping plate is 89.84 °C. According to Figure 9c, the maximum and minimum temperatures of the clamping plate in Scheme 2 are 49.40 °C and 46.80 °C, respectively, which are 1.45 °C and 4.6 °C higher than the maximum and minimum temperatures of the clamping plate in the original structure. The temperature of the clamping plate under Scheme 2 decreases gradually from A2 to B2.

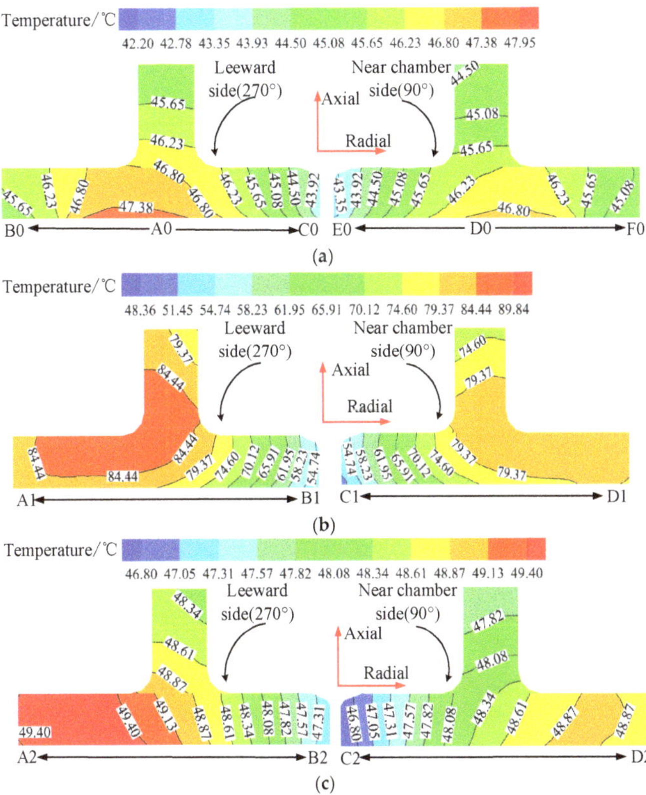

Figure 9. The temperature contour distribution of the clamping plate on the leeward side and the near chamber side: (**a**) original structure; (**b**) Scheme 1; (**c**) Scheme 2.

In order to research the influence of the original structure, Scheme 1 and Scheme 2 on the temperature distribution of the far stator core finger plate in the end of the condenser, the temperature distributions of the far stator core finger plate under three different schemes are given in Figure 10a–c. It can be seen from Figure 10 that the temperature of the finger plate near the chamber is lower than that of the finger plate near the leeward chamber. According to the eddy current loss value of finger plate given in Figure 6, the eddy current loss of the finger plate under the three schemes is similar, but the maximum temperatures of the finger plate under the original structure, Scheme 1 and Scheme 2 are 49.91 °C, 80.11 °C and 51.57 °C, respectively. In Scheme 1, the maximum temperature of the finger plate is 30.2 °C higher than that of the original structure. In Scheme 2, the maximum temperature of the finger plate is 1.66 °C higher than that of the original structure. Through comparative analysis, it can be concluded that the temperature of the finger plate is mainly affected by the heat transfer of other structural parts as well as its own heat source.

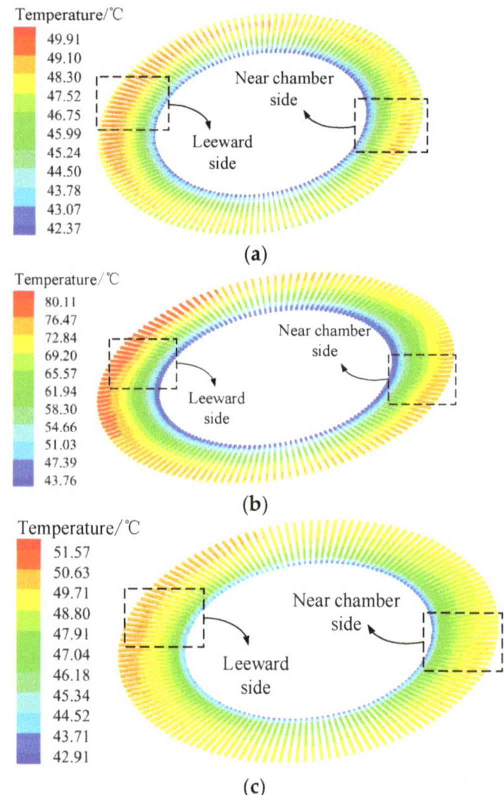

Figure 10. Temperature of far core finger plate under different schemes: (**a**) original structure; (**b**) Scheme 1; (**c**) Scheme 2.

5. Temperature Test and Data Comparison

In order to further verify the correctness of the fluid–solid coupling model of the end of the condenser and the accuracy of the calculation results, the temperature test of the 300 Mvar air-cooled synchronous condenser under rated conditions is carried out. The temperature increase test system of the 300 Mvar condenser is shown in Figure 11. The temperature measurement position of the inner edge of the copper shield of the condenser is shown in Figure 12. The temperature measuring element PT100 is bonded to three positions, A, B and C, of the inner edge of the copper shield.

Figure 11. Temperature increase test system of the 300 Mvar synchronous condenser.

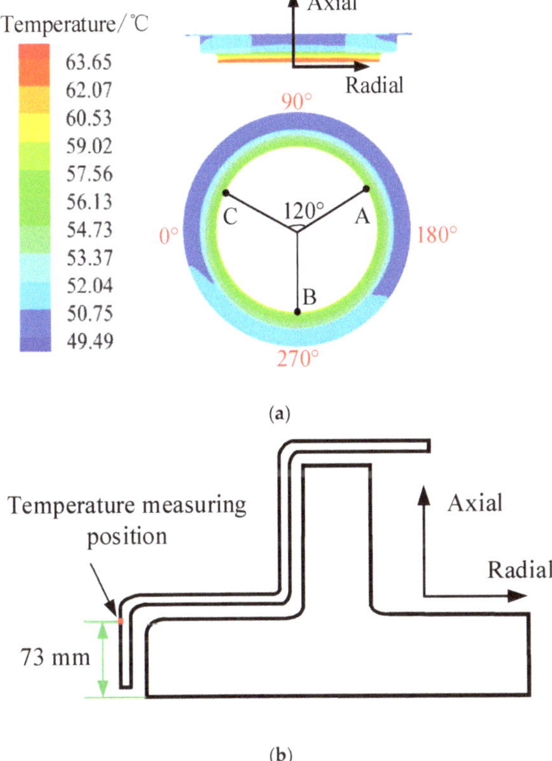

Figure 12. Temperature measurement position of the inner edge of the copper shield: (**a**) the temperature of the copper shield and the circumferential temperature measurement position; (**b**) the axial temperature measurement position of the copper shield.

The temperatures of A, B and C on the inner edge of the copper shield are 57.9 °C, 58.3 °C and 54.2 °C, respectively. The comparison results between the calculated values and the measured values are shown in Table 3. The temperature calculation results are basically consistent with the temperature results of the field test, which can meet the engineering requirements. There is a certain deviation between the copper shield temperature calculation results and the measured temperature values. The reason for the error is that due to the extremely complex structure of the end of condenser, the distance between the stator end windings of the condenser will be different in the assembly process of the condenser, resulting in the uneven distribution of fluid velocity in the end region of the generator. The installation error and unevenness of the copper shield may exist in the actual installation of the project, which will lead to the uneven distribution of the fluid velocity in the ventilation channel between the copper shield and the clamping plate. In addition, the deformation of the end member can cause the uneven distribution of the electromagnetic loss in the end region of the actual condenser, which also leads to the difference between the temperature values measured by the three temperature measuring elements and the calculated values in the inner circle region of the copper shield.

Table 3. Comparison of calculated and measured temperature.

Structure	Position	Measured Value/°C	Calculated Value/°C
	A	57.9	58.7
Copper shield	B	58.3	59.4
	C	54.2	58.6

6. Conclusions

Taking a 300 Mvar air-cooled large synchronous condenser as an example, this paper researches the distribution of magnetic field and eddy current loss in the end of the condenser under three different schemes, and then explores the flow diversion law of the cooling medium and the temperature distribution law of the structure under these three schemes. The temperature of the inner edge of the copper shield in the end of the condenser is measured by the PT100 thermistor. The calculated value of the temperature is consistent with the measured value, and the conclusions are as follows:

(1)   The radial magnetic flux density is the largest in the magnetic flux density component of the upper surface of the lower side of clamping plate under the original structure. Under the three schemes, the magnetic flux density of the inner circle and the outer circle of the clamping plate presents a sinusoidal regular change in the circumferential direction, and the magnetic flux density of the inner circle side is higher than that of the outer circle side. The magnetic flux density of the clamping plate under the original structure is the smallest.

(2)   The temperature near the chamber side is lower than that at the leeward side. The velocity of cooling medium at the inner circle side is higher than that at the outer circle side, but the temperature at the inner circle side is lower than that at the outer circle side.

(3)   The eddy current loss (4036 W) of the clamping plate of Scheme 2 is 3023 W higher than that of the original structure (1013 W), but the maximum temperature and minimum temperature of the clamping plate of Scheme 2 are 1.45 °C and 4.6 °C higher than those of the original structure, respectively. The eddy current loss (234 W) of the finger plate of Scheme 2 is 18.37% higher than that of the original structure (277 W), but the maximum temperature and minimum temperature of the finger plate of Scheme 2 are 1.66 °C and 0.54 °C higher than those of the original structure, respectively.

(4)   When Scheme 2 is adopted, the eddy current loss increases more than that of the original structure, but the temperature increases in the clamping plate, finger plate and other structural parts are less. It can be seen that when the copper shield is removed and the steel clamping plate is replaced by the aluminum clamping plate, the end structure can be simplified and the material of copper shield can be saved. The aluminum clamping plate itself can play a shielding role. Due to the temperature increase being lower, other simple and feasible methods can be adopted to reduce the temperature, such as increasing the fan inlet air volume.

Author Contributions: Conceptualization, X.B. and L.W.; methodology, X.B. and L.W.; formal analysis, L.W.; validation, X.B. and L.W.; investigation, L.W.; data curation, L.W.; writing—original draft preparation, X.B.; writing—review and editing, X.B. and L.W.; project administration, X.B. and L.W.; funding acquisition, L.W., X.B., F.M. and M.Z. All authors have read and agreed to the published version of the manuscript.

Funding: This research was funded only by Natural Science Foundation for Excellent Young Scholars of Heilongjiang Province, grant number YQ2021E038.

Conflicts of Interest: All of our authors declare together that there is no conflict of interest.

References

1. Bian, X.; Liang, Y.P. Circuit Network Model of Stator Transposition Bar in Large Generators and Calculation of Circulating Current. *IEEE Trans. Ind. Electron.* **2015**, *62*, 1392–1399. [CrossRef]
2. Gozdowiak, A. Faulty Synchronization of Salient Pole Synchronous Hydro Generator. *Energies* **2020**, *13*, 5491. [CrossRef]
3. Wang, L.K.; Huo, F.Y.; Li, W.L. A Novel Dual Three-Phase Permanent Magnet Synchronous Motor with Asymmetric Stator Winding. *IEEE Trans. Magn.* **2013**, *49*, 939–945. [CrossRef]
4. Tao, D.J.; Zhou, K.L.; Lv, F.; Dou, Q.P.; Wu, J.X.; Sun, Y.T.; Zou, J.B. Magnetic Field Characteristics and Stator Core Losses of High-Speed Permanent Magnet Synchronous Motors. *Energies* **2020**, *13*, 535. [CrossRef]
5. Ding, S.Y.; Li, Z.J. Analysis of Fluid Field Characteristics and Structure Optimization inside Stator Domain for Air-cooled Turbo-generator. In Proceedings of the 2020 12th IEEE PES Asia-Pacific Power and Energy Engineering Conference (APPEEC), Nanjing, China, 20–23 September 2020; pp. 1–5.
6. Traxler-Samek, G.; Zickermann, R.; Schwery, A. Cooling airflow, losses, and temperatures in large air-cooled synchronous machines. *IEEE Trans. Ind. Electron.* **2010**, *57*, 172–180. [CrossRef]
7. Ding, S.Y.; Zhang, X.F.; Guan, T.Y. Analysis of Heat Transfer Performance for Megawatt Wind Generator in High Altitude Extreme Conditions. In Proceedings of the 2014 17th International Conference on Electrical Machines and Systems (ICEMS), Hangzhou, China, 22–25 October 2014; pp. 2340–2346.
8. Zhang, L.L.; Ling, Y.P.; Bian, X. Analysis of Magnetic Flux Density of Air Gap in Turbo-Generators Using Equivalent Magnetic Network Method. In Proceedings of the 2019 22nd International Conference on Electrical Machines and Systems (ICEMS), Harbin, China, 11–14 August 2019; pp. 1–5.
9. Ling, Y.P.; Bian, X.; Yu, H.H. Analytic Algorithm for Strand Slot Leakage Reactance of the Transposition Bar in an AC Machine. *IEEE Trans. Ind. Electron.* **2014**, *61*, 5232–5240. [CrossRef]
10. Jing, T.T.; Liu, G.H.; Zhou, H.W. Simplified Thermal Modeling of Fault-Tolerant Permanent-Magnet Motor by Using Lumped Parameter Network. In Proceedings of the 2014 IEEE Conference and Expo Transportation Electrification Asia-Pacific (ITEC Asia-Pacific), Beijing, China, 31 August–3 September 2014; pp. 1–5.
11. Zhao, W.X.; Cao, D.H.; Ji, J.H. A Generalized Mesh-Based Thermal Network Model for SPM Machines Combining Coupled Winding Solution. *IEEE Trans. Ind. Electron.* **2021**, *68*, 116–127. [CrossRef]
12. Ling, Y.P.; Yu, H.H.; Bian, X. Finite-Element Calculation of 3-D Transient Electromagnetic Field in End Region and Eddy-Current Loss Decrease in Stator End Clamping Plate of Large Hydrogenerator. *IEEE Trans. Ind. Electron.* **2015**, *62*, 7331–7338. [CrossRef]
13. Demir, Y.; Aydin, M. A Novel Dual Three-Phase Permanent Magnet Synchronous Motor with Asymmetric Stator Winding. *IEEE Trans. Magn.* **2016**, *52*, 8105005. [CrossRef]
14. Tong, W.M.; Dai, S.H.; Wu, S.N. Performance Comparison Between an Amorphous Metal PMSM and a Silicon Steel PMSM. *IEEE Trans. Magn.* **2019**, *45*, 8102705. [CrossRef]
15. Han, J.C.; Zheng, P.; Sun, Y.T. Influence of Electric Shield Materials on Temperature Distribution in the End Region of a Large Water–Hydrogen–Hydrogen-Cooled Turbogenerator. *IEEE Trans. Ind. Electron.* **2020**, *67*, 3431–3441. [CrossRef]
16. Huo, F.Y.; Hai, J.C.; Li, W.L. I Influence of Copper Shield Structure on 3-D Electromagnetic Field, Fluid and Temperature Fields in End Region of Large Turbogenerator. *IEEE Trans. Ind. Electron.* **2013**, *28*, 832–840.
17. Fan, Z.; Yi, H.; Xu, J.; Xie, K.; Qi, Y.; Ren, S.; Wang, H. Performance Study and Optimization Design of High-Speed Amorphous Alloy Induction Motor. *Energies* **2021**, *14*, 2468. [CrossRef]
18. Wang, L.K.; Li, W.L.; Huo, F.Y. Influence of Underexcitation Operation on Electromagnetic Loss in the End Metal Parts and Stator Step Packets of a Turbogenerator. *IEEE Trans. Energy Convers.* **2014**, *29*, 748–757. [CrossRef]
19. Wang, L.K.; Li, W.L. Assessment of the Stray Flux, Losses, and Temperature Rise in the End Region of a High-Power Turbogenerator Based on a Novel Frequency-Domain Model. *IEEE Trans. Ind. Electron.* **2018**, *65*, 4503–4513. [CrossRef]
20. Huo, F.Y.; Li, W.L.; Wang, L.K. Numerical Calculation and Analysis of Three-Dimensional Transient Electromagnetic Field in the End Region of Large Water–Hydrogen–Hydrogen Cooled Turbogenerator. *IEEE Trans. Ind. Electron.* **2014**, *61*, 188–195. [CrossRef]
21. Li, W.L.; Han, J.C.; Huo, F.Y. Influence of the End Ventilation Structure Change on the Temperature Distribution in the End Region of Large Water–Hydrogen–Hydrogen Cooled Turbogenerator. *IEEE Trans. Energy Convert.* **2013**, *28*, 278–288.

 *energies*

Article

Transient System Model for the Analysis of Structural Dynamic Interactions of Electric Drivetrains

Marius Franck *, Jan Philipp Rickwärtz, Daniel Butterweck, Martin Nell and Kay Hameyer

Institute of Electrical Machines (IEM), RWTH Aachen University, Schinkelstrasse 4, 52062 Aachen, Germany; jan.rickwaertz@iem.rwth-aachen.de (J.P.R.); daniel.butterweck@iem.rwth-aachen.de (D.B.); martin.nell@iem.rwth-aachen.de (M.N.); kay.hameyer@iem.rwth-aachen.de (K.H.)
* Correspondence: marius.franck@iem.rwth-aachen.de

Abstract: In electric drivetrains, the traction machines are often coupled to a gear transmission. For the noise and vibration analysis of such systems, linearised system models in the frequency domain are commonly used. In this paper, a system approach in the time domain is introduced, which gives the advantage of analysing the transient behaviour of an electric drivetrain. The focus in this paper is on the dynamic gear model. Finally, the modelling approach is applied to an exemplary drivetrain, and the results are discussed.

Keywords: system model; structural dynamics; multibody simulation; transient electrical machine model; permanent magnet synchronous machine; induction machine; dynamic gear forces; planetary gear

Citation: Franck, M.; Rickwärtz, J.P.; Butterweck, D.; Nell, M.; Hameyer, K. Transient System Model for the Analysis of Structural Dynamic Interactions of Electric Drivetrains. *Energies* **2021**, *14*, 1108. https://doi.org/10.3390/en14041108

Academic Editor: Ryszard Palka

Received: 19 January 2021
Accepted: 13 February 2021
Published: 19 February 2021

Publisher's Note: MDPI stays neutral with regard to jurisdictional claims in published maps and institutional affiliations.

1. Introduction

For the consideration of local loads and acoustics in electric vehicle drivetrains, it is not sufficient to consider the components, such as the motor and gearbox, separately. Due to the interactions of the individual components on the system level, structural dynamic load peaks can occur in combination with the natural oscillation behaviour. Furthermore, the natural frequencies of the coupled system can change when compared to a separate analysis of the components. Consequently, a system approach is essential for mechanical load and noise, vibration, and harshness (NVH) analyses of such systems.

In recent studies, system models for NVH analysis have been frequently discussed, e.g., in [1–3]. Due to the complexity of such system models, simplifications are usually made. In [1], a linearised system approach around a particular operating point was presented in the frequency domain. The electromagnetic forces of a machine were calculated using analytic equations, and the gear excitations were modelled by a load-dependent transmission error formulation [1]. The system model described in [3] used an interpolation approach in the electromagnetic force calculation to reduce the model size. By using this model, a stationary linear interpolation of the resulting force harmonics between different rotational speeds was performed, which was then mapped to a structural dynamic finite element model (FEM) in the frequency domain. In [2], a system model was introduced, where an electromagnetic force model was coupled in the frequency domain to an elastic multi-body simulation (MBS) of a drivetrain. The MBS was solved in the time domain, thus allowing, for instance, the modelling of the non-linear gear dynamics and non-linear bearing characteristics.

The purpose of the models described above is an efficient noise prediction and vibration analysis of the drivetrain. However, possible interactions due to the coupling in the frequency domain of the electric machine with the mechanical system are neglected. In all model descriptions, the feedback of the structural dynamics to the electrical machine and its control is not considered.

163

In this paper, a system model in the time domain is proposed, which allows the analysis of not only the steady-state operating points, but also transient load changes, which can occur, e.g., in situations of driving a vehicle. Therefore, the transient behaviour of an electric machine with an electrical supply system and digital control is modelled using an extended two-axis (dq) model according to [4]. This approach considers, for instance, slotting effects, non-ideal sinusoidal currents due to power electronics and digital control, and non-linearities of the electric machines. For transmission gears, different excitation mechanisms, such as parametric, impulse, and geometric excitations, are described in the literature [5,6]. The models for describing those mechanisms' operational point dependence yield non-linear differential equations [5]. The suggested system model in the time domain allows a consideration of these effects without any operating point linearisation, and therefore, the dynamic gear mesh interaction with the electric machine can be analysed.

The paper is structured as follows. First, a general overview of structural dynamic modelling approaches is given and a particular approach is chosen. In the second step, the excitation mechanisms of transmission gears are summarised and a dynamic gear model is selected. Then, the transient electric machine model used is briefly described. Finally, the modelling domains are coupled to a system model, which is then applied to an exemplary drivetrain, and the results are discussed.

2. Modelling the Structural Dynamics of Electric Drivetrains

The structural dynamic behaviour of drivetrains can be described with various mathematical models. According to [7], these models can be divided into their typical fields of application, as shown in Figure 1. The classification criteria are the system expansion, the complexity, and the frequency range.

For simple systems consisting of basic geometries, the component loads and vibrations can be described with analytical formulas, independently of the frequency range and the system extension. For more complicated geometries, these analytical formulas are not available.

In the lower-frequency range, simulations with multi-body systems (MBSs) offer the opportunity to analyse the vibrational behaviour of large and complicated systems. Here, individual components of the system are modelled by rigid bodies that can be linked with each other and the environment using so-called joint and force elements. Information is obtained about the joint and coupling forces as well as the motion quantities of the individual bodies. For higher frequencies, the assumption that bodies behave rigidly is no longer permissible, since the elastic deformation and the inherent behaviour of the individual bodies are essential.

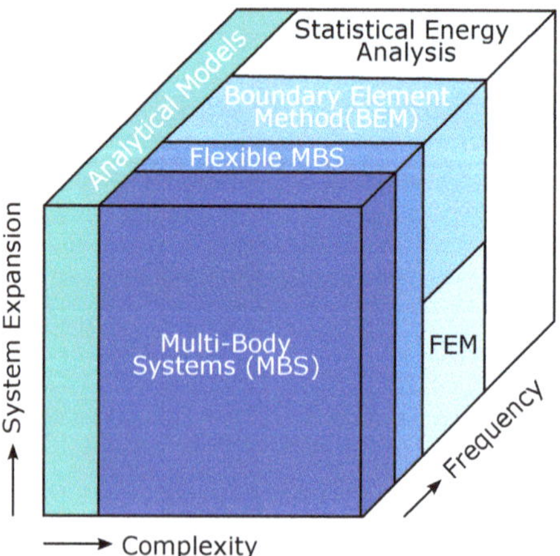

Figure 1. Common simulation approaches and their typical areas of application according to [7].

The methods of flexible multi-body systems extend the valid frequency range of MBSs by considering these elastic deformations of individual bodies [8]. In order to map all of the elastic properties of the system, the finite element method (FEM) is used.

The number of degrees of freedom in the FEM is larger compared to MBSs, which leads to a higher computational effort and limits the applicability with regard to the system size [7]. If one is only interested in the acoustic radiation behaviour of a structure, the model of the boundary element theory (BEM) offers an efficient approach.

In the high-frequency range, the mode density increases, i.e., the number of natural oscillations per frequency interval, which makes the calculation of the individual natural oscillations increasingly difficult [7]. Statistical energy analysis (SEA) is an approach for efficiently generating statements about the structural dynamic behaviour in this area. For use in a system model that considers the structural dynamic interactions, in this paper, the multi-body system approach is chosen. As described later, it can model the non-linearities of a dynamic gear transmission within a force element. The relatively low number of degrees of freedom and, consequently, the lower computational effort allow an analysis and coupling in the time domain. This makes the method suitable for complex systems, such as electric drivetrains.

2.1. Gear Excitations

The dynamic forces in gearboxes that lead to vibrations are decisively determined by the tooth engagement of the gears. According to [5], excitation mechanisms in gears can be divided according to their excitation characteristics into the groups of parameter excitation, impulse excitation, and geometric excitation.

The parameter excitation model describes the force that arises through temporal changes in the system parameters for damping and stiffness. Such changes are mainly caused by the stiffness of the teeth, which depends on the angle of engagement.

According to [5], under certain conditions and under load, the meshing teeth can come in contact with the next unloaded tooth, resulting in an impulsive force. Furthermore, it has been pointed out that these non-ideal contact conditions during meshing can cause tooth flank damage during operation and should, therefore, not occur if a gear unit is carefully designed [5].

The geometric excitation model describes the force when rolling on a surface that is not ideally flat. If a simple single-mass oscillator is considered, this excitation describes a time-varying path $s(t)$ at the base of an oscillator, through which a force is generated. In gear meshing, this results from manufacturing deviations and tooth flank modifications when the involute is rolled off [5].

When considering a gearbox under load, the geometrical excitation can be neglected when compared to the parameter excitation [5]. In addition to the excitation mechanisms presented, the authors of [5] described the frictional force excitation and the excitation due to tilting moments. These can be modelled as additional force excitations. The frictional force excitation in tooth contact has a significantly lower amplitude than the normal tooth force due to the friction coefficient and will, therefore, not be regarded in this paper. The tilting moments are automatically considered if the tooth contact is modelled by a three-dimensional simulation model with all six degrees of freedom [5].

Due to the circumstances described above, the parameter excitation is identified as the main vibration source of a gear under load and is, therefore, considered in more detail in the next section.

2.2. Dynamic Gear Transmission

In order to model the vibration behaviour and the excitations of a gearbox, various model approaches have been described in the literature [5,6]. In [6], a simple torsional vibration model of an unbounded cylindrical gear pair with the two rotational angles of the gear wheels as the degrees of freedom, φ_1 and φ_2, was described. This model is shown in Figure 2. The parameter excitation of the gear mesh is represented by the force elements of the variable spring $c(\varphi(t))$ and damper $k(\varphi(t))$, which are oriented in the direction of the line of contact. The geometric excitation can be modelled using a path source $s(\varphi(t))$.

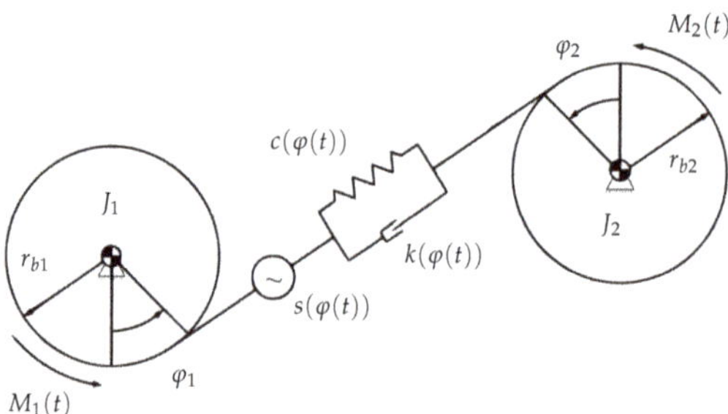

Figure 2. Torsional vibration model with two degrees of freedom [6].

Using the moment of inertia J around the individual rotational axis of the gear wheels and the base circle r_b, the equation of motion for the first gear wheel can be set up [6]:

$$M_1(t) = J_1 \frac{d^2}{dt^2}\varphi_1 + k(\varphi_1(t))[r_{b1}^2 \frac{d}{dt}\varphi_1 + r_{b1}r_{b2}\frac{d}{dt}\varphi_2 + r_{b1}\frac{d}{dt}s(\varphi_1(t))]$$
$$+ c(\varphi_1(t))[r_{b1}^2\varphi_1 + r_{b1}r_{b2}\varphi_2 + r_{b1}s(\varphi_1(t))]. \tag{1}$$

The equation of motion for the second gear wheel follows in the same way [6]. For a helical gear, the torque $M_i(t)$ acting on the two gear wheels can be expressed by the tangential force F_t on the base circle r_b or by the normal tooth force F_N, as shown in Figure 3:

$$M_i(t) = F_{ti}(t) \cdot r_{b_i} = F_{Ni}\cos(\beta)\cos(\alpha_n) \cdot r_{bi}, \quad i = \{1,2\}. \tag{2}$$

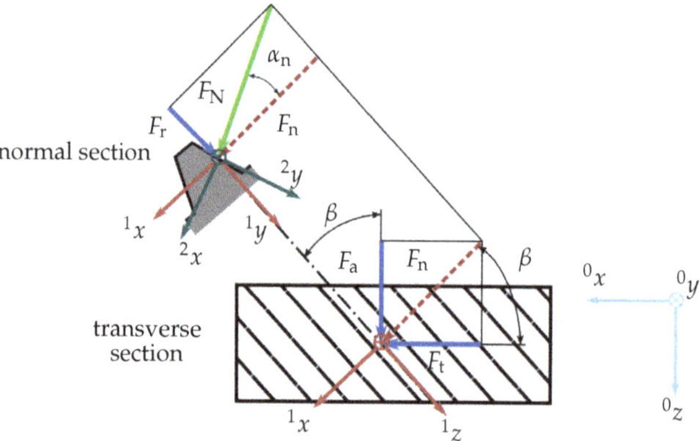

Figure 3. Tooth forces with extension of the coordinate systems used.

Considering the geometrical relations of the gear forces shown in Figure 3, it can be seen that the normal tooth force F_N is composed of the tangential force F_t, the axial force F_a, and the radial force F_r with helix angle β and normal contact angle α_n. Mathematically, the connection of the forces can be described by a coordinate transformation from the global coordinate system Index 0 of the gear wheel to the coordinate system Index 2 in the normal section [6].

$$^0\vec{F} = {^{01}S} \cdot {^{12}S} \cdot {^2\vec{F}} \tag{3}$$

In (3) the rotation matrix ^{01}S results from the elementary rotation matrix by rotating $-\beta$ around the ^0y axis. Furthermore, ^{12}S is obtained from the elementary rotation matrix by rotating $+\alpha_n$ around the ^1z axis.

The tangential, radial, and axial force can also be determined from the tooth normal force by inversion of (3):

$$^2\vec{F} = {}^{01}S^{-1} \cdot {}^{12}S^{-1} \cdot {}^0\vec{F} = {}^{01}S^{T} \cdot {}^{12}S^{T} \cdot {}^0\vec{F}. \tag{4}$$

As described in the previous section, the gear excitation forces due to friction are neglected. Therefore, the three-dimensional tooth force vector $^0\vec{F} = [F_t \quad F_r \quad F_a]^T$ in the global coordinate system can be represented in the local coordinate system by an one-dimensional force, the normal tooth force F_N, with $^2\vec{F} = [F_N \quad 0 \quad 0]^T$. This force description in (4) enables an implementation in a multi-body system environment by a one-dimensional force element using a standard variable spring and damper in this local coordinate system [6]. This modular implementation, as suggested in [6], is used for the gear modelling in the following. It allows the development of complicated dynamic gear models with different gear stages within an MBS environment.

3. Modelling the Transient Behaviour of Electric Machines

In order to analyse the interactions between a machine, the power electronics, the digital control, and the mechanical system, a transient machine model approach was used, as suggested in [4]. As described in [4], a voltage-driven machine model was used to model the dynamic behaviour of the machine. This model approach gives the opportunity to consider any voltage source $u(t)$. Thereby, non-ideal sinusoidal voltages of the power electronics and the digital control strategy can be considered. Figure 4 gives an overview of the schematic model structure.

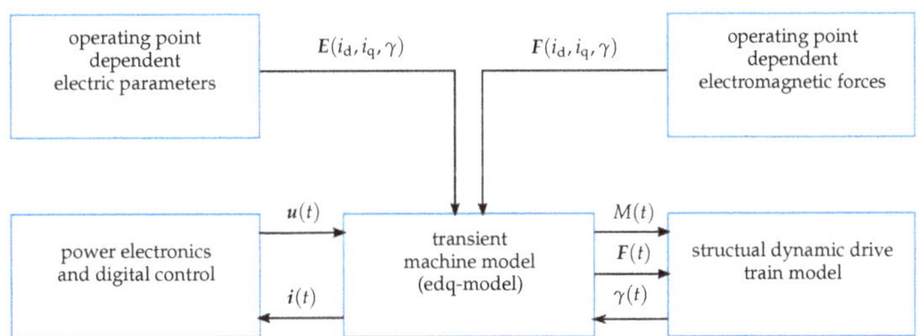

Figure 4. Schematic structure of the extended dq-model (edq-model) according to [4].

The central component of the model is an extended formulation of the voltage equation in the dq axis coordinate system of an electric machine, which allows the inclusion of saturation, cross-coupling, and slotting effects [4]. The interactions between the electric machine model and a mechanical model can be considered through the feedback of the actual rotor angle $\gamma(t)$. The fundamental voltage equation of a three-phase electric machine serves as a starting point:

$$\vec{u} = \mathbf{R}\vec{i} + \frac{d}{dt}\psi, \tag{5}$$

where $\vec{u}$ describes the vector of the three phase voltages, $\mathbf{R}$ is the ohmic resistance matrix, $\vec{i}$ is the vector of the three phase currents, and ψ is the matrix of the magnetic flux linkage. Using the commonly known dq-transformation from, e.g., [9], the stationary reference frame of the three phase quantities is transformed into a rotating reference frame of two phase quantities. From this follows the the fundamental voltage equation in dq axis coordinates in tensor notation [10]:

$$\vec{u}_{dq0} = R_{dq0}\vec{i}_{dq0} + \omega_{el}AL_{dq0}\vec{i}_{dq0} + (\frac{d}{dt}L_{dq0})\vec{i}_{dq0} + L_{dq0}(\frac{d}{dt}\vec{i}_{dq0}) + \omega_{el}A\psi_{f,dq0} + \frac{d}{dt}\psi_{f,dq0}, \tag{6}$$

with

$$A = \begin{bmatrix} 0 & -1 & 0 \\ 1 & 0 & 0 \\ 0 & 0 & 0 \end{bmatrix}. \tag{7}$$

R_{dq0} is the ohmic resistance matrix, $\vec{i}_{dq0}$ is the vector of the phase currents, ω_{el} is the electrical frequency, L_{dq0} is the operating-point-dependent inductance matrix, and $\boldsymbol{\psi}_{f,dq0}$ is the operating-point-dependent excitation flux linkage.

As described in [10], Equation (6) is transformed into partial differential equations through the use of the total derivative, and the equations are solved according to the respective current derivations. Hence, the partial derivatives can be calculated in advance and do not have to be solved in the system simulation itself. The machine parameters L_{dq0} and $\boldsymbol{\psi}_{f,dq0}$ are modelled in this paper as a function of the machine currents $\vec{i}_{dq0}$ and the rotor angle γ. In order to determine these parameter dependencies, a series of electromagnetic finite element simulations are performed, which extract the described parameters for each possible operating point by varying i_d, i_q, and the rotor angle γ.

The same approach is used to extract the electromagnetic forces $F(i_d, i_q, \gamma)$. Together with the supply voltage $u(t)$ from a power electronics model, the extended machine model calculates the resulting transient currents, $i(t)$ and forces $F(t)$, and the torque $M(t)$.

4. The Coupled Transient System Model Approach

In order to analyse the transient behaviour of a drivetrain system, the model domains described above must be coupled to an overall system model. This is done on an exemplary drivetrain, the "KinelectricDrive", which will be presented in the following section.

4.1. The Exemplary Drivetrain

The "KinelectricDrive" is a kinetic–electric power train concept for small vehicles [11]. The name "KineletricDrive" stands for a kinetic–electric drive system that combines the potential of kinetic and electrical energy storage systems on a 48 V basis in one system for automotive applications. The power train and its integration into the vehicle drive are shown schematically in Figure 5.

The central component of the drive system is a high-speed induction machine (IM) with a coupled flywheel. The IM is operated in a technical vacuum to reduce drag losses. In order to realise a maximum short-term mechanical power of the motor of 15 kW in combination with the kinetic energy storage of the flywheel inertia, a maximum speed of up to 45,000 rpm is achieved.

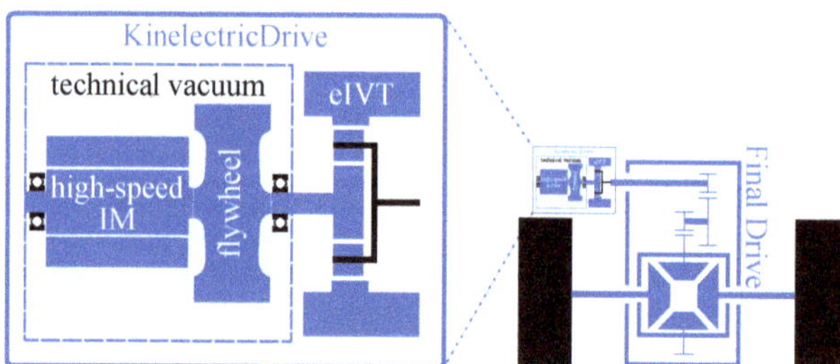

Figure 5. Schematic layout of the "KinelectricDrive" according to [11].

In order to decouple the flywheel speed from the powertrain output speed, an electric infinite variable transmission (eIVT) is employed. It consists of a helical planetary gear set where the ring is powered by a permanent magnet synchronous machine (PMSM). By controlling the ring torque of the planetary gear set together with a speed-controlled machine on the sun gear, the output power at the carrier can be controlled independently of the flywheel speed and its kinetic energy.

4.2. System Model of the "KinelectricDrive"

For the development of a multi-physical transient system model of the "KinelectricDrive", the modelling environment of MATLAB Simulink with the extension toolbox Simscape Multibody was chosen. The focus of this model is on analysing the interactions of the two electrical machines with the rotor dynamics of the drivetrain. Therefore, both machines, the gearbox with the machines' rotors, and the output shaft were modelled in Simulink. The model structure is shown in Figure 6.

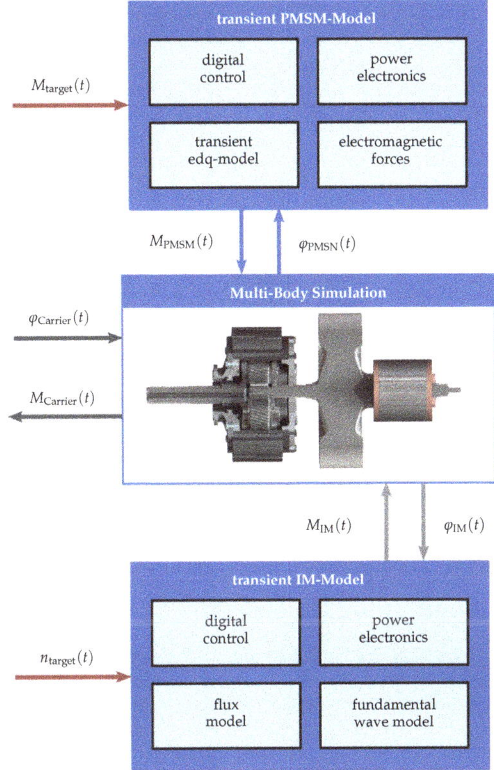

Figure 6. Schematic structure of the proposed system model.

The central component of the system model was the MBS of all rotating components of the "KinelectricDrive". The rotor of the PMSM at the ring, the drive at the planet carrier with the supported planets, the sun wheel with the flywheel, and the rotor of the high-speed IM were modelled as rigid bodies. All bearings were assumed to be ideally rigid and, thus, offered only one rotational degree of freedom. For the consideration of the dynamic gearing forces, as described in Section 2.2, the force element was implemented and extended for use in a planetary gear, as described in the next section.

Table 1 shows selected resulting mechanical parameters of rigid bodies modelled in the MBS. When comparing the mass and inertia characteristics of the rotor of the high-speed IM to those of the rotor of the PMSM in Table 1, it can be seen that the high-speed machine has approximately twice the inertia of the PMSM due to the coupled flywheel. Furthermore, due to the larger ratio of the planetary gearbox of $i_{12} = -5$, the PMSM is designed for a nominal torque that is approximately five times greater than the torque of the IM. Due to these two facts, it is assumed in this paper that the harmonic torques of the induction machine do not contribute significantly to the structural dynamic response of the system. Therefore, a fundamental wave model of the IM was employed as described in [9].

This model was set up together with a field-oriented control and an associated flux model, super-ordinated speed control, and power electronics model.

Table 1. Mechanical parameters of modelled bodies.

Component	Mass m in kg	Moment of Inertia J_z around the Main Axis of Rotation in kgm^2
Rotor of high-speed induction machine (IM) with coupled flywheel	10.3	32.80×10^{-3}
Planet gear wheel	0.1	0.03×10^{-3}
Planet carrier with outputshaft	1.0	1.03×10^{-3}
Rotor of permanent magnet synchronous machine (PMSM)	5.0	16.79×10^{-3}

The PMSM at the ring gear was modelled by using the previously described extended transient dq model, which was parametrised by an electromagnetic FEM. The FEM was solved for a total of 30 i_d, 30 i_q current combinations, and 60 rotor positions with a mechanical resolution of 0.1 degrees. Figure 7a shows the absolute value of the resulting normalised torque for a current in the q-axis of 10% of the maximum value versus the normalised current in the d-axis and the mechanical rotor angle. The slotting effects and saturation of the machines are visible. Figure 7b shows the Fourier decomposition of the resultant torque versus one mechanical revolution for the same q-current component of 10% of the maximum value and a d-current component of 1% of the maximum value.

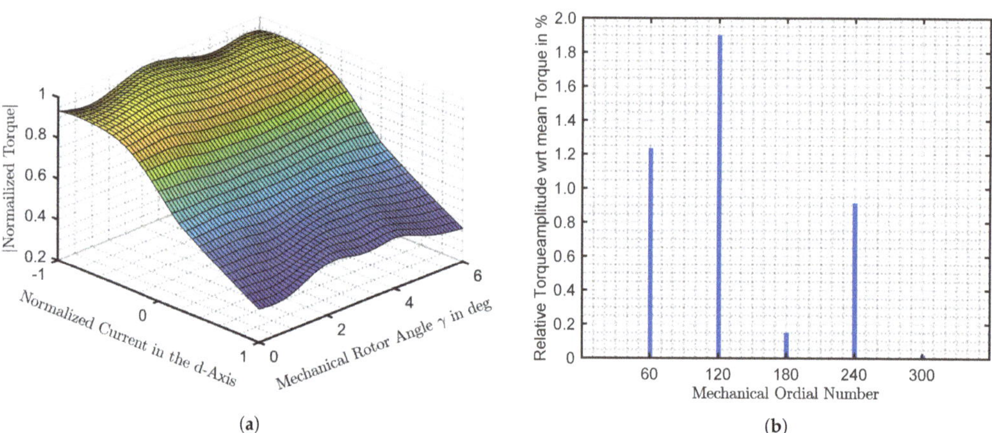

Figure 7. Results of the electromagnetic finite element model: (**a**) Synchronous machine torque as a function of the d-current component and rotor angle; (**b**) Fourier decomposition of the synchronous machine torque for one mechanical rotation.

According to [12], the fundamental ordinal number of the stator slot cogging torque for a three-phase machine with a fractional winding can be determined in a simplified manner according to $\nu_c = N \cdot n_q$ from the truncated denominator n_q of the number of slots per pole per phase q and the number of stator slots N. In the case of the machine under consideration, the number of slots per pole per phase is $q = 0.4$, and therefore, the truncated denominator of $nq = 5$. With the number of stator slots of $N = 24$, the fundamental ordinal number of the stator slot cogging torque is $\nu_c = 24 \times 5 = 120$. This ordinal number is visible in the Fourier decomposition of the results, which confirms the results of the finite element model used. The power electronics and the influence of non-ideal sinusoidal currents are not considered in this work. A standard digital control structure is modelled in the form of a torque control.

As the input variables for the MBS model, the transient torques $M(t)$ of the two machines are transferred. By solving the non-linear equations of motion within the model, the resulting movements $\varphi(t)$ are transferred back to the two machine models. This particular model structure enables the consideration of the transient dynamic behaviour on a system level. The output shaft is modelled in the context of this paper by means of a motion target, whereby the resulting torque at the output can be determined within the MBS model.

4.3. Dynamic Planetary Gear Model

In order to model the dynamic gear forces of the planetary gear, a force element in Simscape Multibody was developed according to the approach described in Section 2.2. Within the element, the resulting bearing forces of the planet wheels were calculated so that the resulting torque at the carrier could be evaluated.

Within the tooth transformation inside the force element, the coordinate transformation in (3) is used to transform the three-dimensional forces in the global system of the planet wheel to the normal section of the gear tooth contact. The reaction forces due to the parameter excitation are transformed back to the global system of the planet wheel according to (4). The coupling is then realised in the normal section of the gear tooth contact using a one-dimensional variable string and a viscous damper.

For the parametrisation of the model, the torque and angle-dependent gear stiffness of the gear pair are required. Such data can be obtained through a gear mesh contact analysis.

The resulting relative contact stiffness, e.g., between the sun and planet gear, for the "KinelectricDrive" is shown in Figure 8. The resulting load and position dependence of the stiffness due to the changing meshing conditions in the tooth contact is visible. This dependency is considered using a two-dimensional look-up table, which is interpolated linearly between the calculated stiffness curves.

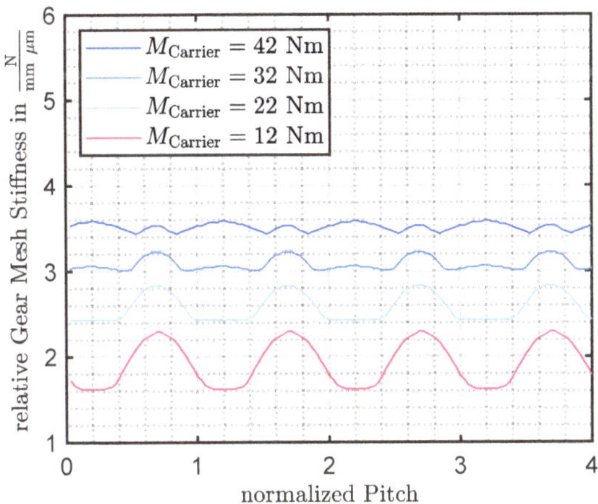

Figure 8. Gear mesh stiffness of the sun and planet wheels.

In addition to the tooth contact stiffness, the damping constant k is required to describe the dynamics of the tooth contact. In this paper, the damping factor, which allows a calculation of the damping constant, was chosen to be $D = 0.05$. It is recommended to parametrise the developed model, particularly the damping factor, using measurement data.

Modelling all three planets as single bodies enables the analysis of the influences of an unequal load distribution between the planets on the excitation behaviour.

5. Results

In this section, selected results of the system model of the "KinelectricDrive" are discussed. Since the entire drivetrain is not modelled in the developed model, but only the "KinelectricDrive" itself, the output speed is set to zero for the following analysis. Thus, only one power flow between the two machines is considered in the following analysis.

A speed-controlled start-up of the IM under load was performed. The target speed value of the IM was ramped up using a constant slope of 2300 rpm/s. The target speed of the IM was 12,000 rpm. The target of the torque load of the PMSM was controlled by an s-function. During the start-up, a maximum torque of the IM of 10.5 Nm was reached. The stationary torque of 10 Nm of the PMSM was obtained after a time of 0.2 s.

The implementation and solution of the modelling approach proposed in the paper led to a stiff numerical problem. For the solution of the model, a variable step solver, *ode23tb* from MATLAB Simulink, was selected. For a convenient evaluation of the results, the numerical values were sampled using a constant sampling frequency of $f_s = 80\,\text{kHz}$. To reduce aliasing effects, an anti-aliasing filter in the form of a lowpass Butterworth filter was employed.

In the first step, the system was modelled with the standard common gear constraint from the Simscape Multibody library with a stationary gear ratio of $i_{12} = -5$. Thereafter, the results were compared to those of the developed dynamic gear model.

Figure 9a shows that the target speed $n_{\text{target,IM}} = 12{,}000\,\text{rpm}$ of the IM is reached at a time of $t = 7\,\text{s}$ after an overshoot of 4% around the target value. Considering the stationary gear ratio of the planetary gear of $i_{12} = -5$, the speed of the PMSM at the time $t = 7\,\text{s}$ is $n_{\text{PMSM}} = \frac{n_{\text{IM}}}{i_{12}} = -2400\,\text{rpm}$. This result is confirmed by the model, which confirms the appropriateness of the implementation. In Figure 9b, the reaction torques calculated from the MBS of the drive are shown. In the reaction torque of the PMSM, the cogging torque modelled by the edq model is visible through a slight superimposed oscillation of around 0.5% of the target value. Due to the ideal rigid gear coupling, the cogging torque of the PMSM acts on the output shaft of the carrier in a similar magnitude of around 0.5% of the mean value.

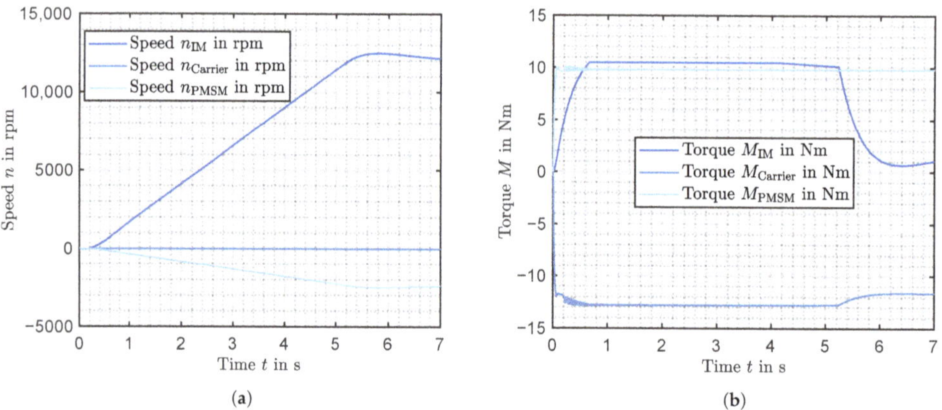

Figure 9. Results of the transient system model using an ideal gear constraint: (**a**) Speeds of the different shafts; (**b**) Reaction torques of the different shafts.

Figure 10 shows the results using the dynamic gear model. The basic course and the steady-state value of the resulting speeds and the reaction torques of the two models are comparable, which speaks for the correct implementation of the force element presented here. When comparing the resulting reaction torque at the carrier output shaft, significantly increased torque oscillations can be determined. In order to analyse the origin of these in more detail, a spectrogram of the torque at the PMSM and at the carrier is calculated in the following.

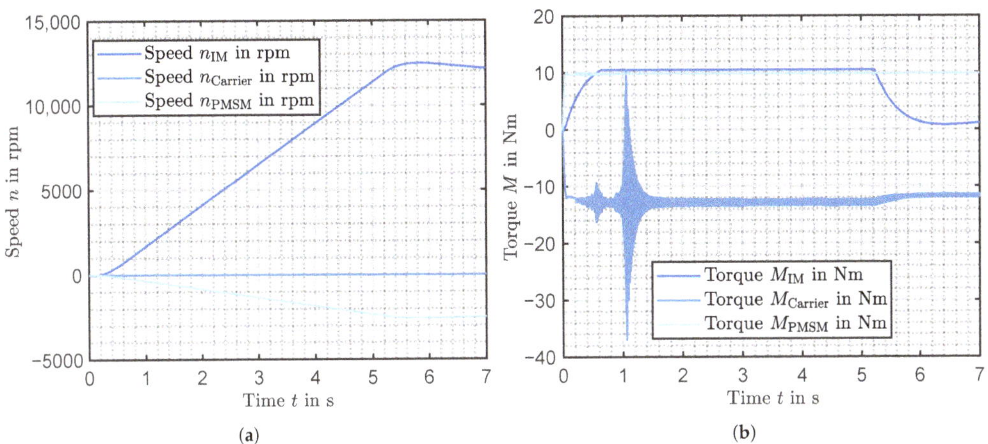

Figure 10. Results of the transient system model using the developed force element: (**a**) Speeds of the different shafts; (**b**) Reaction torques of the different shafts.

Figure 11a shows the spectrogram of the resulting PMSM torque for the studied start-up operation. The significant frequency orders of the cogging torque with the ordinal numbers $\nu = 60; 120; 180; 240$ are highlighted, which correspond to the results from the electromagnetic FEM simulation in Figure 7.

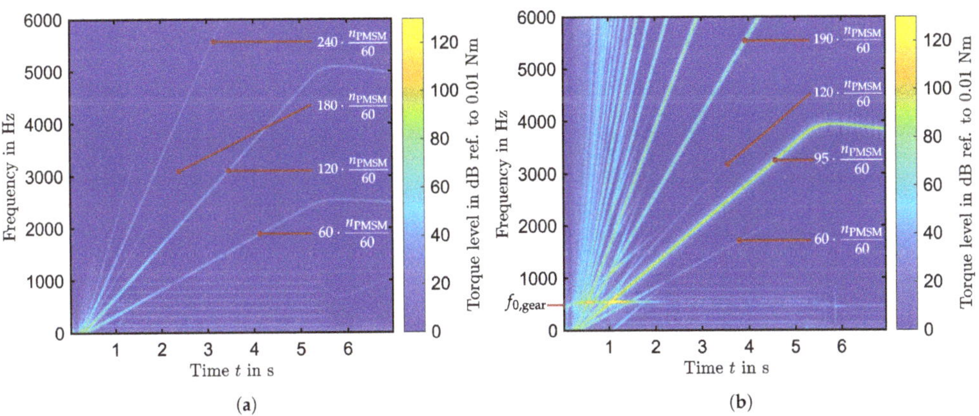

Figure 11. Results of the transient system model using the developed force element: (**a**) Spectrogram of the transient torque of the PMSM; (**b**) Spectrogram of the reaction torque at the carrier shaft.

In Figure 11b, the first orders in the torque with reference to the speed n_{PMSM} of the PMSM at the carrier are highlighted in the spectrum. Among others, the torque orders with the ordinal numbers $\nu = 60; 120$ occur with reference to the PMSM speed. These can be assigned to the cogging torque of the PMSM. Furthermore, orders with the ordinal numbers $\nu = 95; 190$ occur, for instance. These can be related to the gear meshing frequency, which is the frequency of the parameter excitation of the gear. For the considered gearbox, the gear mesh frequency is obtained from the speed of the PMSM through the number of teeth of the ring gear of $z_h = -95$ or from the speed of the IM through the number of teeth of the sun gear, $z_s = 19$:

$$f_{mesh} = z_h \cdot \frac{n_{PMSM}}{60} = z_s \cdot \frac{n_{IM}}{60}. \tag{8}$$

As the torque orders cross the modelled system's first and only natural frequency $f_{0,gear}$ (see Figure 11b), a distinct torque oscillation results due to the resonance. This resonance amplification is particularly evident when the fundamental order of the gear mesh excitation itself is at the natural resonant frequency at time $t \approx 1.1\,$s (see Figure 10b and Figure 11b). Due to the load-dependent stiffness of the gearing contact (see Figure 8), the resonance frequency of the system is not constant, but load dependent. This load dependency can be seen in the spectrogram in Figure 11b, for example, during the start-up process, and when the target speed is reached during the load change as the natural frequency changes. With increasing torque, the gear stiffness increases (see Figure 8). Since the inertia of the system is constant, according to the simplified relation for a one-mass oscillator $\omega_0 = \sqrt{\frac{c}{m}}$, the natural frequency of the system will increase. For example, linearising the system around the operating point, e.g., for the time $t = 0.2\,$s for the starting phase of the system, gives the natural frequency with the inertias and the current effective stiffness as $f_0(0.2\,\text{s}) = 468\,$Hz. For the time $t = 2\,$s, the torque of the IM has reached the steady state (see Figure 10b), and the linearised eigenfrequency around this operating point results in $f_0(2\,\text{s}) = 559\,$Hz.

The proposed model allows the study of load changes, which frequently occur in drive cycles. The transient behaviours of the machines and the controllers are considered. This allows, for example, the study of a controller's parameters for their interaction with the rotor dynamics of a drivetrain. Furthermore, the nonlinearities of the gear mesh contact were included in the simulation. Therefore, the load-dependent changes in the system's natural frequencies and the parameter excitation were taken into account. This overall system model serves as a starting point for analysing the structural dynamic interactions of a drivetrain in detail and for gaining a deeper understanding of the influences on complex drivetrains. A comparison with measured data should be carried out due to the assumption of the damping ratio made here, and this is planned. Finally, this modelling approach gives the possibility of calculating the resulting excitation forces, e.g., the resulting bearing and electromagnetic forces. These can be inputs for a more detailed acoustic and vibration simulation of a drivetrain.

6. Conclusions

In this work, a transient system model for the analysis of structural dynamic interactions in electric drivetrains with gearboxes was developed. The multi-body simulation (MBS) approach was selected. In order to model the dynamic gear forces, a force element in the MBS was implemented and a transient electric machine model was selected. Within this model, the machine controllers were considered. The structural dynamic model was finally coupled to the electric machine model in the time domain.

Compared to commonly used system approaches for noise and vibration analyses of electric drivetrains, in which the coupling is performed in the frequency domain, the presented methodology allows the consideration and analysis of the interactions of the controllers, the electric machine dynamics, and the non-linear structural dynamic response of gears, which are commonly used in electric drivetrains. This particular modelling approach gives the ability to analyse not only steady-state operating points, but also transient load changes and their impact on the system.

The proposed modelling approach was applied to an exemplary drivetrain. Using this drivetrain, a comparison of a controlled start-up of the system with a common gear constraint was made. It was shown that the model provides plausible results. Furthermore, the calculated transient cogging torque of the synchronous machine used and the resulting output torque of the drivetrain were analysed in a spectrum, and excitation orders were assigned to the modelled effects. Finally, the influence of the load dependence of the gear tooth contact on the vibration parameter of the natural frequency of the system was described.

The proposed model can be used as a starting point for a deeper analysis of system interactions and their influences on machine controllers, as well as for detailed vibration and acoustic analyses. A comparison of the results with measurements is planned.

Author Contributions: Conceptualisation, M.F. and D.B.; methodology, M.F.; software, M.F. and J.P.R.; formal analysis, M.N.; writing—original draft preparation, M.F.; writing—review and editing, J.P.R., D.B., M.N. and K.H.; supervision, D.B. and K.H. All authors have read and agreed to the published version of the manuscript.

Funding: This paper was developed in the context of the cooperative project "KinelectricDrive" and was financially supported by the German Federal Ministry of Economic Affairs and Energy (BMWi), reference number 01MY14006B.

Conflicts of Interest: The funders had no role in the design of the study; in the collection, analyses, or interpretation of data; in the writing of the manuscript, or in the decision to publish the results.

Abbreviations

The following abbreviations are used in this manuscript:

NVH	Noise, Vibration, and Harshness
MBS	Multi-Body Simulation
FEM	Finite Element Method
BEM	Boundary Element Method
SEA	Statistical Energy Analysis
IM	Induction Machine
PMSM	Permanent Magnet Synchronous Machine
eIVT	Electric Infinite Variable Transmission

References

1. Holehouse, R.; Shahaj, A.; Michon, M.; James, B. Integrated approach to NVH analysis in electric vehicle drivetrains. *J. Eng.* **2019**, *2019*, 3842–3847. [CrossRef]
2. Rick, S.; Putri, A.K.; Franck, D.; Hameyer, K. Hybrid Acoustic Model of Electric Vehicles: Force Excitation in Permanent-Magnet Synchronous Machines. *IEEE Trans. Ind. Appl.* **2016**, *52*, 2979–2987. [CrossRef]
3. Humbert, L.; Pellerey, P.; Cristaudo, S. Electromagnetic and Structural Coupled Simulation to Investigate NVH Behavior of an Electrical Automotive Powertrain. *SAE Int. J. Altern. Powertrains* **2012**, *1*, 395–404. [CrossRef]
4. Herold, T.; Franck, D.; Lange, E.; Hameyer, K. Extension of a D-Q Model of a Permanent Magnet Excited Synchronous Machine by Including Saturation, Cross-Coupling and Slotting Effects. In Proceedings of the 2011 IEEE International Electric Machines & Drives Conference (IEMDC), IEMDC 2011, Niagara Falls, ON, Canada, 15–18 May 2011.
5. Klocke, F.; Brecher, C. *Zahnrad-und Getriebetechnik: Auslegung—Herstellung—Untersuchung—Simulation*; Hanser: München, Germany, 2017.
6. Dackermann, T.; Miller, S.; Hedrich, L.; Dölling, R. Flexible Gear Model Library—Vibration Excitation Mechanisms and Gear Force Calculation. In Proceedings of the Modelling, Identification and Control/827: Computational Intelligence, Innsbruck, Austria, 16–17 February 2015. [CrossRef]
7. Andreas, E.; Fastl, H.; Kerber, S.; Hobelsberger, J.; Jebasinski, R.; Klerk, D.D.; Moosmayr, T.; Saemann, E.U. *Handbuch Fahrzeugakustik: Grundlagen, Auslegung, Berechnung, Versuch*; ATZ-MTZ Fachbuch, Vieweg + Teubner: Wiesbaden, Germany, 2012.
8. Bauchau, O.A. Flexible Multibody Dynamics. In *Solid Mechanics and Its Applications*; Springer Science + Business Media B.V.: Dordrecht, The Netherlands, 2011; Volume 176.
9. Gerling, D. *Electrical Machines*; Springer Berlin Heidelberg: Berlin/Heidelberg, Germany, 2015; Volume 4. [CrossRef]
10. Herold, T.; Franck, D.; Böhmer, S.; Schröder, M.; Hameyer, K. Transientes Simulationsmodell für lokale Kraftanregungen elektrischer Antriebe. *e i Elektrotechnik Und Informationstechnik* **2015**, *132*, 46–54. [CrossRef]
11. Butterweck, D.; Hombitzer, M.; Hameyer, K. Analysis and Evaluation of Electric Motor Topologies for a Kinematic-Electric 48 Volt Powertrain. In Proceedings of the 26th Aachen Colloquium Automobile and Engine Technology, Aachen, Germany, 9–11 October 2017.
12. Jurisch, F. *Nutrastmomente in Elektrischen Maschinen: Neue Betrachtungsweise und Maßnahmen zur Gezielten Beeinflussung*; Vacuumschmelze GmbH & Co KG: Hanau, Germany, 2003.

 *energies*

Article

Influence of the Preformed Coil Design on the Thermal Behavior of Electric Traction Machines

Benedikt Groschup *, Florian Pauli and Kay Hameyer

Institute of Electrical Machines (IEM), RWTH Aachen University, Schinkelstraße 4, 52062 Aachen, Germany; florian.pauli@iem.rwth-aachen.de (F.P.); kay.hameyer@iem.rwth-aachen.de (K.H.)
* Correspondence: benedikt.groschup@iem.rwth-aachen.de

Abstract: Preformed coils are used in electrical machines to improve the copper slot fill factor. A higher utilization of the machine can be realized. The improvement is a result of both, low copper losses due to the increased slot fill factor and an improved heat transition out of the slot. In this study, the influence of these two aspects on the operational improvement of the machine is studied. Detailed simulation models allow a separation of the two effects. A preform wound winding in comparison to a round wire winding is studied. Full machine prototypes as well as motorettes of the two designs are built up. Thermal finite element models of the stator slot are developed and parameterized with the help of motorette microsections. The resulting thermal lumped parameter model is enlarged to represent the entire electric machine. Electromagnetic finite element models for loss calculation and the thermal lumped parameter models are parameterized using test bench measurements. The developed models show very good agreement in comparison to the test bench evaluation. The study indicates that both, the improvements in the heat transition path and the advantages of the reduced losses in the slot contribute to the improved operational range in dependency of the studied operational point.

Keywords: electrical machines; thermal modeling; preformed coils; insulation systems; pre-shaped conductor

Citation: Groschup, B.; Pauli, F.; Hameyer, K. Influence of the Preformed Coil Design on the Thermal Behavior of Electric Traction Machines. *Energies* **2021**, *14*, 959. https://doi.org/10.3390/en14040959

Academic Editor: Ryszard Palka

Received: 4 January 2021
Accepted: 31 January 2021
Published: 11 February 2021

Publisher's Note: MDPI stays neutral with regard to jurisdictional claims in published maps and institutional affiliations.

1. Introduction

To further increase the thermal operational limits of high torque low speed electrical machines, high slot fill factors are required. The theoretical maximum copper fill factor using round-wire-windings without insulation is given by the value of an orthocyclic winding with 90.7%. This theoretical value is reduced by the wire insulation and a non-perfect arrangement of the conductors within the notch. Several studies focus to approach high slot fill factors with round wires. Authors in [1] use innovative needle winding and indicate realistic copper fill factors of 65% in the machine. Rectangular conductor designs are a well-known approach to increase the copper slot fill factor in comparison to a round wire design. A state-of-the-art approach is using parallel slot designs with rectangular conductors of the same size in a so called hairpin winding design [2–7]. Only few studies focus on parallel teeth such as [3] where rectangular conductors of the same size are inserted into the notch. Studies with rectangular conductors arranged in a parallel tooth design are more seldom to find [8]. A widely studied field is the loss calculation of ac-copper losses within the winding in two dimensional [5] or three dimensional [4] electromagnetic simulations. Comparison of round wire designs and rectangular wire designs can be found in [2,5]. The authors of both publications focus on the loss comparison of the two designs. Thermal aspects are not studied in detail, even if high slot fill factors cause different thermal resistances within the notch. Investigations focusing on thermal aspects of machines with rectangular wire regularly study different innovative cooling concepts. Direct cooling concepts of the conductors within the notch [7,9] or the end winding [6] can be found. In these studies, authors usually focus on the improved heat extraction of the cooling designs.

Energies **2021**, *14*, 959. https://doi.org/10.3390/en14040959

In order to avoid the disadvantage of parallel slots, semi finished coils are used in this study and compressed to fit the geometry of a parallel tooth stator slot. The production process of such coils is described in [10,11].

The semi finished coil has larger radial dimensions than the final tooth of the machine as depicted in Figure 1a. This coil is compressed in a specially designed fixture as described in [10] to match the final size of the slot as shown in Figure 1b. The coil is then impregnated. The final product before inserting into the stator tooth is shown in Figure 1c.

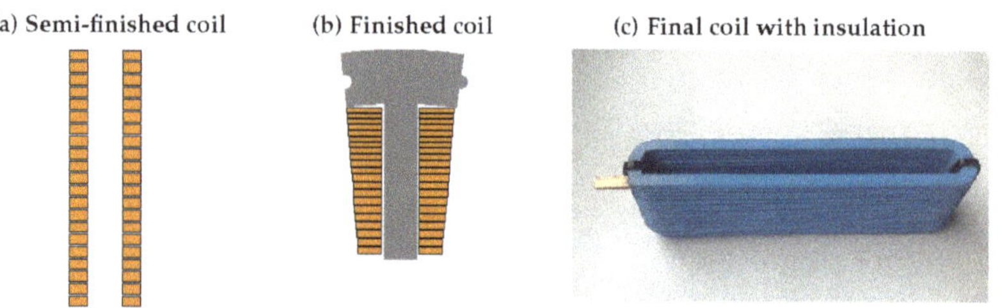

Figure 1. Overview of preformed coil: (**a**) Semi finished coil. (**b**) Finished coil in the stator tooth. (**c**) Images of the final coil with insulation.

Within this study, two different machines are compared to each other. One machine is equipped with classical round wires, the second is equipped with the newly developed preformed coils. In order to have comparable results, the iron core and the entire machine design excluding the coil are kept identical. The machines are designed as directly driven wheel hub drives in an automotive application without using a gearbox. The design requires a relatively low maximum speed of 1200 rpm. The machines are designed with 20 pole pairs, 60 stator slots and a v-shaped permanent magnet arrangement such as depicted in Figure 2.

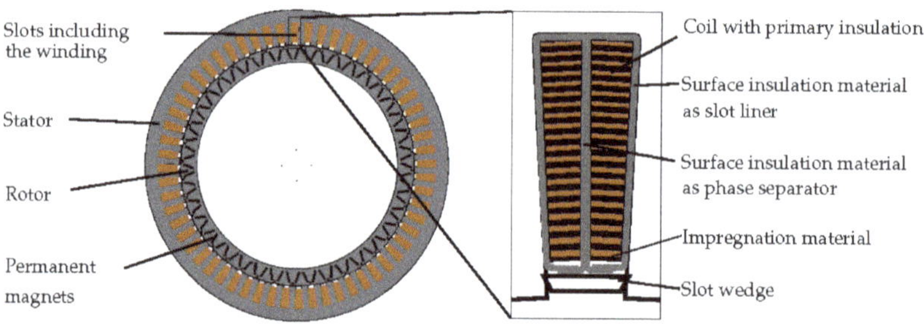

Figure 2. Axial view of the electromagnetic active motor parts and detailed axial view of the slot of the proposed preformed coil design.

A single-tooth winding pattern is used to enable short end windings. The used round wire is the film grade one wire Synwire W210 from SynFlex with a outer diameter of 0.95 mm. The used rectangular wire for the preformed coil has the outer dimensions of 3.4 mm × 1.4 mm with a corner radius of 0.5 mm. The newly developed technology shows influence on three different categories:

- Insulation system,
- thermal behavior, and
- electromagnetic behavior.

A detailed study of the **insulation system** is given in [8]. The insulation system of the round wire winding is designed to hold a temperature of 180 °C (thermal class H) while the interturn insulation of the formed coil in its current setup can hold 120 °C (thermal class E).

An insulation system for the formed coils, which meets the requirements of thermal class H, is however under development.

This study focuses on the enhancement of torque and power density of the drive by analyzing the **thermal behavior**. Microsections of the designs are evaluated to identify the geometrical arrangement of all components within the stator slot, i.e., surface insulation, impregnation, interturn insulation or conductors. The results are used to develop detailed two-dimensional finite element analysis (FEA) models of the stator slot. Lumped parameter thermal network (LPTN) models such as introduced in [12] or [13] are developed for each machine. The FEA models are used to parameterize the thermal resistances of the LPTN models. As all components of the two machines are kept constant between the two machine designs except the winding, only these resistances are adjusted in the LPTN models. The local temperature distribution is analyzed in detail. The LPTN models are validated by thermal transient test bench measurements so that measured temperatures on the test bench, simulated temperatures of the machine LPTN models, and two-dimensional thermal FEA models are coincident in their results. The developed model is used to determine continuous power (S1) for different maximum permitted winding temperatures. The results of this evaluation are validated by test bench measurements of the continuous power (S1).

From a thermal perspective, the **electromagnetic behavior** can be specified by the amount of copper and iron losses that are produced at a specific operational point of the machine, i.e., at defined torque and speed. The approach of substituting the machine winding, which is chosen in the present publication, leaves the iron losses unchanged. The increased copper fill factor leads to a reduction of the copper losses of the machine. From the application of hairpin windings, it is known that large conductor cross sections lead to more pronounced skin- and proximity effects at high speeds [14], which diminishes the advantage of an increased copper fill factor. Electromagnetic finite element simulations and test bench trials are performed to investigate this influence. The efficiency of both machines is simulated and measured for the entire torque-speed range.

2. Simulation Methodology

2.1. Electromagnetic Simulation

The electromagnetic characteristics of both machines are obtained by applying electromagnetic 2D finite element simulations. The torque and the magnetic flux distribution are calcualated in the d-q current reference frame. An exemplary distribution of the magnetic flux density B is given in Figure 3. An exemplary operating point with $\hat{I}_d = -105\,\text{A}$ and $\hat{I}_q = 320\,\text{A}$ is used. The local distribution of the losses within the device under test are obtained by applying this data to the post processing process.

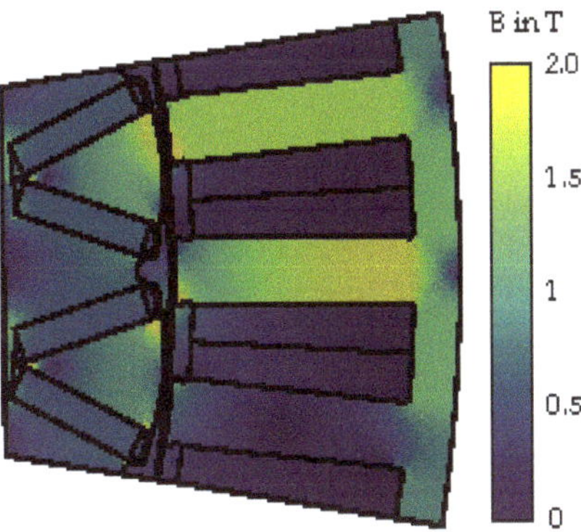

Figure 3. Flux distribution inside the PMSM at an examplary operating point $\hat{I}_d = -105\,\text{A}$ and $\hat{I}_q = 320\,\text{A}$.

The torque-speed map is obtained by applying a maximum torque per ampere (MTPA) control strategy in the base speed region. The copper loss calculation is initially introduced in [8]. Skin and proximity effects are not considered for the machine with conventional round wires due to the small conductor cross section. The copper losses are calculated according to:

$$P_{\text{wind,round}} = \frac{3}{2} \cdot R_{1-1,\text{DC,m,round}} \cdot I^2 ,$$ (1)

where I is the amplitude of the phase current and $R_{1-1,\text{DC,m,round}}$ is the line-to-line resistance. For the preformed coil, skin and proximmity effects are taken into consideration. Therefore, a transient 2D simulation is performed and the ac resistance $R_{1-1,\text{AC,s,pref}}(T,n)$ is obtained as a function of torque and speed. As this calculated resistance only considers the part of the coil, which is placed inside the slots, a factor based on the measured dc resistance $R_{1-1,\text{DC,m,pref}}$ and the simulated dc resistance $R_{1-1,\text{DC,s,pref}}$ is introduced:

$$f_R = \frac{R_{1-1,\text{DC,m,pref}}}{R_{1-1,\text{DC,s,pref}}} .$$ (2)

The copper losses as a function of torque and speed are calculated by multiplying this factor by the simulated ac resistance:

$$P_{\text{wind,pref,s}} = \frac{3}{2} f_R \cdot R_{1-1,\text{AC,s}}(T,n) \cdot I^2 .$$ (3)

Thus, the resistance of the end winding can also be considered. The specific iron losses p_{Fe} are calculated according to the IEM-five-parameter-formula which is introduced in [15–17]:

$$p_{\text{Fe}} = a_1 B^{\alpha} f + a_2 B^2 f^2 (1 + a_3 B^{a_4}) + a_5 B^{1.5} f^{1.5} ,$$ (4)

where B is the magnetic flux density, f is the magnetization frequency. α as well as a_1 are material dependent parameters that define hysteresis losses. The eddy current losses are defined by a_2, excess losses by a_4. The remaining parameters a_3, a_5 specify the amount of non-linear losses in the application. The iron losses P_{Fe} of the drive can be derived by an integration of the specific iron losses p_{Fe} over the mass dm:

$$P_{\text{Fe}} = \int_m p_{\text{Fe}}\, dm .$$ (5)

The loss data to parametrize Equation (4) is taken from the testbench measurements. As only the total losses $P_{\text{loss,tot,m}}$ can be measured, the simulated copper losses $P_{\text{wind,s}}$ are substracted to obtain the ironlosses P_{Fe}:

$$P_{\text{Fe}} = P_{\text{tot,m}} - P_{\text{wind,s}} .$$ (6)

Friction losses are not considered. As both machines are identical except from the winding, the same set of parameters is used to calculate the iron losses. An overview of the resulting losses for each studied operational point (OP) of this parameterization process can be found in Table 1.

Table 1. Overview of losses for studied operational points.

OP	T	n	$P_{Fe,sta,to}$	$P_{Fe,sta,yo}$	$P_{Fe,rot,emp}$	$P_{Fe,rot,yo}$	$P_{wind,round}$	$P_{wind,pref}$
	in Nm	in rpm	in W	in W	in W	in W	in W	in W
1	650	50	32	14	9	2	2699	2277
2	0	800	507	217	155	39	0	0
3	400	800	515	222	152	38	905	894
4	300	1200	461	198	205	51	1275	1440

OP1 is selected at low speed n and high torque T. Due to the low magnetization frequency f in the soft magnetic material, OP1 shows low iron losses in the stator tooth $P_{Fe,sta,to}$, the stator yoke $P_{Fe,sta,yo}$, the rotor embedded magnet pole $P_{Fe,rot,emp}$, and the rotor yoke $P_{Fe,rot,yo}$. The high torque demand in the base speed region leads to high currents in the winding and a high share of copper losses. The round wire design with $P_{wind,round} = 2699\,\mathrm{W}$ shows increased copper losses by 18.5% in comparison to the preformed coil design with $P_{wind,round} = 2277\,\mathrm{W}$. OP2 and OP3 are operational points, both placed at a rotational speed n of 800 rpm. OP2 has no load, is placed in the base speed region, and thus has no copper losses P_{wind}. OP4 is the operational point with the highest studied speed of $n = 1200\,\mathrm{rpm}$. It is interesting to note that the round wire design shows significantly higher winding losses for OP1 (+18.5%), but lower winding losses for OP4 (-11.5%) due to the influence of proximity and skin effects. This difference needs to be kept in mind for the later comparison of the temperature curves. It is interesting to see weather this disadvantage of the preformed coil design is outperformed by the expected improved heat path in the later study.

2.2. Thermal Lumped Parameter Model

For the representation of the thermal behavior of both machines, LPTN models are developed. An assignment of the most important thermal nodes within the network model to the existing geometry is give in Figure 4 using the preformed coil design. The figure includes a longitudinal and a transverse section of the motor as shown in Figure 4a,b respectively.

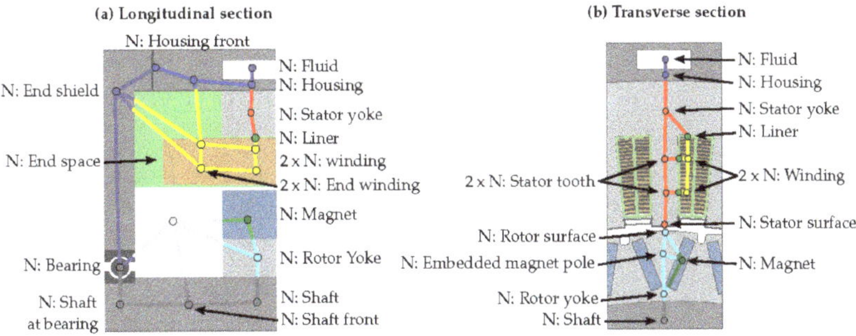

Figure 4. Location of the nodes of the LPTN model: (**a**) Longitudinal section of the geometry. (**b**) Transverse section of the geometry.

The machine is equipped with a water based cooling channel within the housing of the machine. The fluid has a volume flow rate of $7.5\,\mathrm{L\,min^{-1}}$ and a heat transfer coefficient of $h_{ho-fluid} = 4011\,\mathrm{W\,m^{-2}\,K^{-1}}$. The thermal resistance between the housing and the fluid $R_{fluid,ho}$ is calculated to be $1.2\,\mathrm{mW\,K^{-1}}$. A more detailed description of the LPTN model with the most important nodes and resistances being labeled, is shown in Figure 5.

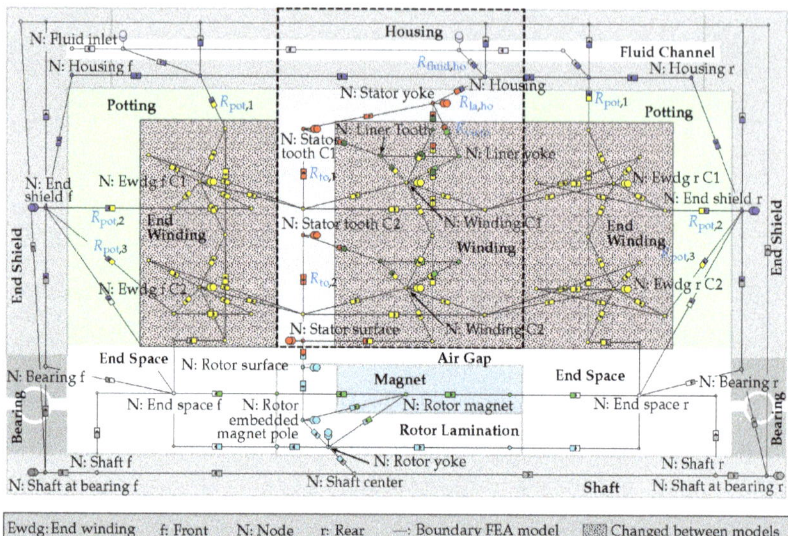

Figure 5. Detailed overview of entiere LPTN model.

Different areas are highlighted in Figure 5. The middle section represents the electromagnetic active length and is the portion that is later modeled in the two-dimensional finite element model. The dotted area marks the thermal resistances and capacities that are changed between the round wire and the preformed coil model. All other parameters of the model are kept constant. The end winding of both machines is potted using a mixture of a mineral-filled epoxy resin Araldit CW 5730 N and a hardener HY 5731 by Bodo Möller Chemie GmbH. Due to the potting, the end winding nodes are directly linked to the end shield and the housing front node of the model. The model is designed symmetrically on both sides of the end winding. The radial connection resistance $R_{pot,1}$ is calculated to be $690\,\mathrm{mK\,W^{-1}}$. The two axial connection resistances $R_{pot,2}$ and $R_{pot,3}$ are calculated to be $1160\,\mathrm{mK\,W^{-1}}$. The interface resistance between the housing and the stator lamination $R_{la,ho}$ shows significant influence on the temperature distribution within the machine. Conductive resistances are calculated using the thickness of the conductive layer d, the area of the conductive layer A and its thermal conductivity k

$$R = \frac{d}{A \cdot k}.\tag{7}$$

The resistance $R_{la,ho}$ is a sum of the conductive share of the housing $R_{la,ho,1}$, the interface $R_{la,ho,2}$, and the conductive share of the lamination $R_{la,ho,3}$. The second resistance with $R_{la,ho,2} = 3.3\,\mathrm{mK\,W^{-1}}$ has the largest share of $R_{la,ho} = 4.7\,\mathrm{mK\,W^{-1}}$ and is frequently discussed in literature. Thus it is described here in detail. The resistance $R_{la,ho,2}$ is influenced by several parameters, such as the surface roughness, the contact pressure between the components, or the material hardness [18]. It can be modeled by an equivalent interface air gap d_{ig} between the components:

$$R_{la,ho} = \frac{d_{ig}}{A_{ig} \cdot k_{air}}.\tag{8}$$

The thermal conductivity of air is assumed to be $k_{air} = 31.7\,\mathrm{mW\,m^{-1}\,K^{-1}}$ and the area of the interface $A_{ig} = 0.095\,\mathrm{m^2}$. Typical literature values for the interface thickness d_{ig} are between $10\,\mu\mathrm{m}$ and $77\,\mu\mathrm{m}$ [18–20] with one outlier in [20] of $95\,\mu\mathrm{m}$. This range is used for the later described parameter identification process using test bench measurements.

The final value is identified to be 10 μm and thus placed at the lower level of identification range. Some other important resistances in the LPTN are the conductive resistances within the stator lamination. For the stator and rotor lamination, a thermal conductivity of $28\,\mathrm{W\,m^{-1}\,K^{-1}} \pm 10\%$ is used for the parameterization. The final value is identified to be $30\,\mathrm{W\,m^{-1}\,K^{-1}}$ leading to a thermal resistance between the stator yoke and the stator tooth of $R_{yo,to} = 7.3\,\mathrm{mK\,W^{-1}}$ for the outer part of the tooth of $R_{to,1} = 12\,\mathrm{mK\,W^{-1}}$ and the inner part of the tooth of $R_{to,2} = 12\,\mathrm{mK\,W^{-1}}$. The winding and the stator tooth of the model is represented by two radially placed cuiboidal elements as introduced in [21]. The two radially aligned cuboidal elements of the active length are labeled with C1 for the radially outside and C2 for the radially inside element. The two cuboidal elements in the active section are supplemented by two elements for each, the front and the rear section of the motor, i.e., N Ewdg f and N Ewdg r.

The losses according to Table 1 are used for the model. The iron losses of the stator tooth $P_{Fe,sta,to}$ are inserted into the stator tooth nodes labeled with C1 and C2. $P_{Fe,sta,yo}$ is applied to stator yoke node. The iron losses $P_{Fe,rot,emp}$ and $P_{Fe,rot,yo}$ are applied to the rotor embedded magnet pole node and the rotor yoke respectively. The copper losses $P_{wind,round}$ and $P_{wind,pref}$ are distributed between the end winding section and the winding section based on the estimated length of the coil

$$P_{wind,i} = P_{wind} \cdot \frac{l_i}{l_i + l_{ewdg}}, \tag{9}$$

with the active length $l_i = 80\,\mathrm{mm}$ and the length of one end winding turn $l_{ewdg} = 34\,\mathrm{mm}$.

2.3. Thermal Finite Element Analysis

In order to achieve a well suited network model for the winding section, thermal two-dimensional finite element analysis (FEA) of the slot are performed. The adiabatic boundary of the FEA model is highlighted in the LPTN model description in Figure 5. The LPTN is run with cut resistances along this boundary for comparison and parameterization. The determination of the exact positioning of the components within the slot as well as the penetration of the impregnation material into the cavities is crucial for an exact representation of the slot. Partial stator models that are referred to as motorettes are build up for this purpose. They are first used for the accelerated aging of the winding as described in [8]. After this process, microsections are processed as depicted in Figure 6. A Leica S9i microscope is used for the measurement of the geometrical arrangement of the components. Four images of different locations of the slot of the preformed coil are shown in Figure 6. The results are used for the development of the FEA model as depcited in Figure 7.

Figure 6. Microsection of a partial motor model of the preformed coil design.

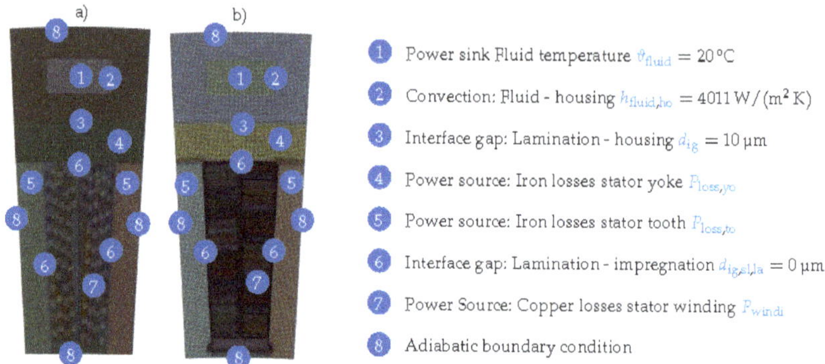

Figure 7. Comparison of mesh settings and boundary conditions for the thermal FEA models. Round wire design (**a**) and preformed coil design (**b**).

No major air pockets can be found in the potting in Figure 6 and thus, no airgap between the lamination stack and the slot liner is modeled (see $d_{\text{ig,sl,la}} = 0\,\mu\text{m}$ in Figure 7). Four layers of the phase-to-ground insulation material Nomex Kapton Nomex (NKN) can be found between the two coils in image 1 of Figure 6, i.e., two overlying layers for each coil. The well visible contour is the capton layer. The NKN has a datasheet thickness of $d_{\text{sl}} = 170\,\mu\text{m}$. This thickness can be confirmed in section 4 of Figure 6. With a measured thickness in section 1 of Figure 6 for four layers of 683 µm, it is obvious that no additional layer of impregnation needs to be inserted in the FEA model between the four layers in the coil divider area. The slot liner is modeled with isotropic conductivity. The value is varied by $\pm20\%$ between the value of $0.15\,\text{W}\,\text{m}^{-1}\,\text{K}^{-1}$ and is identified to be $0.18\,\text{W}\,\text{m}^{-1}\,\text{K}^{-1}$ during the parameterization process using test bench measurements as described later in this study. The wire insulation thickness varies significantly in dependency of the location. The maximum measured value is 119 µm as found in section 1 of Figure 6. The minimum value is highlighted by a circle in section 4. No significant remaining wire insulation can be found at this position. The sharp edge is a parasitic effect of the forming process. This effect can lead to a local error-prone wire insulation in combination with the used spray coating process. Values for two layers of wire insulation between the conductors of one coil between 90 µm and 120 µm are measured (see Section 4). This indicates a thickness between 45 µm and 60 µm for the wire insulation thickness. The cross sectional area of one conductor is known with $4.69\,\text{mm}^2$. The wire insulation thickness with given cross sectional area of the conductor and given slot shape influences the position of the last conductor as measured in section 3 of Figure 6. In order to meet measured distances, an uniformly distributed wire insulation thickness of 52.5 µm is selected for the FEA model.

A heat sink representing the cooling fluid is added to the model with $\vartheta_{\text{fluid}} = 20\,°\text{C}$. The heat transfer coefficient of $h_{\text{fluid,ho}} = 4011\,\text{W}\,\text{m}^{-2}\,\text{K}^{-1}$ and the interface gap between the stator lamination and the housing with $d_{\text{ig}} = 10\,\mu\text{m}$ is selected in accordance to the values of the LPTN. The model is surrounded by adiabatic boundary conditions.

3. Measurement Setup

Test bench measurements are performed for the validation of the study as introduced in [8] and depicted in Figure 8.

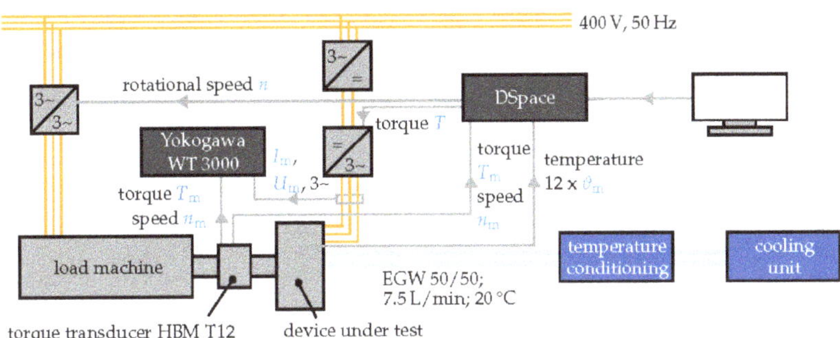

Figure 8. Schematic overview of the test bench setup.

The control system is designed in Matlab-Simulink using d-q lookup tables. The control system is flashed on a dSPace DS1006 system. A dc-link voltage of 300 V for the device under test is used. The tested machine is conditioned with 7.5 L min^{-1} and 20 °C. The two machines are mechanically connected without a gearbox. A HBM T12 torque transducer is placed in the connecting shaft measuring the torque T_m and the rotational speed n_m of the setup. The currents I_m are measured by LEM sensors. The voltages U_m are measured by a voltage transducer. The measured values are transmitted to a Yokogawa-WT-3000 power analyzer for efficiency evaluation. Temperatures are measured by twelve thermocouples type J per motor. The locations of the sensors are shown in Figure 9. Two sensors for each phase are placed in the end winding region (A: Ewdg). The temperature between the housing and the stator lamination is measured by two sensors, placed in a grove of the stator teeth (B: Lam-Hous). The slot center of the machine is equipped with two sensors (C: Slot Center). The last two sensors are placed in the potting region between the end winding and the housing (D: Potting).

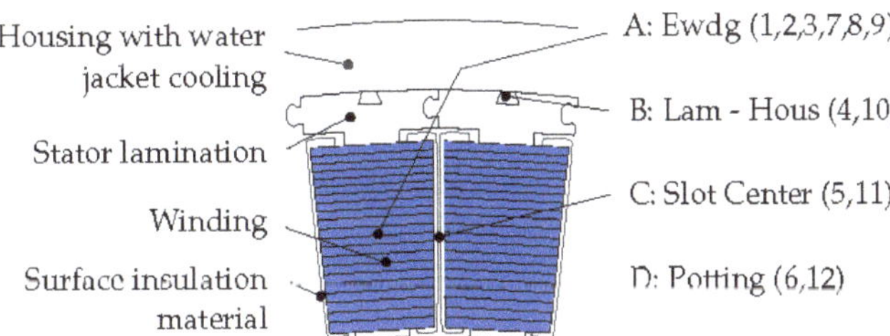

Figure 9. Positioning of the thermocouples type J within the machine.

4. Results

4.1. Comparison of Simulated and Measured Efficiency and Losses

A detailed description of the measurement preparation including results of induced voltage measurements as well as drag torque measurements can be found in [8]. This study also includes a detailed comparison of measured and simulated efficiency. As the losses in the machine are crucial for the thermal modeling, a short review is given. The efficiency of the machines is evaluated using grid points which are placed at steps of 50 N m and 50 rpm. A comparison of the efficiency between measured and simulated losses is given in [8]. The measured and simulated efficiencies fit well in the entire torque speed map. The measured maximum efficiency of the round wire design with 95.1% is similar to the measured efficiency of the preformed coil design with 95.0%. The increased copper fill factor leads to reduced losses of the preformed coil design in comparison to the round design in the base speed region. Similar behavior was already observed for the simulated losses as shown in Table 1. For high magnetization frequencies f, the advantage of the preformed coil is outperformed due to the increased skin and proximity effects occuring in the winding.

4.2. Influence of Winding Design on Measured Transient Temperature Curves

The parameterization of the models is performed in an iterative process. The simulated losses are used for the parameterization. The thermal 2D-FEA model is adjusted by the parameters that are already introduced in the pervious sections. The maximum and the average temperature ($\vartheta_{\mathrm{wind,max}}$ and $\vartheta_{\mathrm{wind,avg}}$) of OP1 are calculated and the LPTN model is calibrated to fit these temperatures. The maximum temperature is chosen, because it is crucial for the lifetime and the operational range. The average temperature is chosen, as copper losses of the LPTN model P_{wind} are scaled with average temperature $\vartheta_{\mathrm{wind,avg}}$. OP1 is selected, as the highest winding temperature is expected in this operational point. After calibrating the FEA with the simplified two-dimensional LTPN, the LPTN is transformed to represent the three-dimensional machine. Transient simulations of this model are performed and compared to the test bench measurements. The parameters are changed until measurements, 3D-LPTN as well as 2D-LPTN and 2D-FEA coincide. A comparison of the steady state temperatures for the four operational points between 2D-FEA and 2D-LPTN is given in Table 2. Due to the usage of operational point OP1 for the parameterization, this point fits with almost no deviation. The other operational points all fit well with a deviation of less than 5.5 °C. It is interesting to note that the LPTN model seems to over-predict the temperatures for the operational points with increasing speed and a larger share of iron losses. As the cuboidal model of the winding in the LPTN is a low-discretizised model in comparison to the FEA some deviations seem to be inavoidable.

Table 2. Comparison of Steady State Temperatures of Two-Dimensional Thermal LPTN and FEA in °C.

	Maximum Temperature ϑ_{max}				Average Temperature ϑ_{avg}			
	OP1	OP2	OP3	OP4	OP1	OP2	OP3	OP4
round wire design: $\vartheta_{\mathrm{LPTN-2D}} - \vartheta_{\mathrm{FEA-2D}}$	0.4	3.1	3.5	5.4	0.3	3.1	3.1	4.5
preformed coil design: $\vartheta_{\mathrm{LPTN-2D}} - \vartheta_{\mathrm{FEA-2D}}$	−0.4	3.1	3.3	4.9	0.0	3.3	3.1	4.4

A comparison of the transient temperature curves of the two machines for all operational points as introduced in Table 1 is given in Figure 10.

A very good fit of 3D-LPTN-simulated and the measured temperatures is achieved. A deviation of not more than 6 K is observed. The previously identified trend regarding the exact representation of the winding temperature of the LPTN model can be confirmed. The LPTN model seems to overpredict temperatures at OP4 and tends to under-predict temperatures at OP1. As an example, the end winding temperature in OP1 for the round machine is underpredicted by the LPTN model by 5.2 K, whereas the end winding temperature of the preformed coil design in OP4 is overpredicted by 4.9 K. The fit of measured and simulated temperatures was possible under the following boundary conditions:

- Only the input of the copper losses P_{wind} is changed between the models of the round wire and the preformed coil as shown in Table 1. All other losses remain constant.
- Only the resistances labeled within the dotted area in Figure 5 are changed between the models.
- The resistances labeled within the dotted area in Figure 5 are changed so that the temperatures between 2D-FEA and 2D-LPTN coincide according to Table 2. The material properties of the impregnation, slot liner and the slot shape are the same in the 2D-FEA and the 2D LPTN.

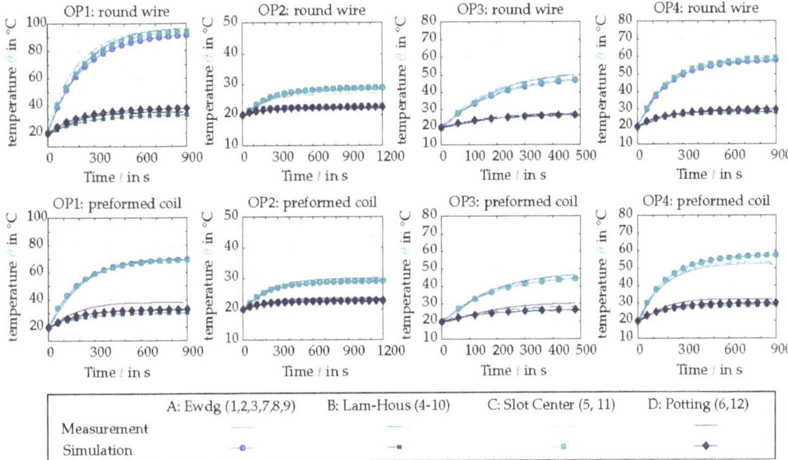

Figure 10. Comparison between LPTN simulation models and test bench measurement data.

It is interesting to compare the temperature curves for the slot center and the end windings for the different operational points. Due to the short end winding, temperatures in the electromagnetic active part of the machine are very similar to the curves of the end winding for each operational point. In OP1, the maximum winding temperature of the round design of around 95 °C is approximately 25 °C higher than the winding temperature of the preformed coil with around 70 °C. The copper losses of the round wire machine are 18.5% higher than those of the preformed coil design. The maximum winding temperature for OP4 is also lower for the preformed coil design by some °C, even if the input losses of the round design are 11.5% lower than those of the preformed coil design. The advantages of the improved heat dissipation paths of the preformed coil design outperform the disadvantage of the increased losses. Following observations can be concluded:

1. The 2D-LPTN and the 2D-FEA show good agreement with a maximum temperature deviation of 5.4 K.
2. The 3D-LPTN and the transient test bench measurements show a very good agreement with a maximum deviation of 6 K.
3. A parameterization of all models was possible under consideration of several boundary conditions that ensure a high degree of comparability.
4. The maximum winding temperature of the preformed coil design at the transient operation of OP1 for 15 min is approximately 25 °C lower than the winding temperature of the round wire design.
5. The disadvantage of the increased winding losses of the preformed coil design at an operational point with maximum speed (OP4: 15 min) is outperformed by the improved heat transfer path.

4.3. Comparison of the Influence of Reduced Losses and Improved Heat Extraction Path

Some more detailed investigations are performed to separate the two different influences on the maximum temperature of the winding, i.e., the reduction of the losses by the preformed coil design in the base speed region and the improved heat transfer path. The boundary conditions of OP1 are used for this investigation. Three different temperature distributions within the slot are depicted in Figure 11.

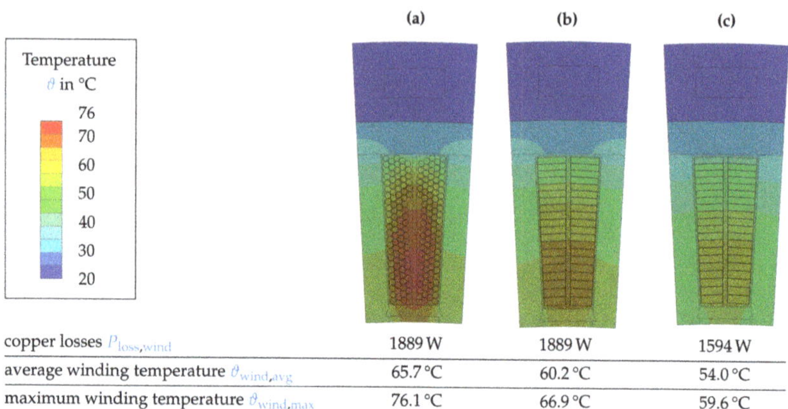

	(a)	(b)	(c)
copper losses $P_{\text{loss,wind}}$	1889 W	1889 W	1594 W
average winding temperature $\vartheta_{\text{wind,avg}}$	65.7 °C	60.2 °C	54.0 °C
maximum winding temperature $\vartheta_{\text{wind,max}}$	76.1 °C	66.9 °C	59.6 °C

Figure 11. Comparison of temperature distribution in the slot using two-dimensional thermal FEA: (a) Round wire design and high copper losses. (b) Preformed coil design with high copper losses. (c) Preformed coil design with low copper losses.

The first setup (a) uses the geometry and the winding losses of the round design according to Table 1 with the conversion of Equation (9). The second setup (b) is an imaginary setup, where the geometry is changed to the settings of the preformed coil, but the high losses of the round wire are used. The third setup (c) is the real design of the preformed coil with the losses of the preformed coil. The average temperature ϑ_{avg} as well as the maximum temperature in the slot ϑ_{max} are evaluated for the investigation. Two effects can be determined:

1. Influence on the average temperature ϑ_{avg} in the slot: The improved heat path (from a to b) shows a reduction of 5.5 K and the reduced losses (from b to c) of 6.2 K. The reduced losses have a slightly higher potential of improvement.
2. Influence on the maximum temperature ϑ_{max} in the slot: The improved heat transfer path (from a to b) reduces the maximum temperature by 9.2 K, the reduced losses (from b to c) by 7.3 K. The improved heat path has a more significant influence on the maximum temperature.

The second described influence is a result of the good thermal heat dissipation from the center of the slot to the side. While setup (a) shows iso-temperature lines with an elliptical shape within the slot, (b) and (c) only show horizontal iso-temperature lines in the slot. The preformed coil design enables a more uniform temperature distribution within the slot.

4.4. Influence of Winding Design on the Operational Limits of the Machine

Simulations as well as test bench measurements are performed to investigate the influence of the winding design on the continuous operational limit S1. In the simulation approach, the validated LPTN models are used. The losses from the electromagnetic simulated torque-speed map are used as heat sources in the LPTN model. Steady state simulations are performed in a grid of 50 N m torque and 40 rpm rotational speed steps. The maximum winding temperature $\vartheta_{\text{S1,s}}$ of the steady state simulation is evaluated. The resulting maximum temperatures are plotted in Figure 12. For the measurement approach, the rotational speed beginning from 100 rpm is investigated in steps of 100 rpm. Each rotational speed is set by the load machine on the test bench. The torque demand for the device under test is increased until a maximum steady state temperature of the hottest sensor in the winding of 120 °C is reached. The operational point is held for a time of $t = 30$ min. Within this period of time, a maximum deviation of the measured temperatures of ±2 K is permitted. The measurement curve is added to the overview in Figure 12.

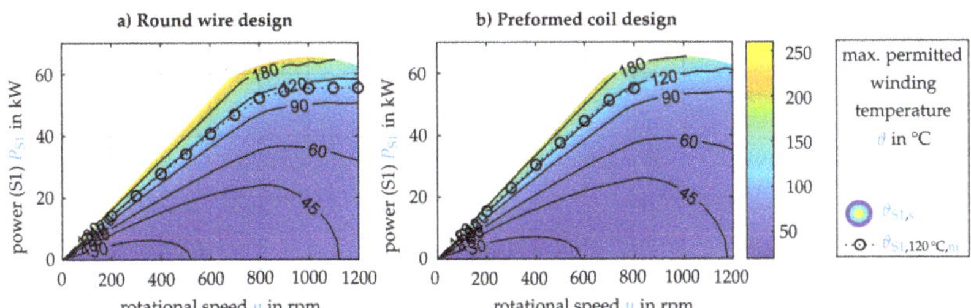

Figure 12. Comparison of continuous operational power (S1) between simulation and measurement: (a) Round wire design (b), Preformed coil design.

The process is repeated for rotational speeds up to 1200 rpm for the round wire design and up to 800 rpm for the preformed coil design. A heat-up of the winding up to 120 °C for speeds larger than 800 rpm is not possible for the preformed coil design with the given lookup tables and restrictions of the test bench setup. The curves of the possible S1-power P_{S1} for a maximum permitted temperature of 120 °C show a good agreement between measurement and simulation. Following observations can be made:

1. For the round wire design, the simulated S1-power $P_{S1,120°C,round,s} = 53.5$ kW is 2.7% larger than the measured S1-power of $P_{S1,120°C,round,m} = 52.1$ kW at $n = 800$ rpm.
2. For the preformed coil design, the simulated S1-power of $P_{S1,120°C,pref,s} = 55.9$ kW is 1.6% larger than the measured S1-power of $P_{S1,120°C,pref,m} = 55.0$ kW at $n = 800$ rpm.
3. According to the measurement, the S1-power $P_{S1,120°C,m}$ of the preformed coil design at $n = 800$ rpm is increased by 4.5% in comparison to the round wire design.
4. According to the simulation, the S1-power $P_{S1,120°C,s}$ of the preformed coil design at $n = 800$ rpm is increased by 5.3% in comparison to the round wire design.
5. Be aware that the given results of the S1-power are calculated based on the temperature limitations caused by the stator winding. Restrictions for the temperature of the permanent magnets in the rotor are likely to occure for high rotational speed [12].

A more significant deviation between measurement and simulation can be observed at maximum speed for the round wire design. At 1200 rpm, the simulation gives a 5.3% higher S1-power than the measurement. One reason for this deviation is the difference between the magnet temperatures between measurement and simulation. The torque of the device under test in the measurement is not controlled in a closed-loop, i.e., the direct and quadrilateral current from the lookup table is controlled in a closed loop. The achievable S1 power is determined at the end of the 30 min time frame. During measurement, the magnet temperature increases. The remanence flux density of the rear earth magnets decrease with increasing temperature. The decreasing rotor flux linkage leads to a decreased permanent magnet torque as shown in [22]. In the simulation, this influence is not considered, because the electromagnetic simulation is performed at room temperature. A more detailed analysis is not possible with the given hardware setup. In order to get a fully validated model, including the permanent magnet temperatures in the rotor, a measurement incorporating a rotor telemetry would be necessary such as performed in [23]. Such investigations could be content of future work.

5. Discussion

A detailed comparison of the conventional round wire design and the innovative preformed coil design is performed in this study. The used losses for the thermal modeling are validated in previous studies [8]. A thermal 2D-FEA model is developed to parameterizte a 2D-LPTN model. The 2-D FEA model is parameterized using microsections of a partial stator prototype. The 2D-LPTN and the 2D-FEA show good agreement with a maximum temperature deviation of 5.4 K. The 2D-LPTN model is extended to a 3D-LPTN model to represent the entire machine. Transient test bench measurements of four operational points are performed for each machine design. The 3D-LPTN and the transient test bench measurements show a very good agreement with a maximum deviation of 6 K. The parameterization of all models is possible under consideration of several boundary conditions that ensure a high degree of comparability. The maximum winding temperature of the preformed coil design at transient operation (OP1: 15 min) is approximately 25 °C lower than the winding temperature of the round wire design. The disadvantage of the increased winding losses of the preformed coil design at a operational point with maximum speed (OP4: 15 min) is outperformed by the improved heat transfer path. A detailed analysis of the local temperature distribution in the slot is performed for steady state temperature distribution of OP1 at 650 N m and 50 rpm. The influence of the loss reduction and the improvement of the heat extraction path on the winding temperature distribution is performed. The reduction of losses has a slightly higher potential of improving the average temperature (6.2 K) in the slot than the improvement of the heat path (5.5 K). The improved heat path has a more significant influence on the maximum temperature (9.2 K) within the slot than the reduction of losses (7.3 K). The preformed coil design leads to a more uniform temperature distribution within the slot. Both effects significantly contribute to the improved performance of the preformed coil design in comparison to the round wire design. The relevance of the two factors depends on the studied operational point. The maximum continuous power of the preformed coil design and the round wire design is compared. Simulations as well as test bench measurements are performed. The improvement of the S1-power is estimated by the simulation model by 4.5% and by the measurements by 5.3% at 800 rpm and a maximum allowed winding temperature of 120 °C. Measurement and simulation show a very good agreement up to 800 rpm by a maximum deviation of 2.7%. For higher speeds, a maximum deviation of up to 4.5% between measurement and simulation of the round wire design is observed. A reason for this difference is the decreasing remanence flux density of the magnets for increasing magnet temperatures.

An investigation of the designs including rotor telemetry measurements could be content of a future study. The interturn insulation of the formed coil in its current setup can only withstand 120 °C (thermal class E). The analysis of the microsections additionally shows some error-prone locations, where no significant insulation thickness can be found. An insulation system for the formed coils, which meets the requirements of thermal class H, is under development.

6. Conclusions

The study shows that both aspects of the newly developed preformed coil design contribute to gain an improved performance in comparison to the round wire design in dependency of the operational conditions. The reduction of losses, especially in the base speed region of the machine, shows high potential of improvement. The improved heat removal path shows high potential for operational points with high copper losses. The thermal studies demonstrate that the improved heat dissipation path especially contributes to an improved heat removal ability from the middle of the notch to the side, i.e., to the stator teeth not to the stator yoke. This new heat path causes new inter-dependencies that should be taken into account during the design process of machines with preformed coil designs.

Author Contributions: Conceptualization, B.G., F.P. and K.H.; methodology, B.G. and F.P.; software, B.G. and F.P.; validation, B.G. and F.P.; formal analysis, B.G. and F.P.; writing—original draft preparation, B.G.; writing—review and editing, B.G., F.P. and K.H.; visualization, B.G.; supervision, K.H.; project administration, K.H.; funding acquisition, K.H. All authors have read and agreed to the published version of the manuscript.

Funding: This research and development project is funded by the German Federal Ministry of Education and Research (BMBF) within the Framework Concept "Serienflexible Technologien für elektrische Antriebe von Fahrzeugen 2". The authors are grateful to the BMBF for the financing of this collaboration and are responsible for the contents of this publication.

Data Availability Statement: Data sharing not applicable.

Acknowledgments: We want to thank our partners of the project "FlexiCoil" for the excellent collaboration concerning this project, i.e., Schaeffler AG, Breuckmann GmbH and the Institute of Metal Forming (IBF) of RWTH Aachen University.

Conflicts of Interest: The authors declare no conflict of interest. The funders had no role in the design of the study; in the collection, analyses, or interpretation of data; in the writing of the manuscript, or in the decision to publish the results.

Abbreviations

The following abbreviations are used in this manuscript:

ac	alternating current
dc	direct current
FEA	Finite Element Analysis
LPTN	Lumped Parameter Thermal Network Model
MTPA	Maximum Torque per Ampere
OP	Operational Point

References

1. Stenzel, P.; Dollinger, P.; Richnow, J.; Franke, J. Innovative needle winding method using curved wire guide in order to significantly increase the copper fill factor. In Proceedings of the 17th International Conference on Electrical Machines and Systems (ICEMS), Hangzhou, China, 22–25 October 2014; pp. 3047–3053.
2. Zhao, Y.; Li, D.; Pei, T.; Qu, R. Overview of the rectangular wire windings AC electrical machine. *CES Trans. Electr. Mach. Syst.* **2019**, *3*, 160–169. [CrossRef]
3. Bianchini, C.; Vogni, M.; Torreggiani, A.; Nuzzo, S.; Barater, D.; Franceschini, G. Slot Design Optimization for Copper Losses Reduction in Electric Machines for High Speed Applications. *Appl. Sci.* **2020**, *10*, 7425. [CrossRef]
4. Aoyama, M.; Deng, J. Visualization and quantitative evaluation of eddy current loss in bar-wound type permanent magnet synchronous motor for mild-hybrid vehicles. *CES Trans. Electr. Mach. Syst.* **2019**, *3*, 269–278. [CrossRef]
5. Berardi, G.; Nategh, S.; Bianchi, N.; Thioliere, Y. A Comparison between Random and Hairpin Winding in E-mobility Applications. In Proceedings of the 46th Annual Conference of the IEEE Industrial Electronics Society (IECON), Singapore, 18–21 October 2020; pp. 815–820.
6. Liu, C.; Xu, Z.; Gerada, D.; Li, J.; Gerada, C.; Chong, Y.C.; Popescu, M.; Goss, J.; Staton, D.; Zhang, H. Experimental Investigation on Oil Spray Cooling with Hairpin Windings. *IEEE Trans. Ind. Electron.* **2020**, *67*, 7343–7353. [CrossRef]
7. Venturini, G.; Volpe, G.; Villani, M.; Popescu, M. Investigation of Cooling Solutions for Hairpin Winding in Traction Application. In Proceedings of the International Conference on Electrical Machines (ICEM), Gothenburg, Sweden, 23–26 August 2020; pp. 1573–1578.
8. Pauli, F.; Groschup, B.; Schröder, M.; Hameyer, K. High Torque Density Low Voltage Traction Drives with Preformed Coils: Evaluation of Operation Limitations. In Proceedings of the 10th International Electric Drives Production Conference (EDPC), Ludwigsburg, Germany, 8–9 December 2020; pp. 1–7.
9. Reinap, A.; Andersson, M.; Marquez-Fernandez, F.J.; Abrahamsson, P.; Alakula, M. Performance Estimation of a Traction Machine with Direct Cooled Hairpin Winding. In Proceedings of the IEEE Transportation Electrification Conference and Expo (ITEC), Detroit, MI, USA, 19–21 June 2019; pp. 1–6.
10. Petrell, D.; Teller, M.; Hirt, G.; Börzel, S.; Schäfer, W. Manufacturing of Conically Shaped Concentrated Windings for Wheel Hub Engines by a Multi-Stage Upsetting Process. In Proceedings of the 2019 9th International Electric Drives Production Conference (EDPC), Esslingen, Germany, 3–4 December 2019; pp. 1–7.
11. Petrell, D.; Teller, M.; Hirt, G.; Börzel, S.; Schäfer, W. Economical Production of Conically Shaped Concentrated Windings Using Forming Technology in Wheel Hub Drives. In Proceedings of the 10th International Electric Drives Production Conference (EDPC), Ludwigsburg, Germany, 8–9 December 2020; pp. 1–7.
12. Groschup, B.; Komissarov, M.; Stevic, S.; Hameyer, K. Operation Enhancement of Permanent Magnet Excited Motors with Advanced Rotor Cooling System. In Proceedings of the IEEE Transportation Electrification Conference and Expo (ITEC), Detroit, MI, USA, 19–21 June 2019; pp. 1–6.
13. Boglietti, A.; Cavagnino, A.; Staton, D.; Shanel, M.; Mueller, M.; Mejuto, C. Evolution and Modern Approaches for Thermal Analysis of Electrical Machines. *IEEE Trans. Ind. Electron.* **2009**, *56*, 871–882. [CrossRef]
14. Du-Bar, C.; Wallmark, O. Eddy Current Losses in a Hairpin Winding for an Automotive Application. In Proceedings of the 2018 XIII International Conference on Electrical Machines (ICEM), Alexandroupoli, Greece, 3–6 September 2018; pp. 710–716.
15. Schmidt, D.; van der Giet, M.; Hameyer, K. Improved iron-loss prediction by a modified loss-equation using a reduced parameter identification range. In Proceedings of the 20th International Conference on Soft Magnetic Materials (SMM), Mpenou, Kos Island, Greece, 18–22 September 2011; p. 421.

16. Eggers, D.; Steentjes, S.; Hameyer, K. Advanced Iron-Loss Estimation for Nonlinear Material Behavior. *IEEE Trans. Magn.* **2012**, *48*, 3021–3024. [CrossRef]
17. Steentjes, S.; Eggers, D.; Leßmann, M.; Hameyer, K. Iron-loss model for the FE-simulation of electrical machines. In Proceedings of the INDUCTICA Technical Conference, Berlin, Germany, 26–28 June 2012; Monkman, G., Ed.; pp. 239–246.
18. Staton, D.; Boglietti, A.; Cavagnino, A. Solving the More Difficult Aspects of Electric Motor Thermal Analysis in Small and Medium Size Industrial Induction Motors. *IEEE Trans. Energy Convers.* **2005**, *20*, 620–628. [CrossRef]
19. Boglietti, A.; Cavagnino, A.; Staton, D.A. Determination of Critical Parameters in Electrical Machine Thermal Models. *IEEE Trans. Ind. Appl.* **2008**, *44*, 1150–1159. [CrossRef]
20. Wang, W.; Zhou, Y.; Chen, Y. Investigation of lumped-parameter thermal model and thermal parameters test for IPMSM. In Proceedings of the 17th International Conference on Electrical Machines and Systems (ICEMS), Hangzhou, China, 22–25 October 2014; pp. 3246–3252.
21. Wrobel, R.; Mellor, P.H. A General Cuboidal Element for Three-Dimensional Thermal Modelling. *IEEE Trans. Magn.* **2010**, *46*, 3197–3200. [CrossRef]
22. Thul, A.; Groschup, B.; Hameyer, K. Influences on the Accuracy of Torque Calculation for Permanent Magnet Synchronous Machines. *IEEE Trans. Energy Convers.* **2020**, *35*, 2261–2268. [CrossRef]
23. Jaeger, M.; Ruf, A.; Hameyer, K.; Tongeln, T.G.V. Thermal Analysis of an Electrical Traction Motor with an Air Cooled Rotor. In Proceedings of the IEEE Transportation and Electrification Conference and Expo (ITEC), Long Beach, CA, USA, 13–15 June 2018; pp. 467–470.

Article

Selected Aspects of Decreasing Weight of Motor Dedicated to Wheel Hub Assembly by Increasing Number of Magnetic Poles

Piotr Dukalski [1] and Roman Krok [2,*]

[1] Łukasiewicz Research Network—KOMEL Institute of Electric Drives and Machines, 40-203 Katowice, Poland; piotr.dukalski@komel.lukasiewicz.gov.pl
[2] Department of Electrical Engineering and Computer Science, Faculty of Electrical Engineering, Silesian University of Technology, 44-100 Gliwice, Poland
* Correspondence: Roman.Krok@polsl.pl

Abstract: Decreasing the mass of a wheel hub motor by improving the design of a motor's electromagnetic circuit is discussed in this paper. The authors propose to increase the number of magnetic pole pairs. They present possibilities of mass reduction obtained by these means. They also analyze the impact of design changes on losses and temperature distribution in motor elements. Lab tests of a constructed prototype, as well as elaborated conjugate thermal-electromagnetic models of the prototype motor and modified motor (i.e., motor with increased number of magnetic poles) were used in the investigation. Simulation models were verified by tests on the prototype. Results of calculations for two motors, differing by the number of pair poles, were compared over a wide operational range specific to the motor application in the electric traction. A detailed analysis of the operational range for these motors was also made.

Keywords: wheel hub motor; electric drive; permanent magnet synchronous motor

Citation: Dukalski, P.; Krok, R. Selected Aspects of Decreasing Weight of Motor Dedicated to Wheel Hub Assembly by Increasing Number of Magnetic Poles. *Energies* **2021**, *14*, 917. https://doi.org/10.3390/en14040917

Academic Editor: Armando Pires
Received: 25 December 2020
Accepted: 1 February 2021
Published: 9 February 2021

Publisher's Note: MDPI stays neutral with regard to jurisdictional claims in published maps and institutional affiliations.

1. Introduction

Design of an electric motor dedicated for assembly in wheel hubs opens up a whole new vista of opportunities for the automobile industry. The new possibilities of such a solution include the elimination of elements of the drive mechanisms used to transfer the torque between the electric motor and the wheel, increasing the efficiency of the entire drive, as well as allowing for more dynamic driving and turning, and new designs of hybrid drives. Placing electric motors into wheels increases the amount of space available inside the vehicle where we may put additional batteries, so that vehicle range is significantly improved. Removing the power unit from the car's body also allows for a more aerodynamic car design.

The use of this type of drive creates many challenges for motor designers. It should be remembered that electric motors mounted in wheels constitute an additional unsprung mass of the vehicle, which may affect the driving comfort and the vehicle's steerability [1–9]. Therefore, the mass of the electric motor in such a drive solution should be as small as possible.

Additionally, if the driving characteristics of a car with wheel hub motors are to be comparable to ICE (internal combustion engine), or electric cars (with centrally mounted electric motors with gearboxes), the motor in the wheel hub must be characterized by a high maximum torque necessary for adequate acceleration, high long-term torque overload, enabling movement with the required speed on the slopes, and a sufficiently high rotation speed at maximum speed (to overcome the motion resistance at this speed). The electric motor should also be highly efficient in the entire operating range.

Designing an electric motor meeting all of these requirements is not easy due to the fact that the dimensions and weight of the motor are limited due to its location in the car.

The design of the motor for such application requires electromagnetic calculations and estimation of power losses, which are the reason for heating of various elements of the motor (especially the insulation of windings and permanent magnets), as well as the calculation of their temperatures.

The technical development of such electric motors requires large financial resources and multidisciplinary teamwork. The design guidelines set very high requirements for the operating parameters of the electric motor, while maintaining a relatively low weight. For this reason, the cooling efficiency must be extremely high. Appropriately selected materials should be used, which should meet the requirements in many physical aspects. The issue of optimization of the wheel hub motor structure and the selection of appropriate materials has been discussed by many researchers [10]. There are publications in which various types of electric motors of various designs are selected for use in vehicle drive wheels [11–18]. In industry, technologically advanced structures are most often permanent magnet synchronous motors with external rotors [19–25]. This is due to a number of advantages, including the ability to control such motors in which the field weakening method can be used. Another advantage of this design is the relatively high torque due to the large diameter of the air gap. Another factor is the shape of the motor (motor geometry); it is best suited for multi-pole magnetic circuit system. Researchers also undertake research on this subject; however, these are usually theoretical studies [26] or studies of electric motors with relatively low power, in which many design issues (construction, thermal problems) are not taken into account. It should be emphasized that, when developing electromagnetic circuits for electric motors with compact construction and high power densities, attention should be paid to many design features, such as dimensional limitations, careful selection of materials (ensuring adequate structural strength and thermal conductivity), and technological possibilities. In wheel hub motors, all of these aspects influence the temperature distribution of the electric motor components to a much greater degree than in standard electric motors, as they are motors with a relatively high power density in relation to their weight.

The development of wheel hub motors requires testing on prototypes. These tests allow you to solve many important questions related to, among others with unsprung mass, the solution of effective seals, the operation of electric motor elements related to different thermal expansion, taking into account the influence of additional power losses affecting the temperatures of the motor elements.

The authors of the article, based on their own research to date, indicate the direction of development of the construction of wheel hub motors. The goal is to reduce the mass of the motor by reducing the mass of the electromagnetic circuit. The prototype wheel hub motor SMzs200S32 produced in the Łukasiewicz Research Network—KOMEL Institute of Electric Drives and Machines, was used for the simulation and laboratory work. This electric motor weighs 36 kg. The assumed operating parameters were achieved, but the designers decided to make changes to reduce weight. In the performed tests and simulations, the impact of the proposed design changes on the temperature distribution in the electric motor was assessed. The simulation models used in this study were verified with measurements.

In the article, the authors focus on reducing the mass of the electromagnetic circuit, which is the heaviest part of the mass of the prototype wheel hub motor under consideration. The article does not discuss possible changes to the mechanical structure of the electric motor in order to reduce its weight.

Many scientific studies concern the problem of improving the design of permanent magnet motors with concentrated windings, which are commonly used in wheel hub motors, but only simulation models are regularly used in research. One of the possible methods of improving the motor structure is the appropriate selection of the number of slots depending on the number of pairs of magnetic poles [27,28]. This results in a reduction of losses in permanent magnets [29–33], as well as a reduction in the dimensions of electromagnetic circuits. Therefore, the total weight of the motor is also reduced. In their research, the authors used the existing SMzs200S32 motor prototype. Laboratory tests of this motor were used to calibrate the electromagnetic/thermal model of the motor generated in Ansys Motor-CAD software (Ansys Canonsburg, Canonsburg, PA, USA). A new motor design was then developed, featuring an increased number of pole pairs. The simulation results obtained for this model were compared with the results for the prototype motor. In scientific publications and available technical studies, very little work on the directions of further development of electric motors in wheel hubs is based on laboratory measurements of prototype electric motors. Very high use of the electric motor's electromagnetic circuit, resulting from the need to obtain high power with significantly limited dimensions, requires solving a number of issues related to the construction of seals, ensuring sufficiently high strength of structural elements, the use of a new generation of insulation materials with relatively high thermal conductivity, and the development of new production technologies. The authors, developing the coupled electromagnetic-thermal model for simulation studies, based on the prototype wheel hub motor built, in which all the previously presented issues were solved, which had a significant impact on the electromagnetic field, generated power losses and the temperature field. Then, the model was validated using temperature measurements with sensors installed in many places, both in the stator and in the rotor. Considerations regarding the improvement of the design of the considered type of engines, which are very often found in the literature, based only on mathematical models without prototypes and measurements, may only set certain trends. The procedure proposed by the authors leads to the development of a proposal for a new engine solution with a significantly reduced mass, supported by a verified thermal model. This method allows for a more accurate approximation of design solutions to be put into production.

2. Motor Construction

Multi-pole motors with external rotors are often used for assembly in EV wheels. This is due to the character of space in the wheel, where the motor is placed. The electromagnetic circuit of such a motor is toroidal, so that additional space found inside this toroid may be used. The cross-section of the three-dimensional (3D) model of the discussed SMzs200S32 motor is shown in Figure 1.

Most of the hull is the rotating element. It contains a magnetic core with mounted permanent magnets. The stationary element is the anchor disc with the supporting structure, in which there is a labyrinth cooling system. The stator's magnetic core with winding is mounted on the supporting structure (Figure 1).

The motor is dedicated for assembly in a 17" wheel rim.

A rotor position sensor is required to control the motor. Typically, an incremental encoder is used.

The space in which the electric motor must fit is limited by the dimensions of the wheel rim (outer diameter and motor length), while the inner diameter depends on how the space inside the toroid is used. In the case of the presented structure, this space houses the vehicle brake drum.

The space containing the electromagnetic circuit is limited by the dimensions of the support structure, anchor shell, and rotor, which must be thick enough to ensure the required mechanical strength (Figure 1).

Pictures of the prototype SMzs200S32 motor are shown in Figure 2. For research purposes, this motor was equipped with a number of PT100 temperature sensors, placed in various elements of the stator and rotor (permanent magnets). Additionally, a small wireless temperature recorder was developed. It is installed on the rotor surface and the sensor mounted on the magnet is connected to the recorder. Temperature can be registered continuously and data are sent wirelessly [34]. The cross-section of the motor with the positions of the temperature sensors is shown in Figure 3.

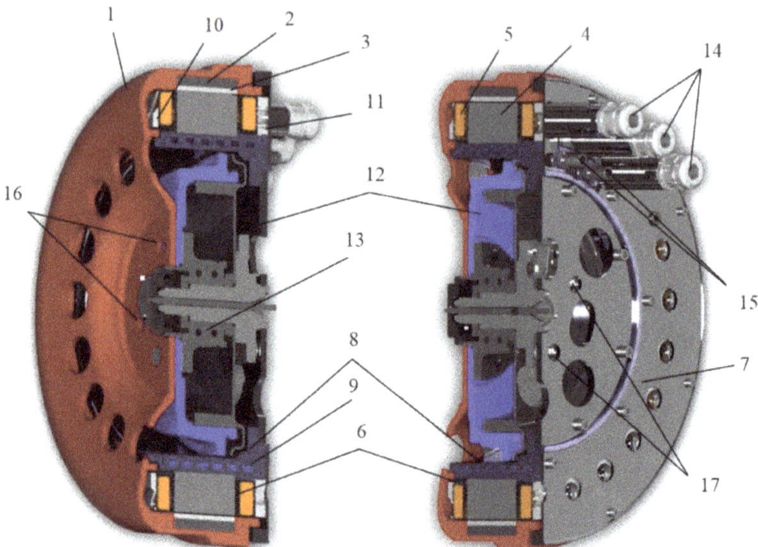

Figure 1. Models The cross-section of disassembled three-dimensional (3D) model of SMzs200S32 motor, manufactured by Łukasiewicz Research Network–KOMEL Institute, and dedicated for assembly in wheel hub of car: 1—rotor, 2—rotor's magnetic core, 3—magnet, 4—stator's magnetic core, 5—stator winding coil ends, 6—resin, 7—permanent anchoring shield, 8—supporting structure, 9—casing with coolant ducts, 10—radiator of coil outhang, drive end, 11—radiator of coil outhang, non-drive drive, 12—brake drum, 13—bearing assembly, 14—entry for supply wires, 15—cooling system ports, 16—rotor assembly openings, 17—stator assembly openings.

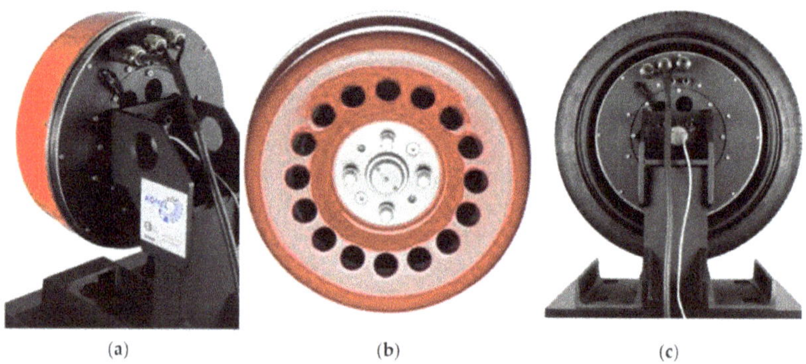

(a)   (b)   (c)

Figure 2. Prototype of SMzs200S32 motor manufactured by Łukasiewicz Research Network—Institute of Electrical Drives and Machines KOMEL (Ł-KOMEL): (**a**) angled view; (**b**) front view; (**c**) rear view.

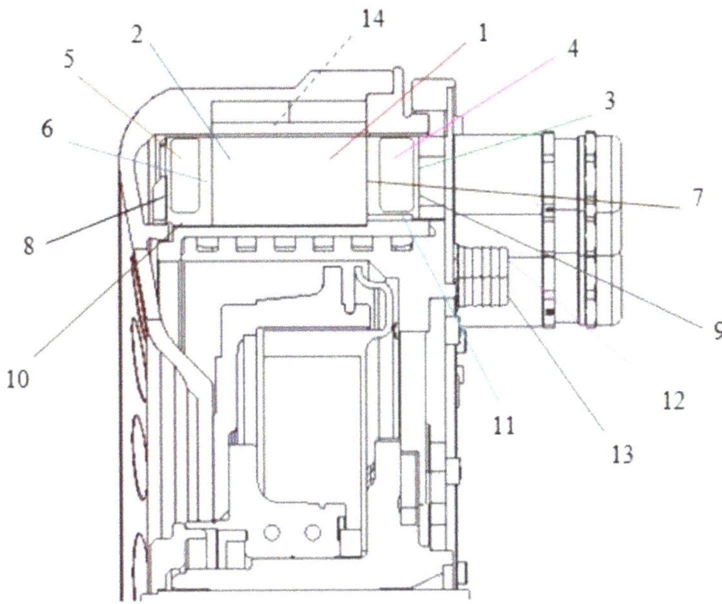

Figure 3. Cross-section of SMzs200S32 manufactured by Ł-KOMEL; siting of PT-100 temperature sensors in different elements of motor is shown, they are partly indicated as drive end (D) and non-drive end (N) sensors: 1—winding in slot N, 2—winding in slot D, 3—neutral point of winding, 4—winding ends N, 5—winding ends D, 6—laminations D, 7—laminations N, 8—radiator disk D, 9—radiator disk N, 10—casing of cooling system D, 11—casing of cooling system N, 12—water at inlet point, 13—water at outlet point, 14—permanent magnets. D—drive side; N—non-drive side.

Motor at the test stand was supplied from Sevcon Gen4 Size8 inverter dedicated to EV supply. During tests, the inverter was supplied with nominal voltage $V = 350\ V_{DC}$ (about 235 V at motor terminals). Since the rotor position angle sensor had to be used, a magnetic absolute encoder with analog Sin–Cos output was applied. Control of the motor drive was conducted with the help of a computer-aided measurement and control device with dedicated control.

The motor operates in two zones: constant torque zone and field weakening zone. The rated operational parameters of this motor are presented in Table 1.

Table 1. Rating parameters of SMzs200S32 motor, Sevcon Gen4 Size8 inverter supply.

Parameter	Value	Unit
Inverter's rated voltage U_{DC}	350	V
Motor's rated current I_{DC}	173	A
Rated torque T_m	450	Nm
Rated rotational speed n_n	1050	rpm
Maximum speed n_{max}	1500	rpm
Maximum torque T_{max} (within rotational speed range $n = 0$–400 rpm)	750	Nm
Maximum efficiency η_{max}	95	%
Motor mass m	36	kg

The operating parameters of the electric motor make it possible to use it in various car drive systems, where two, four, or more electric motors can be used. Taking the example of a Fiat Panda car with two motors, assuming a car weight of 1600 kg and driving at a speed of 150 km/h, each motor must be able to produce a torque of 150 Nm. When going up a

10% gradient at 105 km/h (950 rpm), each motor must generate a torque of approximately 300 Nm. The rated duty point allows the car to be driven continuously (i.e., until the supply battery is fully discharged) at a speed of more than 115 km/h on a road with a slope of 15.5%. This is a very abstract operating point because continuous operation under these conditions (road gradient, speed) is highly unlikely.

In the case of investigated drive, the maximum torque is limited by peak value of inverter current (seen from motor terminals). This specific inverter was used in tests. Further discussion will relate to simulation of motor operation within full operational range, this limitation will be ignored.

The wheel hub motor SMzs200S32 meets the operating parameters assumed by the constructors, but the authors planned to reduce its weight. For this purpose, an electromagnetic circuit with an increased number of pole pairs has been proposed. The further part of the article presents the procedure for modifying the design of the electromagnetic circuit using the calculation models generated in the Ansys Motor-CAD program.

3. Computational Models

If we aim to decrease the motor mass, then, apart from the use of lightweight and resistant material for the supporting structure, housing, and anchoring shield, the mass of the electromagnetic circuit should be decreased as much as possible (this mass usually constitutes more than 50% of the total motor mass). In the design, we should try to obtain a reasonably high number of poles, because this will decrease the volume of the electromagnetic circuit on account of lower magnetic flux within a single pole. Thus, the width of the stator yoke and the rotor of the electromagnetic circuit may be significantly diminished. This will result in a better ratio of core volume (or mass) to rated torque. Use of winding with concentrated coils is also useful, because winding ends may be reduced.

When we consider possibilities of lessening mass of the electromagnetic circuit by increasing the number of pole pairs, different existing limitations must be examined. Increasing the number of poles in SPM (Surface Permanent Magnet Motor) motors is easy from a technological standpoint, but it may be restricted by minimum dimensions of a single magnet (technological possibilities of mounting small magnets). The other limitation is maximum output frequency of supply inverter. We must also keep in mind that, when yoke thickness is reduced, the mechanical strength of the rotor also goes down.

An additional problem emerges when we try to increase the number of stator slots in proportion to the increased number of poles. For instance, with the number of poles $2p = 32$ and number of slots $Qs = 48$, the number of slots per pole per phase is $q = 0.5$. If we raise the number of poles up to $2p = 56$, then in order to maintain $q = 0.5$, the number of slots should be increased to $Qs = 84$ (Figure 4). Since the outer diameter of the machine is constant, such a large number of slots may not be feasible due to the technological limitations of production machines, the strength of the thin stator teeth, or the workability of the winding. In addition, the cost of producing a winding will also increase, as the number of coils is much greater. The area of the active part of the slot winding also decreases, because, with the increase in the number of slots, the share of slot insulation is greater.

In this situation, it is possible to reduce the number of sockets per pole per phase, but it should be remembered that, for motors with $q < 0.5$, the spatial distribution of magnetomotive force mmf is deformed, which causes increased eddy current losses in permanent magnets [35].

The winding temperature limited by the permissible winding insulation temperature is a standard torque limiting factor in electrical machines. In the case of multipole motors, and in particular when the number of gaps per phase per q pole is equal to or less than 0.5, the increased power losses in permanent magnets may lead to a situation, where the magnet temperature will limit the maximum load of the motor (especially when the motor is running with field weakened to increase the speed range [36]). Under field weakening conditions, magnets are exposed, not only to an increase in temperature caused by the magnitude and frequency of the supply current, but also to an external magnetic field.

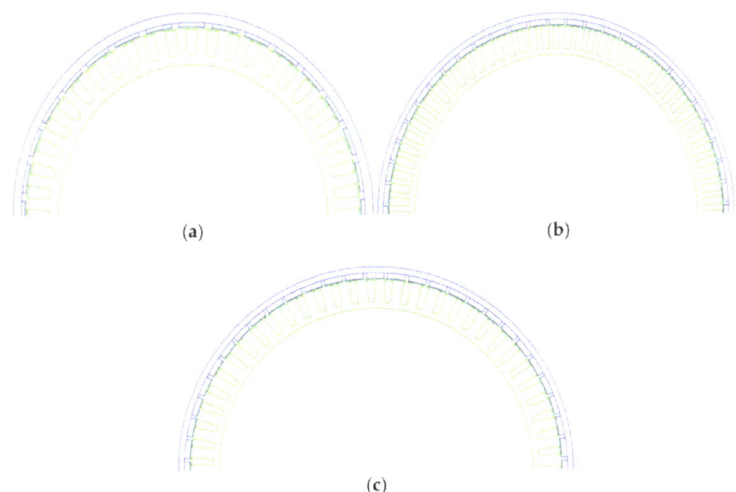

Figure 4. Models of cross-sections of magnetic circuits: (**a**) Q = 48, 2p = 32, q = 0.5; (**b**) Q = 84, 2p = 56, q = 0.5; (**c**) Q = 63, 2p = 56, q = 0.375.

By increasing the number of pole pairs, we obtain a toroidal motor with an increased internal diameter. As a result, the weight is reduced and additional space is gained inside the motor, which can be used, for example, for mounting a car brake drum (Figure 1).

The used Ansys Motor-CAD software (Ansys Canonsburg, Pensylwania, USA) greatly supports the design of electrical machines, because work simulations can be carried out in the entire working range of torque and speed.

At the same time, it is distinguished by very high computing speeds. It is possible to solve conjugated electromagnetic and thermal fields. It uses a combination of advanced analytical equations and calculations based on 2D FEM. The temperature can be determined in steady and transient conditions; the program uses advanced models in the form of thermal networks [37–42].

Model 1 of the motor with the number of pole pairs 2p = 32, where the number of slots per pole per phase is q = 0.5, is a representation of the prototype SMzs200S32 motor. Then the model of Motor #2 was developed; with the number of poles increased to 2p = 56, the number of slots per pole per phase also changed: q = 0.375. Modification of the electromagnetic circuits was carried out to maintain the outer diameter of the rotor and the width of the air gap. The width of the magnet in Motor #2 also increased from 4 to 5 mm. This was at the expense of the rotor yoke. As a result, the inside diameter of the stator and the diameter of the air gap were increased. The calculated operating parameters of both the motors and the masses of individual elements of the electromagnetic circuits are presented in Table 2. The weight of the winding and the mass of the magnetic core were reduced as assumed. The flux density in the rotor's magnetic core increased with increasing width of the permanent magnet (at the expense of the core width). The weight of the winding was reduced by 0.9 kg, the mass of the stator magnetic core by 2.3 kg, the mass of the rotor magnetic core by 2.1 kg, and the mass of permanent magnets increased by 0.66 kg.

The simulation models and the calculated flux density distribution from the permanent magnets in the magnetic core are shown in Figure 5.

Table 2. Calculated weight of different elements of discussed electromagnetic circuits.

Calculated Parameter	Motor #1	Motor #2
Number of slots Q	48	63
Number of pole pairs 2p	32	56
Number of slots per pole and phase q	0.5	0.375
Calculated length of turn (mm)	204	190
Winding weight (kg)	4.5	3.6
Stator core weight (kg)	9.5	7.2
Rotor core weight (kg)	4.4	2.3
Permanent magnets weight (kg)	1.6	2.26
Supporting structure weight (kg)	4	4.3
Total weight of electromagnetic circuit (kg)	20	15.3

The calculations were made for the given power supply parameters, which are dictated by the permissible operating parameters of the inverter U_{DCmax} = 350 V, I_{max} = 350 A. The control system performs work in two zones, the work zone with constant torque and the work zone weakening the magnetic field from permanent magnets.

Thermal calculations were made using two models in the thermal of the Motor-CAD module.

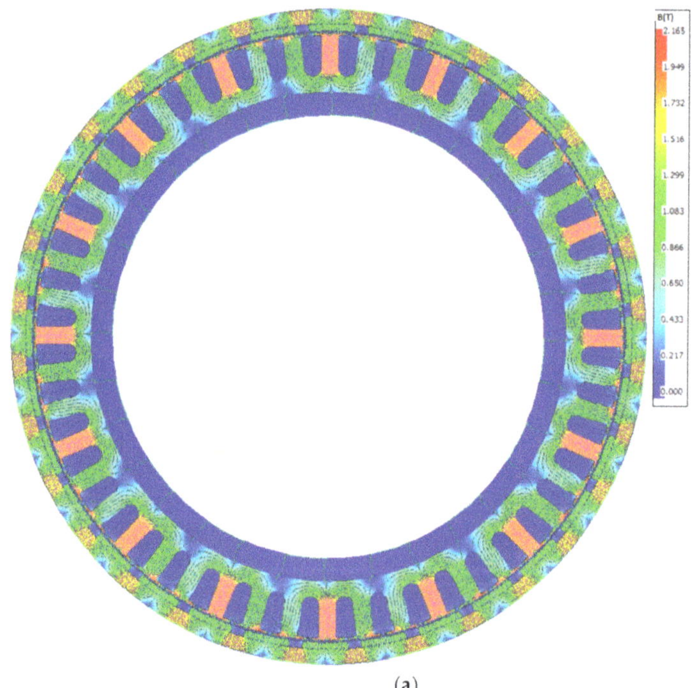

(a)

Figure 5. *Cont.*

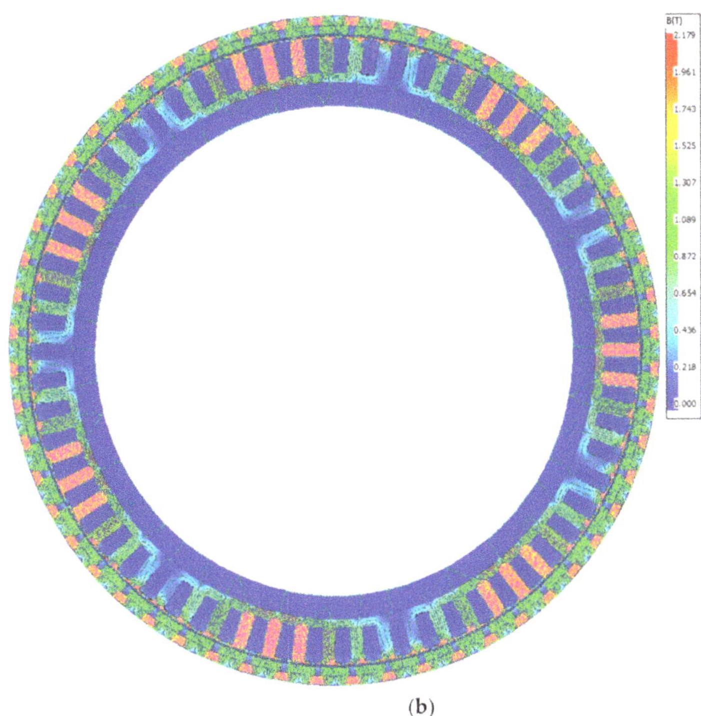

(b)

Figure 5. Two-dimensional (2D) FEM models constructed in Ansys Motor-CAD software: (**a**) Motor #1 (**b**) Motor #2.

As the stator inner diameter increases, the diameter of the stator support structure increases. It was assumed that the thickness of the support structure must be kept so that its internal and external diameter change with the increase of the internal diameter of the stator.

The dimensions of the cooling channels remain unchanged. To facilitate further analysis, both models have identical lengths of winding ends. Note that in Motor #2, the calculated length of the turns is smaller and the teeth of the stator core are narrower. In practice, this can lead to a reduction in the length of the winding ends and a reduction in the length of the entire motor, which will also reduce the weight of the motor further. Determining the length of winding ends requires carrying out technological tests on the physical execution of the prototype winding, which is currently not the subject of research.

The simulation model uses the parameters of the actual materials used in the construction of SMzs200S32: electrical sheets used for stator and rotor laminates, permanent magnets, and aluminum alloy.

To sum up, the mass of the electromagnetic circuit of engine 2 is 4.7 kg less (by 23.5%), while the mass of its supporting structure is 0.3 kg greater than that of motor no. 1. The total mass of Motor #2 in relation to Motor #1 was significantly reduced—by 4.4 kg (by 12%).

The Bertotti method was used for calculating the magnetic core power losses in the elaborated models [43–46]. This lets us account for hysteresis, eddy current, and excess losses:

$$\Delta P_{Fe}(t) = k_h B_m{}^2 f + \sigma \frac{b^2}{12}\left(\frac{dB}{dt}(t)\right)^2 + k_e \left(\frac{dB}{dt}(t)\right)^{\frac{3}{2}}$$

where: k_h—hysteresis factor, B_m—flux density, f—frequency, σ—conductivity, b—thickness of single electrical sheet, k_e—excess loss coefficient.

The first part of the formula relates to hysteresis losses (due to the hysteresis loop of the ferromagnetic material of the magnetic core). The second part of the formula covers eddy current losses (eddy currents are induced in the magnetic core). The third part relates to excess losses, caused by interaction of the external magnetic field and local magnetic fields generated by eddy currents.

In analysis, copper losses are split into constant losses and losses due to skin effect, i.e., in the program named as AC winding losses. Copper losses are calculated on the basis of a well-known relationship:

$$\Delta P = I^2 R(v)$$

where: I_{RMS}—value of supply root mean square current, $R(v)$—winding resistance as a function of temperature

AC Losses on AC windings can be determined by two methods—full FEA and hybrid FEA method. The full FEA method uses an accurate model where the induced eddy currents and then losses are calculated separately for each conductor. Obviously, this is the most time consuming method. Hybrid FEA uses the flux density levels calculated by FEM for each slot, and then the losses are calculated analytically. This method is quick, but much less accurate at relatively high frequencies and relatively large conductor cross-sections. In our case, the mixed method was adopted. For the operating point with maximum speed and load, the losses were calculated using both methods, and then the correction factor was calculated according to the formula:

$$k_{\frac{Full}{Hybrid}} = \frac{\Delta P_{Cu\ Full}}{\Delta P_{Cu\ Hybrid}}$$

where: AC winding losses calculated by Full FEA method—AC winding losses calculated by hybrid FEA method.

Since further analysis takes into account a wide load/speed range, AC winding losses are calculated with hybrid FEA method and each result is then multiplied by correction factor.

One operating point, motor SMzs200S32, was selected to verify the calculations. The operational parameters as well as losses and temperature distribution were tested. The work point was analyzed with the current consumption root mean square $I_{RMS} = 109$ A and the rotational speed 950 rpm. With two motors mounted on the 17-inch wheels of the Fiat Panda, this point corresponds to driving at 100 km/h on a steep road (10% gradient). The values of coolant flow (10 L/min), coolant temperature (23.8 °C) and ambient temperature (28.3 °C) used in the calculations are identical to those measured in the laboratory tests. Table 3 summarizes the different operating parameters of the three motors: the real SMzs200S32 prototype motor, the Motor #1 simulation model, which is a representation of the prototype, and the Motor #2 simulation model, which is a version of the motor with an increased number of magnetic poles. Motor #1 and Motor #2 are shown in Figures 6 and 7 with the calculated temperatures marked.

Table 3. Comparison of measured parameters of SMzs200S32 motor with values calculated with the help of models (Motor #1 and Motor #2).

Parameter	SMzs200S32	Motor #1	Motor #2
Number of magnetic poles	32	32	56
Number of slots	48	48	63
Conductors per slot	10	10	6
Phase resistance at 20 (°C)	0.026	0.025	0.016

Table 3. *Cont.*

Parameter	SMzs200S32	Motor #1	Motor #2
Supply current I_{RMS} (A)	109	109	121
Drive supply voltage U_{DC} (V)	350	350	350
Rotational torque T_m (Nm)	300	311	311
Rotational speed n_n (rpm)	950	950	950
Mechanical power P_m (kW)	29.9	30.9	31.1
Supply frequency (Hz)	253.6	253.3	443.4
Power factor	0.72	0.73	0.67
Torque ripple (%)	-	21	3.8
Cogging torque (Nm)	15	12.4	1.8
Efficiency (%)	93.2	93.4	92.5
Steady-state temperatures (°C)			
In slot (bottom) R	76.5	76.5	54.9
In slot (bottom) L	75	76.5	54.9
Coil outhang (center) R	76.8	76.5	64.9
Coil outhang (center) L	72.3	76.6	64.6
Top of tooth R	65	67.7	64
Top of tooth L	65.8	67.7	64
Yoke laminations R	43.4	40	40.1
Yoke laminations L	51.6	40	40.1
Radiator shield R	41.4	53.3	47.5
Radiator shield L	49.5	34	33
Temperature of supporting structure R	33.2	42.7	39
Temperature of supporting structure L	39.6	33	32
Magnet temperature	67.1	64.9	57.5
Coolant (water)	23.8	23.8	23.8
Ambient temperature	28.3	28.3	28.3

Based on the data from Table 3 and Figure 6, it is possible to see the importance of the position of the temperature sensor in a given element of the electric motor. In the case of a physically manufactured motor with a "compact" design, it is rather difficult to locate a large number of temperature sensors and their terminals. In addition, the stator of the motor, which houses almost all sensors, is embedded in epoxy. This excludes the possibility of replacing the sensor or changing the location. Since the gap-filling factor is high (approximately 75% with insulation) and due to the winding technology, the temperature sensors in the slots are placed on its bottom during the winding process. On the other hand, the sensors at the winding ends are placed in the middle. When the measured and calculated steady-state temperatures for the operating point in question are compared, we can see that model Motor #1 provides sufficient accuracy. The temperature differences between the prototype motor and Motor #1 result from the simplification of the shapes, possible slight differences between the physical parameters of the materials adopted in the model and the actual ones occurring in the motor (this may be due to technological differences) and the inaccuracy of determining the contact resistance between motor elements. The temperature comparison shows that an increase in the number of pole pairs (motor 2) lowers the temperature of the winding and magnets.

Motor cross-sections in Ansys Motor-CAD software (Ansys Canonsburg, Pensylwania, USA) (Figures 6 and 7) are simplified in comparison with real cross-sections. Test verification of the model confirmed sufficient accuracy of temperature distribution representation in motor.

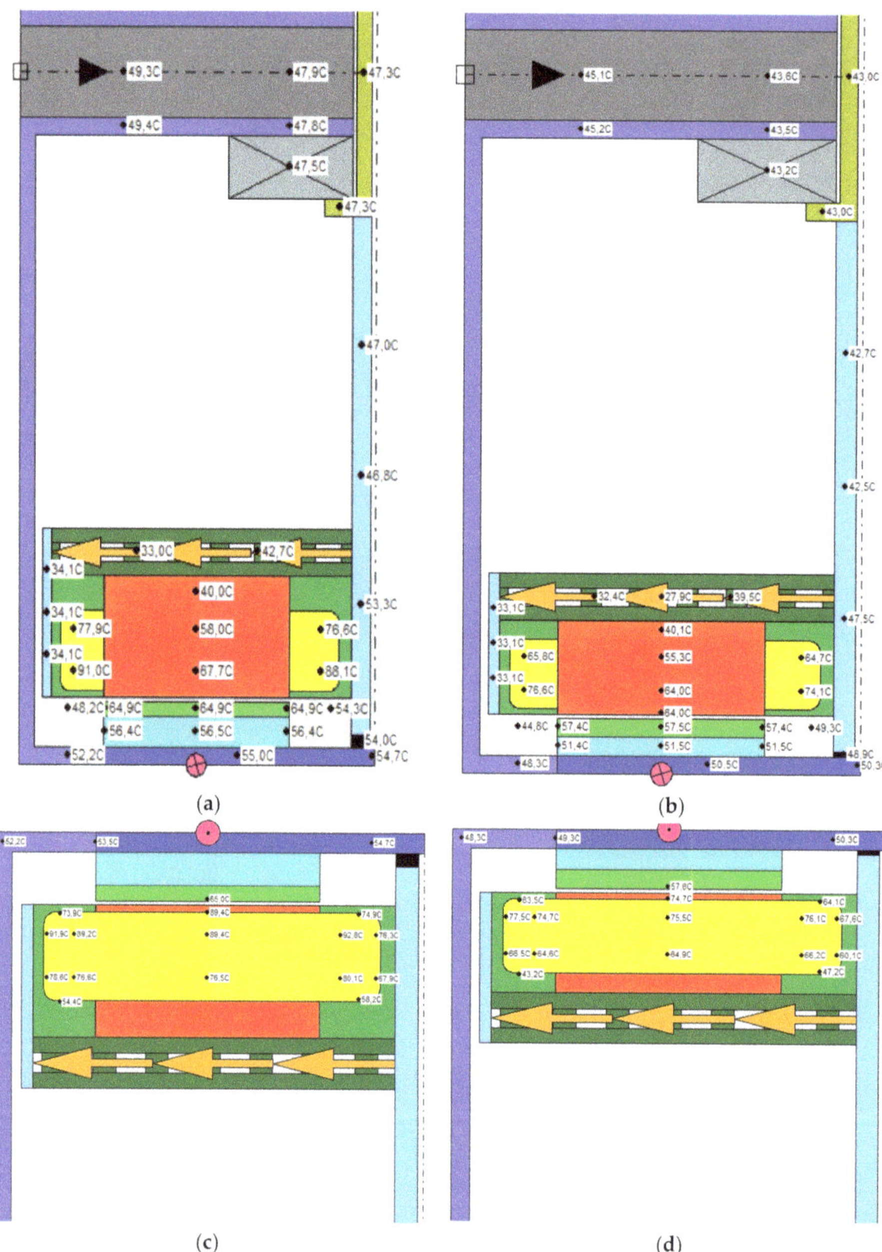

Figure 6. Results of calculations for Motor #1 model and Motor #2 model, speed n = 950 rpm, supply current T_m = 311 Nm: (**a**) Model of longitudinal cross-section through stator tooth Model #1. (**b**) Model of longitudinal cross-section through stator tooth Model #2. (**c**) Model of longitudinal cross-section through stator slot Model #1. (**d**) Model of longitudinal cross-section through stator slot Model #2.

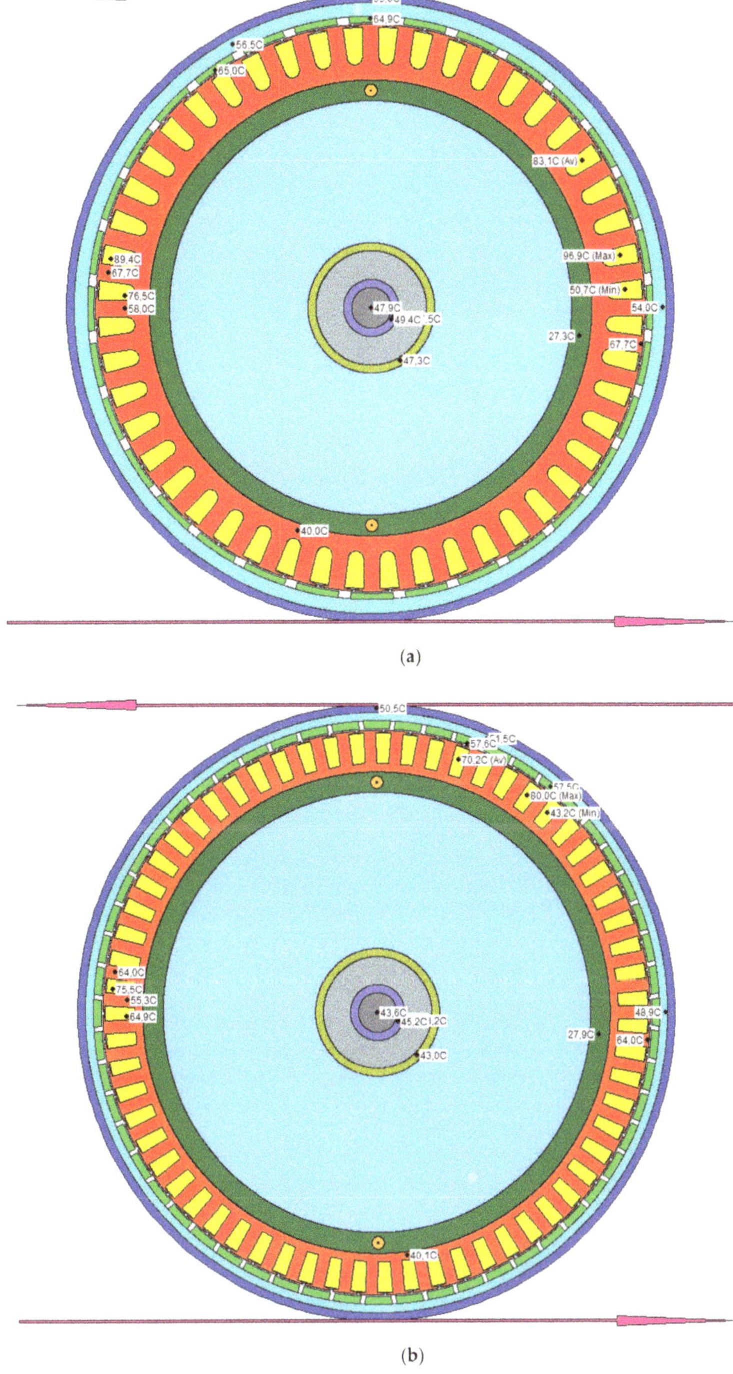

(**a**)

(**b**)

Figure 7. Results of calculations for Motor #1 model and Motor #2 model—transverse cross-section, speed n = 950 rpm, supply current T_{m} = 311 Nm. (**a**) Model #1 (**b**) Model #2.

4. Discussion Calculation Results—Comparison of Operational Parameters of Investigated Motors

During subsequent analysis, we calculated operational parameters over a wide range of rotational speed and supply current I_{RMS}. In order to take into account differences between two designs (and these differences have impact on temperature), comparison of operational parameters was conducted with temperatures calculated for current I_{RMS} = 150 A and speed n = 1000 rpm. In all bar graphs, results for Motor #1 and Motor #2 are marked in blue and yellow, respectively. Rotational torque versus speed is shown in Figure 8.

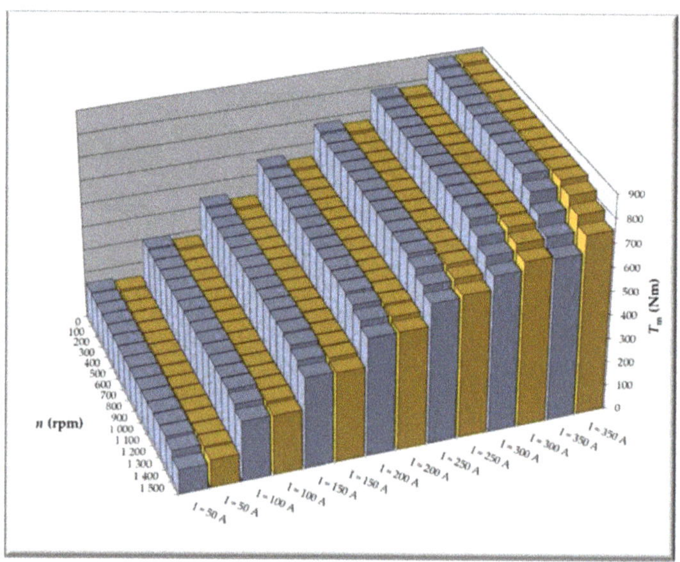

Figure 8. Bar chart of calculated rotational torque characteristics for Motor #1 and Motor #2; supply current is treated as input quantity and is identical for both motors.

In the first control zone, Motor #2 attains somewhat smaller rotational torques than Motor #1 (for identical supply currents). When the current is increased, then torque becomes comparable or higher. In the second control zone, when the field is weakened, Motor #2 is characterized by rotational torque equal or slightly higher than that of Motor #1.

Keeping in mind the planned later analysis, which would compare the different types of losses for the two motors (copper losses, losses in the magnetic core, and losses on permanent magnets), and these losses depend on the value of the supply current, we made calculations with the operating torque as the input quantity. The calculated torque characteristics are shown in Figure 9, and the corresponding supply current characteristics in Figure 10.

Basing on characteristics shown in Figure 9, we may say that, for higher rotational torques, operational range of Motor #1 is slightly less than that of Motor #2. This is because the supply voltage of the electric drive for both motors is the same U_{DC} = 350 V. Motor #1 starts to work in the magnetic field weakening control zone in relation to Motor #2 slightly earlier (as shown in Figure 10, Motor #1 currents increase a little earlier at higher rotational speeds), while with increasing load torque, more current power is field weakening, and not generating torque.

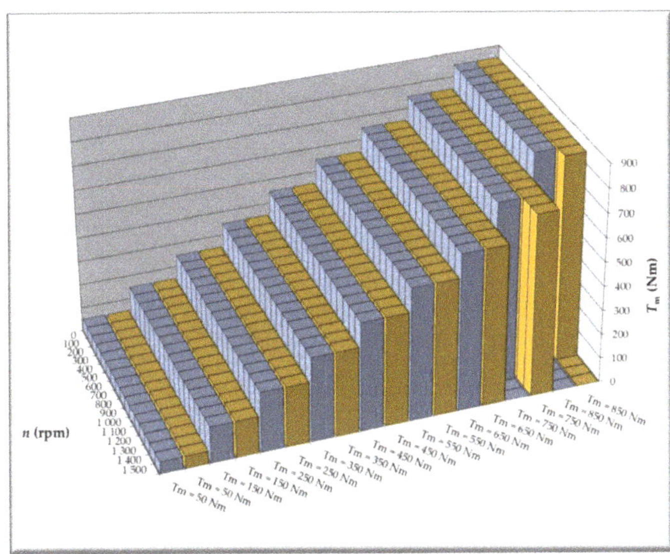

Figure 9. Bar chart of calculated rotational torque for Motor #1 and Motor #2; rotational torque is treated as input quantity and is identical for both motors.

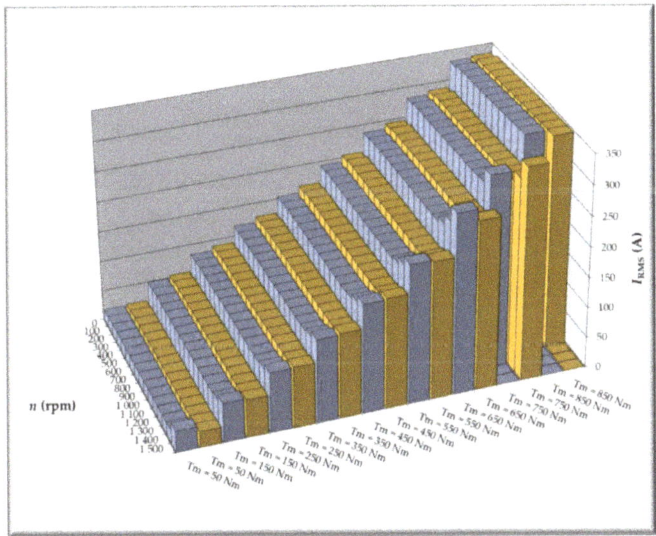

Figure 10. Bar chart of calculated supply current for Motor #1 and Motor #2; rotational torque is treated as input quantity and is identical for both motors.

Keeping in mind that these motors are dedicated to an electric car drive (modified Fiat Panda car) with two motors, the less extensive range of rotational torque may be of minimal importance in the first control zone, since it may affect car acceleration to some extent. In the second control zone, especially at maximum speeds, rotational torques of 750 Nm range practically do not occur.

Characteristics of supply currents versus assumed rotational torques are shown in Figure 10. The supply currents of two motors are not identical, and their disparity increases as rotational speed goes up and we enter the second control zone. As in the case of calculation results shown in Figure 8, we may state that equalization of the rotational torque in the first control zone has led to an increase in the supply current of Motor #2. In the second control zone, and as the field becomes more weakened, Motor #1 must be supplied with a higher current in order to maintain rotational torque at the required level.

Total winding losses are shown in Figure 11, while AC winding losses are shown in Figure 12.

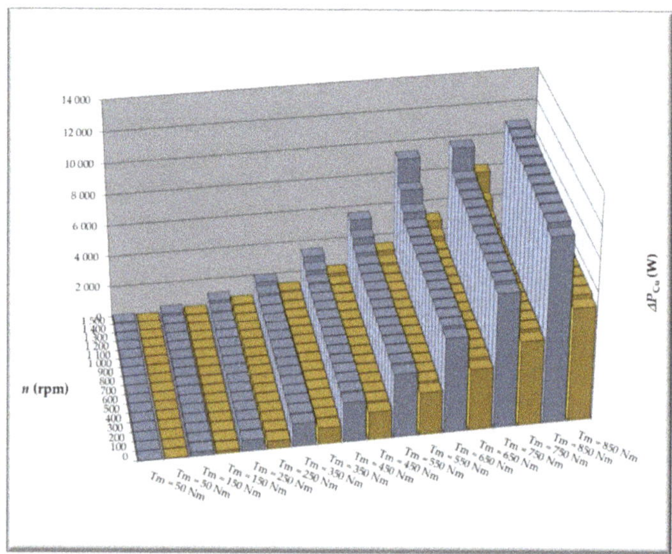

Figure 11. Bar chart of calculated total power losses in windings for Motor #1 and Motor #2; rotational torque is treated as input quantity and is identical for both motors.

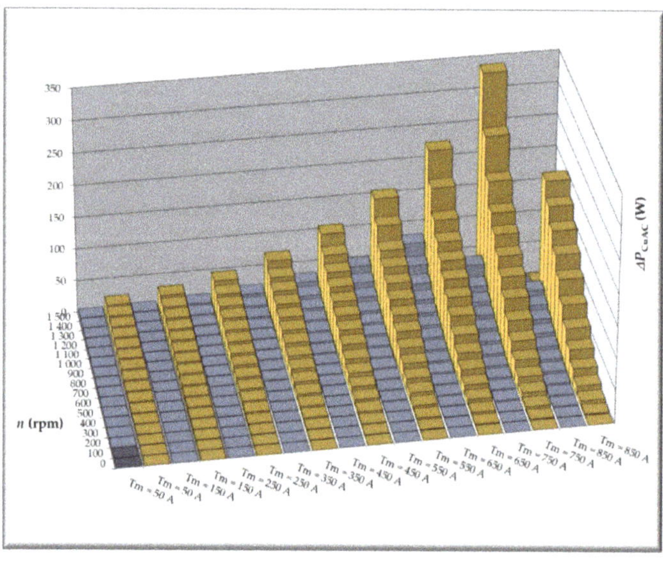

Figure 12. Bar chart of calculated AC winding power losses for Motor #1 and Motor #2; rotational torque is treated as input quantity and is identical for both motors.

Characteristics shown in Figure 11 confirm the fact that Motor #1 is characterized by higher winding losses over the entire operational range. This is due to greater winding resistance and higher winding operational temperatures. Power losses associated with current displacement effects, i.e., AC winding losses, are shown separately in Figure 12. In this case, higher losses are clearly generated in Motor #2 windings; this is supplied at a much higher frequency since the number of pole pairs is increased from $p = 16$ to $p = 28$, and speed range remains unchanged. Even though AC winding losses are higher, their contribution to total losses (Figure 11) is insignificant.

Calculated total losses in the stator core are shown in Figure 13. Losses in Motor #2 core are greater than those of Motor #1 on account of higher values of flux density (produced by permanent magnets) in the electromagnetic circuit (see Figure 5) and higher supply currents (Figure 11). We may observe in these curves that core losses increase as rotational speed increases and then they stabilize at more or less constant levels; this is due to field weakening and operation in the second control zone. The field weakening zone for Motor #1 is commenced somewhat earlier.

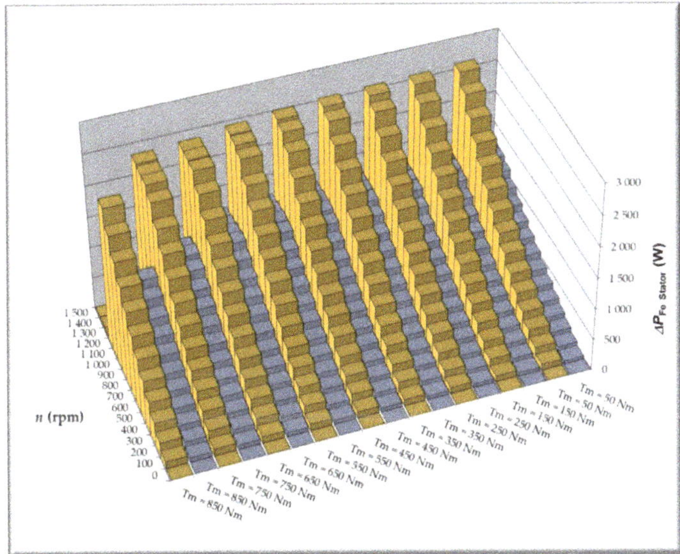

Figure 13. Bar chart of calculated stator core losses for Motor #1 and Motor #2; rotational torque is treated as input quantity and is identical for both motors.

Characteristics of calculated stator total losses are presented in Figure 14; this is the sum of losses shown in Figures 11 and 13. It must be noted that these losses determine temperature distribution in the stator core. The stator core is heated by its own losses and winding losses, which are transferred through the core to the coolant. In the case of lowest loads (rotational torque $T_m = 50$ Nm) losses of Motor #2 are higher over the entire range of rotational speed. These are mostly losses generated in the stator's magnetic core; throughout the entire operational range, these losses are higher in Motor #2 (Figure 14). As the load increases, winding losses increase in both motors, but winding losses in Motor #1 increase at a much higher rate (see Figure 11), while stator core losses do not vary much with changes in the current. When charts for different losses are compared, we observe that while load increases, the area where Motor #2 losses are lower than Motor #1 losses moves away from lower speeds towards maximum speed. In the case of rotational torque equal to $T_m = 450$ Nm, total losses generated in the motor stator are higher in Motor #1 over the entire rotational speed range. When load is increased, difference in stator's total losses also grows.

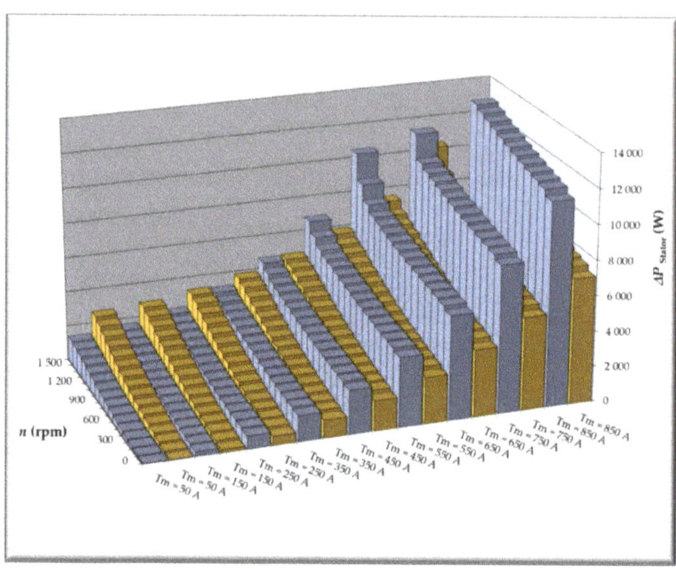

Figure 14. Bar chart of calculated stator total losses for Motor #1 and Motor #2; rotational torque is treated as input quantity and is identical for both motors.

Permanent magnet losses also have significant impact on the motor's operational range and these losses will be investigated next. PM losses are shown in Figure 15.

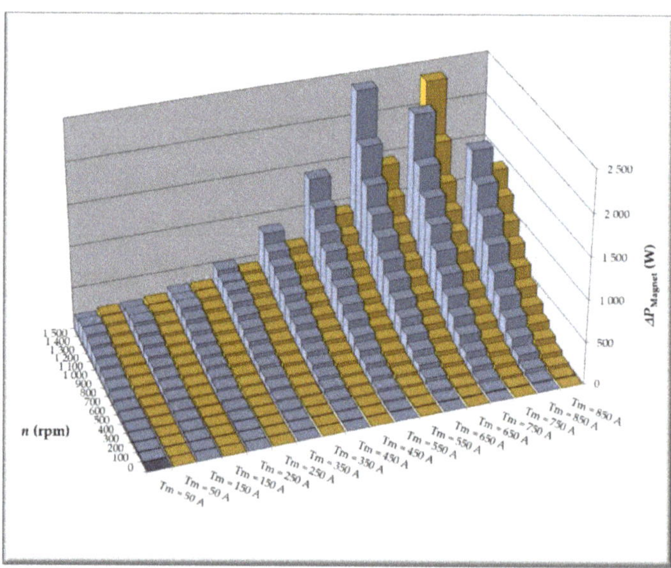

Figure 15. Bar chart of calculated permanent magnet losses for Motor #1 and Motor #2; rotational torque is treated as input quantity and is identical for both motors.

Examining the bar chart presented in Figure 15, we may observe that over (nearly) the entire operational range, PM power losses are smaller in Motor #2; the difference is increased as rotational speed rises (this is true with exception of operating points at speed n = 1500 rpm and at loads up to T_m = 150 Nm). Publications on the subject make it evident that decrease in number of slots per pole per phase results in increased PM losses [47]. Still, it must be pointed out that Motor #2 differs from Motor #1 by the number of magnetic pole pairs positioned upon diameter of similar size; this is related to significant decrease in PM angular length, so, finally, PM losses are lessened.

Characteristics of calculated rotor core losses are presented in Figure 16.

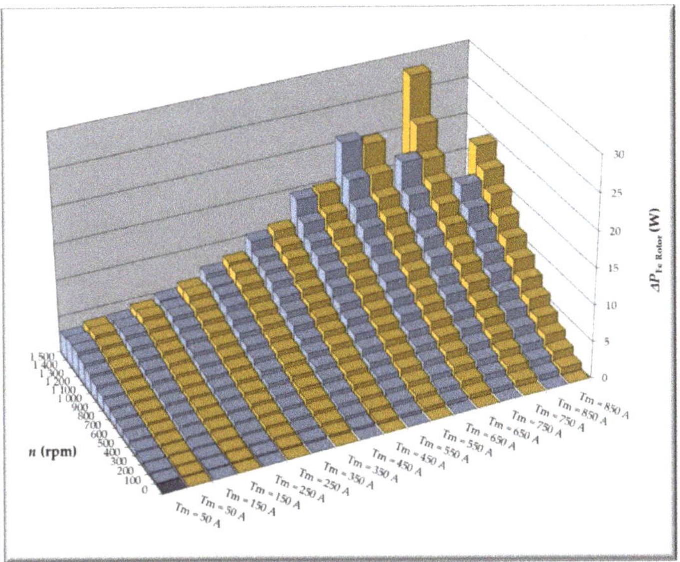

Figure 16. Bar chart of calculated rotor magnetic core losses for Motor #1 and Motor #2; rotational torque is treated as input quantity and is identical for both motors.

Rotor magnetic core losses are higher in Motor #2; still, value of these losses in relation to remaining rotor power losses, including PM losses, is very small.

Temperature rise in motor elements is influenced by the ratio of existing power losses to the area of heat removal. Design of Motor #2, with respect to Motor #1, is characterized by greater stator diameter and diameter of structure containing a water-cooling system. Thus, while lengths of both motors are identical, the area of heat removal is greater in Motor #2.

Bar charts showing the ratio of stator total losses to area of structure, where stator is positioned, are presented in Figure 17. If we compare losses shown in Figure 14 charts, with charts showing the ratio of these losses to the heat removal area (Figure 17), then we see that Motor #2 is much more promising; this is especially noticeable in the case of maximum speed at load torque equal to T_m = 450 Nm and T_m = 350 Nm.

A similar comparison was performed for winding losses in Figure 18. Total slot area is much greater in Motor #2 (SQ = 3389.4 cm^2) than in Motor #1 (SQ = 235.6 cm^2).

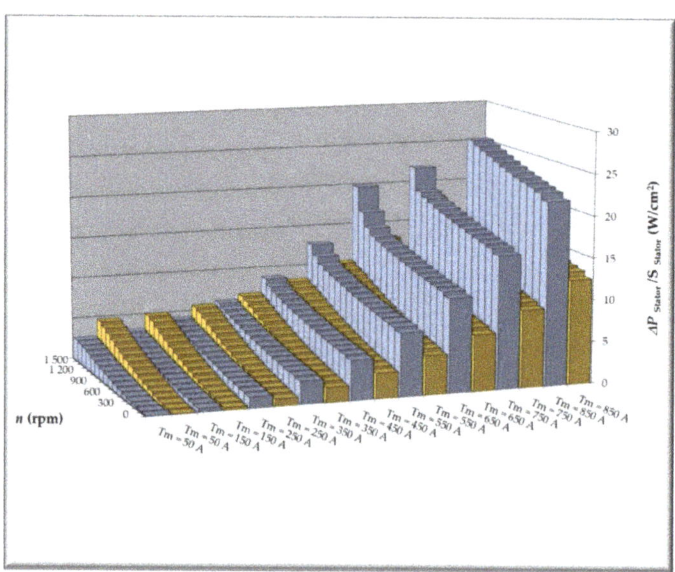

Figure 17. Bar chart of the calculated ratio of stator total losses to the area of the supporting structure containing the water cooling system for Motor #1 and Motor #2; rotational torque is treated as input quantity and is identical for both motors.

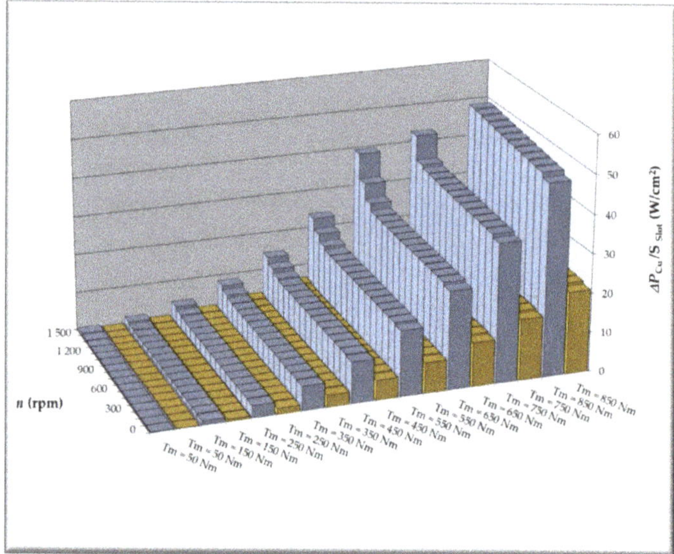

Figure 18. Bar chart of calculated ratio of stator winding losses to slot area for Motor #1 and Motor #2; rotational torque is treated as input quantity and is identical for both motors.

Ratio of power losses generated in permanent magnets to the rotor surface area where magnets are mounted is shown in Figure 19. The ratio of PM loss to heat removal area again underlines the advantages displayed by Motor #2, with respect to Motor #1.

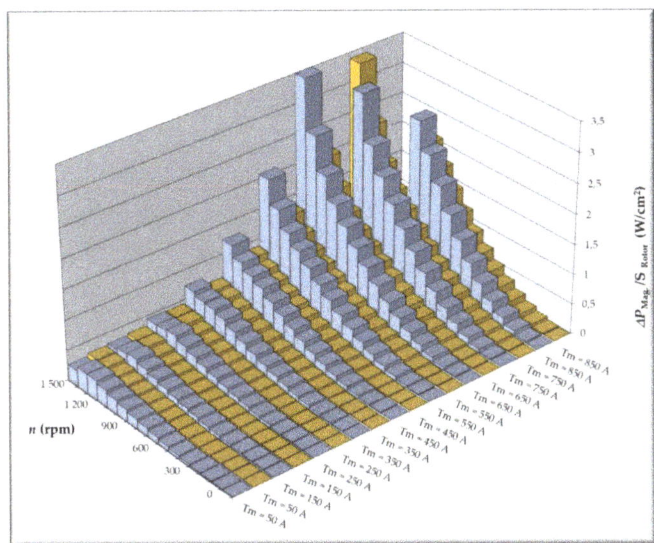

Figure 19. Bar chart of calculated ratio of PM losses to rotor's inner surface area for Motor #1 and Motor #2; rotational torque is treated as input quantity and is identical for both motors.

In order to compare properties of these two motors, and taking into account the fact that cooling systems of the two motors differ since dimensions of structural elements are not identical, calculations were conducted for the S1 duty cycle, assuming that steady-state winding temperature cannot exceed $T_{Cu} = 150\,°C$, while steady-state temperature of permanent magnets cannot exceed $T_{mag.} = 100\,°C$. Ansys Motor-CAD (Ansys Canonsburg, Pensylwania, USA) analysis used here was based on the electromagnetic circuit model as well as the thermal model. The conjugated thermo-electromagnetic model was solved by the iterative method for speed, ranging from $n = 0$ rpm to $n = 1500$ rpm. Calculated torque-speed curves of the motors are shown in Figure 20; these have been worked out adopting the previously stated assumptions related to allowable steady-state temperatures for winding and permanent magnets.

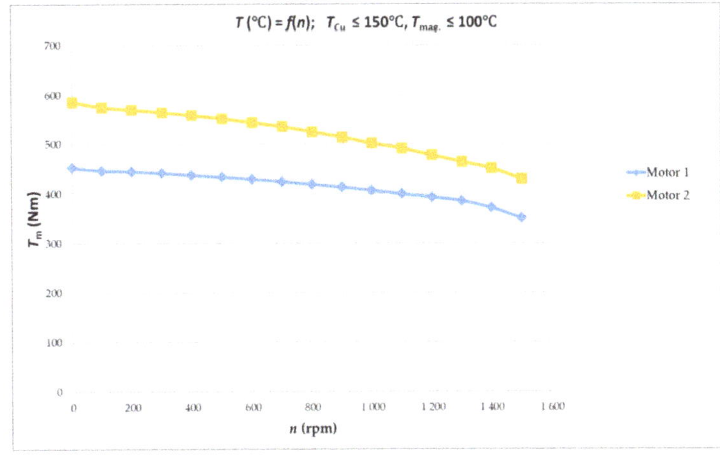

Figure 20. Calculated torque-speed curves with assumed steady-state temperatures of winding and magnet equal to, respectively: $T_{Cu} \leq 150\,°C$, $T_{mag.} \leq 100\,°C$.

The presented curves demonstrate clearly the advantage of Motor #2. Under the assumed conditions (maximum allowable temperatures of winding and magnets), the continuous torque of this motor is greater over the entire speed range than torque generated by Motor #1. In accordance with calculated power losses, difference in rotational torque becomes smaller as speed increases. Curves of maximum steady-state temperatures in winding, magnets, and magnetic core teeth of the stator are shown in Figure 21.

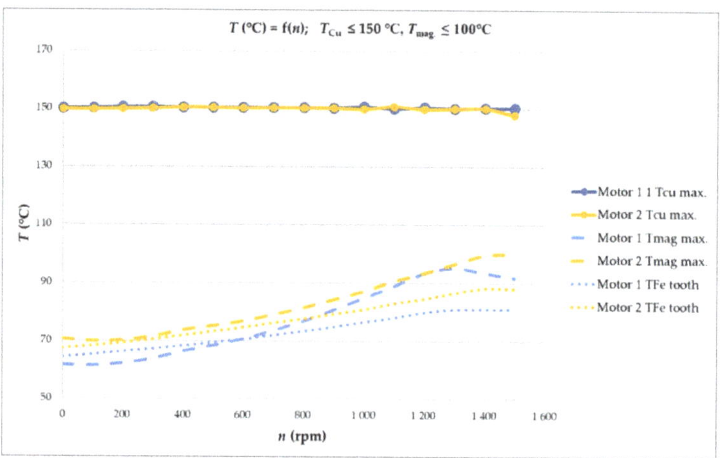

Figure 21. Maximum steady-state temperatures in winding, magnets, and magnetic core teeth of the stator vs. rotational speed; allowable steady-state temperatures for winding and magnets have been assumed as: $T_{Cu} \leq 150\,°C$, $T_{mag.} \leq 100\,°C$.

The calculated curves show that winding temperature determines the limits of the motor's continuous operation. The exception is the operational point of Motor #2, at speed $n = 1500$ rpm, where winding temperature does not attain allowable temperature $T_{Cu} < 150\,°C$, while permanent magnet temperature is equal to allowable temperature $T_{mag.} = 100\,°C$. When rotational speed rises above c. $n = 1200$ rpm, the temperatures of magnet and stator's core teeth stop increasing, and in case of Motor #1, they even start to decrease. Calculated winding losses, stator core losses and permanent magnet losses for limit motor load are shown in Figure 22. When we analyze the curves of calculated steady-state temperatures and corresponding power losses, we may observe that Motor #2 may operate continuously with much higher power losses in windings and core, generating higher rotational torque. At the same time, winding temperature is maintained at $T_{Cu} = 150\,°C$. When rotational speeds are higher and motors operate with weakened magnetic field, this impact is more perceptible in Motor #1, where magnetic core and permanent losses are noticeably decreased, which results in a decrease of magnet temperature. When speed is maximum, winding losses are almost identical in both motors; the same is true for magnet losses. Losses in the stator core are greater in Motor #2.

Supply current (RMS) curves for Motor #1 and Motor #2 are shown in Figure 23.

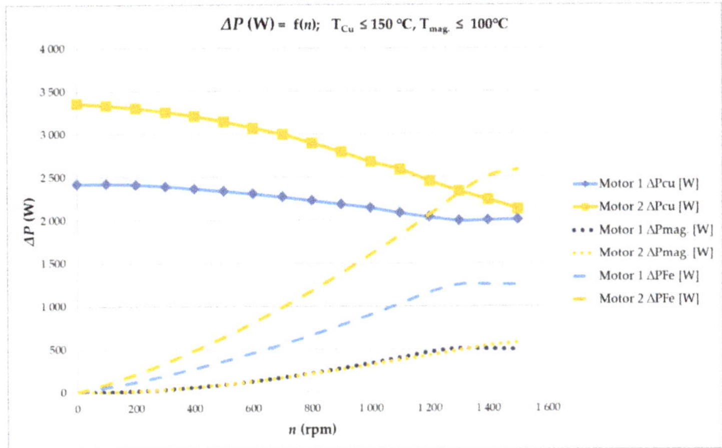

Figure 22. Calculated power losses in winding, magnets, and magnetic core teeth of the stator vs. rotational speed; allowable steady-state temperatures for winding and magnets have been assumed as: $T_{Cu} \leq 150\,°C$, $T_{mag.} \leq 100\,°C$.

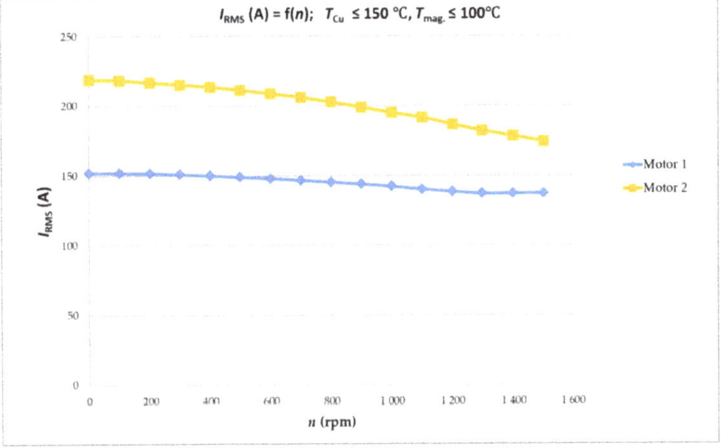

Figure 23. Supply current (I_{RMS} value) vs. rotational speed; allowable steady-state temperatures for winding and magnets have been assumed as: $T_{Cu} \leq 150\,°C$, $T_{mag.} \leq 100\,°C$.

5. Conclusions

The proposed electric motors intended for installation in a vehicle wheel must meet a number of requirements. One requirement is to reduce the weight of the motor so that the unsprung mass can be reduced. The research described was aimed at checking the possibility of reducing the mass by increasing the number of magnetic poles and increasing the width of the magnet at the expense of the rotor yoke. Model tests fully confirmed our expectations. Increasing the number of poles from 32 to 56 reduced the mass of the motor's electromagnetic circuit: from 20.0 to 15.3 kg, i.e., by 23.5%. Models used for electromagnetic and thermal calculations were generated in Ansys Motor-CAD software (Ansys Canonsburg, Pensylwania, USA). The models are conjugated, so that interaction of electromagnetic field and thermal field may be taken into account. Motor #1 model is a representation of prototype SMzs200S32 motor designed and built in Łukasiewicz Research Network—Institute of Electrical Drives and Machines KOMEL. Motor #2 model is a modification of Motor #1 model, number of magnetic poles is increased. Number of slots per phase per pole was also modified, from q = 0.5 to q = 0.375.

The calculations show that, in spite of a significant decrease in mass of the electromagnetic circuit, Motor #2 is characterized by improved thermal parameters.

By increasing the number of pairs of magnetic poles and the volume of the permanent magnet, the allowable constant torque of Motor #2 increased. Despite the significant increase in the power frequency and the consequent increase in the losses in the magnetic core, the reduction of power losses in the winding and the geometric changes in the cooling system were much more important. The presented characteristics show that the total losses in the stator are largely lower in Motor #2. The higher losses in the stator are taken by Motor #2 at high rotational speeds and low and medium torques, which is also visible in the operating point parameters in Table 3, where Motor #2 has a slightly lower efficiency. By illustrating this operating point on the stator accumulative loss graphs, you can see that this is the area where Motor #2 has slightly higher accumulative losses. Observing the changes in the stator total loss ratio in both motors, it can be seen that, at higher rotational speeds, the effect of high frequency is significant at higher rotational speeds and lower supply currents. The effect of the frequency losses decreases as the overall winding losses increase. The total losses in the winding change with the current intensity and only negligibly with the frequency (due to the relatively small AC losses in the winding).

The temperature of the stator core drops; this is due to a decrease in the winding losses transferred to the core and an increase in the heat dissipation surface (caused by an increase in the diameter of the stator support structure). The losses of permanent magnets are smaller in Motor #2, regardless of the increase in frequency, due to the much smaller angular dimension of the magnets. In addition to the calculated power losses converted into heat, the temperature of the motor components is influenced by the dimensions of the cooling system. Design modifications introduced in the motor increase the heat dissipation area in the stator (increased internal diameter of the stator) and permanent magnets (increased rotor diameter on which the magnets are mounted). In addition, it should be noted that when we reduce the width of the stator and the rotor yoke, the thermal resistance between the winding and the cooling system housing and between the permanent magnets and the rotor outer surface is also reduced. The reduced stator yoke leads to a reduction in the distance between the coil terminals and the cooling system, which in turn leads to a reduction in the thermal resistance of the epoxy resin layer interposed between these elements. As a consequence, the heat dissipation from the winding part, which is characterized by the highest temperature in standard electric machines, improves [48,49].

Motor-CAD iterative calculations allowed the comparison of the torque vs. speed curves (maximum continuous torque), assuming that the winding and magnet temperatures did not exceed T_{Cu} = 150 °C and $T_{mag.}$ = 100 °C, respectively. Motor #2, showing greater total losses at high rotational speeds, is characterized by a higher torque (in the S1 duty cycle) in the entire rotational speed range.

An additional advantage resulting from the increased number of pole pairs and the internal diameter of the stator is the distance of the motor structure from the brake drum (it is an additional source of heat during braking).

It should be noted that the power losses generated in the permanent magnets of the motor and the magnetic core can be further reduced by using permanent magnet segmentation and electrical sheets with a lower loss (we took M300-35A sheets in our calculations).

Apart from reducing the mass of the electromagnetic circuit, it is also necessary to analyze the possibilities of reducing the mass of the motor by reducing the mass of other elements, for example, by using light and durable materials [50].

The authors are currently exploring further options for reducing engine weight by improving cooling efficiency. The developed designs will allow better use of the electromagnetic circuit, which should result in a reduction of the engine weight while improving operating parameters.

Author Contributions: P.D.—development of computational models, running simulation, designing electromagnetic circuits, analysis of simulation results, preparation of the article. R.K.—analysis of simulation results, participation in the creation of proposals, editorial article. All authors have read and agreed to the published version of the manuscript.

Funding: The research is co-financed under the Program of the Ministry of Science and Higher Education "Implementation Doctorate" (POLAND). The research is a continuation of the project "Innovative Solutions for Direct Drive of Electric Vehicles", financed by National Centre for Research and Development under the LIDER VII program, in accordance with the agreement: LIDER/24/0082/ L-7/15/NCBR/2016 (POLAND). The project received the Research Award (main award) 25th Siemens Award Competition for scientists and research teams (Poland).

Institutional Review Board Statement: Not applicable.

Informed Consent Statement: Not applicable.

Data Availability Statement: Not applicable.

Conflicts of Interest: The authors declare no conflict of interest.

References

1. Ślaski, G.; Gudra, A.; Borowicz, A. Analysis of the influence of additional unsprung mass of in-wheel motors on the comfort and safety of a passenger car. *Arch. Autom. Eng. Arch. Motoryz.* **2014**, *65*, 51–64.
2. Parczewski, K.; Romaniszyn, K.; Wnęk, H. Influence of electric motors assembly in hubs of vehicle wheels on the dynamics of movement, especially on surfaces with different adhesion coefficient. *Combust. Eng.* **2019**. [CrossRef]
3. Dukalski, P.; Będkowski, B.; Parczewski, K.; Wnęk, H.; Urbaś, A.; Augustynek, K. Analysis of the influence of assembly electric motors in wheels on behaviour of vehicle rear suspension system. *Mater. Sci. Eng.* **2018**, *421*. [CrossRef]
4. Dukalski, P.; Będkowski, B.; Parczewski, K.; Wnęk, H.; Urbaś, A.; Augustynek, K. Dynamics of the vehicle rear suspension system with electric motors mounted in wheels. *Maint. Reliab.* **2019**, *21*, 125–136. [CrossRef]
5. Frajnkovic, M.; Omerovic, S.; Rozic, U.; Kern, J.; Connes, R.; Rener, K.; Biček, M. Structural Integrity of In-Wheel Motors. *SAE Tech. Paper* **2018**, *1829*, 2018. [CrossRef]
6. Biček, M.; Connes, R.; Omerović, S.; Gündüz, A.; Kunc, R.; Zupan, S. The Bearing Stiffness Effect on In-Wheel Motors. *Sustainability* **2020**, *12*, 4070. [CrossRef]
7. Parczewski, K.; Wnek, H. Comparison of overcoming inequalities of the road by a vehicle with a conventional drive system and electric motors placed in the wheels. In Proceedings of the Conference Transport Means 2020, Palanga, Lithuania, 2 October 2020.
8. Li, G.; Wang, Y.; Zong, C. Driving State Estimation of Electric Vehicle with Four-wheel-hub-motors. *Qiche Gongcheng Automot. Eng.* **2018**, *40*, 150–155.
9. Wanner, D.; Kreusslein, M.; Augusto, B.; Drugge, L. Single wheel hub motor failures and their impact on vehicle and driver behavior. *Veh. Syst. Dyn.* **2016**, *54*, 1–17. [CrossRef]
10. Luo, Y.; Tan, D. Lightweight design of an in-wheel motor using the hybrid optimization method. *Inst. Mech. Eng. Part D J. Automob. Eng.* **2013**, *227*, 1590–1602. [CrossRef]
11. Rahim, N.; Ping, H.; Tadjuddin, M. Design of an in-wheel axial flux brushless dc motor for electric vehicle. In Proceedings of the 2006 International Forum on Strategic Technology, Ulsan, Korea, 18–20 October 2006; pp. 16–19.
12. Shin, P.S.; Kim, H.D.; Chung, G.B.; Yoon, H.S.; Park, G.S.; Koh, C.S. Shape Optimization of a Large-Scale BLDC Motor Using an Adaptive RSM Utilizing Design Sensitivity Analysis. *IEEE Trans. Magn.* **2007**, *43*, 1653–1656. [CrossRef]
13. Seo, I.M.; Kim, H.K.; Hur, J. Design and analysis of modified spoke type BLDC motor using a ferrite permanent-magnet. In Proceedings of the 17th International Conference on Electrical Machines and Systems (ICEMS), Hangzhou, China, 22–25 October 2014; pp. 1701–1705.
14. Zhu, J.; Cheng, K.E.; Xue, X.; Zou, Y. Design of a novel high-torque-density in-wheel switched reluctance motor for electric vehicles. In Proceedings of the 2017 IEEE International Magnetics Conference (INTERMAG), Dublin, Ireland, 24–28 April 2017.
15. Artetxe, G.; Paredes, J.; Prieto, B.; Martinez-Iturralde, M.; Elosegui, I. Optimal pole number and winding designs for low speed–high torque synchronous reluctance machines. *Energies* **2018**, *11*, 128. [CrossRef]
16. Zhu, X.; Shu, Z.; Quan, L.; Xiang, Z.; Pan, X. Design and Multicondition Comparison of Two Outer-Rotor Flux-Switching Permanent-Magnet Motors for In-Wheel Traction. *IEEE Trans. Ind. Electron.* **2017**, *64*, 6137–6148. [CrossRef]
17. Rechkemmer, S.K.; Zhang, W.; Sawodny, O. Modeling of a Permanent Magnet Synchronous Motor of an E-Scooter for Simulation with Battery Aging Model. *IFAC PapersOnLine* **2017**, *50*, 4769–4774. [CrossRef]
18. Gao, P.; Gu, Y.; Wang, X. The Design of a Permanent Magnet In-Wheel Motor with Dual-Stator and Dual-Field-Excitation Used in Electric Vehicles. *Energies* **2018**, *11*, 424. [CrossRef]

19. Łebkowski, A. Design, Analysis of the Location and Materials of Neodymium Magnets on the Torque and Power of In-Wheel External Rotor PMSM for Electric Vehicles. *Energies* **2018**, *11*, 2293. [CrossRef]
20. Freitag, G.; Schramm, M. Electric wheel hub motor with high recuperative brake performance in automotive design. *World Electr. Veh. J.* **2012**, *5*, 510–513. [CrossRef]
21. Biček, M.; Lampič, G.; Zupan, S.; Obrul, B.; Gotovac, G.; Štefe, B.; Valentinčič, J. High Torque "In-Wheel" Motors for Rescue Vehicles. In Proceedings of the Innovative Automotive Technology—IAT 2012, Dolenjske Toplice, Slovania, 12–13 April 2012.
22. Lampič, G.; Detela, A.; Valentinčič, J. Management of innovative technology of Elaphe* "in-wheel" electric motors—A case study. In Proceedings of the 9th International Conference on Management of Innovative Technologies MIT'2007, Fiesa, Slovenia, 8–10 October 2007.
23. Elaphe. Available online: http://in-wheel.com/en/ (accessed on 19 October 2019).
24. Kostic Perovic, D. Making the Impossible, Possible—Overcoming the Design Challenges of In Wheel Motors. *World Electr. Veh. J.* **2012**, *5*, 514–519. [CrossRef]
25. Protean Electric. Available online: https://www.proteanelectric.com/ (accessed on 19 October 2019).
26. Ziehl-Abegg. Available online: https://www.ziehl-abegg.com/de/en/product-range/automotive/in-wheel-hub-motors/ (accessed on 19 October 2019).
27. Libert, F.; Soulard, J. Investigation on Pole-Slot Combinations for Permanent-Magnet Machines with Concentrated Windings. In Proceedings of the 16th International Conference on Electrical Machines, Lodz, Poland, 5–8 September 2004.
28. Peng, M.-T.; Flack, T. Design and Analysis of an In-Wheel Motor with Hybrid Pole-Slot Combinations. *IEEE Trans. Magn.* **2016**, *52*, 1–10. [CrossRef]
29. Guo, Q.; Zhang, C.; Li, L.; Gerada, D.; Zhang, J.; Wang, M. Design and implementation of a loss optimization control for electric vehicle in-wheel permanent-magnet synchronous motor direct drive system. *Appl. Energy* **2017**, *204*, 1217–1332. [CrossRef]
30. Yamazaki, K.; Shina, M.; Kanou, Y.; Miwa, M.; Hagiwara, J. Effect of Eddy Current Loss Reduction by Segmentation of Magnets in Synchronous Motors: Difference Between Interior and Surface Types. *IEEE Trans. Magn.* **2009**, *45*, 10. [CrossRef]
31. Martin, F.; El-Hadi Zaïm, M.; Tounzi, A.; Bernard, N. Improved Analytical Determination of Eddy Current Losses in Surface Mounted Permanent Magnets of Synchronous Machine. *IEEE Trans. Magn.* **2014**, *50*, 6.
32. Huang, W.-Y.; Bettayeb, A.; Kaczmarek, R.; Vannier, J.-C. Optimization of Magnet Segmentation for Reduction of Eddy-Current Losses in Permanent Magnet Synchronous Machine. *IEEE Trans. Energy Convers.* **2010**, *25*, 2.
33. Upadhayay, P.; Patwardhan, V. Magnet Eddy-Current Losses in External Rotor Permanent Magnet Generator. In Proceedings of the International Conference on Renewable Energy, Research and Applications, Madrid, Spain, 20–23 October 2013.
34. Będkowski, B.; Dukalski, P.; Jarek, T.; Wolnik, T. Tests of Electrical Motor for Installation in the Wheel Hub of an Electric Car. *IOP Conf. Series Mater. Sci. Eng.* **2020**, *841*, 012003. [CrossRef]
35. Toda, H.; Xia, Z.; Wang, J. Rotor Eddy-Current Loss in Permanent Magnet Brushless Machines. *IEEE Trans. Magn.* **2004**, *40*, 4. [CrossRef]
36. Soong, W.L.; Ertugrul, N. Field-Weakening Performance of Interior Permanent-Magnet Motors. *IEEE Trans. Ind. Appl.* **2002**, *38*, 5. [CrossRef]
37. Motor-CAD Help © 2019 Motor Design Ltd. Available online: https://www.motor-design.com/ (accessed on 17 May 2020).
38. Ahmad, A.S.; Youguang, G. Predicting the Behavior of Induction Machine Using Motor-CAD and MATLAB. In Proceedings of the 12th International Conference on Compatibility, Power Electronics and Power Engineering, Doha, Qatar, 10–12 April 2018. [CrossRef]
39. Jurkovic, M.; Žarko, D. Optimized Design of a Brushless DC Permanent Magnet Motor for Propulsion of an Ultra Light Aircraft. *Automatika* **2012**, *53*, 244–254. [CrossRef]
40. Abdullah, A.T.; Ali, A.M. Thermal analysis of a three-phase induction motor based on motor-CAD, flux2D, and matlab. *J. Electr. Eng. Comput. Sci.* **2019**, *15*. [CrossRef]
41. Jie, D. Thermal performance analysis of motor based on motor-CAD. *AIP Conf. Proc.* **2019**, *2073*, 020049.
42. Moreno, Y.; Almandoz, G.; Egea, A.; Madina, P.; Escalada, P.M.A.A.J. Multi-Physics Tool for Electrical Machine Sizing. *Energies* **2020**, *13*, 1651. [CrossRef]
43. Kowal, D.; Sergeant, P.; Dupré, L.; Vandenbossche, L. Comparison of Iron Loss Models for Electrical Machines with Different Frequency Domain and Time Domain Methods for Excess Loss Prediction. *IEEE Trans. Magn.* **2015**, *51*, 1. [CrossRef]
44. Yamazaki, K.; Fukushima, N. Iron-Loss Modeling for Rotating Machines: Comparison Between Bertotti's Three-Term Expression and 3-D Eddy-Current Analysis. *IEEE Trans. Magn.* **2010**, *46*, 3121–3124. [CrossRef]
45. Zhao, H.; Luo, Y.; Wang, H.; Peter, B. A Complete Model for Iron Losses Prediction in Electric Machines Including Material Measurement. Data Fitting, FE Computation and Experimental Validation Electrical Review. *Przegląd Elektrotechniczny* **2012**, *88*, 52–55.
46. Bertotti, G. General properties of power losses in soft ferromagnetic materials. *IEEE Trans. Magn.* **1988**, *24*, 621–630. [CrossRef]
47. Ouamara, D.; Dubas, F. Permanent-Magnet Eddy-Current Losses: A Global Revision of Calculation and Analysis. *Math. Comput. Appl.* **2019**, *24*, 67. [CrossRef]

48. Będkowski, B.; Madej, J. The innovative design concept of thermal model for the calculation of the electromagnetic circuit of rotating electrical machines. *Maint. Reliab.* **2015**, *17*, 481–486. [CrossRef]
49. Krok, R. *Sieci Cieplne w Modelowaniu Pola Temperatury w Maszynach Elektrycznych Prądu Przemiennego*; Wydawnictwo Politechniki Śląskiej, Monografia Habilitacyjna: Gliwice, Poland, 2010.
50. Kleine, A.; Rosefort, M.; Koch, H. Lightweight Construction for Electric Mobility Using Aluminium. *Light Metals* **2014**, *2014*, 331–335.

Article

Nonlinear Digital Simulation Models of Switched Reluctance Motor Drive

Xing Wang [1], Ryszard Palka [2] and Marcin Wardach [2,*

1   Clean Energy Conversion and Power Generation Research Centre, China University of Mining and Technology, Xuzhou 221116, China; wxstarwx@163.com
2   Department of Power Systems and Electrical Drives, West Pomeranian University of Technology, 70-313 Szczecin, Poland; Ryszard.Palka@zut.edu.pl
*   Correspondence: marwar@zut.edu.pl

Received: 15 November 2020; Accepted: 17 December 2020; Published: 19 December 2020

Abstract: The paper deals with nonlinear simulation models of a drive consisting of the four-phase 8/6 doubly salient switched reluctance motor (SRM), the four-phase dissymmetric bridge power converter and the closed-cycle rotor speed control strategy carried out by the pulse width modulation (PWM) with variable angle and combined control scheme with the PI algorithm. All presented considerations are based on a MATLAB-SIMULINK platform. The nonlinear mathematical model of the analyzed SRM drive was obtained as a combination of the two dimensional (2D) finite element model (FEM) of the motor and the nonlinear model of the electrical network of the power supply circuit. The main model and its seven sub-modules, such as the controller module, one phase simulation module, rotor position angle transformation module, power system module, phase current operation module, "subsystem" module, and electromagnetic torque of one phase operation module, are described. MATLAB functions store the magnetization curves data of the motor obtained by the 2D FEM electromagnetic field calculations, as well as the data of magnetic co-energy curves of the motor calculated from the magnetization curves. The 2D specimen insert method is adopted in MATLAB functions for operating the flux linkage and the magnetic co-energy at the given phase current and rotor position. The phase current waveforms obtained during simulations match with the tested experimentally phases current waveforms at the same rotor speed and the same load basically. The simulated rotor speed curves also agree with the experimental rotor speed curves. This means that the method of suggested nonlinear simulation models of the analyzed SRM drive is correct, and the model is accurate.

Keywords: motor control; switched reluctance motor; MATLAB; simulation

---

1. Introduction

The switched reluctance motor (SRM) drive, described comprehensively in [1], has been developed for mining equipment [2], electric vehicles [3], high speed equipment [4], different generators [5,6], wind generators [7], high speed and high power applications [8,9], flywheel energy storage applications [10], linear transportation [11], automotive applications [12], and so on, due to favorable conditions in the design of the motor and power supply, ease of four quadrants operation, high start torque with low start current, high efficiency within vast rotor speed ranges, high dependability with the independence of magnetic paths for each phase and the independence of circuits in each phase. The performance of switched reluctance motors is better than that of the hysteresis direct torque control of permanent-magnet synchronous motor system [13]. The SRM drive generally requires closed-cycle rotor speed control, where the opening angle and the turn-off angle of the power transistors, the chopping limitation of phase current, and the average phase voltage are the rotor speed control parameters.

Digital industrial controllers based on field programable gate array technologies are developed to implement artificial intelligence for industrial control applications of advanced drives [14]. The neural estimators are implemented for control of two-mass system with digital field programable gate array hardware [15]. The electric propulsion systems managed and controlled by a digital electronic control unit are designed for light electric vehicles [16]. A model with an adaptive control system for the flexible joint drive system based on the accommodative sliding neuro-fuzzy speed controller has been presented in [17]. The model of the communication network with time-delay was effectively handled by Smith predictor [18]. The description, modeling and control of discrete event systems for machines control logic project are developed in [19].

As with the developed novel mechatronics system [20–23], it is essential to integrate the double salient reluctance motor and power supply of single polarity when creating simulation models of the SRM drive. It is also important to set up an appropriate control scheme and strategy, together with the regulation algorithm which can be used for comparing and selecting the closed-loop rotor speed control algorithms. Such an approach can improve the control parameters and increase drive efficiency.

Because the magnetic characteristics of the motor parts, especially the iron core, are nonlinear and the power network also has nonlinear characteristics, the analyzed SRM drive system is also nonlinear. Considering the above, the simulation models of the SRM drive system should also take into consideration all nonlinear mathematical dependencies. For the active simulation of the SRM a flexible modeling technique within the SABER environment has been developed [24]. Thanks to this model, comprehensive modeling of a vast range of control policies and proper switching control angles could be performed. The model can reasonably predict the flux-linkage characteristics of the typical two-phase impelled 6/4 SRM drive [25]. An improved active model of a three-phase SRM includes modeling of influence of iron loss on current waveform, transient torque and hysteresis, eddy-current and extra iron losses [9,26]. An active modeling strategy of SRMs for synchronous two-phase excitation has been presented in [27]. To carry out a simulation model using MATLAB-SIMULINK software, the flux linkage data required for the model were measured experimentally by expanding the procedure for gauging single phase flux characteristics, and the gauged flux linkage data have been used. Modeling of SRMs using an adaptive neural fuzzy inference system (ANFIS) can be applied to study switched reluctance motor dynamic performance by torque and current models [28]. It has been shown that the reduplicative process is very helpful simulation tool in developing a servo-type four-quadrant SRM drive system [29]. Taking into account the machine nonlinearities the paper, [30] presents the flux linkage new analytical expression of a SRM as a function of the rotor position and current. According to the presented model, it may be concluded that a Fourier series representation can estimate the rotor position dependency on the flux linkage. The paper [31] states that a nonlinear model of SRM is characterized by state variables which are constant in the steady-state. This in turn facilitates the high-speed simulation, control and design of the machine. Nonlinearities of SRM resulting from the magnetic saturation and a non-sinusoidal self-inductance profile, causes complications in deriving this model. The difficulties were conquered by introducing a variable representation which approaches motor variables by the time-varying quotient vector and inner product of a vector of basic functions. In high-performance systems for the real-time control, models based on invertible equation can be very helpful. This expression is useful when the set current is derived of the torque command and denoting the torque-phase current relationship [32]. Paper [33] presents a novel rapid nonlinear simulation method for SRMs. This method uses the relationship among the magnetization curves of the machine estimated based on the curve matching the parameters of the Torrey model.

Many problems connected with the SRM design and its applications (mechanical, acoustic and thermal properties, losses, etc.) were discussed in [34], where additionally some significant comments and recommendations for the SRM power electronics are given. Additionally, in paper [35] a procedure similar to the one proposed in this paper for the improvement of the low-speed performance of SRMs is presented.

The main goal of the paper was to develop the nonlinear simulation models of the advanced SRM drive using the MATLAB platform combined with the 2D electromagnetic field FE analysis and SIMULINK software. The nonlinear electrical meshwork model of the 4-phase asymmetric bridge SRM power supply circuit and the closed-cycle rotor speed control strategy have been realized by the variable angle pulse width modulation control scheme together with the PI algorithm. Finally, the simulation solution of phase current waveforms and rotor speed curves were compared with the measured waveforms for a real four-phase 8/6 poles SRM motor.

This developed procedure is similar to the algorithm given in [36], where the multi-objective finite element model (FEM) optimization of a PMSM motor was performed with the help of MATLAB and ANSYS-Maxwell scripting capabilities. This result was used after that in paper [37] for the synthesis of the dead-beat current controller for this motor.

2. System Components and Mathematical Model

The analyzed SRM drive prototype consists of a four-phase 8/6 poles SRM and its four-phase asymmetric bridge power supply. The cross-section of the SRM under consideration has been shown in Figure 1. One phase winding of the motor is composed by two coils placed on the opposite stator poles. The motor windings are supplied by the current from the bridge power supply.

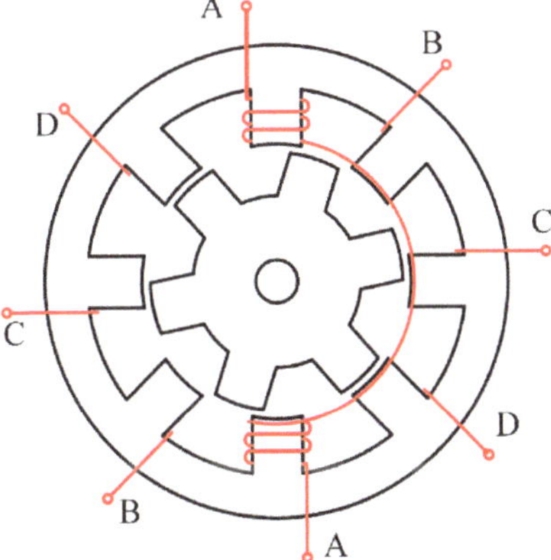

Figure 1. Cross-section of the four-phase 8/6 poles switched reluctance motor (SRM).

The topology of power circuit is shown in Figure 2. It consists of two major switches and two diodes in every phase of the power supply system.

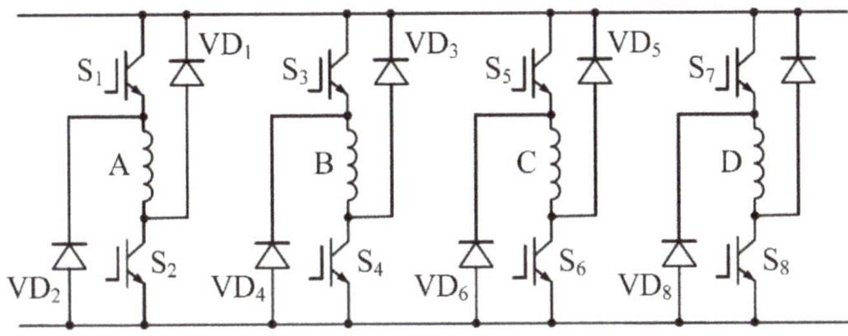

Figure 2. Topology of the power converter main circuit.

The alterable angle PWM combined control scheme with the PI algorithm is used for the closed-cycle rotor speed drive control. The opening angle of the major switches, θ_1, and the turn-off angle of the main switches, θ_2, at the two rotor speed values are given in Table 1, where (θ_1 and $\theta_2 = 0$ are defined as the rotor position while the axis of the rotor slot is in accordance with that of the stator pole of the conducting phase).

Table 1. θ_1 and θ_2 at the different rotor speed ranges.

n (r/min)	θ_1	θ_2
100~500	2.5°	22.5°
500~1000	0°	22.5°

The developed mathematical model of this SRM drive is in fact the nonlinear model of the electrical circuit of the power supply system combined with the two 2D FE motor models. Neglecting mutual inductances, it becomes [1]:

$$
\begin{bmatrix} U_A(\theta_A) \\ U_B(\theta_A - \theta_r/4) \\ U_C(\theta_A - \theta_r/2) \\ U_D(\theta_A - 3\theta_r/4) \end{bmatrix} = \begin{bmatrix} R_A + \frac{\partial \psi_A}{\partial i_A} \cdot D & 0 & 0 & 0 \\ 0 & R_B + \frac{\partial \psi_B}{\partial i_B} \cdot D & 0 & 0 \\ 0 & 0 & R_C + \frac{\partial \psi_C}{\partial i_C} \cdot D & 0 \\ 0 & 0 & 0 & R_D + \frac{\partial \psi_D}{\partial i_D} \cdot D \end{bmatrix} \begin{bmatrix} i_A \\ i_B \\ i_C \\ i_D \end{bmatrix} + \omega \cdot \begin{bmatrix} \frac{\partial \psi_A}{\partial \theta_A} \\ \frac{\partial \psi_B}{\partial (\theta_A - \theta_r/4)} \\ \frac{\partial \psi_C}{\partial (\theta_A - \theta_r/2)} \\ \frac{\partial \psi_D}{\partial (\theta_A - 3\theta_r/4)} \end{bmatrix}, \quad (1)
$$

where:

$Di_k = di_k/dt$, $(k = A, B, C, D)$,
U_A, U_B, U_C, U_D—voltages of phases A, B, C and D,
θ_A—the rotor position of phase A,
θ_r—one rotor period,
R_A, R_B, R_C, R_D—resistances of phases A, B, C and D,
$\Psi_A, \Psi_B, \Psi_C, \Psi_D$—flux linkages of phases A, B, C and D,
i_A, i_B, i_C, i_D—currents in phases A, B, C and D,
ω—the rotor angular velocity,
t—the time,

and

$$
\psi_k = \frac{NI}{3S} \sum \left(A_i \Delta_i - A_j \Delta_j \right)_k, \, k = A, B, C, D, \quad (2)
$$

where:

N—each phase turn numbers of winding,
L—active length of motor iron core,
S—stator pole area at one side,
Δ_i and Δ_j—area of the split cell at the right and left side of the stator poles,
A_i and A_j—magnetomotive force vector of the split cell at the right and left side of the stator poles, separately.

The typical equation for the magnetic vector potential (together with the homogeneous boundary conditions) describes the magnetic field distribution within the SRM:

$$
\begin{cases} \frac{\partial}{\partial x}\left(\gamma \frac{\partial A}{\partial x} \right) + \frac{\partial}{\partial y}\left(\gamma \frac{\partial A}{\partial y} \right) = -J \\ A|_{D_2,d_2} = 0 \end{cases}, \quad (3)
$$

where:

γ—the reciprocal of permeability,
J—stator phase winding current density,

D_2—outer diameter of stator,

d_2—diameter of rotor bore,

A—magnetic vector potential.

Figure 3 shows the computational cross-section of the analyzed motor.

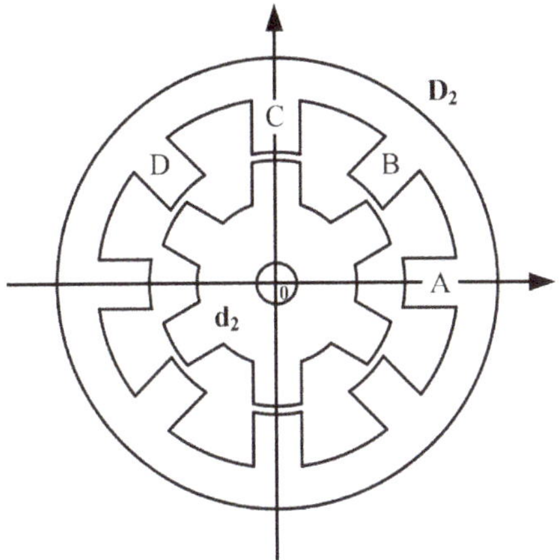

Figure 3. Analyzed cross-section of the motor.

The machine phase voltages can be expressed using equations:

(1) During the supply part of cycle, when the major switches in one phase are open

$$U_k = U_S - 2U_T, k = A, B, C, D, \tag{4}$$

where U_S is the DC supply voltage of the power supply, and U_T is the on-state voltage drop of the major switch.

(2) During the course of commutation, the major switches in one phase circuit are closed

$$U_k = -U_S - 2U_D, k = A, B, C, D, \tag{5}$$

where, U_D is the diode on-state voltage drop.

(3) While the power supply does not work, the current in phase is naturally continuous and one phase major switch is turned off,

$$U_k = -U_T - U_D, k = A, B, C, D, \tag{6}$$

The system equation of the mechanical motion can be written as [1,12,35]:

$$T_e = J\frac{d\omega}{dt} + H\omega + T_L, \tag{7}$$

$$\frac{d\theta}{dt} = \omega, \tag{8}$$

where J is the sum of inertia moments of the motor and loads, T_e is the electromagnetic torque of the motor, T_L is the torque of the loads, H is the viscous coefficient of friction, t is the time period and θ is the rotor position.

3. Simulation Models

The nonlinear simulation model of the SRM drive developed by MATLAB-SIMULINK platform is given in Figure 4. In this model, the turn-on angle of the major switches in the power converter, θ_1, is fixed at "*Xon*", the turn-off angle of the major switches in the power supply system, θ_2, is fixed at "*Xoff*", which is shown for different rotor speeds ranges in Table 1. "*Us*" is the DC supply voltage of the power supply, "θ" is the rotor position, "ω" is the angular speed of the motor, "n^*" is the set speed, "*n*" is the practical rotor speed, "*D*" is the duty ratio of the pulse width modulation signal, "*n*" is the given rotor speed, "*Te*" is the total electromagnetic torque of the motor, "*Ta*", "*Tb*", "*Tc*", and "*Td*" are the electromagnetic torque in *A phase*, *B phase*, *C phase*, *D phase*, separately, "*TL*" is the load, "*J*" is the sum of all system inertia moments, "*H*" is the viscous coefficient of friction, "*ia*", "*ib*", "*ic*", and "*id*" are the phase current in *A phase*, *B phase*, *C phase*, *D phase*, separately. The "*PI controller*" module is responsible for the closed-cycle rotor speed control. The expanding module of "*PI controller*" is shown in Figure 5, where, "*Kp*" is the proportion coefficient, "K_I" is the integral coefficient, the input "*in1*" is the difference among the given rotor speed and the real rotor speed, the output "*out1*" is the duty ratio of the pulse width modulation semaphore, and the "*saturation*" is used to limit the duty ratio between 0.000 and 1.000.

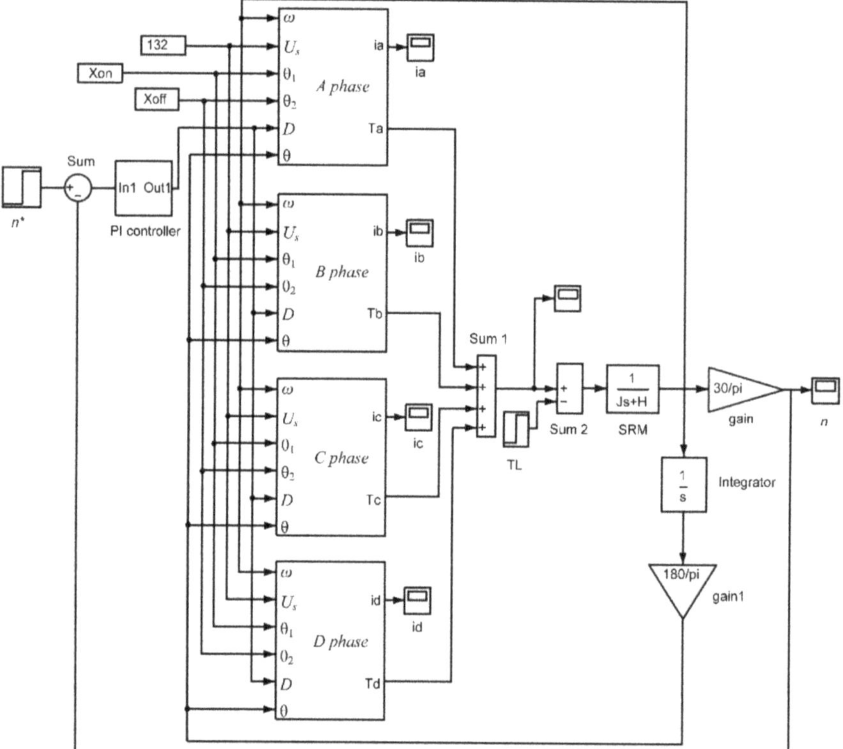

Figure 4. Nonlinear simulation main model of the switched reluctance motor drive.

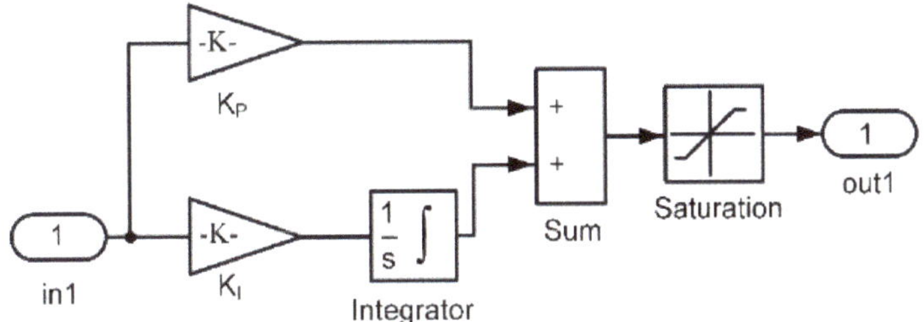

Figure 5. Controller module.

One phase simulation module of the main model is shown in Figure 6. It contains the power system module *"converter"*, the phase current computation module *"calculating i"*, the one phase electromagnetic torque calculation module *"calculating T"*, the rotor place angle conversion module *"θ conversion"*. The phase difference of the analyzed SRM is 15°. The rotor position angle conversion module is used to convert the real rotor position of every phase as the relative rotor position. The expanding rotor position angle conversion module is indicated in Figure 7. The input *"In1"* is the absolute rotor place angle, and the output *"out1"* is the relative rotor position of the certain phase. *"Constant"* is one rotor period, θ_r, such as one rotor period of four-phase 8/6 is 60°. *"Constant1"* is 0° for *A* phase, 15° for *B* phase, 30° for *C* phase, and 45° for *D* phase.

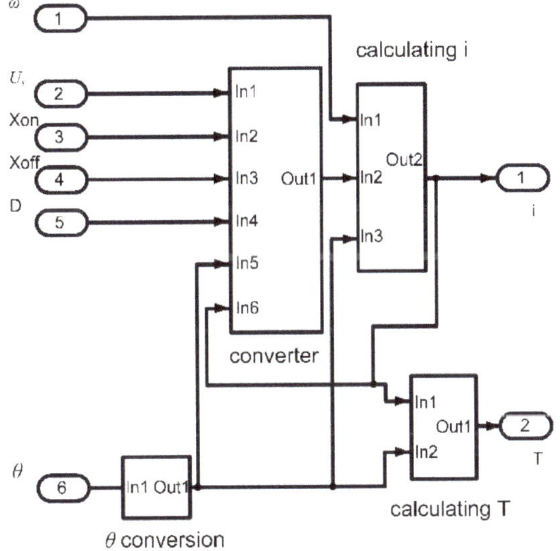

Figure 6. One phase simulation module.

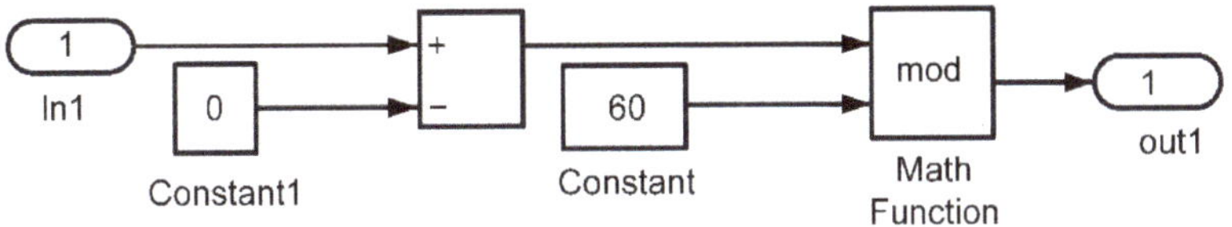

Figure 7. Rotor position conversion module.

Figure 8 presents the module of the expanding power converter, where the import of module *"In1"~"In6"*, are the DC supply voltages, respectively. This module contains also turn-on and turn-off angles of the major switches, a pulse width modulation duty ratio, position angle of the machine rotor, phase currents respectively, and an output "*out1*", which is the phase voltage "U_k". Proposed "*MATLAB Function*" is used to calculate the phase voltages, and contains switches "turn-on" and "turn-off" rule. This rule is based on the PWM duty ratio. The "*MATLAB Function*" includes also opening angle and major switches turn-off angle. While $\theta_1 \le \theta \le \theta_2$, if the PWM semaphore is "1", the phase voltage is expressed by Equation (4), if the PWM semaphore is "0", the phase voltage is denoted as Equation (6). While $\theta > \theta_2$, the phase voltage is expressed by Equation (5) if the phase current exists, and the phase voltage is zero if the phase current does not exist. While $\theta < \theta_1$, the voltage is zero if the phase current does not exist, and the voltage is expressed by Equation (5) if the current in phase exists.

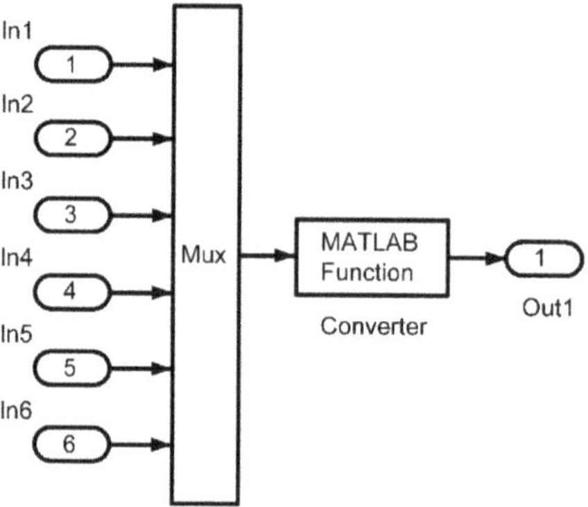

Figure 8. Power converter module.

From Equation (1), we have:

$$\frac{di_k}{dt} = \frac{U_k - R_k i_k - \frac{\partial \psi_k}{\partial \theta} \omega}{\partial \psi_k / \partial i_k}. \tag{9}$$

Based on the Equation (9), the simulation model of the phase current calculation module is shown in Figure 9. The import of module, "*In1*"~"*In3*", are the rotor angular velocity, the phase voltage, the converted relative rotor place, respectively, and the export of module "*out1*" is the instantaneous phase current. "*Subsystem*" is the module for calculating $\partial \Psi / \partial i$ and $\partial \Psi / \partial \theta$, which are gained by the magnetization curves of the motor.

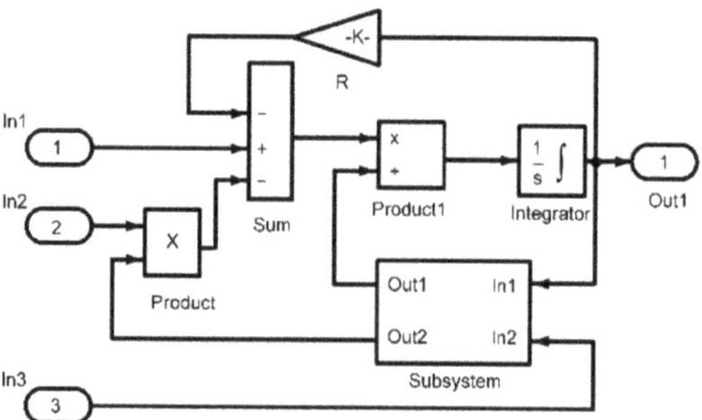

Figure 9. Phase current calculation module.

Finally, we have:

$$\begin{cases} \frac{\partial \psi}{\partial i} = \frac{\psi(i+\Delta i,\theta)-\psi(i,\theta)}{\Delta i} \\ \frac{\partial \psi}{\partial \theta} = \frac{\psi(i,\theta+\Delta\theta)-\psi(i,\theta)}{\Delta\theta} \end{cases}, \qquad (10)$$

Based on Equation (10), the expanding "*Subsystem*" module is shown in Figure 10.

In the module, "*in1*" is the instantaneous current value, "*in2*" is the instantaneous converted relative rotor position value, "*out1*" is $\partial\Psi/\partial i$, and "*out2*" is $\partial\Psi/\partial\theta$. In the "*Subsystem*" the current incremental quantity Δi is calculated from "*Constant1*" and "*Constant2*", but the rotor place incremental quantity $\Delta\theta$ is calculated from "*Constant3*" and "*Constant4*". Figure 11 shows magnetization map of analyzed motor, computed by the electromagnetic 2D FEM. These curves are stored in MATLAB Functions by data, such as "*MATLAB Fcn1*", "*MATLAB Fcn2*", "*MATLAB Fcn3*" presented in Figure 10. In order to compute the flux linkage at the given rotor position and phase current, in view of the magnetization curves, the "MATLAB Function" with the 2-D specimen insert method has been adopted.

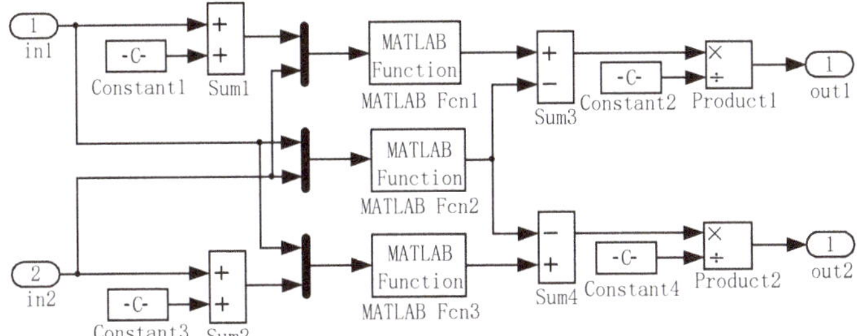

Figure 10. Expanded "*Subsystem*" module.

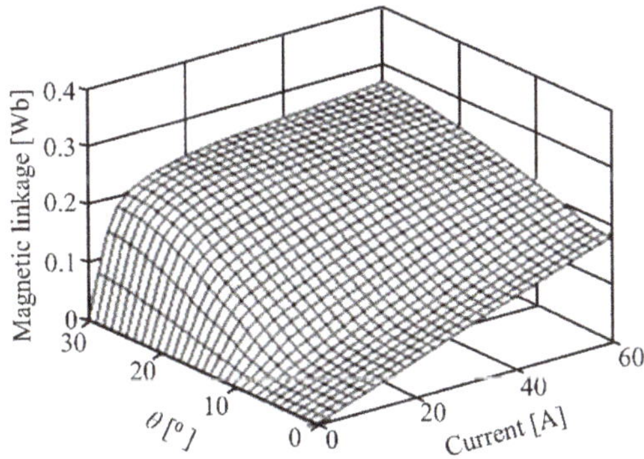

Figure 11. Magnetization map.

According to Equations (9) and (10) and Figures 6–10, the 2D FE model of the electric machine and the nonlinear electrical meshwork model of the power system have been integrated. The magnetic co-energy of the motor can be computed as follows:

$$W' = \int_0^{i_k} \psi_k di. \qquad (11)$$

The magnetic co-energy curves of the motor are calculated by Equation (11) based on the magnetization curves in Figure 11. They are shown in Figure 12.

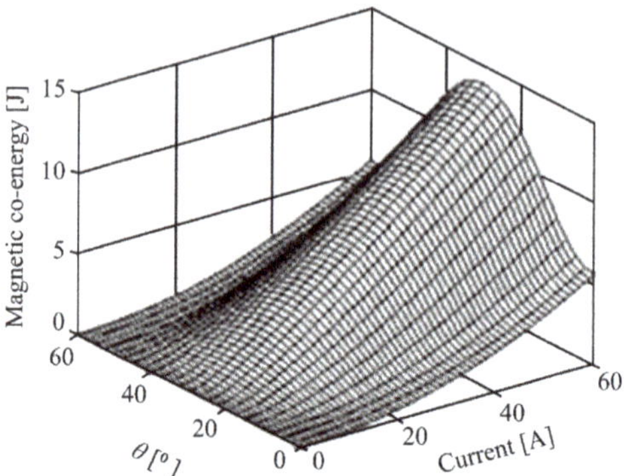

Figure 12. Magnetic co-energy surfaces.

The equation for calculating electromagnetic torque generated by each phase of the double salient reluctance machine can be written as

$$T_k = \frac{\partial W'}{\partial \theta}\Big|_{i_k=const} = \frac{W'(\theta + \Delta\theta) - W'(\theta)}{\Delta\theta}\Big|_{i_k=const}. \tag{12}$$

Based on the Equation (12), the simulation model of one phase electromagnetic torque computation module is shown in Figure 13.

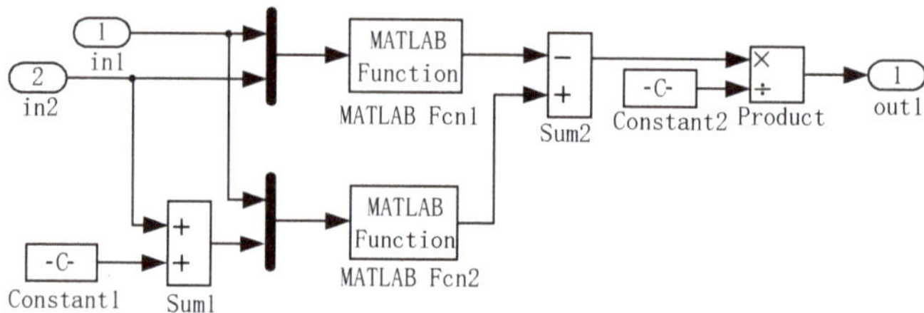

Figure 13. One phase electromagnetic torque calculation module.

In the module, "*In1*" is the instantaneous current value, "*in2*" is the instantaneous converted relative rotor place value, and the export of module "*out1*" is the one phase electromagnetic torque instantaneous value. "*Constant1*" and "*Constant2*" are the rotor position increment, $\Delta\theta$.

The magnetic co-energy curves data of the motor in Figure 12 is stored in MATLAB Functions by data, such as "*MATLAB Fcn1*", "*MATLAB Fcn2*" in Figure 13. During calculating for computing magnetic co-energy at the given rotor place and phase current, in view of the magnetic co-energy curves in MATLAB Functions the 2D specimen insert method has been adopted.

4. Simulated and Measured Results

The developed SRM (see Appendix A) drive prototype with the closed-cycle control speed conducted with use of the variable angle of the PWM combined control scheme with the PI algorithm is simulated by the proposed nonlinear simulation models on MATLAB-SIMULINK, and it is also tested experimentally at the same conditions. The photography of the SRM with the torque/speed meter and the load is shown in Figure 14, where the motor is on the left, the torque/speed meter is in the middle, and the load is on the right. The picture of the 4-phase asymmetric bridge power unit with the Intel 8XC196KC MCU controller (high performance member of the MCS 96 microcontroller family) is indicated in Figure 15, where on the right the power converter is shown, the Intel controller is on the left, and the isolation/enlargement and current/voltage protected circuit board for major switches in the power system is shown in the middle.

In the prototype, the DC voltage of the power supply is 132 V, the frequency of the PWM semaphore is 5.0 kHz, the major switches opening and turn-off angles in the power electronics is fixed based on the rotor speed ranges included in Table 1. The rating output torque is 0.5 Nm and the rating rotor speed is 900 rpm.

Figure 14. Measuring stand for determination of SRM-drive system properties.

Figure 15. Power converter with the controller.

Figure 16 shows phase current waveforms obtained during simulating research and Figure 17 presents current waveforms which are result from experimental tests.

The comparison between simulated and experimental peak value of currents is given in Table 2. As it results from this comparison, the maximum error between the calculated and measured values is not more than 8% for low speeds and not more than 6% for high speeds.

Table 2. Comparison of simulated and experimental phase current peak values.

	Simulated (A)	Tested (A)
300 rpm, 0.5 Nm	6.0	6.5
600 rpm, 0.5 Nm	4.6	5.0
600 rpm, 1.0 Nm	9.0	9.6
900 rpm, 1.0 Nm	7.0	7.0

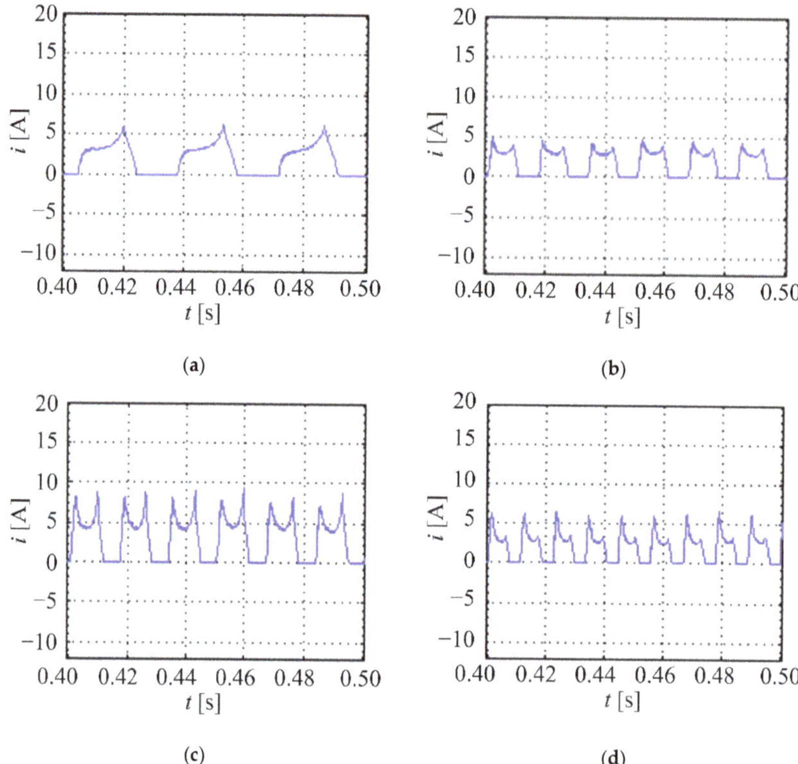

Figure 16. Phase current waveforms of the analyzed SRM machine: (**a**) the rotor speed is n = 300 rpm and the load is T = 0.5 Nm; (**b**) n = 600 rpm and T = 0.5 Nm; (**c**) n = 600 rpm and T = 1.0 Nm; (**d**) n = 900 rpm and T = 1.0 Nm.

Based on Table 1, the opening angle is fixed at 2.5° in Figures 16a and 17a, the opening angle is fixed at 0° in Figures 16b–d and 17b–d. The turn-on angle in Figures 16a and 17a are later than those in Figures 16b–d and 17b–d, so that the ascending rate of the phase current in the former case is lower than the ascending rate of the phase current in the latter. At the same rotor speed, the load in Figures 16c and 17c is larger than the load in Figures 16b and 17b, so that the duty ratio of the PWM semaphore in Figures 16c and 17c is larger than that in Figures 16b and 17b, the phase current in Figures 16c and 17c is bigger than that in Figures 16b and 17b.

Figures 16 and 17 prove that at the same rotor speed and load, simulated phase current waveforms agree basically with the experimental ones.

The curves of simulated rotor speeds are given in Figure 18 and the curves of experimentally tested rotor speeds are given in Figure 19. The comparison of simulated and tested rotor speeds is given in Table 3.

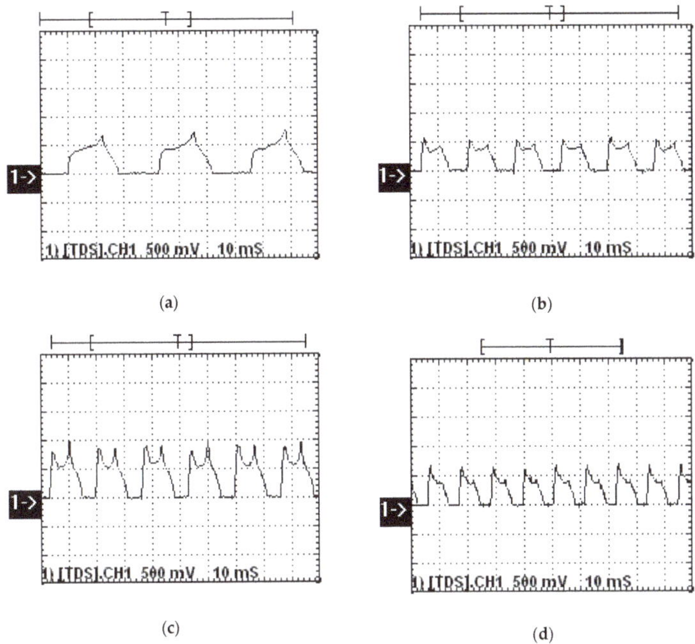

Figure 17. Tested phase current waveforms (ordinate: 5.0 A/div., abscissa: 10.0 ms/div.): (**a**) the rotor speed is *n* = 300 rpm and the load is *T* = 0.5 Nm; (**b**) *n* = 600 rpm and *T* = 0.5 Nm; (**c**) *n* = 600 rpm and *T* = 1.0 Nm; (**d**) *n* = 900 rpm and *T* = 1.0 Nm.

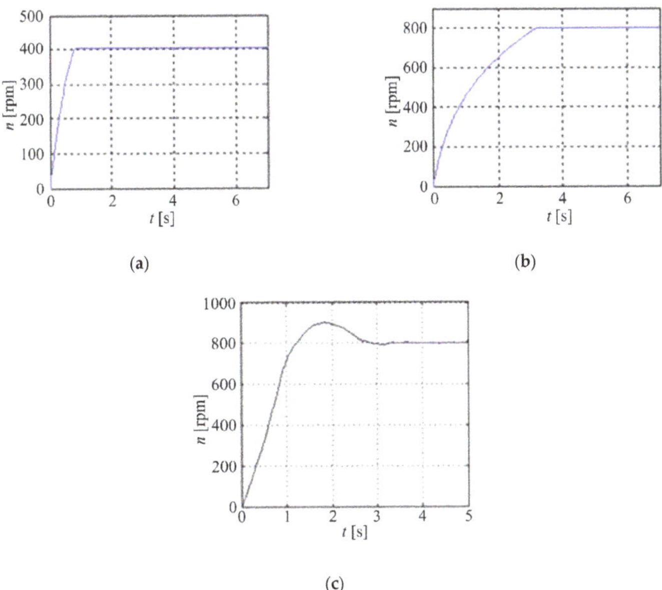

Figure 18. Simulated rotor speed curves: (**a**) the given rotor speed is *n* = 400 rpm, and the load is *T* = 0.05 Nm; (**b**) *n* = 800 rpm and *T* = 0.05 Nm; (**c**) *n* = 800 rpm and *T* = 0.50 Nm.

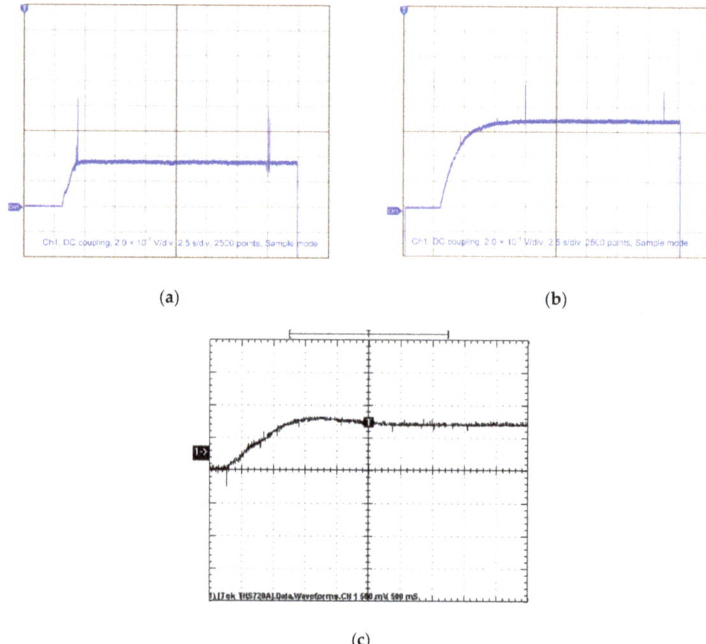

(a)   (b)

(c)

Figure 19. Curves of the tested rotor speed: (**a**) the given rotor speed is $n = 400$ rpm, and the load is $T = 0.05$ Nm (ordinate: 230 rpm/div., abscissa: 2.5 s/div.); (**b**) $n = 800$ r/min and $T = 0.05$ Nm (ordinate: 230 rpm/div., abscissa: 2.5 s/div.); (**c**) $n = 800$ r/min and $T = 0.50$ Nm (ordinate: 570 rpm/div., abscissa: 0.5 s/div.).

Table 3. Comparison simulated and tested rotor speed curves.

	400 rpm, 0.05 Nm		800 rpm, 0.05 Nm		800 rpm, 0.50 Nm	
	Simulated	Tested	Simulated	Tested	Simulated	Tested
Rising time (s)	0.70	0.70	2.60	2.70	1.00	1.00
Settling time (s)	0.80	0.80	3.10	3.10	3.30	3.25
Overshoot (%)	0.00	0.00	0.00	0.00	+12.50	+14.29

As can be seen, the curves of the simulated rotor speeds agree with the tested experimentally ones at the same certain rotor speed and load, and the closed-cycle rotor speed control can be implemented by the PI controller which works with high accuracy. All this proves the correctness of the adopted assumptions and the good functioning of the proposed method of SRM control.

5. Conclusions

The presented nonlinear simulation model of the 4-phase 8/6 poles SRM drive has been developed on a MATLAB-SIMULINK platform, which is combined with the magnetization curves data computed by the 2D electromagnetic field FE analysis (Figures 11 and 12). The nonlinear electrical meshwork model of the 4-phase asymmetric bridge power electronics main circuit and the closed-cycle rotor speed control strategy have been realized by the variable angle pulse width modulation combined control scheme with the PI algorithm. It includes one main module and seven sub-modules, such as: controller, one phase simulation module, rotor place angle conversion module, power electronics module, phase current computation module, "subsystem" module and one phase electromagnetic torque computation module (Figure 4). The magnetization curves data of the motor are stored in the

MATLAB functions, the magnetic co-energy curves data of the motor calculated from the magnetization curves data are also stored in other MATLAB functions. The two dimensions specimen insert method was adopted in MATLAB functions for computing the flux linkage and the magnetic co-energy curves at the given rotor position and the given phase current. The waveforms of the simulated current agree very well with the waveforms of the experimentally tested ones at the same rotor speed and load (Figures 16 and 17). The simulated rotor speed curves also agree basically with the curves of the experimentally tested rotor speeds (Figures 18 and 19). It was proved that the method of the proposed nonlinear simulation models of the analyzed SRM drive is correct, and the final model is accurate. The main achievement of the work is the creation of a full stable algorithm for controlling SRM machines. The limitations of the proposed algorithm are caused by the accuracy of the SR machine FEM model and the quality of the determined torque and magnetic flux values. On the hardware side, the limitations are the speed and accuracy of the hardware used. The results obtained in this paper enable better control of SR machines in different working states. The proposed method can be applied to develop other simulation models of SRM drives with other double salient reluctance motors and different topologies of the control system and power electronics. Simulation models proposed in the paper can contribute to develop semi-physical simulation platforms which can be used for actual time simulation and—as a consequence—facilitate development of the SRM drive. In the future, research will consist in improving the proposed method and automating the connection of individual computational programs: FEM, MATLAB and SIMULINK.

Author Contributions: Conceptualization, X.W., R.P. and M.W.; methodology, R.P. and M.W.; software, X.W.; validation, X.W., R.P. and M.W.; formal analysis, R.P. and M.W.; data processing, X.W.; writing—original draft preparation, X.W.; writing—review and editing, R.P. and M.W. All authors have read and agreed to the published version of the manuscript.

Funding: This research was funded by Xuzhou Science and Technology Plan Project, grant number KH17004.

Conflicts of Interest: The authors declare no conflict of interest.

Appendix A

Table A1. Parameters of tested Switch Reluctance Machine.

SRM Parameters	
Phase number	4
Number of stator pole	8
Number of rotor pole	6
Stator bore diameter	78.0 mm
Stator outer diameter	130.0 mm
Air gap length	0.15 mm
Rotor bore diameter	30.0 mm
Length of laminated iron core	135 mm
Number of turns per pole	7
Type of iron	DR510–50

References

1. Miller, T.J.E. *Switched Reluctance Motors and Their Control*; Oxford University Press: Oxford, UK, 1993.
2. Chen, H.; Gu, J.J. Implementation of three-phase switched reluctance machine system for motors and generator. *IEEE/ASME Trans. Mechatron.* **2010**, *15*, 421–432. [CrossRef]
3. Radimov, N.B.; Rabinovici, R. Switched reluctance machines as three-phase ac autonomous generator. *IEEE Trans. Magn.* **2006**, *42*, 3760–3764. [CrossRef]
4. Calverley, S.D.; Jewell, G.W.; Saunders, R.J. Aerodynamic losses in switched reluctance machine. *IEEE Proc. Electr. Power Appl.* **2000**, *147*, 443–448. [CrossRef]
5. Krishnan, R.; Blanding, D. High reliability SRM driver system for aerospace application. *Proc. IEEE APEC* **2003**, *2*, 1110–1115.

6. Chang, Y.; Liaw, C. On the design of power circuit and control scheme for switched reluctance generator. *IEEE Trans. Power Electron.* **2008**, *23*, 445–454.

7. Echenique, E.; Dixon, J.; Cárdenas, R.; Peña, R. Sensorless control for a switched reluctance wind generator, based on current slopes and neural networks. *IEEE Trans. Ind. Electron.* **2009**, *56*, 817–825. [CrossRef]

8. May, H.; Canders, W.-R.; Palka, R.; Holub, M. Optimisation of the feeding of switched reluctance machines for high speed and high power applications. *Stud. Appl. Electromagn. Mech.* **2002**, *22*, 489–494.

9. Yan, W.; Chen, H.; Liu, X.; Ma, X.; Lv, Z.; Wang, X.; Palka, R.; Chen, L.; Wang, K. Design and multi-objective optimisation of switched reluctance machine with iron loss. *IET Electr. Power Appl.* **2019**, *13*, 435–444. [CrossRef]

10. Holub, M.; Palka, R.; Canders, W.-R. Control of Switched Reluctance Machines for Flywheel Energy Storage Applications. *Electromotion* **2005**, *12*, 185–191.

11. Nie, R.; Chen, H.; Liu, J.; Zhao, W.; Wang, X.; Palka, R. Compensation analysis of longitudinal end effect in three-phase switched reluctance linear machines. *ET Electr. Power Appl.* **2020**, *14*, 165–174. [CrossRef]

12. Krishnamurthy, M.; Edrington, C.S.; Emadi, A.; Asadi, P.; Ehsani, M.; Fahimi, B. Making the Case for Applications of Switched Reluctance Motor Technology in Automotive Products. *IEEE Trans. Power Electron.* **2006**, *21*, 659–675. [CrossRef]

13. Gulez, K.; Adam, A.A.; Pastaci, H. A novel direct torque control algorithm for ipmsm with minimum harmonics and Torque Ripples. *IEEE/ASME Trans. Mechatron.* **2007**, *12*, 223–227. [CrossRef]

14. Monmasson, E.; Idkhajine, L.; Cirstea, M.N.; Bahri, I.; Tisan, A.; Naouar, M.W. FPGAs in industrial control applications. *IEEE Trans. Ind. Informat.* **2011**, *7*, 224–243. [CrossRef]

15. Orlowska-Kowalska, T.; Kaminski, M. FPGA implementation of the multilayer neural network for the speed estimation of the two-mass drive system. *IEEE Trans. Ind. Informat.* **2011**, *7*, 436–445. [CrossRef]

16. Bertoluzzo, M.; Buja, G. Development of electric propulsion systems for light electric vehicles. *IEEE Trans. Ind. Informat.* **2011**, *7*, 428–435.

17. Orlowska-Kowalska, T.; Szabat, K. Damping of torsional vibrations in two-mass system using adaptive sliding neuro-fuzzy approach. *IEEE Trans. Ind. Informat.* **2008**, *4*, 47–57. [CrossRef]

18. Lai, C.; Hsu, P. Design the remote control system with the time-delay estimator and the adaptive smith predictor. *IEEE Trans. Ind. Informat.* **2010**, *6*, 73–80.

19. Polic, A.; Jezernik, K. Closed-loop matrix based model of discrete event systems for machine logic control design. *IEEE Trans. Ind. Informat.* **2005**, *1*, 39–46.

20. Filev, D.P.; Chinnam, R.B.; Tseng, F.; Baruah, P. An industrial strength novelty detection framework for autonomous equipment monitoring and diagnostics. *IEEE Trans. Ind. Informat.* **2010**, *6*, 767–779. [CrossRef]

21. Kong, K.; Bae, J.; Tomizuka, M. Control of rotary series elastic actuator for ideal force-mode actuation in human–robot interaction applications. *IEEE/ASME Trans. Mechatron.* **2009**, *14*, 105–118. [CrossRef]

22. Lu, L.; Chen, Z.; Yao, B.; Wang, Q. Desired compensation adaptive robust control of a linear-motor-driven precision industrial gantry with improved cogging force compensation. *IEEE/ASME Trans. Mechatron.* **2008**, *13*, 617–624. [CrossRef]

23. Liu, H.; Meusel, P.; Hirzinger, G.; Jin, M.; Liu, Y.; Xie, Z. The modular multisensory DLR-HIT-hand: Hardware and software architecture. *IEEE/ASME Trans. Mechatron.* **2008**, *13*, 461–469.

24. Michon, M.; Calverley, S.D.; Powell, D.J.; Clark, R.E.; Howe, D. Dynamic model of a switched reluctance machine for use in a saber based vehicular system simulation. In Proceedings of the IEEE 40th IAS Annul Meeting, Hong Kong, China, 2–6 October 2005; pp. 2280–2287.

25. Farshad, M.; Faiz, J.; Lucas, C. Development of analytical models of switched reluctance motor in two-phase excitation mode: Extended miller model. *IEEE Trans. Magn.* **2005**, *41*, 2145–2155. [CrossRef]

26. Charton, J.T.; Corda, J.; Stephenson, J.M.; Randall, S.P. Dynamic modelling of switched reluctance machines with iron losses and phase interactions. *IEEE Proc.-Electr. Power Appl.* **2006**, *153*, 327–336. [CrossRef]

27. Jain, A.K.; Mohan, N. Dynamic modeling, experimental characterization, and verification for SRM operation with simultaneous two-phase excitation. *IEEE Trans. Ind. Electron.* **2006**, *53*, 1238–1249. [CrossRef]

28. Ding, W.; Liang, D. Modeling of a 6/4 switched reluctance motor using adaptive neural fuzzy inference system. *IEEE Trans. Magn.* **2008**, *44*, 1796–1804. [CrossRef]

29. Husain, I.; Hossain, S.A. Modeling, simulation, and control of switched reluctance motor drives. *IEEE Trans. Ind. Electron.* **2005**, *52*, 1625–1634. [CrossRef]

30. Khalil, A.; Husain, I. A Fourier series generalized geometry-based analytical model of switched reluctance machines. *IEEE Trans. Ind. Appl.* **2007**, *43*, 673–684.
31. Loop, B.; Essah, N.; Sudhoff, S. A basis function approach to the nonlinear average value modeling of switched reluctance machines. *IEEE Trans. Energy Convers.* **2006**, *21*, 60–68. [CrossRef]
32. Vujicic, V.P. Modeling of a switched reluctance machine based on the invertible torque function. *IEEE Trans. Magn.* **2008**, *44*, 2186–2194. [CrossRef]
33. Xia, C.; Shi, T.; Xue, M. A new rapid nonlinear simulation method for switched reluctance motors. *IEEE Trans. Energy Convers.* **2009**, *24*, 578–586. [CrossRef]
34. Burkhart, B.; Klein-Hessling, A.; Ralev, I.; Weiss, C.P.; De Doncker, R.W. Technology, Research and Applications of Switched Reluctance Drives. *CPSS Trans. Power Electron. Appl.* **2017**, *2*, 12–27. [CrossRef]
35. Sovicka, P.; Rafajdus, P.; Vavrus, V. Switched reluctance motor drive with low-speed performance improvement. *Electr. Eng.* **2020**, *102*, 27–41. [CrossRef]
36. Caramia, R.; Piotuch, R.; Palka, R. Multiobjective FEM optimization of BLDC motor based on Matlab and Maxwell scripting capabilities. *Arch. Electr. Eng.* **2014**, *63*, 115–124. [CrossRef]
37. Palka, R.; Piotuch, R. Usage of FEM for synthesis of dead-beat current controller for permanent magnet synchronous motor. *COMPEL Int. J. Comput. Math. Electr. Electron. Eng.* **2019**, *38*, 1386–1400. [CrossRef]

Publisher's Note: MDPI stays neutral with regard to jurisdictional claims in published maps and institutional affiliations.

Article

Effect of Demagnetization on a Consequent Pole IPM Synchronous Generator

Roberto Eduardo Quintal Palomo [1] and Maciej Gwozdziewicz [2],*

[1] Faculty of Engineering, Universidad Autónoma de Yucatán, Mérida 97000, Mexico;
    roberto.quintal@correo.uady.mx
[2] Department of Electrical Machines, Drives and Measurements, Wroclaw University of Science
    and Technology, 50-105 Wroclaw, Poland
* Correspondence: maciej.gwozdziewicz@pwr.edu.pl

Received: 13 November 2020; Accepted: 29 November 2020; Published: 2 December 2020

Abstract: The design and analysis of a permanent magnet synchronous generator (PMSG) are presented. The interior permanent magnet (IPM) rotor was designed asymmetric and with the consequent pole approach. The basis for the design was a series-produced three-phase induction motor (IM) and neodymium iron boron (Nd-Fe-B) cuboid magnets were used for the design. For the partial demagnetization analysis, some of the magnets were extracted and the results are compared with the finite element analysis (FEA).

Keywords: synchronous generator; permanent magnets; demagnetization

---

1. Introduction

In the last three decades, the advances in material sciences have brought about higher-energy density products ($BH_{\max}$) in rare-earth permanent magnet materials, like neodymium iron boron (Nd-Fe-B) and samarium-cobalt (Sm-Co) [1,2]. This characteristic allowed electrical machines to have higher efficiency and to be more compact [3]. Nonetheless, special attention must be taken in the design of electrical machines with Nd-Fe-B magnets since they have temperature-dependent magnetization curves with lower coercivity than other materials [4]. Further, some novel applications like magnetic refrigeration or refrigeration by the magnetocaloric effect require the magnetic material to work at extreme temperatures, as stated in [5]. For that reason, Nd-Fe-B magnets can suffer irreversible demagnetization or partial demagnetization at high temperatures (>80 °C), and this is even below the Curie temperature, which is much higher, as explained in [6].

In Figure 1, three operating points over the 80 °C curve are shown. The difference between these depends on the magnet shape. Notice that the slope of the line crossing the *B-H* curve (load line) is calculated as

$$\mu_L = \frac{B_d}{\left(H_d - \frac{NI}{l_{PM}}\right)} = P_{sys}\frac{l_{PM}}{a_{PM}} \tag{1}$$

where μ_L is the permeance coefficient, B_d is the operating point's magnetic flux density, H_d is the permanent magnet's field intensity, N is the number of turns in the coil, I is the current in amperes, l_{PM} is the length of the permanent magnet, P_{sys} is the system's permeance, and a_{PM} is the permanent magnet's area transverse to the magnetic flux. These are important figures of merit when designing a permanent magnet machine since, as explained before, the volume (mass) of rare earth's permanent magnet impacts directly on the machine cost.

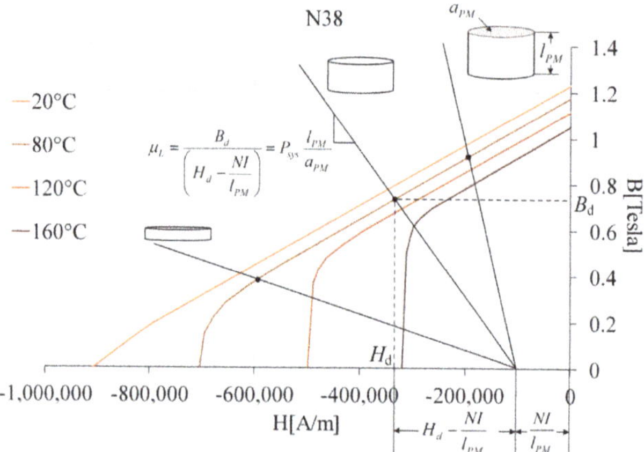

Figure 1. Three operating points for three different permeance coefficients.

In Figure 2, a simplified magnetic circuit C_1 is shown. Note that the slope of the load line when the current $I = 0$ depends on the geometry of the magnet (a_{PM}, l_{PM}) and the geometry of the air gap (a_g, l_g). This is a very important relation since the operating point should not lie close to the *B-H* curve's knee, as this would mean that the magnet is at demagnetization risk because, as shown in Figure 1, the load line will move to the left as a response to the armature reaction; in other words, the opposing magnetic field produced by the machine coils will move the operating point further to the left perhaps after the knee point, causing demagnetization. This is a very similar effect to the one described for small ferromagnetic material samples in cylindrical shapes from [7–9].

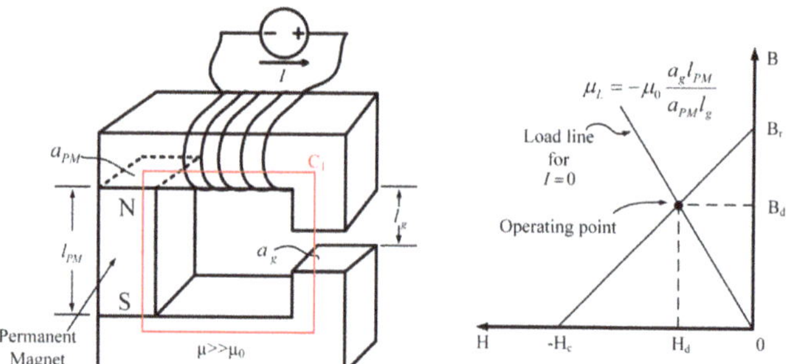

Figure 2. C-core and load line when $I = 0$ A.

This paper shows the design and measurements of an interior permanent magnet (IPM) rotor for a permanent magnet synchronous generator (PMSG). The finite element analysis (FEA) is conducted based on the designed geometry and then validated with measurements performed in the laboratory. Finally, some results from the FEA with a partially demagnetized rotor are presented and discussed.

2. Design of the PMSG

The design of the PMSG was based on the series-produced three-phase induction motor (IM) of 1.5 kW, the Sh90-L4 [10]. The housing, stator geometry, and windings are the same as the IM. The parameters of the IM are presented in Table 1. The air gap of the PMSG is 0.5 mm and therefore the rotor has 81 mm diameter with the same 30 mm diameter shaft. The Sh90L-4 IM has a 0.3 mm air gap but it is common for permanent magnet machines to have a bigger air gap.

Table 1. Parameters of the induction motor (IM), which the permanent magnet synchronous generator (PMSG) is based on.

Phases	3
Nominal power	1.5 kW
Nominal speed	1410 rpm
Nominal frequency	50 Hz
Nominal current	3.5 A
Nominal torque	10.16 Nm
Number of poles	4
Stator Slots	36
Shaft diameter	30 mm

The magnets available for the design exploration were 18 by 4 by 30 mm rectangular cuboid N38SH Nd-Fe-B magnets, where the rating SH refers to a suggested maximum operating temperature of 150 °C (see [11]). This is due to the magnetic material powder's content of dysprosium (Dy) which makes the magnet more resistant to demagnetization but at the same time more expensive, since Dy is even more expensive than neodymium.

In Figure 3, the W shape of the consequent poles is shown. The W shape was selected due to its field-focusing abilities and low total harmonic distortion (*THD*) of the *back EMF*, as studied in [12].

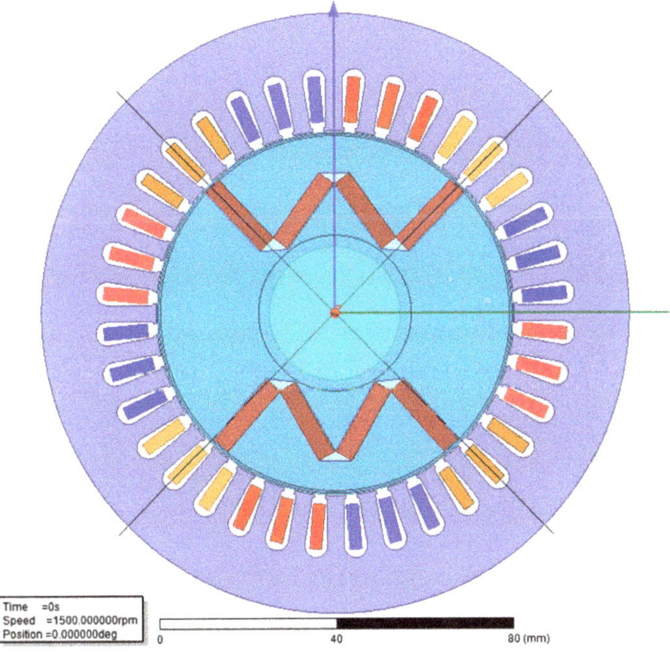

Figure 3. Transverse view of the PMSG geometry.

Notice in Figure 3 that the upper and lower poles are not symmetrical. This was done on purpose to lower even further the *THD* of the *back EMF* and cogging torque. Further, notice that the rotor has 4 poles but only two W shapes are visible. This is because of the "consequent pole" approach [13], sometimes also called induced pole in the literature [14], that was used in the design.

A detailed explanation of the design process and optimization of the W shape's angles was published in [15]. One of the most important issues of the process was the design space delimited by the shaft (inner circle) and the dotted line (Figure 4). This limit is given by the rotor diameter of 81 mm (outer circle) minus a bridge or small part of the steel sheet lamination that will hold the magnets inside the rotor. The minimum size for this bridge is given by the mechanical strength of the steel sheet from which the rotor laminations are cut, but also from the saturation due to the magnetic flux density since the flux should go through the stator and not through the bridge. For this project, a 0.5 mm bridge was chosen as the constraint for the design space, but also because the rotor laminations were cut by laser on a computer numerical controlled (CNC) machine whose precision is about ±0.1 mm. This distance is later analyzed by stress calculation in the FEA software to ensure a proper safety factor.

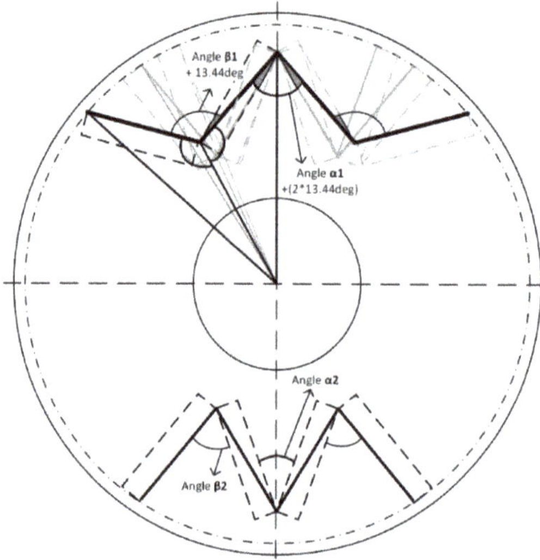

Figure 4. The PMSG parametrized rotor design space limitation.

Moreover, in Figure 4, a relation between angle and angle is shown. These are angles between the magnets which form the W shape. In order to obtain the maximum flux density through the stator, the tip of the outer magnets in the W shape should be coincident with the dotted line.

The rotor was built and tested in the laboratory in order to validate the FEA results. In Figure 5, the isometric view of the rotor is shown. Notice that it uses 4 threaded rods with hex nuts at both ends. Two of these rods are for holding the steel sheets together, and the other two are for holding the end plates at each end of the rotor stack. The space between the tip of the outer magnets and the air gap is determined by two factors. The first one is the precision of the laser cutter with which the prototype was manufactured. The second is the mechanical strength at high-speed rotation at which the generator may operate (1.2 times the maximum speed of 3000 rpm according to the norm IEC 60034 part 1). The stress on that space of the steel lamination, called the bridge, must be calculated to ensure a proper safety factor.

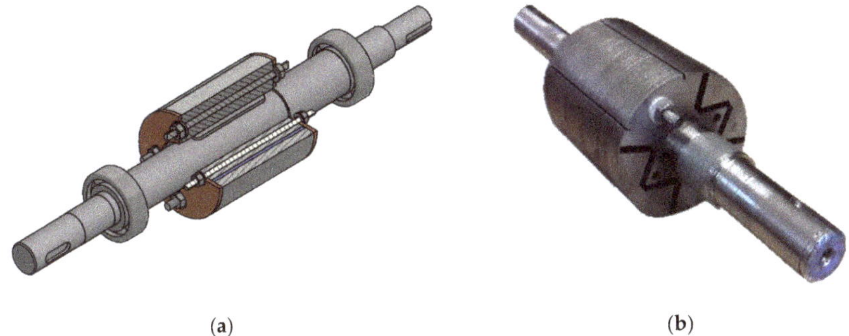

(a)   (b)

Figure 5. Isometric view of: (**a**) the designed rotor and (**b**) the built rotor without magnets.

In Figure 6, the results of mechanical stress analyses are shown. During the first analysis, the rotor speed was 1.2 times the maximum speed (n_{max} = 3000 rpm) without any torque on the rotor shaft. During the second analysis, the rotor speed was equal to the rated speed and the shaft torque was 2 times the rated shaft torque (t_n = 16 Nm). The maximum yield strength of M400-50A steel is 325 MPa. A safety factor equal to 2 allows a maximum strength of 162.5 MPa. The maximum mechanical stress during the analyses was about 43 MPa. This clearly demonstrates that the bridge will withstand the centrifugal forces of the generator spinning inside the nominal speed range and shaft torque even two times greater than the nominal one.

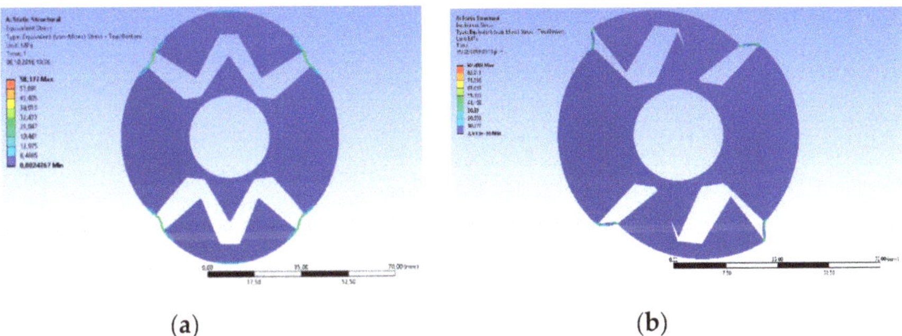

(a)   (b)

Figure 6. Results of the mechanical FEA: (**a**) with rotor speed $n = 2.0\,n_n$ and shaft torque $t = 0$, (**b**) with rotor speed $n = 1.0\,n_n$ and shaft torque $t = 2\,t_n$.

In Figure 7, the test setup is presented. The measurements were conducted in the Laboratory of Electric Machines at Wroclaw University of Science and Technology in Poland. The generator was driven by the servomotor which can work both as a motor or as a brake.

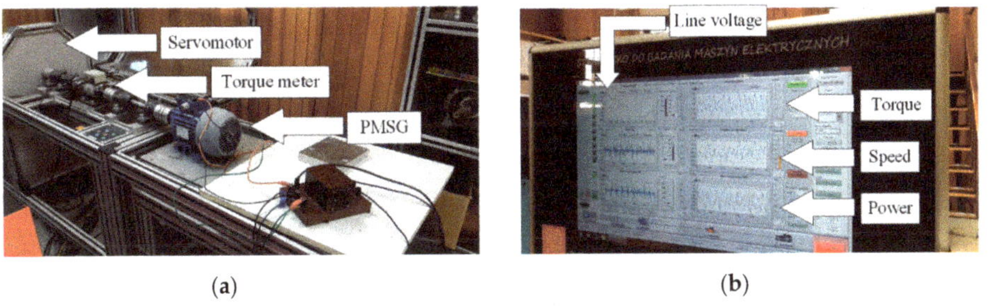

(a)   (b)

Figure 7. Laboratory: (**a**) test bench and (**b**) measurement's display interface.

In Figure 8, the *back EMF* of the PMSG is shown. Notice the similarities between the FEA (dotted lines) and the measurements conducted at nominal speed.

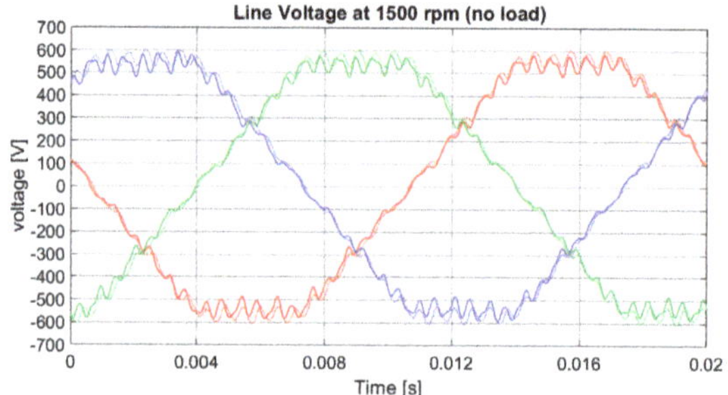

Figure 8. *Back EMF* of the healthy PMSG; dotted lines are FEA results.

FEA results predicted 90% efficiency when operating at 1500 rpm and a nominal input torque of 16 Nm. This can be confirmed by the measurements shown in Figure 9.

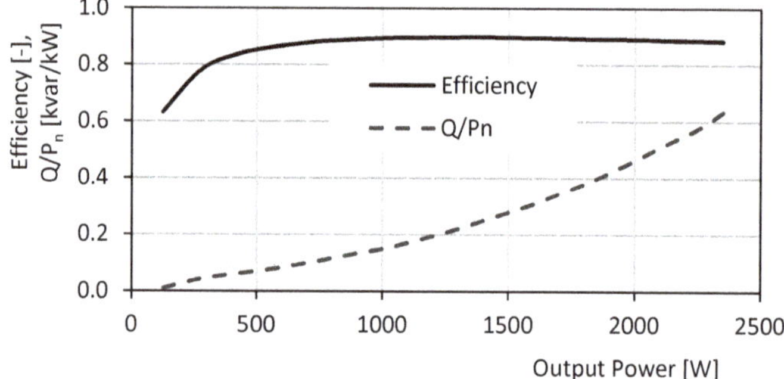

Figure 9. Efficiency and reactive power ratio measurements at 1500 rpm.

In Figure 10, the calculated efficiency map of the generator is shown. Notice that the calculated maximum efficiency is about 90%. This was corroborated by laboratory measurements also shown in Table 2. These measurements were conducted only with a three-phase resistor (56 ohm). The FEA results appear to be in good agreement with the measurements conducted in the laboratory. Once again, the similarities between the measurements and FEA transient results are evident.

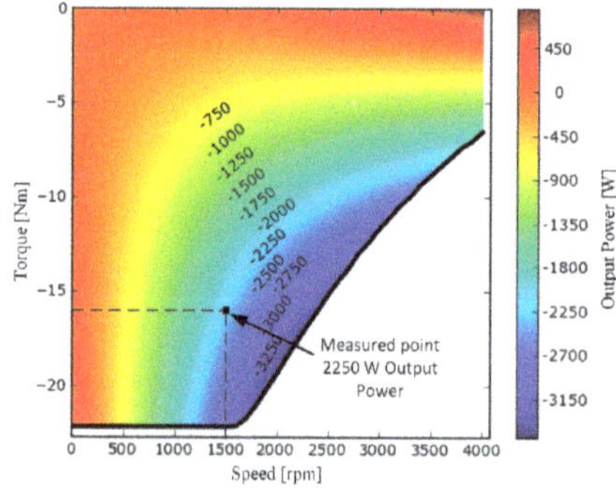

Figure 10. Efficiency and reactive power ratio measurements at 1500 rpm.

Table 2. Results of the investigated generator at 1500 rpm and 56 ohm load.

	P_{mech} [W]	Line Voltage [V_{rms}]	Current per Phase [A_{rms}]	$P_{electrical}$ [W]	Efficiency [%]
FEA	2530	320	3.603	2275	89.0
Measurements	2530	316	3.661	2250	88.9

In Figure 11, the PMSG was loaded with a resistive load of 45 Ω per phase, providing 291 V_{rms} line voltage and 4.16 A phase current with an output power of 2.35 kW.

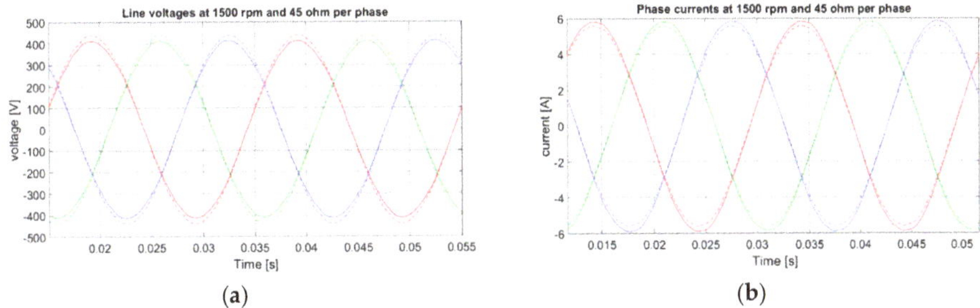

Figure 11. Loaded PMSG: (**a**) line voltages and (**b**) currents; dotted lines are FEA results.

Once the validation was conducted, the PMSG FEA was used for the analysis of partial demagnetization's effects. First, an experiment (transient simulation) was designed in order to demagnetize the PMSG in the time-stepping FEA.

3. Demagnetization Analysis

Once the validation was conducted, the PMSG FEA was used for the analysis of partial demagnetization's effects. First, an experiment (transient simulation) was designed in order to demagnetize the PMSG in the time-stepping FEA. The results are analyzed and presented. Then, some validation of the demagnetization was conducted. The demagnetization model embedded in the FEA software was first presented in [16], where it is explained as a formulation of the Newton–Raphson method using the true Jacobian matrix.

3.1. Recoil Line

The phenomenon of demagnetization can be explained better with Figure 12. In Figure 12, the recoil line is shown in red. First, the magnet is at the operating point A, then a demagnetizing current moves the operating point to B; notice that the load line moves parallel to the original but further to the left thanks to the armature reactance (NI/l_{PM}). Now, the operating point is below the knee of the B-H curve for 80 °C. When the opposing magnetic field is turned off, the magnet's operating point returns to C. This means that the magnet has lost its original BH_{max}. Note that the recoil line is parallel to the linear part of the B-H curves.

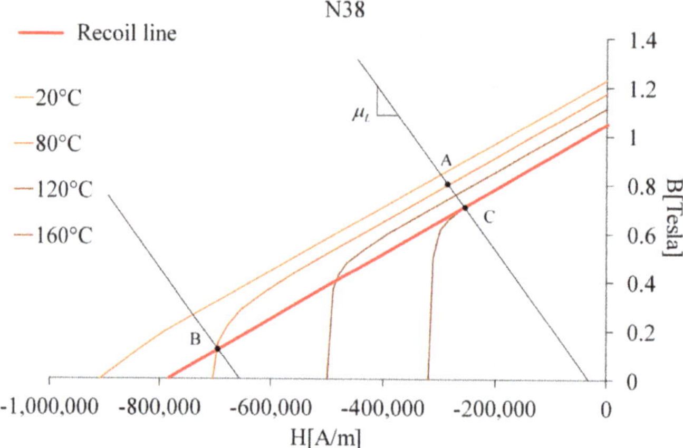

Figure 12. Recoil line of the N38 magnet at 80 °C.

Another failure mechanism that will obtain the same results (operating point C) is to augment the temperature from 80 to 160 °C. This is clear in Figure 12 because the operating point C coincides with the 160 °C *B-H* curve.

In Figure 13, a qualitative depiction of the hysteresis curves (*B-H* curves) for two different temperatures is shown. Notice the change in the remanence B_r and coercivity H_c when the temperature rises. In the magnet's data sheet, this rate of change (reversible temperature coefficients) is expressed as ΔB_r and ΔH_c, both expressed in %/°C. These figures of merit are useful for the non-linear model of the magnetic material in the FEA since they are used to calculate the new *B-H* curve when the temperature changes.

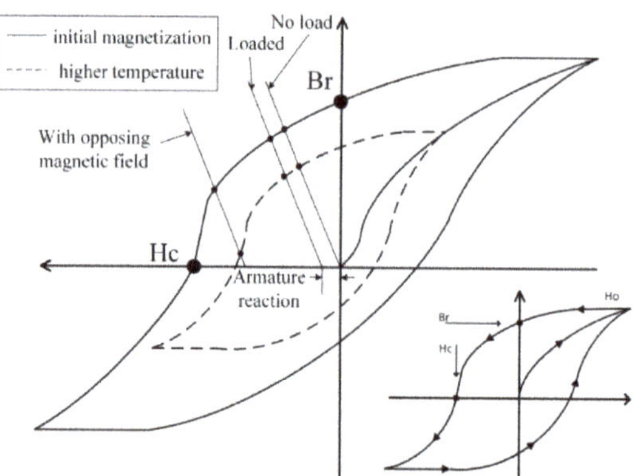

Figure 13. Qualitative depiction of the *B-H* curves under different temperatures.

3.2. Three-Phase Short-Circuit Demagnetization

The FEA was connected to the circuit shown in Figure 14, in order to simulate a three-phase short-circuit fault that would generate the currents necessary to drive the PMSG magnets' operating point below the knee point of the magnetization curves. This was first studied by Professor Jahns in [17] and more recently in [18]. In this last reference, the authors used the non-linear *B-H* characteristics of the magnets as a data for the FE simulation to improve the accuracy of the demagnetization analysis.

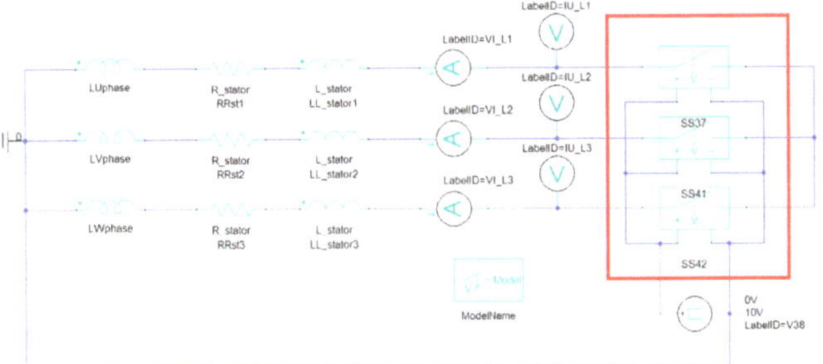

Figure 14. Maxwell's Circuit Editor schematic; red rectangle shows the short circuit position.

In this paper, a temperature-dependent demagnetization model of permanent magnets for FEA (first used in [19]) was used in the transient simulation, although the temperature was kept constant for each experiment, as shown in Table 3.

Table 3. *Back EMF* at 1500 rpm before and after the three-phase short circuit.

	Before	After
60 °C	416 V$_{rms}$	415 V$_{rms}$
70 °C	413 V$_{rms}$	412 V$_{rms}$
80 °C	410 V$_{rms}$	408 V$_{rms}$
90 °C	407 V$_{rms}$	404 V$_{rms}$
100 °C	403 V$_{rms}$	399 V$_{rms}$
110 °C	400 V$_{rms}$	392 V$_{rms}$
120 °C	396 V$_{rms}$	382 V$_{rms}$
130 °C	392 V$_{rms}$	367 V$_{rms}$
140 °C	387 V$_{rms}$	329 V$_{rms}$
150 °C	382 V$_{rms}$	272 V$_{rms}$
160 °C	376 V$_{rms}$	206 V$_{rms}$

For each temperature shown in Table 3, a transient simulation was performed and the *back EMF* value in V$_{rms}$ was calculated before the short circuit and after the short circuit. In Figure 15, the amplitude of the *back EMF* before and after the short circuit is shown. The difference is bigger at 150 °C because the knee point in the demagnetization curve is closer to the operating point, as shown in Figure 1. Further, notice the difference between the initial *back EMF* (before the short circuit) with magnets at 60 °C in Figure 15a and at 150 °C in Figure 15b. This difference is due to the different magnetization curves calculated by the model for each temperature.

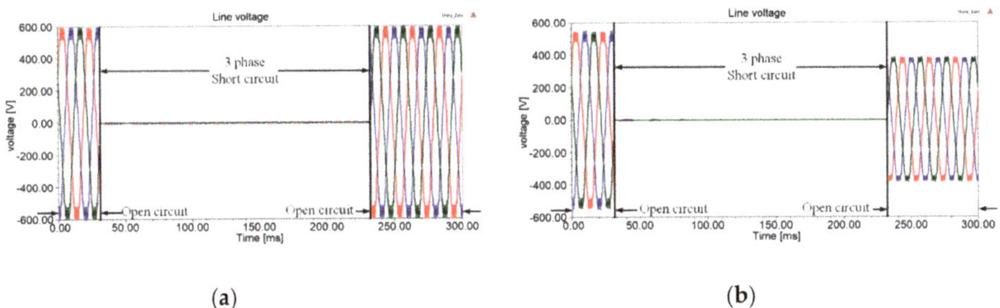

(a)   (b)

Figure 15. *Back EMF* before, during, and after the three-phase short circuit for (**a**) 60 °C and (**b**) 150 °C.

In Figure 16, the results of the first test at 60 °C are shown. Note that the tests were made with a constant speed input. This means that before the short circuit, during the short circuit, and after the short circuit, the speed is kept constant at 1500 rpm. In Figure 16b, the torque at the shaft is presented, and the PMSG behaves as an electromagnetic brake during the three-phase-to-ground fault, opposing the input torque with about a 5 Nm average.

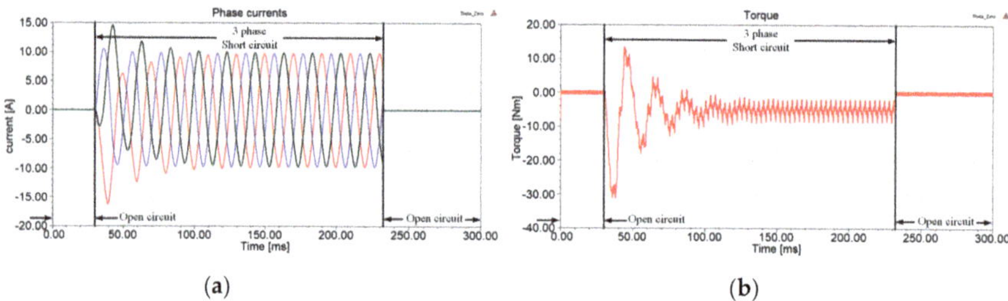

Figure 16. Open circuit–short circuit–open circuit test at 60 °C: (**a**) currents and (**b**) torque.

In Figure 17, the results of the same test (open circuit–short circuit–open circuit) for a higher temperature of 150 °C are presented. Here, the opposing torque is only of about 2 Nm in the steady state (between 150 and 230 ms).

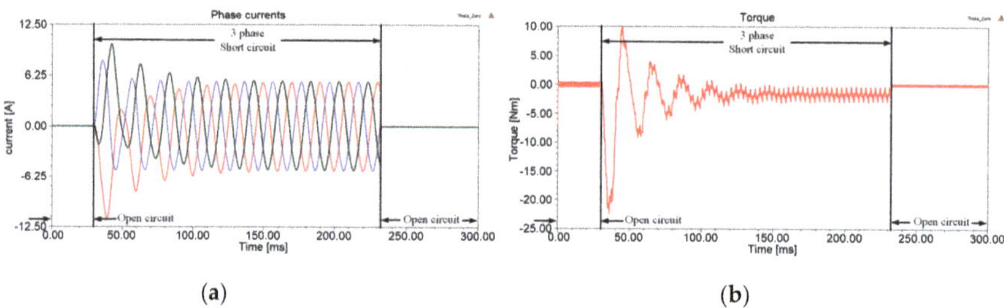

Figure 17. Open circuit–short circuit–open circuit test at 150 °C: (**a**) currents and (**b**) torque.

In Figure 18, the magnetic flux density distribution inside the magnets at 150 °C is shown. Notice in Figure 17b that the red circles indicate where the highest demagnetization has happened. If we take into account that the direction of rotation is counter-clockwise, then the results are in good agreement with the ones reported in [20,21]; this means the trailing edge of the magnets is the one that gets partially demagnetized first.

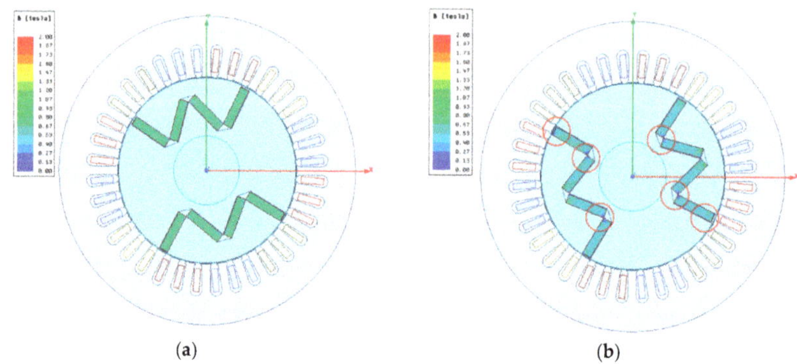

Figure 18. Magnetic flux density distribution inside the magnets: (**a**) at 150 °C before the fault and (**b**) after the fault.

In Figure 19, the results for three analyses are summarized. This means two other similar tables to Table 3 were produced for 500 and 2500 rpm. Notice how after 130 °C, the difference in the back EMF before and after the short circuit becomes bigger than 50 V_{rms}, all the way to 170 V_{rms} at 160 °C. This indicates a very high demagnetization rate.

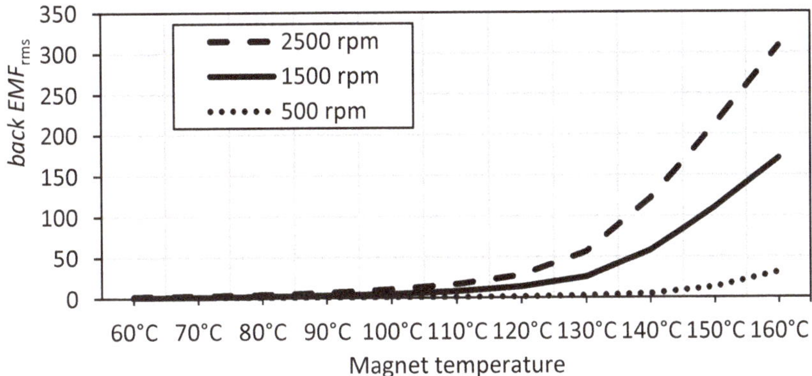

Figure 19. *Back EMF* difference (before and after the fault) at 3 different speeds.

Further, in Figure 19, the other lines corresponding to the same analysis but for a different constant speed indicate that the speed of operation of the PMSG, before and during the fault, also has an impact on the level of demagnetization that the IPM machine will get. A bigger ΔV_{rms} is worse. From Figure 19, it can be concluded that the PMSG can be operated up to 120 °C without the risk of high demagnetization in case of a fault, or even at higher temperatures but at low speed. Keep in mind that all the tests shown in this section were conducted at constant speed, namely no acceleration occurring during the fault.

In order to analyze the partial demagnetization of the W-shape PMSG, an experiment was conducted in the laboratory by taking out two of the cuboid magnets from the rotor. This is a common procedure widely used in the literature, e.g., [22–24]. The magnets that were taken out are shown in Figure 20. As mentioned in Section 2, the cuboid magnets are only 30 mm long; therefore, to fill the 90 mm rotor stack, three of them must be inserted in each hole of the W shape. The difference between those two measured cases is small. The difference becomes more evident when the V_{rms} is analyzed. For the healthy case, 394 V_{rms} *back EMF* is obtained. For the demagnetized rotor, only 382 V_{rms} is obtained.

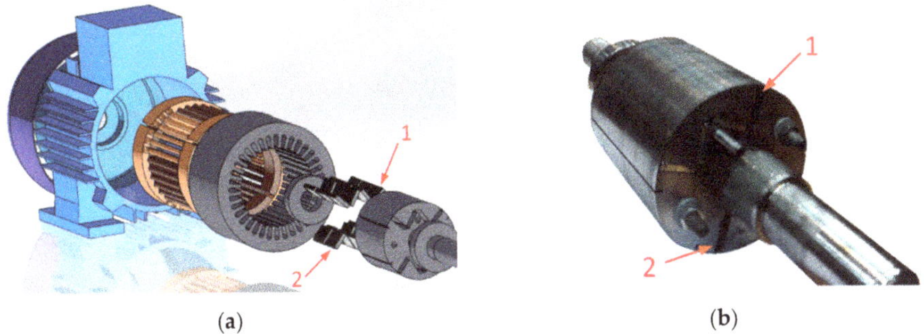

Figure 20. Back EMF measurements with: (**a**) healthy magnets and (**b**) partially demagnetized magnets.

4. Conclusions

The authors developed a novel PMSG on the basis of an IM (Sh90-L4). The PMSG was designed as an IPM with the same stator of the IM (overlapping, distributed three-phase windings). In order to save costs, an approach in the literature called "consequent pole" was adopted. Specifically, this means the usage of both sides of the magnets to avoid using double the number of magnets.

Analyses of the effects of the partial demagnetization of the W-shape PMSG are presented. By virtue of the FEA material's non-linear properties, the demagnetization by short-circuit currents is studied. It was found that the demagnetization risk is higher when operating at speeds higher than nominal.

The magnet thickness was chosen not only to produce the necessary magnetic flux density in the air gap but also to avoid possible demagnetization, by having operating points close to the *B-H* curve's knee. As was shown in the article, the selection of the N38SH material for the Nd-Fe-B magnets allowed for a continuous operation with magnet temperatures up to 100 °C with almost no risk of demagnetization.

These results demonstrate the need to simulate possible failure modes when designing a PMSG, especially one with Nd-Fe-B.

This work was part of the PhD thesis by the first author, were demagnetization effects of the PMSG with a voltage-source converter (VSC) was analysed [25].

Future work should include more experimental verifications of the demagnetization and measurements of the demagnetized rotor in order to better detect and assess this kind of fault.

Author Contributions: Methodology, software, validation, writing and visualization R.E.Q.P. Conceptualization, resources, project administration and data curation M.G. All authors have read and agreed to the published version of the manuscript.

Funding: R.E.Q.P. thanks the funding from the Mexican Council of Science and Technology CONACYT and the Department of Research, Innovation and Higher Education of the Yucatan State SIIES grant number 312140. Calculations have been carried out using resources provided by Wroclaw Centre for Networking and Supercomputing (http://wcss.pl), grant No. 400.

References

1. Wang, A.; Li, S. Investigation of E-Core Modular Permanent MagnetWind Turbine. *Energies* **2020**, *13*, 1751. [CrossRef]
2. Eriksson, S. Permanent Magnet Synchronous Machines. *Energies* **2019**, *12*, 2830. [CrossRef]
3. Gutfleisch, O.; Willard, M.A.; Brück, E.; Chen, C.H.; Sankar, S.G.; Liu, J.P. Magnetic Materials and Devices for the 21st Century: Stronger, Lighter, and More Energy Efficient. *Adv. Mater.* **2011**, *23*, 821–842. [CrossRef] [PubMed]
4. Cao, Z.; Li, W.; Li, J.; Zhang, X.; Li, D.; Zhang, M. Research on the Temperature Field of High-Voltage High Power Line Start Permanent Magnet Synchronous Machines with Different Rotor Cage Structure. *Energies* **2017**, *10*, 1829. [CrossRef]
5. Zhang, H.; Gimaev, R.; Kovalev, B.; Kamilov, K.; Zverev, V.; Tishin, A. Review on the materials and devices for magnetic refrigeration in the temperature range of nitrogen and hydrogen liquefaction. *Phys. B Condens. Matter* **2019**, *558*, 65–73. [CrossRef]
6. Zverev, V.I.; Gimaev, R.R.; Tishin, A.M.; Mudryk, Y.; Gschneidner, K.A.; Pecharsky, V.K. The role of demagnetization factor in determining the "true" value of the Curie temperature. *J. Magn. Magn. Mater.* **2011**, *323*, 2453–2457. [CrossRef]
7. Joseph, R.I. Ballistic Demagnetizing Factor in Uniformly Magnetized Cylinders. *J. Appl. Phys.* **1966**, *37*, 4639. [CrossRef]
8. Kobayashi, M.; Ishikawa, Y. Surface magnetic charge distributions and demagnetizing factors of circular cylinders. *IEEE Trans. Magn.* **1992**, *28*, 1810–1814. [CrossRef]
9. Chen, D.-X. Demagnetizing factors of long cylinders with infinite susceptibility. *J. Appl. Phys.* **2001**, *89*, 3413–3415. [CrossRef]

10. Indukta Electrical Machine Factory. Sh90-L4 Electrical Parameters. Available online: https://www.cantonigroup.com/celma/en/page/offer/details/1/146/Sh90L-4 (accessed on 11 September 2020).
11. Supermagnete. What Temperatures Can Magnets Withstand? Frequently Asked Questions. Available online: https://www.supermagnete.de/eng/faq/What-temperatures-can-magnets-sustain (accessed on 3 September 2020).
12. Gwozdziewicz, M. Limitation of torque ripple in medium power line start permanent magnet synchronous motor. *Przegląd Elektrotechniczny* **2017**, *1*, 3–6. [CrossRef]
13. Tapia, J.; Leonardi, F.; Lipo, T. Consequent-pole permanent-magnet machine with extended field-weakening capability. *IEEE Trans. Ind. Appl.* **2003**, *39*, 1704–1709. [CrossRef]
14. Ugale, R.T.; Pramanik, A.; Baka, S.; Dambhare, S.; Chaudhari, B.N. Induced pole rotor structure for line start permanent magnet synchronous motors. *IET Electr. Power Appl.* **2014**, *8*, 131–140. [CrossRef]
15. Gwozdziewicz, M.; Quintal-Palomo, R.E. Induced pole permanent magnet synchronous generator. In Proceedings of the 2017 International Symposium on Electrical Machines (SME), Naleczow, Poland, 18–21 June 2017; Institute of Electrical and Electronics Engineers (IEEE): New York, NY, USA, 2017; pp. 1–4.
16. Fu, W.N.; Ho, S.L. Dynamic Demagnetization Computation of Permanent Magnet Motors Using Finite Element Method with Normal Magnetization Curves. *IEEE Trans. Appl. Supercond.* **2010**, *20*, 851–855. [CrossRef]
17. Welchko, B.; Jahns, T.; Soong, W.; Nagashima, J. IPM synchronous machine drive response to symmetrical and asymmetrical short circuit faults. *IEEE Trans. Energy Convers.* **2003**, *18*, 291–298. [CrossRef]
18. Choi, G.; Jahns, T.M. Interior permanent magnet synchronous machine rotor demagnetization characteristics under fault conditions. In Proceedings of the 2013 IEEE Energy Conversion Congress and Exposition, Denver, CO, USA, 15–19 September 2013; pp. 2500–2507. [CrossRef]
19. Zhou, P.; Lin, D.; Xiao, Y.; Lambert, N.; Rahman, M.A. Temperature-Dependent Demagnetization Model of Permanent Magnets for Finite Element Analysis. *IEEE Trans. Magn.* **2012**, *48*, 1031–1034. [CrossRef]
20. Kim, K.-T.; Lee, Y.-S.; Hur, J. Transient Analysis of Irreversible Demagnetization of Permanent-Magnet Brushless DC Motor with Interturn Fault Under the Operating State. *IEEE Trans. Ind. Appl.* **2014**, *50*, 3357–3364. [CrossRef]
21. Kim, H.K.; Hur, J. Dynamic Characteristic Analysis of Irreversible Demagnetization in SPM- and IPM-Type BLDC Motors. *IEEE Trans. Ind. Appl.* **2017**, *53*, 982–990. [CrossRef]
22. Roux, W.L.; Harley, R.G.; Habetler, T.G. Detecting rotor faults in permanent magnet synchronous machines. In Proceedings of the 4th IEEE International Symposium on Diagnostics for Electric Machines, Power Electronics and Drives, SDEMPED 2003, Atlanta, GA, USA, 24–26 August 2003; Institute of Electrical and Electronics Engineers (IEEE): New York, NY, USA, 2003; Volume 22, pp. 198–203.
23. Yang, Z.; Shi, X.; Krishnamurthy, M. Vibration monitoring of PM synchronous machine with partial demagnetization and inter-turn short circuit faults. In Proceedings of the 2014 IEEE Transportation Electrification Conference and Expo (ITEC), Dearborn, MI, USA, 15–18 June 2014; pp. 1–6. [CrossRef]
24. Park, Y.; Fernandez, D.; Bin Lee, S.; Hyun, D.; Jeong, M.; Kommuri, S.K.; Cho, C.; Reigosa, D.D.; Briz, F. Online Detection of Rotor Eccentricity and Demagnetization Faults in PMSMs Based on Hall-Effect Field Sensor Measurements. *IEEE Trans. Ind. Appl.* **2019**, *55*, 2499–2509. [CrossRef]
25. Quintal Palomo, R. Operation and Faults Analysis of energy-Saving Permanent Magnet Synchronous Generator for Small Wind Turbine. Ph.D Thesis, Wroclaw University of Science and Technology, Wroclaw, Poland, 2020.

Publisher's Note: MDPI stays neutral with regard to jurisdictional claims in published maps and institutional affiliations.

Article

Faulty Synchronization of Salient Pole Synchronous Hydro Generator

Adam Gozdowiak

Faculty of Electrical Engineering, Wroclaw University of Science and Technology, 50-370 Wroclaw, Poland;
adam.gozdowiak@pwr.edu.pl

Received: 10 July 2020; Accepted: 16 October 2020; Published: 20 October 2020

Abstract: This article presents the simulation results of hydro generator faulty synchronization during connection to the grid for various voltage phase shift changes in a full range ($-180°$; $180°$). A field-circuit model of salient pole synchronous hydro generator was used to perform the calculation results. It was verified using the measured no-load and three-phase short-circuit characteristics. This model allowed observing the physical phenomena existing in the investigated machine, especially in the rotor which was hardly accessible for measurement. The presented analysis shows the influence of faulty synchronization on the power system stability and the construction components which are the most vulnerable to damage. From a mechanical point of view, the most dangerous case was for the voltage phase shift equal to $-120°$, and this case was analyzed in detail. Great emphasis was placed on the following physical quantities: electromagnetic torque, stator current, stator voltage, rotor current, current in rotor bars, and active and reactive power. The physical quantities existing during faulty synchronization were compared with a three-phase sudden short-circuit state. From this comparison, we selected the values of physical quantities that should be taken into account during design of new hydro generators to withstand the greatest possible threats during long-term work.

Keywords: electrical machine; hydro generator; faulty synchronization; finite element method; field-circuit modeling

---

1. Introduction

Synchronization of a generator with a power system must be carried out carefully. It is a dynamic process that requires the coordinated operation of many components such as mechanical, electrical, and human. The voltage and frequency of the disconnected generator must be closely matched to the voltage and frequency existing in the network bus. The instantaneous value of the voltage induced in the armature winding must be close to the instantaneous value of the network bus voltage. The interconnection of large numbers of synchronous generators operating in parallel constitutes the power system. These generators are hydro generators (possessing a salient-pole rotor) and turbogenerators (with a cylindrical rotor). These machines are connected by transmission lines supplying the network loads. A disconnected generator can be paralleled with the network by driving it at synchronous speed and adjusting its excitation current so that its terminal voltage is equal to the network bus voltage.

Failure of the synchronizing procedure results in out-of-phase synchronization, mainly caused by the following [1–4]:

- Failure in wiring during commissioning Wiring errors lead to particular out-of-phase angles. Polarity errors at a voltage transformer can cause synchronizing at 180°.
- Delay during breaker closure This can occur if the breaker physically closes slower than anticipated and the systems go beyond the designed safe conditions before the breaker closes. The closing process cannot be stopped when the breaker coil is energized, and out-of-phase synchronization

can happen. During this abnormal condition, transformers, generators, and associated equipment can be damaged.

- Flash-over in breaker's contacts A breaker is designed to sustain the voltage that occurs before synchronizing and in the case of inequality of the generator and the network voltage phase. Several factors can reduce the electrical strength of the breaker's insulation. This results in arcing between contacts before galvanic closing. The following phenomena favor the flash-over in a breaker: pollution, low pressure, humidity, and decomposition of insulation.
- Wrong setting of synchronous system This emerges from a human mistake.
- Problem in manual synchronization In addition to automatic synchronization, the vast majority of generators have the possibility of manual synchronization. An operator may not predict how fast the phase angle difference converges and energizes the breaker close coil in advance or with delay. Sometimes, the operator does not take into account the closing mechanism delay of the generator breaker and, therefore, the main contacts do not make a close to $0°$ angle difference.

A hydro generator is a synchronous generator. Synchronous speed comes from interaction between two poles coming from a stator and rotor that create a rotating magnetic field. Interaction between these two poles occurs when direct current flows in the excitation (rotor) winding and three-phase voltages are applied to the armature (stator) winding. Rotor speed is determined by the number poles and the grid frequency. The rotor is coupled to a water turbine by a mechanical shaft which supplies the mechanical energy for transformation to electrical energy.

During synchronizing, before closing the breaker between the power system and generator terminals, the frequency of the voltage induced in the armature is adjusted by the angular velocity of the rotating magnetic field (the rotor speed), whereas, after synchronizing, when the breaker is closed, the frequency of the power system regulates the speed of the rotating magnetic field. The position and speed of the rotor must be closely matched at the instant the hydro generator is connected to the power system in order to eliminate the transient torque required to bring the rotor into synchronism. If the frequency of voltage induced on the stator (coming from angular velocity) is significantly different from the frequency in the power system, then large transient torque will appear. Stability can be achieved by accelerating or decelerating the rotating masses (rotor and turbine) until the rotor speed matches the power system frequency. If the voltage phase angle difference is significant (rotor position is off), then transient torque required to set the rotor position into phase with the power system can be even higher.

Transient torques appearing during faulty synchronization can cause instantaneous and cumulative fatigue damage to the hydro generator and water turbine over their lifetime. Instantaneous stator current associated with this abnormal state can exceed the three-phase short circuit. Huge current windings cause the appearance of large forces in end-winding.

The consequences of faulty synchronization are the following:

- Damage to the hydro generator rotor and water turbine because of mechanical stresses caused by rapid acceleration or deceleration of rotating masses.
- Damage to step-up transformer and stator windings caused by high currents.
- Disturbances such as power and voltage deviations.
- Catastrophic failure and reduced lifetime of generator construction elements.

IEEE Standards C50.12 and C50.13 [5,6] define the following limits which guarantee that a generator is for service without inspection or repair after synchronizing:

- Phase shift angle difference between generator-side voltage and the power system: $\pm10°$.
- Generator-side voltage relative to the power system: 1.0–1.05 U_N.
- Frequency difference: ±0.067 Hz.

The synchronizing can be done by an operator using manual means or automated control systems. The synchronizing system is dedicated to the following:

- Closing the breaker as close to $0°$ angle difference as possible. The operator must predict how fast the phase angle difference is coming and energize the breaker close coil in advance to account for the closing mechanism delay.
- Controlling the governor to match speed.
- Controlling the excitation current value to match voltage at the stator terminals.

Faulty synchronization is an abnormal operating condition and it can be detected by protection devices which give a signal to a device able to disconnect the generator from the power system. Protection devices such as reverse power and loss-of-field protections have time delays to avoid unwanted trips during transient operation of the generator. In this case, these times are around several seconds, whereas the generator breakers have a certain operating time which does not exceed 100 ms [7].

The faulty synchronization of synchronous generators has been the subject of many research works in the latest years. In particular, these studies referred to turbo generators. All of them were based on models with d–q axes considering two damper winding on both axes [8,9]. The parameters taken into account to create these models were extracted using Canay's approach. Simplified d–q models do not reflect the saturation effect of magnetizing steel. The effect of saturation on the rotor shaft torques during faulty synchronization is significant. Reflecting the saturation effect shows that the torsional moment on the shaft is higher [10]. Neglecting this effect affects erroneous conclusions. Currently, the most accurate calculation method for electrical machines is the finite element method. This method reflects the real distribution of generator construction elements and allows observing the physical phenomena existing inside the generator [11]. Additionally, it can be helpful in the choice of the best material for conductive parts and to optimize the magnetic path for the flux. The above statement applies only to large cylindrical-rotor generators. The studies on faulty synchronization for hydro generators with power of 1–10 MVA were omitted.

The literature lacks information on the effect of the damping cage on faulty synchronization for hydro generators, whereas, in the case of turbogenerators, there is a significant influence of the location of rotor wedges on the damping of excitation current oscillation during out-of-phase synchronization [11]. So far, the distribution of the current induced in the rotor bars of hydro generators has not been discussed, and it has not been compared to the most dangerous state, which is sudden short-circuit fault. Most often, articles [8,10,12] focused on showing the maximum values of stator current and electromagnetic torque, i.e., in such physical quantities that can lead to irreversible damage. The influence of induced current in the rotor bars on the possibility of damaging damping bars was ignored.

In this work, the impact of voltage phase shift (synchronizing angles) on hydro generator stability was investigated. A two dimensional field-circuit model was used in calculation. This model was previously verified by comparisons of the calculated and measured no-load and three-phase short-circuit characteristics during running tests. The simulations were carried out for different synchronizing angles to evaluate the instantaneous and peak shaft torques. In addition, the following quantities were also determined: stator current, stator voltage, field current, induced current on rotor bars, and active and reactive power. In this article, the waveforms of physical quantities of the most dangerous case are shown from the mechanical point of view. Finally, a typical comparative study between faulty synchronization and three-phase short circuit faults of a hydro generator is presented.

2. Description of Field-Circuit Model and Main Rated Data of Hydro Generator

A field-circuit model of a hydro generator was utilized in computation. The circuit equations (based on Kirchhoff's laws) for rotor and stator windings were coupled with field equations used to describe the temporal–spatial distribution of the electromagnetic field [13].

The vector magnetic potential A was used to describe the temporal–spatial distribution of the time-varying electromagnetic field. In this way, the partial differential equations were solved. The mathematical description is expressed in Equation (1) for the low-frequency range. The time-varying electric and magnetic fields are calculated using Equation (2) for the fully coupled dynamic physics solution implemented in Ansys Maxwell software. The current density vector is expressed by yjr vector magnetic potential A and scalar electrical potential V.

$$\nabla \times \frac{1}{\mu}(\nabla \times A) = J, \tag{1}$$

$$\sigma \frac{\partial A}{\partial t} + \sigma \nabla V - \sigma v \times \nabla \times A = J, \tag{2}$$

where J is the current density vector, μ is the magnetic permeability, σ is the electric conductivity, and v is the velocity vector of the environment moving relative to the electromagnetic field.

In a two-dimensional model, the vector magnetic potential has only one component. When knowing the mean values of the potential in the cross-section of the winding conductors (s_i) and the effective machine length l_e, the mean value of the flux associated with k-th winding can be expressed by Equation (3). The electromagnetic state of the k-th winding is described by the Kirchoff equation (Equation (4)).

$$\Psi_k = \sum_i \frac{l_e}{s_i} \int_{s_i} A_i ds_i, \tag{3}$$

$$u_k = R_k i_k + \frac{d\Psi_k}{dt}, \tag{4}$$

where u_k, i_k, Ψ_k are the instantaneous values of voltage, current, and flux coupled to the k-th winding.

The two-dimensional model of the hydro generator was created in Ansys Maxwell software dedicated to finite element analysis of electromagnetic field distribution. The investigated hydro generator (type: GCV-1610M, produced in December 2019) was a vertical salient pole machine with static excitation (the slip rings were on the shaft). The ratings of the generator are presented in Table 1. The stator core and rotor poles were laminated. The rotor winding was wound around the poles. The rotor possessed a damper cage located in the pole shoes. These bars were short-circuited by ring-shaped segments in all poles. The bars and ring-shaped segments were made of copper.

Table 1. Main rated data of investigated hydro generator.

Data Name	Symbol	Unit	Value
Apparent power	S_N	kVA	4500
Stator voltage	U_{SN}	V	6000
Stator current	I_{SN}	A	433
Speed	n_N	rpm	600
Torque	T_N	kNm	71.6
Power factor	$cos\varphi_N$	-	0.80
Frequency	f_N	Hz	50
Excitation current	I_{FN}	A	295
Efficiency at T_N and $cos\varphi_N$	η_N	%	97.1 [1]
Generator moment of inertia	J_G	kgm^2	2548

[1] Total losses were not included in the loss margin (the losses were measured).

The field-circuit model of the hydro generator reflected nonlinear magnetizing curves of rotor and stator cores, possibilities of inducing current in damper bars, rotor movement, and external electrical circuit with voltage sources. However, the skin effects in stator coils and eddy currents in the stator and rotor laminations were neglected. The two-dimensional model was reduced to half of the cross-section because of geometrical symmetry and electromagnetic periodicity. The investigated region, together with the finite element mesh, is shown in Figure 1. The analyzed area contained 15,712 finite elements and 31,372 nodes. This model had two boundary conditions to solve the electromagnetic field equations. The first Dirichlet boundary condition was located at the outer stator diameter (edge Γ_2), where the vector magnetic potential was equal to 0. The periodic condition of the magnetic potential was set at edge Γ_1. The numbering of rotor poles and damper bars refer to calculation results, where the waveform of current in each bar is shown. The main dimensions of the geometry are presented in Table 2.

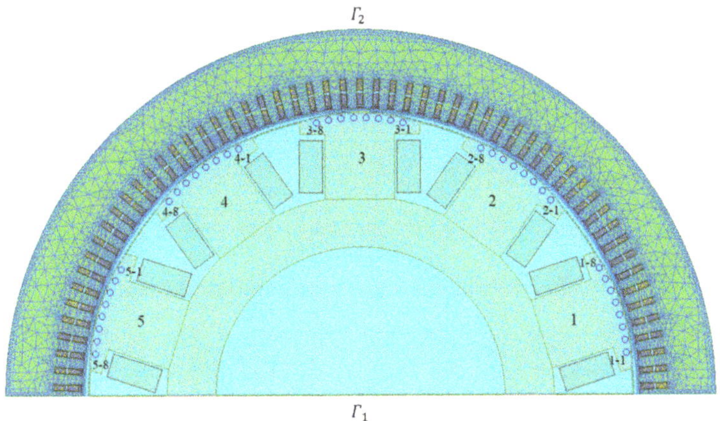

Figure 1. The field-circuit model of the investigated hydro generator with a visible finite element mesh.

Table 2. Main geometry dimensions of the investigated hydrogenerator.

Data Name	Unit	Value
Stator outer diameter	mm	1850
Stator inner diameter	mm	1450
Airgap	mm	11
Stator core length (with ventilation ducts)	mm	770
Ventilation ducts width	mm	10
Number of ventilation ducts	-	12
Stator slots	-	114
Number of turns per phase in stator	-	95
Rotor shoe width	mm	310
Rotor shoe height	mm	55
Rotor body width	mm	180
Rotor body height	mm	165
Number of rotor turns per pole	-	60
Number of rotor bars per pole	-	8
Rotor bar diameter	mm	16
Height of shorted segment of rotor bars	mm	25
Width of shorted segment of rotor bars	mm	35
Distance between two rotor bars in pole	mm	34

Figure 2 presents the scheme of the electrical circuit which contained stator and rotor windings, as well as the distribution of rotor damper bars. The armature winding circuit was extended by resistance (R_S) and inductance (L_{Sew}), representing the end-winding part. Additionally, there were added resistances and inductances representing the transformer (R_{TR}, L_{TR}) and the power system (R_{PR}, L_{PR}). The stator winding was supplied with three-phase voltage sources shifted with respect to each other by 120° (U_1, U_2, U_3). The field part of the excitation winding circuit was extended by resistance (R_F) and inductance (L_{Few}), representing the end-winding part. The excitation winding was supplied by direct current (DC) voltage source U_F. The excitation voltage was modeled as a constant voltage source; therefore, no actuation of the automatic voltage regulator was taken into account. R_{Bew} and L_{Bew} contained fragments of segments of the short-circuiting rotor cage between bars in one pole, whereas R_{B2ew} and L_{B2ew} were part of the ring-shaped segments between rotor poles. Switching the circuit breaker (S$_1$, S$_2$, and S$_3$) on or off enabled the hydro generator to be analyzed in various scenarios of faulty synchronization.

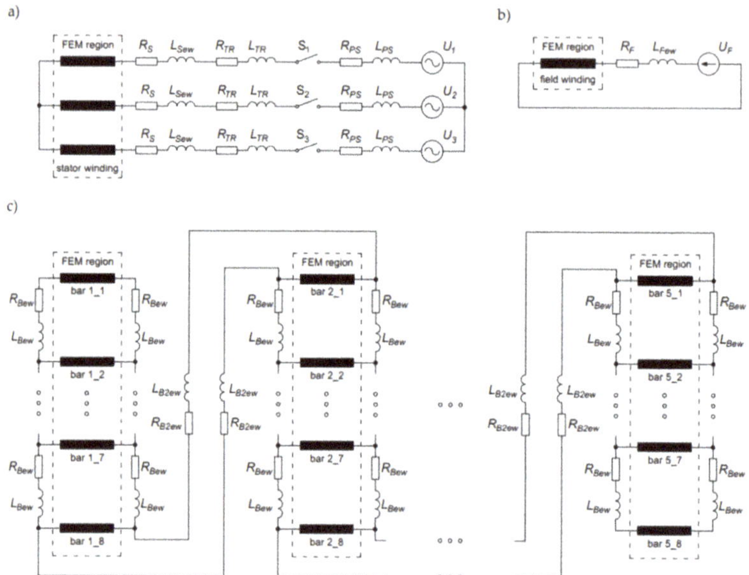

Figure 2. Schemes of stator winding (**a**), rotor winding (**b**), and distribution of damper bars (**c**) associated with the field model of the hydro generator.

During the simulation, it was assumed that the generator was connected to the power system, which possessed a short-circuit power equal to 15,000 MVA (i.e., a strong system) [14,15]. Computed inductances and resistances used in the simulations are shown in Table 3. The names of parameters refer to the circuit model shown in Figure 2.

Table 3. Computed resistances and reactances of the power system and unit transformer.

Symbol	Unit	Value
R_{PS}	mΩ	0.24
L_{PS}	μH	7.60
R_{TR}	Ω	0.12
L_{TR} [1]	mH	1.53

[1] Main data of unit transformer: S_{TR} = 4.5 MVA, U_{TR} = 6 kV/21 kV, $u_{k\%}$ = 6%, and ΔP_O = 1.5%.

The created field-circuit model of the hydro generator allowed calculating waveforms of electromagnetic quantities in steady and transient states. This model was verified experimentally on the basis of no-load and three-phase short-circuit curves calculated and compared with the measurements obtained during running tests. These curves are presented in Figure 3. The computed reactances and time constants of the generator are presented in Table 4.

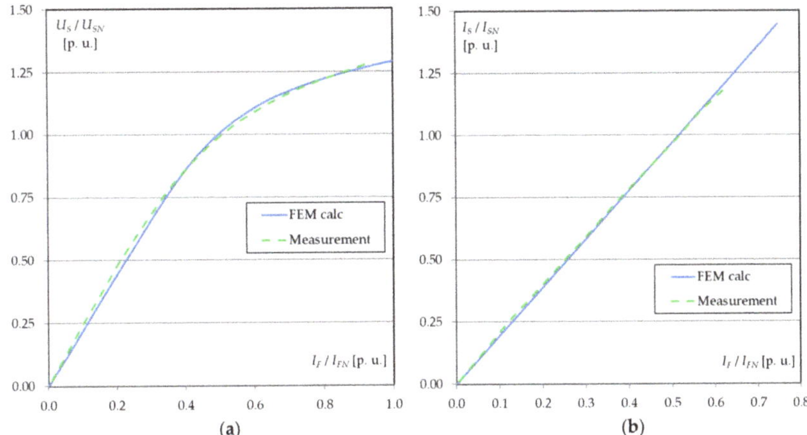

Figure 3. Comparison of measured curves with those calculated for no-load (**a**) and three-phase short circuit (**b**).

Table 4. Computed reactances and time constants of the investigated hydrogenerator.

Symbol	Unit	Value
X_d	p.u.	1.22
X_q	p.u.	0.70
X'_d	p.u.	0.27
X''_d	p.u.	0.19
X''_q	p.u.	0.17
X_2	p.u.	0.18
X_0	p.u.	0.09
T'_d	s	0.710
T''_d	s	0.045
T''_q	s	0.056
T_a	s	0.080

On the basis of the no-load and three-phase short-circuit characteristics, it could be concluded that the measurements did not differ significantly from the calculation results obtained from the field-circuit model of the hydro generator. Therefore, this model was utilized to calculate the faulty synchronization for different values of voltage phase shift.

3. Analysis of Faulty Synchronization

The calculations of faulty synchronization were prepared for different voltage phase shift angles in the range of [−180°; 180°] with a 5° step. This was a case when the generator and power system voltages were in sequence, amplitude, and frequency, but not in phase. When the voltage phase shift was in the range of [−180°; 0°], the power system voltage lagged with respect to generator voltage (Figure 4a); however, in the range of [0°; 180°], the power system voltage led with respect to generator voltage (Figure 4b). The study was extended to include the effect of increased generator voltage up to 1.05 U_N according to IEEE Standard C50.12 [5].

The investigated hydro generator was modeled in a single-machine system, as shown in Figure 5. In the simulation, the generator was connected to the power system at instant $t = 0.02$ s. During the simulation, the following physical quantities were obtained: electromagnetic torque, stator current, stator voltage, field current, currents in rotor bars, and active and reactive power in the power system bus. The frequency during the simulations was 50 Hz. The rotor speed was equal to 600 rpm before synchronization.

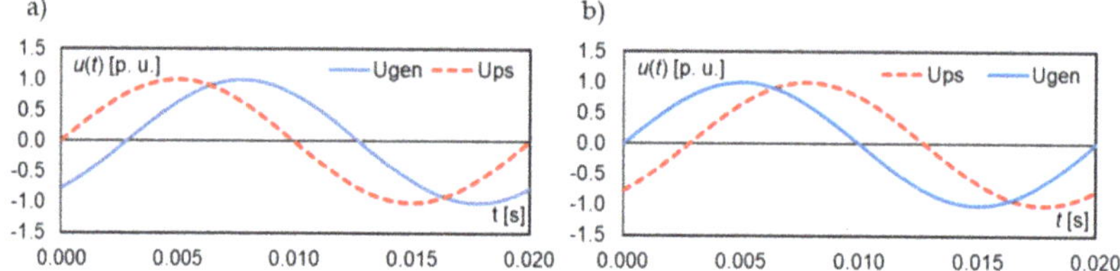

Figure 4. Two cases of voltage waveforms during faulty synchronization: (**a**) power system voltage (U_{PS}) lagging with respect to generator voltage (U_{GEN}); (**b**) power system voltage leading with respect to generator voltage.

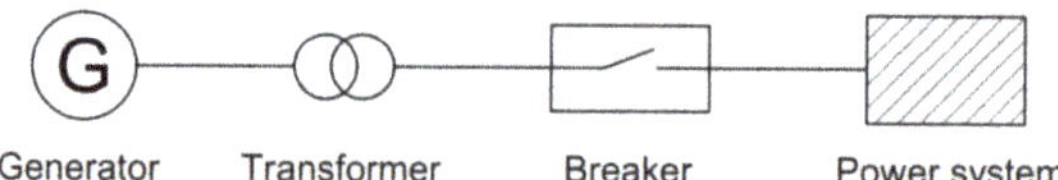

Figure 5. Single-line diagram of simulated case.

The simulations were carried out for different voltage phase shift angles to determine the instantaneous and peak shaft torques. Figure 6a shows the maximum amplitude of electromagnetic torque. The maximum value was for $-120°$ of voltage phase shift, when the power system lagged with respect to generator voltage. If we took into account the absolute sum of two maximum electromagnetic torques with opposite directions (Figure 6b), then the maximum value (peak-to-peak) was obtained for $180°$ of voltage phase shift. The significant differences between cases of $U = 1.00U_N$ and $U = 1.05U_N$ were visible when the voltage phase shift was higher than $60°$ and below $-60°$. The statement of computed extreme values is shown in Table 2.

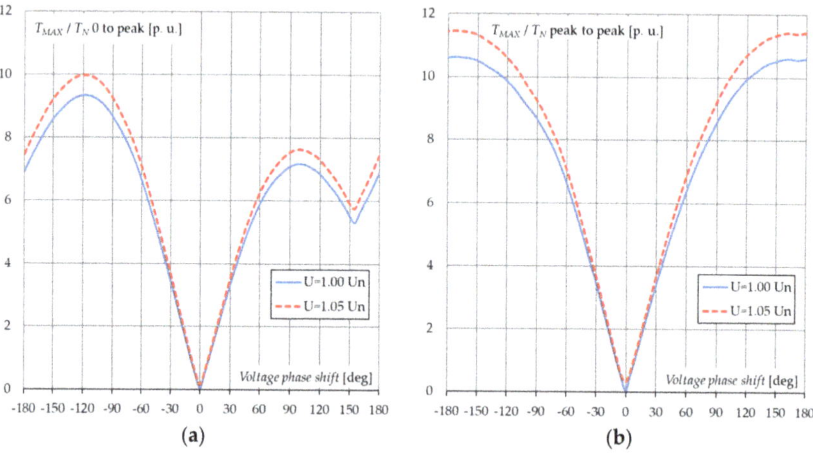

Figure 6. Maximum amplitude of electromagnetic torque (**a**) and absolute sum of two maximum electromagnetic torques with opposite directions (**b**).

Figure 7a,b present the maximum amplitudes of stator current and the minimum amplitude of stator voltage, respectively. When the voltage phase shift was equal to $180°$, the maximum value of stator current and the minimum value of stator voltage were observed. For the stator current, significant differences between cases of $U = 1.00U_N$ and $U = 1.05U_N$ were visible when the angle was higher than 90° and below −90°, whereas, for stator voltage, there were no visible differences in the whole range.

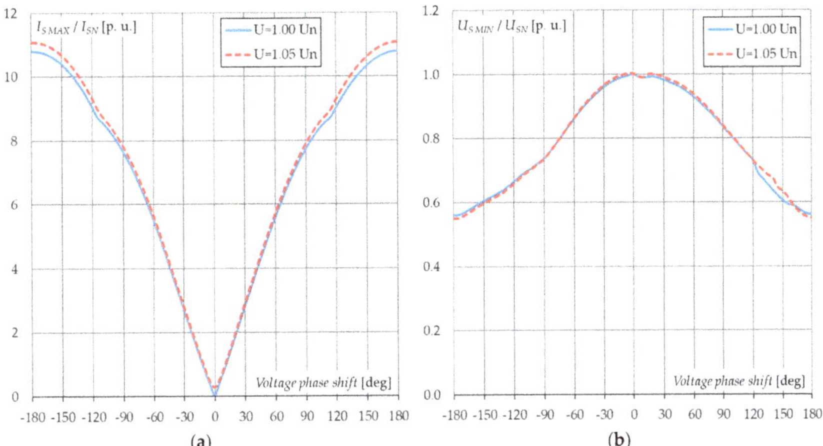

Figure 7. Maximum amplitude of stator current (**a**) and minimum amplitude of stator voltage (**b**) during the first period of faulty synchronization.

The maximum amplitude of induced current in the excitation winding is presented in Figure 8a. The maximum value was for the angle of 180°. Differences between cases of $U = 1.00U_N$ and $U = 1.05U_N$ were visible in the whole range. Figure 8b shows the maximum amplitude of the induced current in rotor bars in the active part and end-winding part (shorted ring segment between bars). The induced current in the end-winding part was almost twofold higher than the induced current in the active part. The maximum value of induced current in the active part was observed for 130°, whereas that in the end-winding part was observed for −130°. According to basic knowledge of hydro generator design, the selected diameter of rotor bars was equal to 16 mm (cross-section = 201 mm^2), whereas the cross-section of the segments of the short-circuiting rotor cage between rotor bars was equal to 500 mm^2. It follows from this that the end-winding part of rotor bars was able to withstand currents 2.5-fold higher than those existing in the rotor bar active part.

Figures 9 and 10 show the distribution of maximum values of induced current in each rotor bar and each end-winding part, respectively. These distributions were determined for a voltage phase shift of 180°. The bar numeration was as follows: the first number denotes the number of rotor poles (according to numeration in Figure 1), followed by the number of bars in the pole (direction: counterclockwise), and the double number (after the number of poles) denotes the segment between two rotor bars (e.g., Segment 1-12). In the case of a segment between two poles, the numeration was as follows: the first number denotes the number of poles, followed by the number of bars, the second number of poles, and the number of bars in the second pole (e.g., Segment 1-8-2-1).

The maximum amplitudes existing in the active part of rotor bars were close to the extremes of the rotor poles, whereas the lower values were in bars located in the center of poles. The opposite situation existed in the case of induced current in the end-winding part of rotor bars.

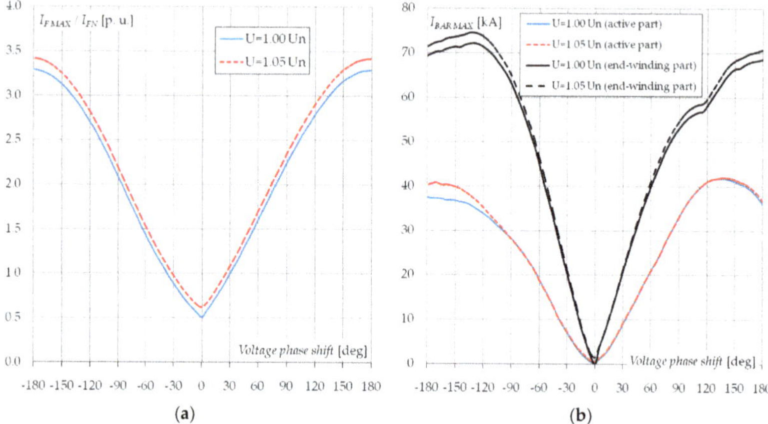

Figure 8. Maximum amplitude of field current (**a**) and rotor bar current (**b**).

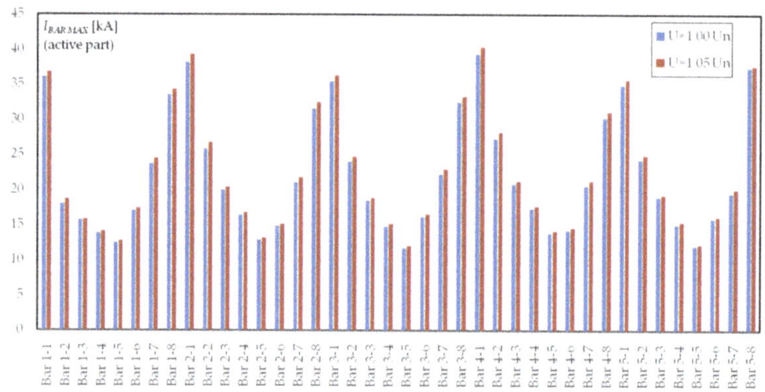

Figure 9. Distribution of maximum values of currents induced in the active part of the rotor bar at 180° of voltage phase shift.

Figure 11a presents the generated and absorbed active power by the investigated hydro generator. Due to the changing nature of the electric circuit, along with the change in voltage phase shift, there were local extremes in the graph. The maximum value of absorbed active power was equal to 5.8–6.6 P_N, where a higher value refers to a higher value of stator voltage at the terminals during synchronization. A higher active power was absorbed when the voltage phase shift during faulty synchronization was in the range of [−180°; 0°], i.e., when the power system voltage lagged with respect to generator voltage, whereas the maximum generated active power occurred in the opposite situation, i.e., when the power system voltage led with respect to generator voltage. The maximum value of generated active power did not exceed 7.2 P_N.

Figure 11b presents the generated and absorbed reactive power by the investigated hydro generator. Due to the changing nature of the electric circuit, along with the change in voltage phase shift, there were local extremes in the graph, as in the case with active power. The maximum value of absorbed reactive power was equal to 5.4–6.4 Q_N. A higher reactive power was absorbed when the voltage phase shift during faulty synchronization was in the range of [−120°; 120°], whereas the maximum generated reactive power did not exceed 9.5 Q_N. The highest values of generated reactive power were in the ranges of [−120°; −120°] and [120°; 180°].

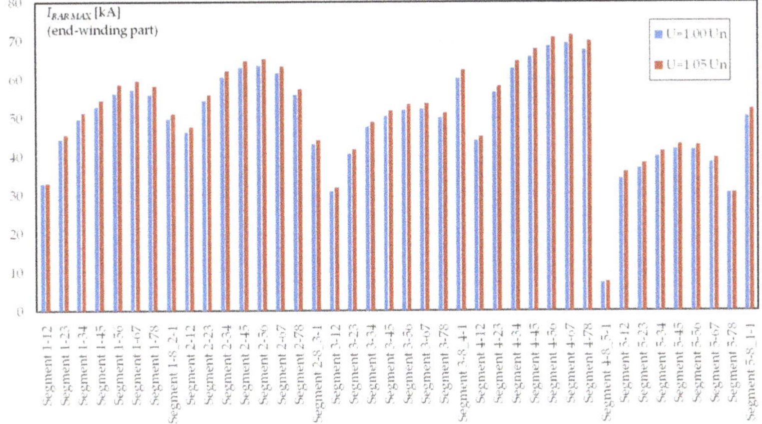

Figure 10. Distribution of maximum values of currents induced in the rotor's circuited segments at 180° of voltage phase shift.

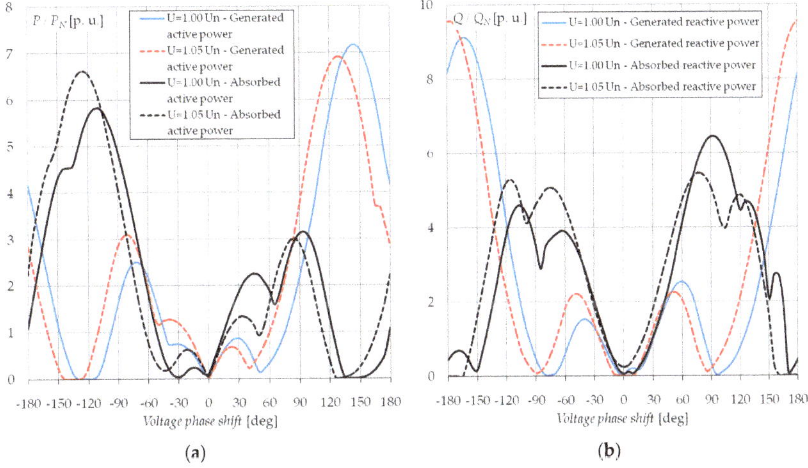

(a)

(b)

Figure 11. Generated and absorbed active (**a**) and reactive power (**b**).

4. Analysis of Out-of-Phase Synchronization of −120°

Simulations were carried out for out-of-phase synchronization of −120°. A negative sign denotes that the power system voltage lagged with respect to generator voltage (as shown in Figure 4a). This was the worst case of examined faulty synchronization, because the highest electromagnetic torque appeared. The value of voltage at the terminal before synchronization was equal to 1.05 U_N (maximum acceptable value according to [5,6]). The connection of the hydro generator to the power system took place at $t = 0.02$ s. The simulation step was equal to 0.2 ms.

Figure 12 presents the waveform of speed. At the beginning, there were fluctuations which disappeared after 3.5 s, and the investigated hydro generator returned to a steady state. The speed fluctuations depended on electromagnetic torque and the moment of inertia for all rotating components. In this study, it was assumed that the moment of inertia of the water turbine as twofold higher than the moment of inertia of the hydro generator.

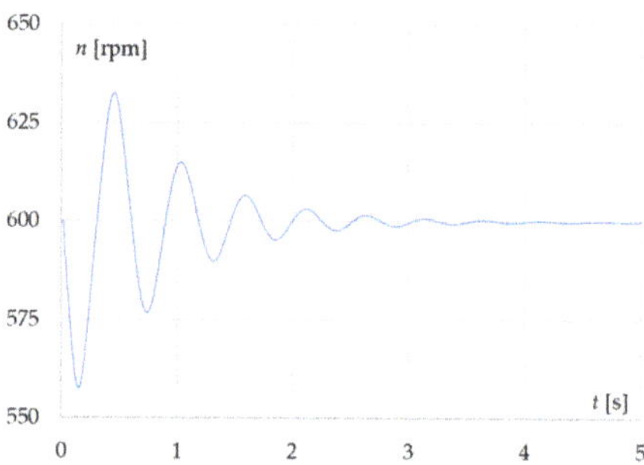

Figure 12. Waveform of speed during out-of-phase synchronization of −120°.

The calculated maximum value of electromagnetic torque was 9.94 T_N (Figure 13), and the largest values were turned in one direction. Significantly lower values were observed in the opposite direction.

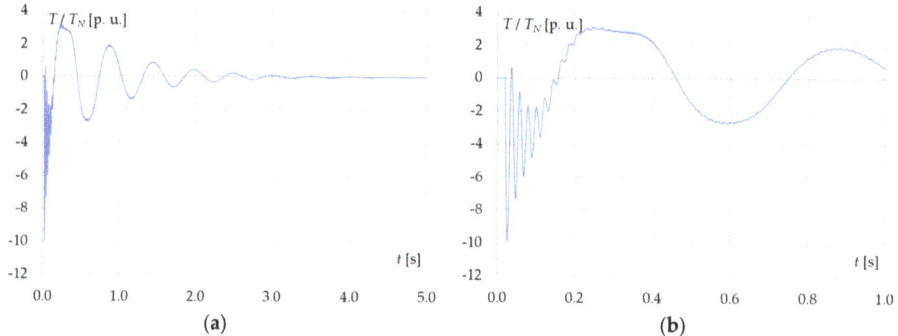

Figure 13. Waveform of electromagnetic torque during out-of-phase synchronization of −120° in (**a**) the scope of the considered calculation time and (**b**) the first second.

Figure 14 presents the stator current. The calculated maximum value was equal to 8.90 I_{SN}. Stator currents reduced to zero after seven periods and then increased and oscillated with a period of ca. 0.3 s. A significant increase in stator current caused a drop in terminal voltage to a value of 0.66 U_{SN} (seen in Figure 15). After one period, the stator voltage was rebuilt and returned to its value before synchronization within 3.5 s.

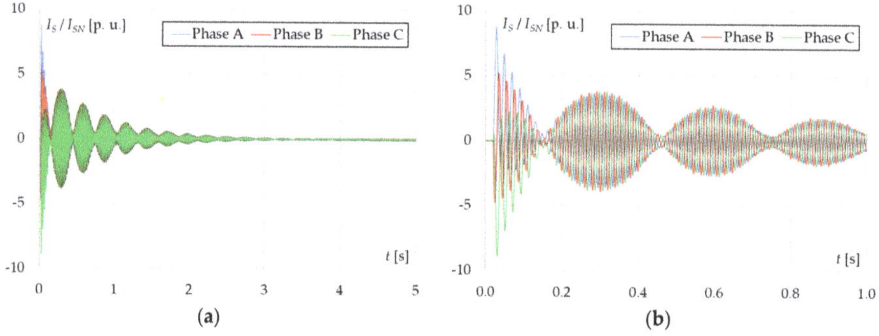

Figure 14. Waveforms of stator current during out-of-phase synchronization of −120° in (**a**) the scope of the considered calculation time and (**b**) the first second.

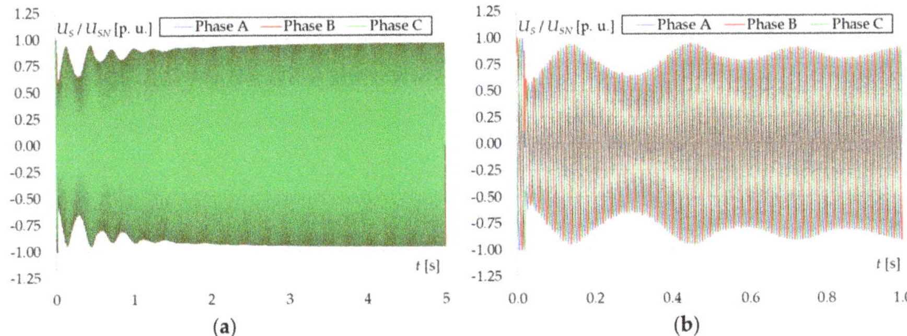

Figure 15. Waveforms of stator voltage during out-of-phase synchronization of −120° in (**a**) the scope of the considered calculation time and (**b**) the first second.

During the simulations, these was a lack of excitation regulation in order to show how the field current changed during faulty synchronization. The maximum computed value of field current was equal to 2.81 I_{FN} (Figure 16). When the maximum value was reached, the field current returned to the value before synchronization, which allowed inducing a value of 1.05 U_{SN} on the stator terminals during the no-load state.

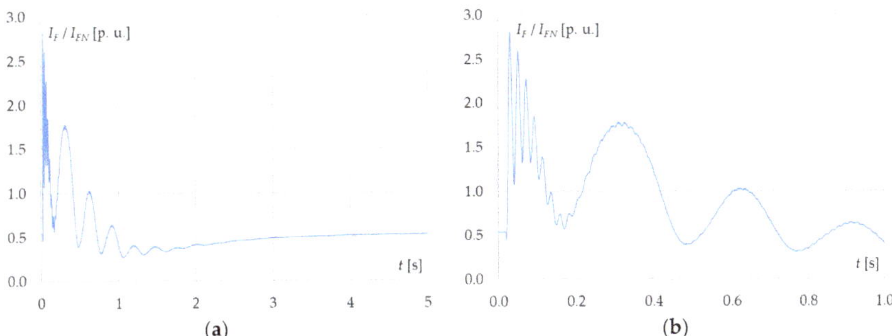

Figure 16. Waveform of field current during out-of-phase synchronization of −120° in (**a**) the scope of the considered calculation time and (**b**) the first second.

Figure 17 shows the distribution of maximum values of currents induced in each rotor bar. The maximum amplitudes of currents were close to extreme of rotor poles. The maximum current appeared in the rotor bar No. 2-1. The waveform of this current is presented in Figure 18.

Figure 19 shows the distribution of maximum values of each current induced in the rotor's circuited segments. The maximum values of currents were in segments located in the center of the pole (poles No. 1 and 2) and close to the extremes of rotor poles (poles No. 3, 4, and 5). The maximum current appeared in a segment between rotor bars No. 3-8 and No. 4-1 (Segment 3-8_4-1). The waveform of this current is presented in Figure 20.

Waveforms of active and reactive power are presented in Figure 21a,b, respectively. An active power above zero denotes that power was absorbed from the power system. The same situation applied to reactive power, whereby a reactive power above zero denotes that power was absorbed from the power system, and a value below zero means that active or reactive power was generated to the grid. The computed active power was huge due to the fact that, at the first moment of analyzed faulty synchronization (voltage phase shift equal to −120°), the generator speed decreased and there was a need to absorb active power from the power system to keep the machine in synchronism. The huge stator currents forced a drop in stator voltage and there was a need to absorb reactive power in order to magnetize the rotor and stator cores.

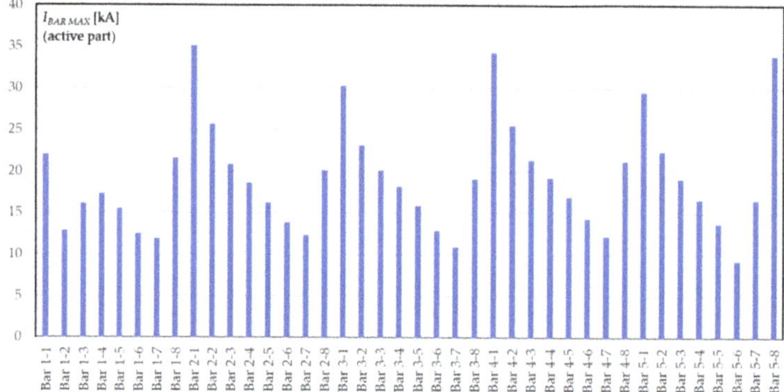

Figure 17. Distribution of maximum values of currents induced in the active part of the rotor bar during out-of-phase synchronization of −120°.

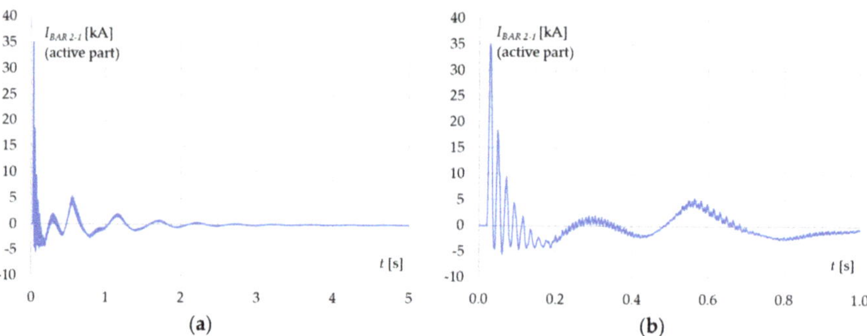

Figure 18. Waveform of induced current in bar No. 2-1 during out-of-phase synchronization of −120° in (**a**) the scope of the considered calculation time and (**b**) the first second.

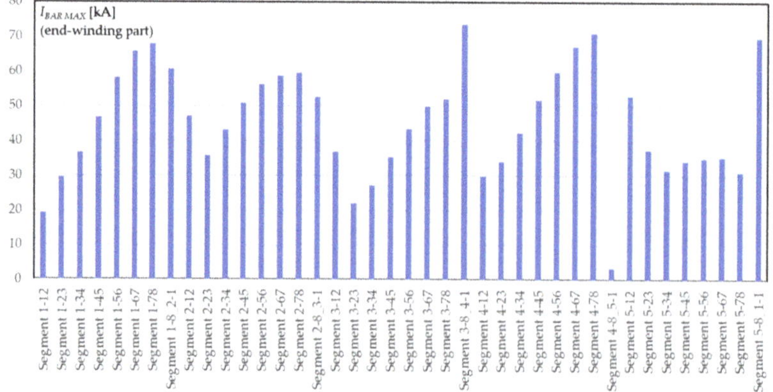

Figure 19. Distribution of maximum values of currents induced in rotor's circuited segments during out-of-phase synchronization of −120°.

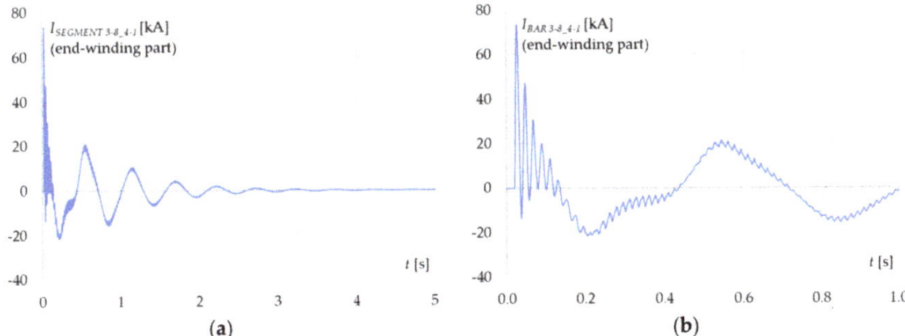

Figure 20. Waveforms of induced current in rotor's circuited segment 3-8_4-1 during out-of-phase synchronization of −120° in (**a**) the scope of the considered calculation time and (**b**) the first second.

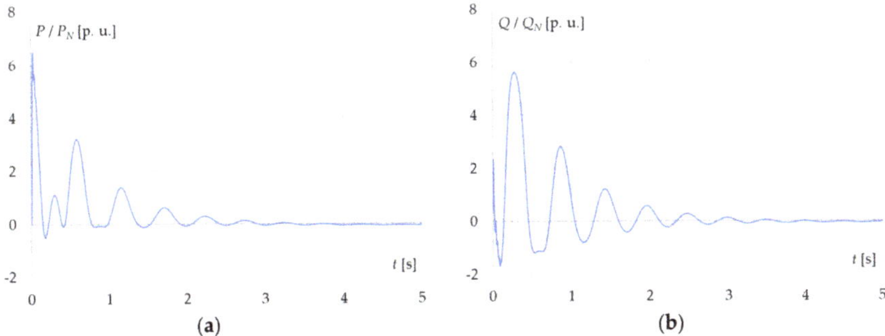

Figure 21. Waveforms of active (**a**) and reactive power (**b**) during out-of-phase synchronization of −120°.

Figures 22–24 present the flux lines, current densities in rotor bars, and flux densities in cores, respectively, for selected time steps. For a time instant of 0.02 s, physical phenomena were shown just before the simulated faulty synchronization. The flux lines presented the symmetrical poles, where the saturation in cores were within the limits (in the stator core below 1.9 T; in the rotor core below 1.5 T) and currents flowing through the rotor bars were very small and did not exceed 65×10^4 A/m^2 (the maximum value of current in the bar was below 130 A).

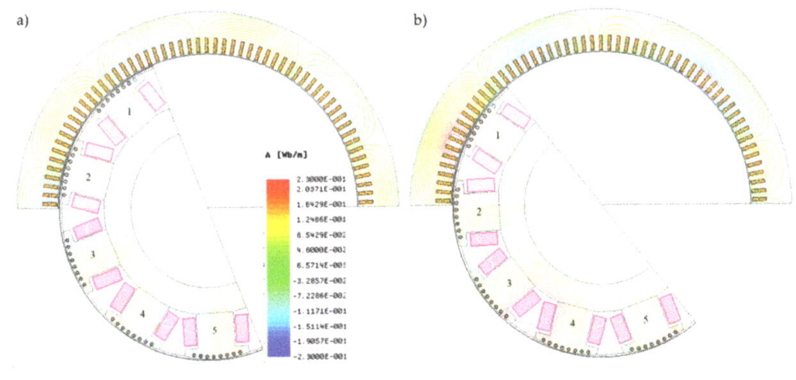

Figure 21. *Cont.*

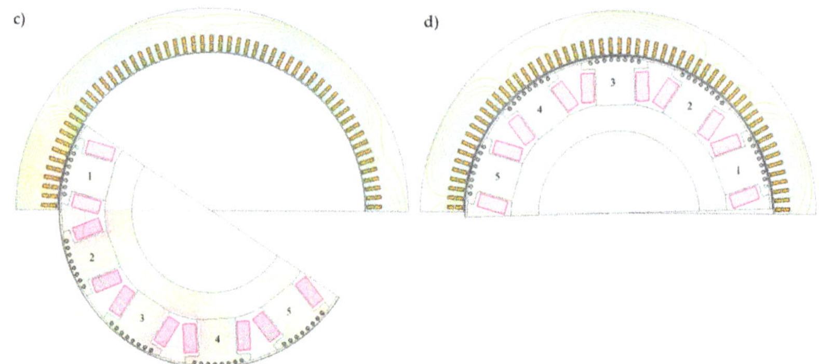

Figure 22. Magnetic field lines during out-of-phase synchronization of −120° at (**a**) 0.02 s, (**b**) 0.025 s, (**c**) 0.03 s, and (**d**) 0.302 s.

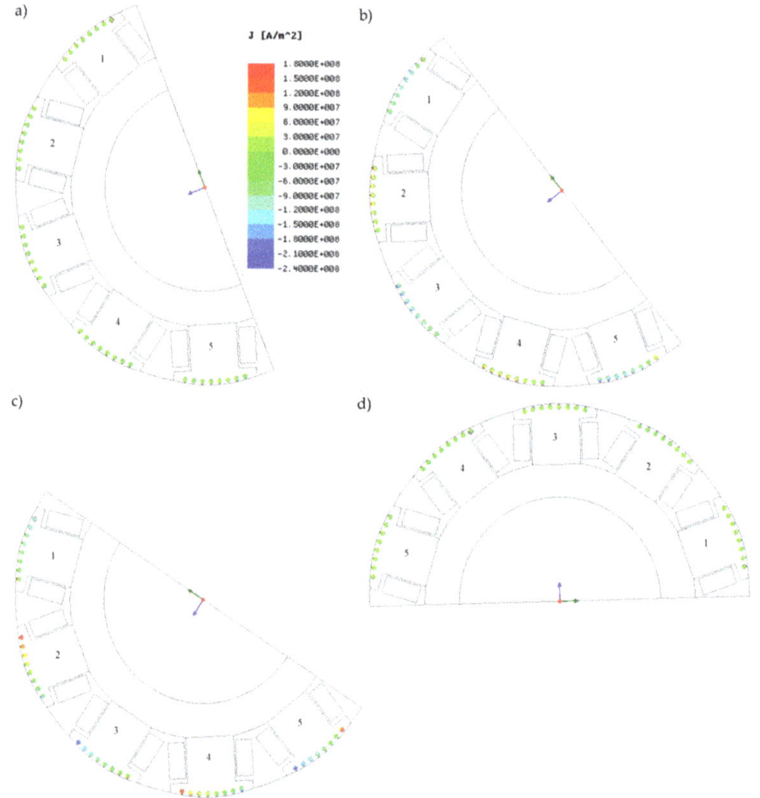

Figure 23. Current densities in rotor bars during out-of-phase synchronization of −120° at (**a**) 0.02 s, (**b**) 0.025 s, (**c**) 0.03 s, and (**d**) 0.302 s.

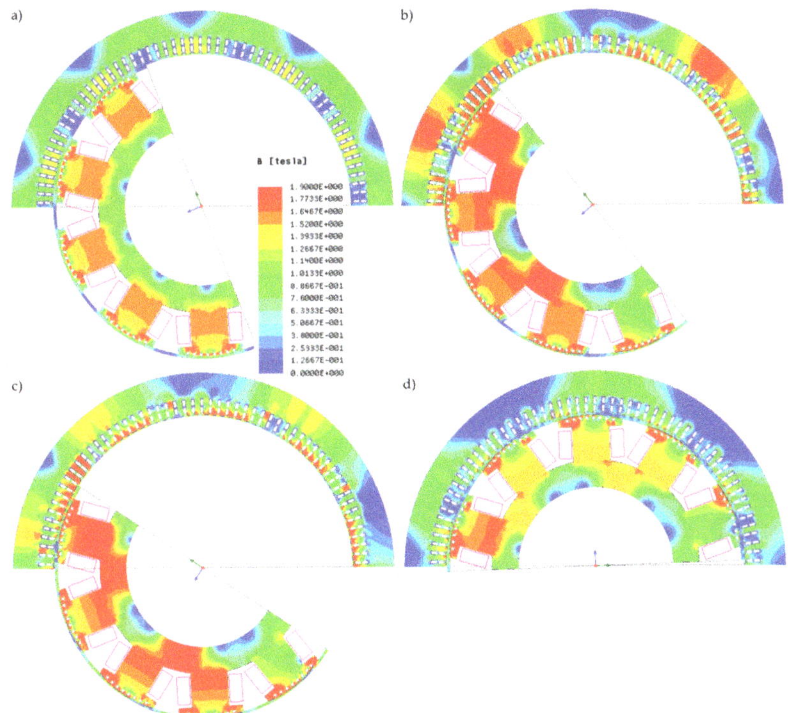

Figure 24. Flux densities during out-of-phase synchronization of $-120°$ at (**a**) 0.02 s, (**b**) 0.025 s, (**c**) 0.03 s, and (**d**) 0.302 s.

After a 1/4 period of faulty synchronization, the highest saturation of cores appeared, especially in the rotor pole and yoke, which significantly exceeded the acceptable value (1.5 T). The saturation in the stator yoke was exceeded as well. The maximum computed value was equal to 1.9 T, which is above the acceptable value for a salient pole synchronous machine (1.4 T). The current densities in rotor bars did not exceed 120×10^6 A/m^2.

At a time of 0.03 s, the maximum value of current densities in the rotor appeared and was equal to 180×10^6 A/m^2. Huge saturation existed only in the rotor pole and yoke. Saturation in the stator teeth and yoke was within the limits. The magnetic field inside the machine was significantly distorted.

At a time of 0.302 s, the most visible field distortion was observed. There was a lack of pole symmetries. Saturation in the stator core was the same as before the faulty synchronization, whereas that in the rotor pole slightly exceeded 1.5 T. The current densities in rotor bars were no higher than 50×10^6 A/m^2.

5. Comparison of Faulty Synchronization with Sudden Short-Circuit State

Hydro generators are designed to withstand the thermal and stress effects in severe emergencies. One such state is sudden three-phase short circuit. Emerging electrodynamic forces on the end-winding part of the stator winding during sudden short circuit were taken into account in the calculation in order to predict the lifetime of this construction element. The computed value of electromagnetic torque was used to estimate the stresses on the shaft. The thermal resistance of the conductive rotor parts was assessed on the basis of verified analytical equations. The temperature increase as a function of the inducted current in the rotor bar and circuited segments did not cause overheating of these construction parts before the expiry of the protection operation.

Simulations were carried out for the three-phase short-circuit fault which occurred suddenly. Before the analyzed state, the investigated hydro generator worked with no load. The three-phase stator winding was short-circuited at 0.1 s. The terminal voltage before the shorted circuit was equal to U_N.

The calculated maximum value of electromagnetic torque (Figure 25a) was equal to 8.39 T_N and was less than the maximum value obtained from the faulty synchronization (9.35 T_N), whereas the absolute sum of the two maximum electromagnetic torques with opposite directions was equal to 14.16 T_N and was significantly higher than the value obtained from faulty synchronization (10.6 T_N).

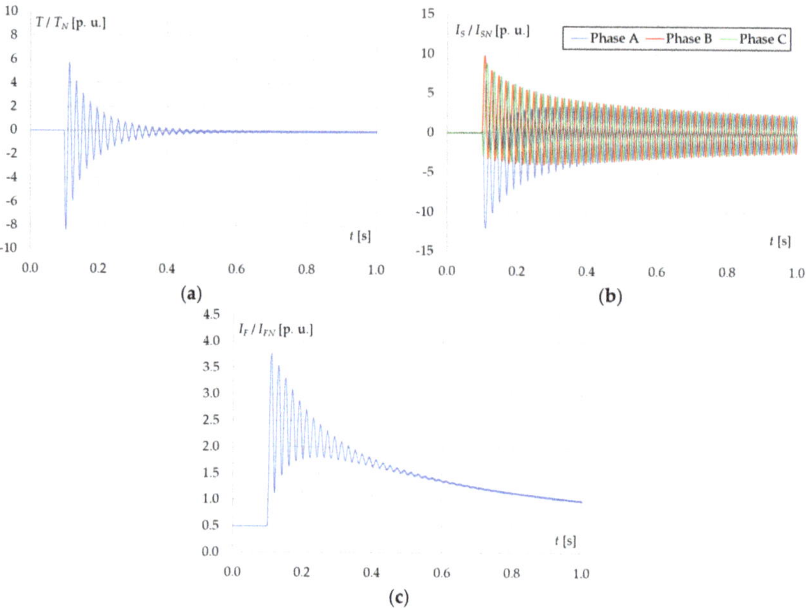

Figure 25. Waveforms of electromagnetic torque (**a**), stator current (**b**), and field current (**c**) during the first second of sudden short circuit.

Figure 25b,c present the waveforms of stator current and field current, respectively. The amplitudes of stator and field current were less than those obtained during faulty synchronization.

The same situation was observed in the case of induced current in rotor bars and circuited segments, which produced values lower than those for faulty synchronization. The distributions of maximum values of these currents are shown in Figures 26 and 27. The maximum values of induced current in the rotor bar were close to the extremes of rotor poles, while smaller values were noted in the bars located in the center of the rotor pole. A different situation was observed in the case of current in the rotor's circuited segment. The maximum values were located in the center of rotor pole, whereas smaller values were found in the extremes of rotor poles. The waveforms of the maximum current in the rotor bar (bar No. 5-8) and in the circuited segment (segment No. 4-45) are presented in Figure 28a,b, respectively.

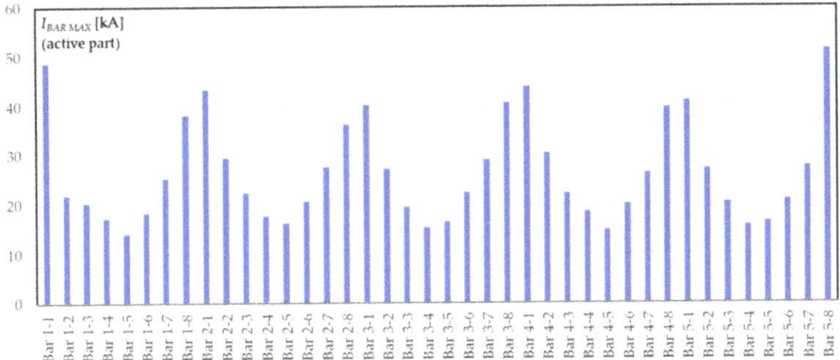

Figure 26. Distribution of maximum values of currents induced in the active part of the rotor bar during the sudden short circuit.

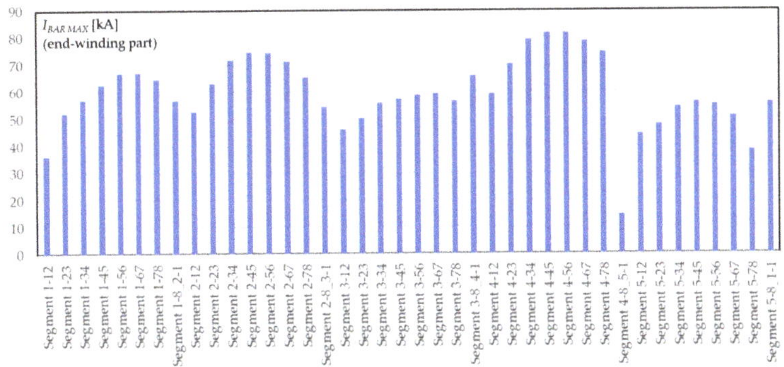

Figure 27. Distribution of maximum values of currents induced in the rotor's circuited segments during the sudden short circuit.

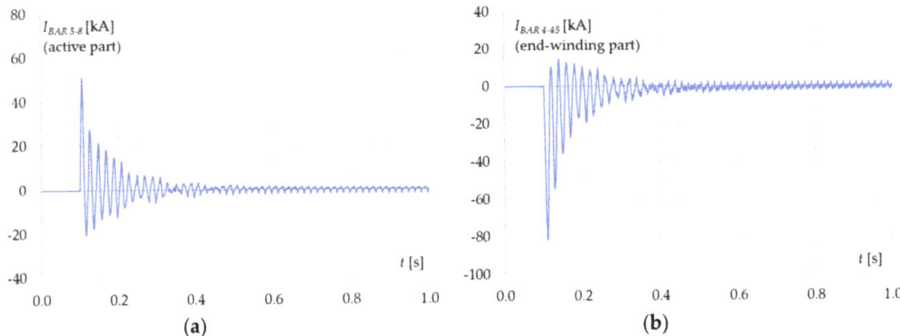

Figure 28. Waveforms of induced current in bar No. 5-8 (**a**) and circuited segment No. 4-45 (**b**) during the first second of sudden short circuit.

A comparison of faulty synchronization (for U_N) with the three-phase sudden short circuit is shown in Table 5. Only the amplitude of electromagnetic torque ($T_{MAX\ 0-PEAK}$) obtained from the analyzed faulty synchronization was higher (by ca. 11%) than the value coming from the sudden short circuit state. A different situation occurred in the case of the absolute sum of the two maximum electromagnetic torques with opposite directions, where a higher value was obtained for the three-phase sudden short circuit.

Table 5. Comparison of maximum values of physical quantities existing during faulty synchronization and sudden short circuit.

Physical Quantities	Unit	Faulty Synchronization	Sudden Short Circuit
$I_{S\,MAX}/I_{SN}$	p.u.	10.79	12.09
$T_{MAX\,0-PEAK}/T_N$	p.u.	9.35	8.39
$T_{MAX\,PEAK-TO-PEAK}/T_N$	p.u.	10.60	14.16
$I_{F\,MAX}/I_{FN}$	p.u.	3.42	3.80
$I_{ROTOR\,BAR\,IN\,ACTIVE\,PART}$	kA	41.79	51.41
$I_{ROTOR\,BAR\,IN\,ACTIVE\,PART}$	kA	72.15	81.71

6. Conclusions

In this article, we analyzed the impact of voltage phase shift during faulty synchronization on physical quantities such as electromagnetic torque, stator current, terminal voltage, field current, induced current in the rotor conductive part, and active and reactive power.

The utilization of the finite element method allowed observing electromagnetic phenomena inside the investigated generator, which until now were unknown. Most published studies used simplified computational models which commutate only the maximum instantaneous values of the stator current and electromagnetic torque. The influence of local core saturations and the uneven distribution of induced current in the rotor bars were ignored.

The greatest thermal hazards existed for a voltage phase shift of 180°. For this angle, the stator current, field current, and induced current in the rotor bars and circuited segments were the highest. The greatest electrodynamics forces on the stator end-winding part existed for this angle as well.

The greatest mechanical hazards existed for the voltage phase shift of −120° (when the power system voltage lagged with respect to generator voltage). In this state, the greatest electromagnetic torque appeared and caused a severe torsional moment on the shaft. The amplitude of electromagnetic torque was higher than that obtained from the three-phase sudden short circuit, whereas, in the case of the absolute sum of the two maximum electromagnetic torques with opposite directions, the maximum value was noted for a phase shift angle equal to 180° during the sudden short circuit.

Larger values of induced current in the rotor conductive parts were observed for the three-phase sudden short circuit but they disappeared relatively quickly (ca. 0.3 s, Figure 26). The induced current decay in the rotor bars was significantly slower and lasted ca. 2 s (Figures 18 and 20). This potentially caused strong heating of the rotor during faulty synchronization.

The stator current and electromagnetic torque did not exceed the rated values during faulty synchronization when the voltage phase shift was less than 10°.

Funding: Computations were carried out using resources provided by the Wroclaw Center for Networking and Supercomputing (http://wcss.pl), grant No. 400.

Conflicts of Interest: The author declares no conflict of interest the results.

References

1. Strang, W.M.; Mozina, C.J.; Beckwith, B.; Beckwith, T.R.; Chhak, S.; Fennell, E.C.; Kalkstein, E.W.; Kozminski, K.C.; Pierce, A.C.; Powell, P.W.; et al. Generator synchronizing industry survey results IEEE power system relaying committee report. *IEEE Trans. Power Deliv.* **1996**, *11*, 174–183. [CrossRef]
2. Thompson, M.J. Fundamentals and advancement in generator synchronizing systems. In Proceedings of the 65th Annual Conference for Protective Relay Engineers, College Station, TX, USA, 2–5 April 2012; pp. 203–214.

3. Ranjbar, A.H.; Gharehpetian, G.B. Transient stability of synchronous generator in out-of-phase synchronization. In Proceedings of the 5th International Conference on Electrical and Electronics Engineering, Bursa, Turkey, 5–9 December 2007; pp. 1–4.

4. Belyaev, N.A.; Khrushchev, Y.; Svechkarev, S.; Prokhorov, A.V.; Wang, L. Generator to grid adaptive synchronization technique based on reference model. In Proceedings of the IEEE Eindhoven PowerTech, Eindhoven, The Netherlands, 29 June–2 July 2015; pp. 1–5.

5. IEEE Standard C50.12-2005. *IEEE Standard for Salient-Pole 50 Hz and 60 Hz Synchronous Generators and Generator/Motors for Hydraulic Turbine Applications Rated 5 MVA and Above*; IEEE: Piscataway, NJ, USA, 2005.

6. IEEE Standard C50.13-2005. *IEEE Standard for Cylindrical-Rotor 50 Hz and 60 Hz Synchronous Generators Rated 10 MVA and Above*; IEEE: Piscataway, NJ, USA, 2006.

7. IEEE Standard C37.102-2006. *IEEE Guide for AC Generator Protection*; IEEE: Piscataway, NJ, USA, 2006.

8. Krause, P.C.; Hollopeter, W.C.; Triezenberg, D.M.; Rusche, P.A. Shaft torque during out of phase synchronization. *IEEE Trans. Power Appar. Syst.* **1977**, *96*, 1318–1323. [CrossRef]

9. Faried, S.O.; Billinton, R.; Aboreshaid, S.; Fotuhi-Firuzabad, M. Stochastic evaluation of turbine-generator shaft torsional torques during faulty synchronization. *IEE Proc. Gener. Transm. Distrib.* **1996**, *143*, 487–491. [CrossRef]

10. Kabir, J.H.; Karim, I.; Chowdhury, A.H. Effect of generator saturation on shaft torques during faulty synchronization. *Electr. Power Syst. Res.* **1998**, *44*, 85–91. [CrossRef]

11. Weili, L.; Purui, W.; Yong, L.; Yi, X.; Dong, L.; Xiaochen, Z.; Jianjun, Z. Influence of rotor structure on field current and rotor electromagnetic field of turbine generator under out-of-synchronization. *IEEE Trans. Magn.* **2017**, *53*, 1–4. [CrossRef]

12. Aboreshaid, S.; Al-Dhalaan, S. Stochastic evaluation of turbine-generator shaft fatigue due to system faults and faulty synchronization. In Proceedings of the 2000 IEEE Power Engineering Society Winter Meeting, Singapore, 23–27 January 2000; pp. 186–191.

13. Salon, S.J. *Finite Element Analysis of Electrical Machines*; Springer: Berlin/Heidelberg, Germany, 1995.

14. Klempner, G.; Kerszenbaum, I. *Handbook of Large Turbo-Generator Operation and Maintenance*, 3rd ed.; Wiley: Hoboken, NJ, USA, 2018.

15. IEEE 492–1999. *IEEE Guide for Operation and Maintenance of Hydro-Generators*; IEEE: Piscataway, NJ, USA, 1999.

Publisher's Note: MDPI stays neutral with regard to jurisdictional claims in published maps and institutional affiliations.

Article

Analysis of Open-Circuit Fault in Fault-Tolerant BLDC Motors with Different Winding Configurations

Mariusz Korkosz *, Jan Prokop, Bartlomiej Pakla, Grzegorz Podskarbi and Piotr Bogusz

The Faculty of Electrical and Computer Engineering, Rzeszow University of Technology, Al.
Powstańców Warszawy 12, 35-959 Rzeszow, Poland; jprokop@prz.edu.pl (J.P.); b.pakla@prz.edu.pl (B.P.);
g.podskarbi@prz.edu.pl (G.P.); pbogu@prz.edu.pl (P.B.)
* Correspondence: mkosz@prz.edu.pl; Tel.: +48-178-651-389

Received: 27 August 2020; Accepted: 9 October 2020; Published: 13 October 2020

Abstract: In this study, tests were carried out on a brushless permanent magnet DC motor with different winding configurations. Three configurations were compared: star, delta and combined star–delta. A mathematical model was constructed for the motor, taking into account the different winding configurations. An analysis of the operation of the motor in the different configurations was performed, based on numerical calculations. The use of different winding configurations affects the properties of the motor. This is significant in the case of the occurrence of various fault states. Based on numerical calculations, an analysis of an open-circuit fault in one of the phases of the motor was performed. Fast Fourier Transform—FFT analysis of the artificial neutral-point voltage was used for the detection of fault states. The results were verified by tests carried out under laboratory conditions. It was shown that the winding configuration has an impact on the behaviour of the motor in the case of an open circuit in one of the phases. The classical star configuration is the worst of the possible arrangements. The most favourable in this respect is the delta configuration. In the case of the combined star–delta configuration, the consequences of the fault depend on the location of the open circuit.

Keywords: brushless direct current motor with permanent magnet (BLDCM); winding configurations; star; delta; star–delta; fault states; open circuit (OC); FFT; neutral-point voltage

---

1. Introduction

Brushless motors with permanent magnets offer the most efficient conversion of electrical energy. The use of permanent magnets significantly increases the efficiency of conversion and allows for the construction of technologically advanced electrical machines [1–3]. In drive systems that are not required to operate with flux weakening, it is possible to use six-step commutation [4,5], which is simpler to implement than the vector control of PMSMs (permanent magnet synchronous machines) [6,7]. This applies to low-power systems in particular, such as drives for pumps and fans. [8–10]. In three-phase motors, a star or delta winding configuration is used [11]. For three-phase motors with permanent magnets, the star configuration is the most common. However, in some cases, to obtain a more favourable torque-to-volume ratio, a delta configuration is used. This type of winding configuration is successfully used in motors designed for small unmanned aerial vehicles (UAVs). Apart from the aforementioned configurations, it is also possible to use a combined star–delta connection [12–14]. This configuration was analysed in [13,14] for a three-phase induction motor. It was shown in those studies that, for a three-phase induction motor, use of the combined star–delta winding configuration reduces losses [13,14] and noise [14].

In sensorless rotor position detection systems for brushless permanent magnet motors, the neutral-point voltage is used [15]. This can also be used in diagnostics of the operation of a drive system [16,17]. In the case of a star winding configuration, for diagnostic purposes, the neutral point of the windings is additionally derived [18]. This is the most favourable solution in terms of diagnostics, but it cannot be applied to, for example, a delta winding configuration. This diagnostic method cannot be used in typical commercial systems, which also lack a fourth wire.

The purpose of this paper is to analyse various fault states in the two basic winding configurations—star and delta—and, as a novel element, a combined star–delta configuration. For a brushless permanent magnet DC motor, how the configuration affects the motor's properties was determined. A state of symmetrical operation was analysed, in addition to the behaviour of the drive system in a selected fault state (an open circuit in one of the phases of the motor). An original mathematical model of the BLDC motor was constructed, taking into account the analysed winding configurations. Based on the numerical model, the content of higher harmonics in the neutral-point voltage was determined. It was shown that the diagnostics of the motor's operating state can be performed based on the Fast Fourier Transform (FFT) analysis of the neutral-point voltage alone, irrespective of the configuration. The selected results of laboratory tests are presented (line currents, contents of higher harmonics of the neutral-point voltage, torque–speed characteristics, overall efficiency and levels of vibrations and noise) for symmetrical operation and for the analysed fault state.

2. Analysis of Winding Configurations

2.1. Star, Delta and Star–Delta Winding Configurations

The analysed device was a brushless permanent magnet motor with 24 slots and 20 permanent magnets placed on a rotor. With two coil groups per phase, the stator windings could be connected in a star (Figure 1b), delta (Figure 1c), or combined star–delta configuration (Figure 1d). A diagram of the electrical connections of the converter—with indications of the neutral-point voltage, u_0, used for diagnostics of the operating state of the drive system—is given in Figure 1a.

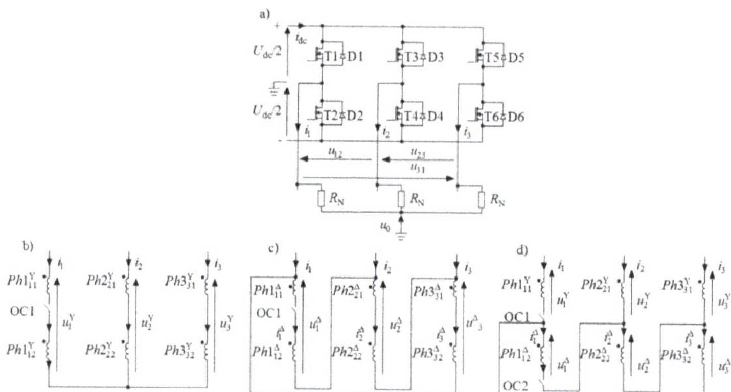

Figure 1. Diagram of (**a**) supply to brushless direct current (BLDC) motor, and configuration of windings in (**b**) star (Y), (**c**) delta (Δ), (**d**) combined star–delta (YΔ) configurations.

In Figure 1b–d, additional switches, labelled OC1 and OC2, indicate the locations of the open circuit in phase *Ph*1. The analysed test construction was designed to drive an unmanned aerial vehicle (UAV) [19]. At the construction stage, the laboratory version was adapted to allow for the analysis of different configurations. Connections of all coil groups of particular phases of the motor were provided. With respect to the UAV drive, the goal of the design was to obtain the maximum torque-to-volume ratio while maintaining the overall efficiency at as high a level as possible. The use of a delta configuration allowed for satisfactory results to be obtained. A view of the wound stator is shown in Figure 2a, and the rotor is shown in Figure 2b.

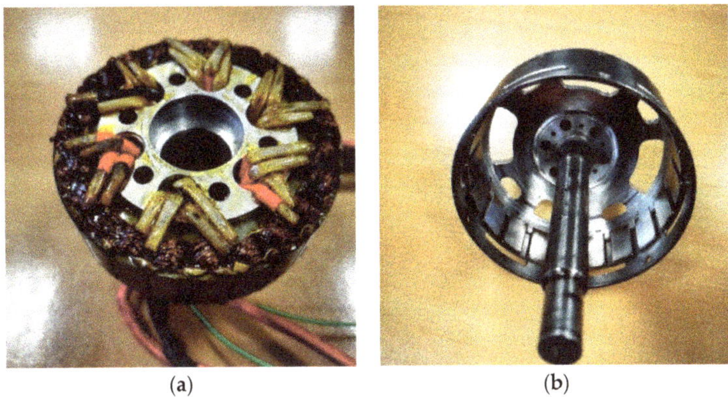

(a)   (b)

Figure 2. Physical model of the analysed BLDC motor design: (**a**) wound stator with terminals; (**b**) rotor.

2.2. Finite Element Method (FEM) Analysis

Using a program for performing numerical calculations in 2D FEM analysis [20], a simulation model was constructed. The 2D Finite Element Method (FEM) model allowed for the analysis of different stator winding configurations. Figure 3 shows an example distribution of magnetic flux lines and current densities for the three winding configurations: star (Figure 3a), delta (Figure 3b) and combined star–delta (Figure 3c).

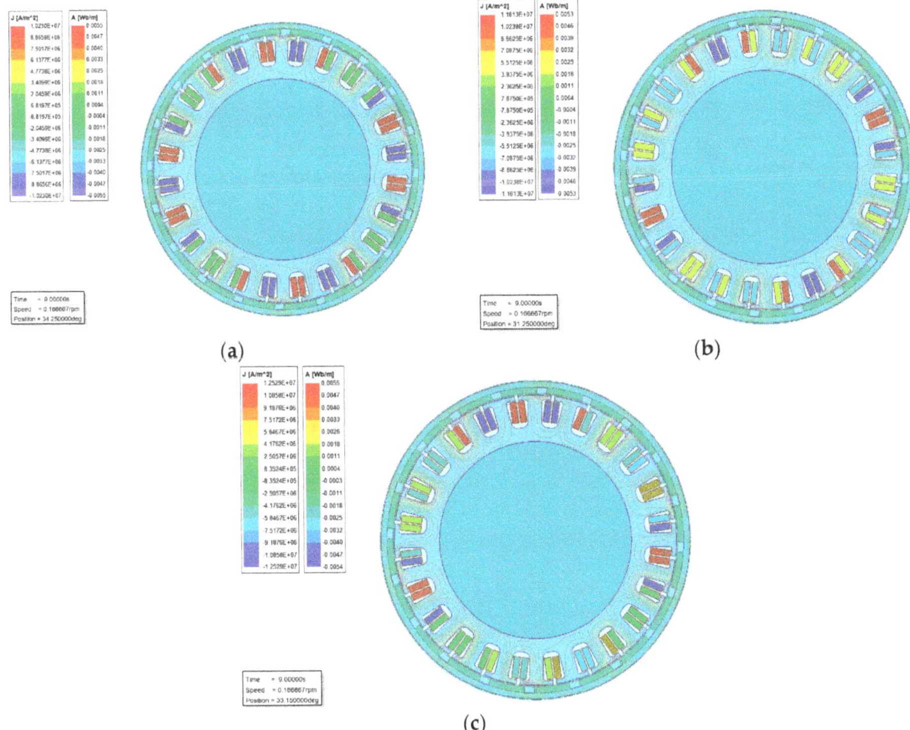

(a)   (b)
(c)

Figure 3. Distribution of current densities and magnetic flux: (**a**) star; (**b**) delta; (**c**) star–delta configurations.

The distribution of current densities varies depending on the analysed winding configuration. The magnetic radial forces depend on the air-gap flux density distribution. This distribution is significantly affected by the choice of winding configuration.

2.3. Static Characteristics

For the comparison of the analysed configurations, their static torque characteristics were determined. The star winding configuration was taken as the base configuration. For this configuration, the value of the supply current I_Y was 10 A. In determining the currents for the other two configurations, it was assumed that the total copper loss P_{Cu} was identical for each connection—i.e., $P_{CuY} = P_{Cu\Delta} = P_{CuY\Delta}$. Based on this assumption, the following relationships were obtained for the currents in particular configurations: $I_Y = \sqrt{3}I_\Delta = \sqrt{\frac{3}{2}}I_{Y\Delta}$. The static torque characteristics were determined both numerically and under laboratory conditions. The stand for the laboratory measurement of static characteristics is shown in Figure 4. Figure 5a shows the results of numerical calculations. The calculations were performed for only one electrical period, which corresponds to a mechanical angle of 36°. The characteristics obtained under laboratory conditions are shown in Figure 5b. In view of the electrical and magnetic asymmetries occurring in the tested prototype, the results for five electrical periods (corresponding to a mechanical angle of 180°) are shown.

Comparing the obtained static torque characteristics, it was observed that with, the same copper losses, the star and delta winding configurations produced practically the same effect. The torque characteristics were almost identical. This was also confirmed by the laboratory measurements (Figure 5b). In the case of the combined star–delta configuration, a smaller torque was obtained in both the numerical calculations (Figure 5a) and the laboratory measurements (Figure 5b). The torque characteristics obtained under laboratory conditions were similar to each other; a greater difference was observed for the numerical results in the case of the Y∆ configuration. Table 1 provides the values obtained (numerically) for the torque constant K_T for each configuration.

Figure 4. Stand for measurement of static characteristics.

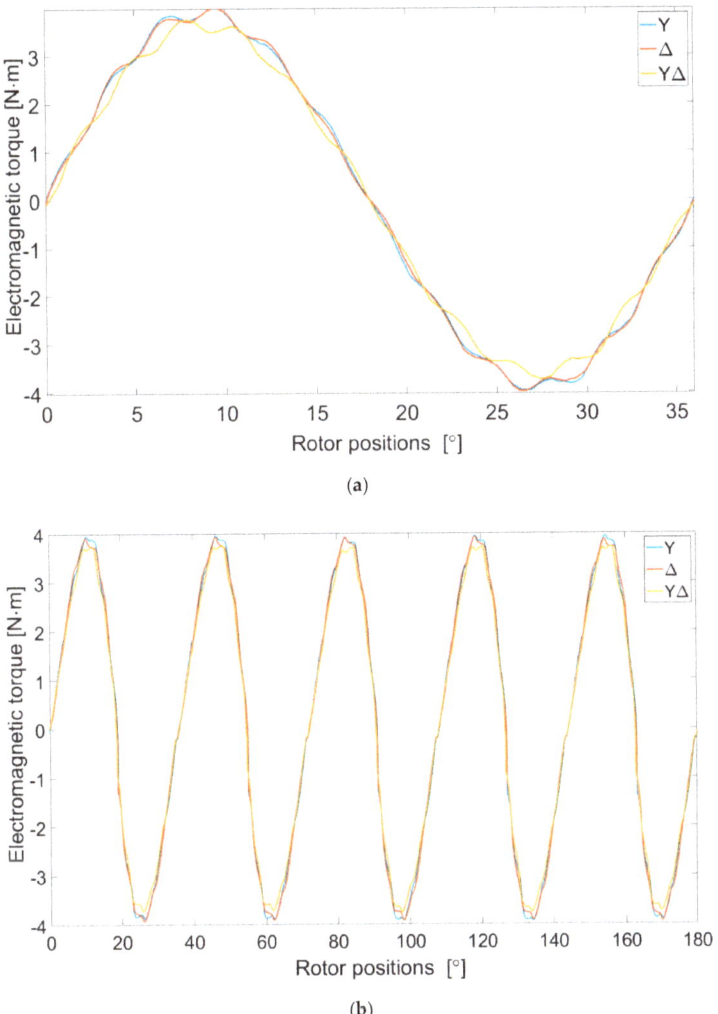

(a)

(b)

Figure 5. Electromagnetic torque vs. rotor position: (**a**) numerical calculations; (**b**) measurements.

Table 1. Torque constants for the analysed configurations.

Y	Δ	YΔ
$K_T^Y = 0.371$ N·m/A	$K_T^\Delta = \sqrt{\frac{1}{3}}K_T^Y$	$K_T^{Y\Delta} = \sqrt{\frac{2}{3}}K_T^Y$

Assuming the same copper losses and identical winding parameters, the highest value of the torque constant was obtained for the star configuration, and the lowest was obtained for the delta configuration (by a factor of $\sqrt{3}$). In the YΔ case, the torque constant fell to approximately 81.6% of the value for the star configuration.

2.4. Back Electromotive Force (BEMF)

Back electromotive force computed numerically for the analysed configurations are shown in Figure 6. These are line-to-line voltages determined for the rotational speed $n = 1000$ rev/min.

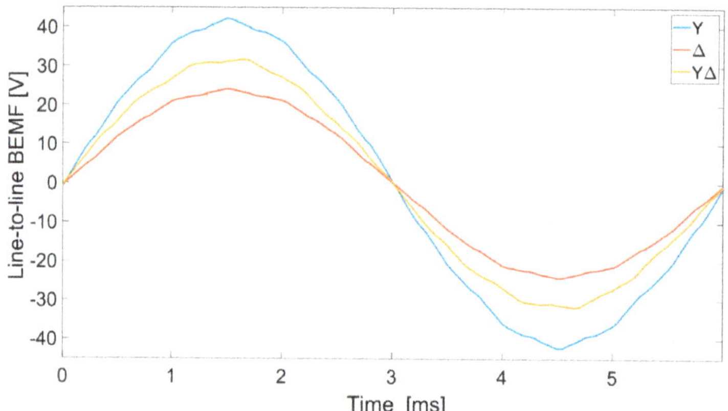

Figure 6. Waveforms of line-to-line BEMF.

Given identical winding parameters, the highest BEMF value is obtained for the star configuration. The BEMFs of the other configurations also differ in shape. The BEMF of the delta configuration is closest to a trapezoidal shape (third harmonic 26%; fifth harmonic 5%).

BEMF constants for all configurations are given in Table 2.

Table 2. BEMF constants for the analysed configurations.

Y	Δ	YΔ
$K_E^Y = 29.08$ V/1000 r/min	$K_E^\Delta = 16.8$ V/1000 r/min	$K_E^{Y\Delta} = 22.27$ V/1000 r/min
$K_E^Y = 100\%$	$K_E^\Delta = 0.563\ K_E^Y$	$K_E^{Y\Delta} = 0.766\ K_E^Y$

The BEMF constant K_E is related to the torque constant K_T. For the delta configuration, the BEMF constant was smaller by a factor of approximately $\sqrt{3}$ compared to the star configuration. This results from the relationship between the phase and line voltages. However, In the case of the YΔ configuration, this relationship to the star configuration was not maintained. The BEMF constant $K_E^{Y\Delta}$ was 76.6% of K_E^Y. In the case of the torque constant, the relationship was $K_T^{Y\Delta} = \sqrt{\frac{2}{3}}K_T^Y = 81.6\%\ K_T^Y$.

3. A Mathematical Model of the Analysed Brushless Direct Current Motor with Permanent Magnet (BLDCM)

Mathematical models are presented for BLDC machines with star (Y), delta (Δ) and star–delta (YΔ) winding configurations. The following simplifying assumptions were adopted in the model of the three-phase BLDC machine:

- Symmetric cylindrical stator and permanent magnet type rotor, linear magnetic circuit;
- Cogging and reluctance torques are neglected;
- Phenomena relating to eddy currents and magnetic hysteresis are neglected—in particular, zero losses in the stator and rotor cores are assumed.

3.1. No-Constraints Phase Voltages, Three-Phase BLDC Star–Delta Model

The general structure of the mathematical model of the three-phase BLDC motor with combined star–delta (YΔ) windings can be written in the following form:

$$\begin{bmatrix} u_1^Y \\ u_2^Y \\ u_3^Y \end{bmatrix} = \begin{bmatrix} R^Y & 0 & 0 \\ 0 & R^Y & 0 \\ 0 & 0 & R^Y \end{bmatrix} \begin{bmatrix} i_1^Y \\ i_2^Y \\ i_3^Y \end{bmatrix} + \begin{bmatrix} L^Y & M^Y & M^Y \\ M^Y & L^Y & M^Y \\ M^Y & M^Y & L^Y \end{bmatrix} \frac{d}{dt} \begin{bmatrix} i_1^Y \\ i_2^Y \\ i_3^Y \end{bmatrix} + \begin{bmatrix} L^{Y\Delta} & M^{Y\Delta} & M^{Y\Delta} \\ M^{Y\Delta} & L^{Y\Delta} & M^{Y\Delta} \\ M^{Y\Delta} & M^{Y\Delta} & L^{Y\Delta} \end{bmatrix} \frac{d}{dt} \begin{bmatrix} i_1^\Delta \\ i_2^\Delta \\ i_3^\Delta \end{bmatrix} + \begin{bmatrix} e_1^Y \\ e_2^Y \\ e_3^Y \end{bmatrix} \quad (1)$$

$$\begin{bmatrix} u_1^\Delta \\ u_2^\Delta \\ u_3^\Delta \end{bmatrix} = \begin{bmatrix} R^\Delta & 0 & 0 \\ 0 & R^\Delta & 0 \\ 0 & 0 & R^\Delta \end{bmatrix} \begin{bmatrix} i_1^\Delta \\ i_2^\Delta \\ i_3^\Delta \end{bmatrix} + \begin{bmatrix} L^\Delta & M^\Delta & M^\Delta \\ M^\Delta & L^\Delta & M^\Delta \\ M^\Delta & M^\Delta & L^\Delta \end{bmatrix} \frac{d}{dt} \begin{bmatrix} i_1^\Delta \\ i_2^\Delta \\ i_3^\Delta \end{bmatrix} + \begin{bmatrix} L^{\Delta Y} & M^{\Delta Y} & M^{\Delta Y} \\ M^{\Delta Y} & L^{\Delta Y} & M^{\Delta Y} \\ M^{\Delta Y} & M^{\Delta Y} & L^{\Delta Y} \end{bmatrix} \frac{d}{dt} \begin{bmatrix} i_1^Y \\ i_2^Y \\ i_3^Y \end{bmatrix} + \begin{bmatrix} e_1^\Delta \\ e_2^\Delta \\ e_3^\Delta \end{bmatrix} \tag{2}$$

$$J\frac{d\omega_\mathrm{m}}{dt} + D\,\omega_\mathrm{m} + T_\mathrm{L} = T_\mathrm{e} \tag{3}$$

where total electromagnetic torque T_e is given by:

$$T_\mathrm{e} = \frac{1}{\omega_\mathrm{m}} \sum_{i=1}^{3} \left(e_i^Y i_i^Y + e_i^\Delta i_i^\Delta \right) \tag{4}$$

The following symbols are used in Equations (1)–(4) for $i = 1, 2, 3$: u_i^Y, u_i^Δ—phase voltages; i_i^Y, i_i^Δ—phase currents; R^Y, R^Δ—phase resistances; $L^Y, M^Y, L^\Delta, M^\Delta, L^{Y\Delta} = L^{\Delta Y}, M^{Y\Delta} = M^{\Delta Y}$—self and mutual inductances; e_i^Y, e_i^Δ—phase BEMFs voltages; J—rotor moment of inertia; ω_m—mechanical angular speed of rotor; D—rotor damping of viscous friction coefficient; T_L—load torque.

The phase BEMF voltages for $i = 1, 2, 3$ in Equations (1), (2) and (4) are defined as follows:

$$e_i^Y = \omega\frac{\partial \psi_i^{Y\mathrm{PM}}(\theta)}{\partial\theta} = \omega K_\mathrm{E}^Y f_i^Y\left(\theta - (i-1)\frac{2\pi}{3}\right) e_i^\Delta = \omega\frac{\partial \psi_i^{\Delta\mathrm{PM}}(\theta)}{\partial\theta} = \omega K_\mathrm{E}^\Delta f_i^\Delta\left(\theta - (i-1)\frac{2\pi}{3}\right) \tag{5}$$

where θ—electrical rotor angle; $\omega = d\theta/dt = p\,\omega_\mathrm{m}$—electrical angular speed; p—machine pole pairs; $\psi_i^{Y\mathrm{PM}}(\theta)$, $\psi_i^{\Delta\mathrm{PM}}(\theta)$—permanent magnetic fluxes linking the stator windings; K_E^Y, K_E^Δ—BEMF constant of one phase; $f_i^Y()$, $f_i^\Delta()$—phase trapezoidal functions, profile BEMF. The phase BEMF functions can be expressed as ideal trapezoidal waveforms or as the summation of all harmonics in the Fourier series—for example, they can be expressed as follows:

$$e_i^Y = \omega \sum_{v=1}^{\infty} E_{iv}^Y \cos(v\theta + \theta_{i0}^Y), \; e_i^\Delta = \omega \sum_{v=1}^{\infty} E_{iv}^\Delta \cos(v\theta + \theta_{i0}^\Delta) \tag{6}$$

where E_{iv}^Y, E_{iv}^Δ and θ_{i0}^Y, θ_{i0}^Δ are the amplitude (at 1 rad/s speed) and initial phase of the harmonic BEMF.

The total electromagnetic torque T_e in Equation (4) is produced by an interaction between the rotor's permanent magnets and the stator's energized windings, and can be expressed as follows:

$$T_\mathrm{e} = \sum_{i=1}^{3} \left(i_i^Y K_\mathrm{T}^Y f_i^Y\left(\theta - (i-1)\frac{2\pi}{3}\right) + i_i^\Delta K_\mathrm{T}^\Delta f_i^\Delta\left(\theta - (i-1)\frac{2\pi}{3}\right)\right) \tag{7}$$

where K_T^Y, K_T^Δ—torque constants of one phase.

Additional constraints on voltages and currents are imposed by the arrangement of motor phase windings in a star (Y), delta (Δ) or combined star–delta (YΔ) configuration.

3.2. Star, Delta and Star–Delta Line Voltage Models

The relationship of phase currents i_i^Y ($i = 1, 2, 3$) in a star (Y) configuration and phase voltages u_i^Δ ($i = 1, 2, 3$) in a delta (Δ) configuration can be written as follows:

$$\begin{aligned} \text{Star (Y)} &: \sum_{i=1}^{3} i_i^Y = 0; \\ \text{Delta } (\Delta) &: \sum_{i=1}^{3} u_i^\Delta = 0 \end{aligned} \tag{8}$$

The relationship between the line-to-line and phase voltages and line and phase currents in the star (Y) and delta (Δ) configurations can be written as follows:

$$\text{Star (Y)}:\ \begin{bmatrix} u_{12} \\ u_{23} \\ u_{31} \end{bmatrix} = \begin{bmatrix} 1 & -1 & 0 \\ 0 & 1 & -1 \\ -1 & 0 & 1 \end{bmatrix}\begin{bmatrix} u_1^Y \\ u_2^Y \\ u_3^Y \end{bmatrix};\ \begin{bmatrix} i_1 \\ i_2 \\ i_3 \end{bmatrix} = \begin{bmatrix} i_1^Y \\ i_2^Y \\ i_3^Y \end{bmatrix} \tag{9}$$

$$\text{Delta (Δ)}:\ \begin{bmatrix} u_{12} \\ u_{23} \\ u_{31} \end{bmatrix} = \begin{bmatrix} u_1^\Delta \\ u_2^\Delta \\ u_3^\Delta \end{bmatrix};\ \begin{bmatrix} i_1 \\ i_2 \\ i_3 \end{bmatrix} = \begin{bmatrix} 1 & 0 & -1 \\ -1 & 1 & 0 \\ 0 & -1 & 1 \end{bmatrix}\begin{bmatrix} i_1^\Delta \\ i_2^\Delta \\ i_3^\Delta \end{bmatrix} \tag{10}$$

Including the constraint Equations (9) and (10), the final equations of the BLDC motor models can be written for the Y, Δ and YΔ configurations. The model line-to-line voltages and the line currents, phase BEMF voltages and total electromagnetic torques can be written as follows.

A. Model for star (Y) winding BLDC motor

$$\begin{bmatrix} u_{12} \\ u_{23} \\ u_{31} \end{bmatrix} = \begin{bmatrix} 2R^Y & 0 & R^Y \\ R^Y & 2R^Y & 0 \\ 0 & R^Y & 2R^Y \end{bmatrix}\begin{bmatrix} i_1^Y \\ i_2^Y \\ i_3^Y \end{bmatrix} + \begin{bmatrix} 2(L^Y - M^Y) & 0 & L^Y - M^Y \\ L^Y - M^Y & 2(L^Y - M^Y) & 0 \\ 0 & L^Y - M^Y & 2(L^Y - M^Y) \end{bmatrix}\frac{d}{dt}\begin{bmatrix} i_1^Y \\ i_2^Y \\ i_3^Y \end{bmatrix} + \begin{bmatrix} e_1^Y - e_2^Y \\ e_2^Y - e_3^Y \\ e_3^Y - e_1^Y \end{bmatrix} \tag{11}$$

$$\begin{bmatrix} i_1 \\ i_2 \\ i_3 \end{bmatrix} = \begin{bmatrix} i_1^Y \\ i_2^Y \\ i_3^Y \end{bmatrix};\ e_i^Y = \omega K_E^Y f_i^Y\left(\theta - (i-1)\frac{2\pi}{3}\right);\ T_e^Y = \sum_{i=1}^{3}\left(i_i^Y K_T^Y f_i^Y\left(\theta - (i-1)\frac{2\pi}{3}\right)\right) \tag{12}$$

B. Model for delta (Δ) winding BLDC motor

$$\begin{bmatrix} u_{12} \\ u_{23} \\ u_{31} \end{bmatrix} = \begin{bmatrix} R^\Delta & 0 & 0 \\ 0 & R^\Delta & 0 \\ 0 & 0 & R^\Delta \end{bmatrix}\begin{bmatrix} i_1^\Delta \\ i_2^\Delta \\ i_3^\Delta \end{bmatrix} + \begin{bmatrix} L^\Delta & M^\Delta & M^\Delta \\ M^\Delta & L^\Delta & M^\Delta \\ M^\Delta & M^\Delta & L^\Delta \end{bmatrix}\frac{d}{dt}\begin{bmatrix} i_1^\Delta \\ i_2^\Delta \\ i_3^\Delta \end{bmatrix} + \begin{bmatrix} e_1^\Delta \\ e_2^\Delta \\ e_3^\Delta \end{bmatrix} \tag{13}$$

$$\begin{bmatrix} i_1 \\ i_2 \\ i_3 \end{bmatrix} = \begin{bmatrix} 1 & 0 & -1 \\ -1 & 1 & 0 \\ 0 & -1 & 1 \end{bmatrix}\begin{bmatrix} i_1^\Delta \\ i_2^\Delta \\ i_3^\Delta \end{bmatrix};\ e_i^\Delta = \omega K_E^\Delta f_i^\Delta\left(\theta - (i-1)\frac{2\pi}{3}\right);\ T_e^\Delta = \sum_{i=1}^{3}\left(i_i^\Delta K_T^\Delta f_i^\Delta\left(\theta - (i-1)\frac{2\pi}{3}\right)\right) \tag{14}$$

C. Model for star–delta (YΔ) winding BLDC motor

$$\begin{bmatrix} u_{12} \\ u_{23} \\ u_{31} \end{bmatrix} = \begin{bmatrix} R & -R^Y & -R^Y \\ -R^Y & R & -R^Y \\ -R^Y & -R^Y & R \end{bmatrix}\begin{bmatrix} i_1^\Delta \\ i_2^\Delta \\ i_3^\Delta \end{bmatrix} + \begin{bmatrix} L & M & M \\ M & L & M \\ M & M & L \end{bmatrix}\frac{d}{dt}\begin{bmatrix} i_1^\Delta \\ i_2^\Delta \\ i_3^\Delta \end{bmatrix} + \begin{bmatrix} e_1^Y - e_2^Y + e_1^\Delta \\ e_2^Y - e_3^Y + e_2^\Delta \\ e_3^Y - e_1^Y + e_3^\Delta \end{bmatrix} \tag{15}$$

$$\begin{bmatrix} i_1 \\ i_2 \\ i_3 \end{bmatrix} = \begin{bmatrix} i_1^Y \\ i_2^Y \\ i_3^Y \end{bmatrix} = \begin{bmatrix} 1 & 0 & -1 \\ -1 & 1 & 0 \\ 0 & -1 & 1 \end{bmatrix}\begin{bmatrix} i_1^\Delta \\ i_2^\Delta \\ i_3^\Delta \end{bmatrix};\ T_e = T_e^Y + T_e^\Delta \tag{16}$$

The equivalent motor parameters in Equation (15) are determined by the following relationships:

$$R = 2R^Y + R^\Delta;\ L = 2(L^Y - M^Y) + L^\Delta + 2(L^{Y\Delta} - M^{Y\Delta});$$
$$M = -L^Y + M^Y + M^\Delta - L^{Y\Delta} + M^{Y\Delta} \tag{17}$$

The phase BEMF voltages K_T^Y, K_T^Δ ($i = 1, 2, 3$) in Equation (15) and the torques T_e^Y, T_e^Δ in Equation (16) are defined in Equations (12) and (14).

Equations (3) and (11)–(16) constitute the mathematical models of BLDC motors with star (Y), delta (Δ) and star–delta (YΔ) winding configurations.

4. Comparative Analysis, Simulation and Experimental Verification

All numerical calculations and laboratory tests were carried out in a fixed operating state for different rotational speeds. The numerical results are presented here for only a speed of 500 r/min. The supply voltage was varied, taking account of the BEMF constants k_V for particular configurations (Table 2). In the laboratory tests, to obtain the required operating point, the value of the load torque was varied.

4.1. Star Configuration—Transient Analysis in Healthy Mode and under Open Phase Fault

4.1.1. Waveforms of Electromagnetic Torque and Currents

The star winding configuration is shown in Figure 1b. Opening the switch OC1 breaks the supply in the phase *Ph*1. The effects of the open circuit state on the produced electromagnetic torque and line currents are shown in Figures 7, 8a, and 9a. The current waveforms obtained under laboratory conditions for the analysed operating states are shown in Figures 8b and 9b.

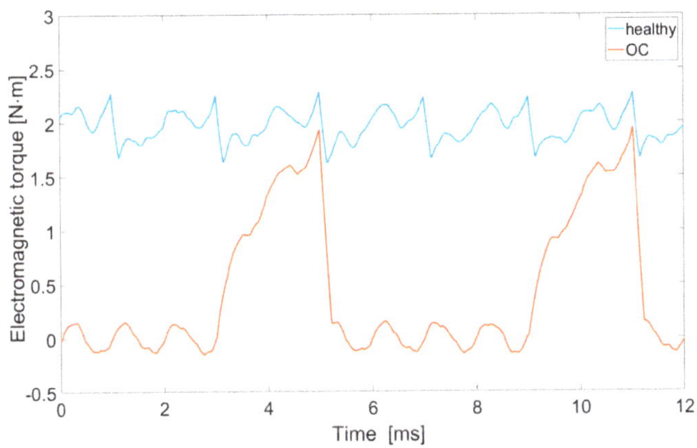

Figure 7. Star (Y) configuration—waveforms of electromagnetic torque.

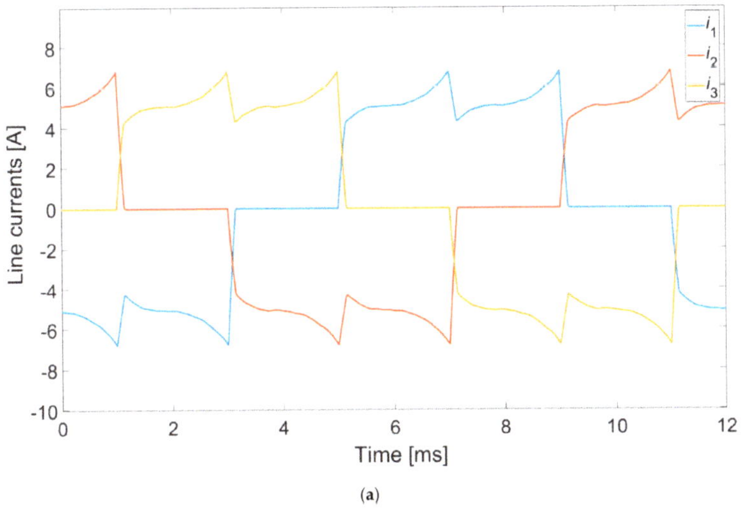

(a)

Figure 8. *Cont.*

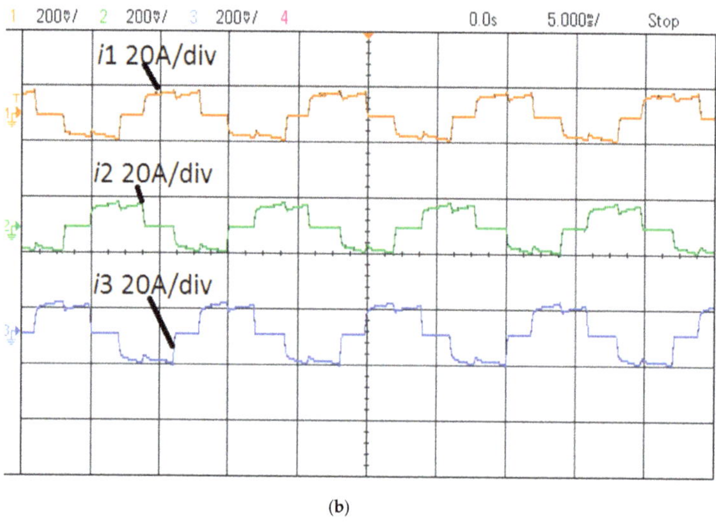

(**b**)

Figure 8. Star (Y) configuration—waveforms of line currents in healthy mode: (**a**) numerical calculations; (**b**) measurements.

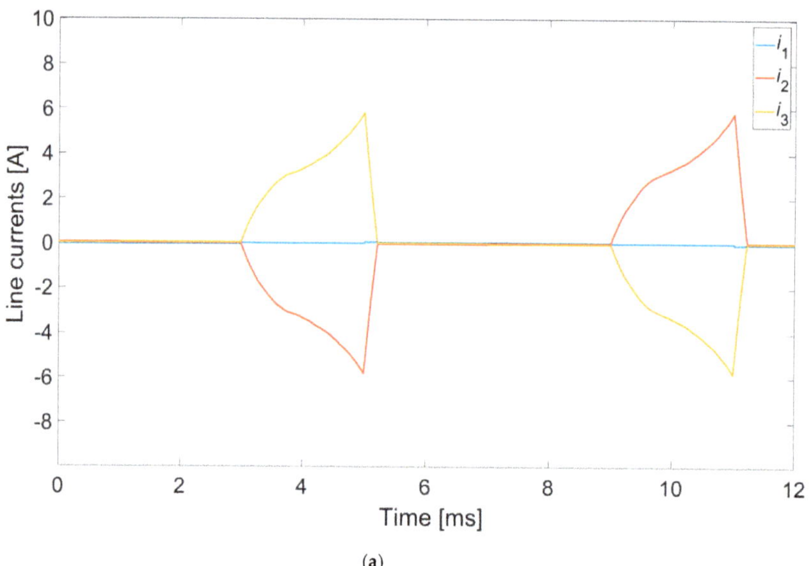

(**a**)

Figure 9. *Cont.*

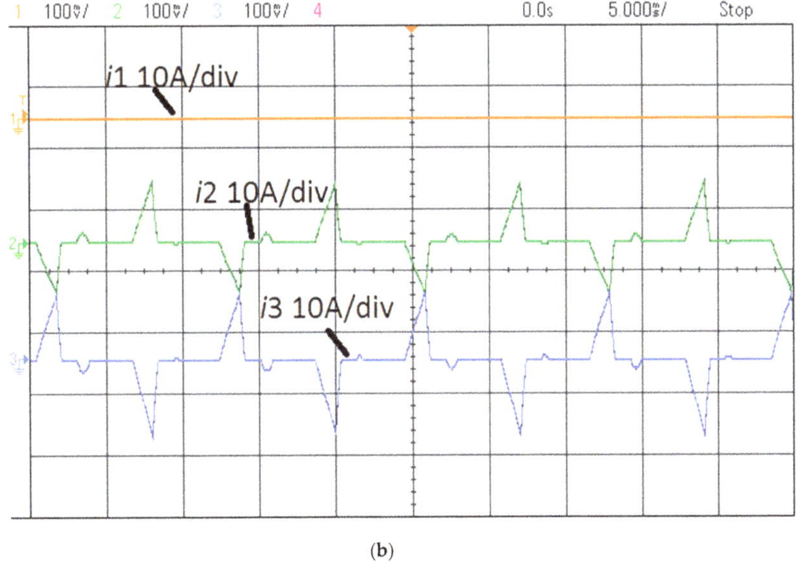

(b)

Figure 9. Waveforms of line currents in the star configuration with open OC1 switch: (**a**) numerical calculations; (**b**) measurements.

Opening the switch OC1 (Figure 1b) causes the whole of phase *Ph*1 to be switched off. In this case, the motor operates in two phases only. This means that four out of six control sequences cannot be applied. If the fault occurs during operation, the motor can continue to function, but only at a small load torque (below 20% of the rated value).

4.1.2. FFT of u_0

Waveforms of the voltage u_0 were recorded for the analysed states (Figure 1a). Using harmonic analysis (FFT), significant harmonics present in the voltage signal, u_0, were identified (Figure 1a). Figure 10 shows the content of higher harmonics for the star configuration under normal operation (numerical calculations in Figure 10a; measurements in Figure 10b). The content of higher harmonics of the voltage signal in fault state OC1 is shown in Figure 11a (numerical calculations) and Figure 11b (measurements).

The fundamental frequency $f1$ of the voltage signal u_0 depends on the rotational speed n and on the design parameter p and the number of pole pairs of the rotor.

$$f_1 = \frac{p \cdot n}{60} \tag{18}$$

In a state of electrical and magnetic symmetry, the voltage signal u_0 contains only the third harmonic and its odd multiples (the 9th, 15th, etc.), as can be seen in Figure 10a. In practice, most electrical machines have electrical and magnetic asymmetries to a greater or lesser degree, usually arising during the technological process. Under laboratory conditions, the appearance of other harmonics that were multiples of three was observed (the 6th, 12th, 18th, etc.; see Figure 10b).

In the fault state OC1, all odd harmonics appeared in the voltage signal u_0 (the first, third, fifth, seventh, etc.). This is visible in Figure 11a. In real conditions, not all odd harmonics were present—for example, the fifth harmonic was absent (Figure 11b). Additionally, the even harmonic $f2$ appeared in the spectrum. The appearance in the voltage signal of the harmonic $f1$ with an amplitude comparable to that of $f3$ indicates an open circuit in one of the phases of the motor.

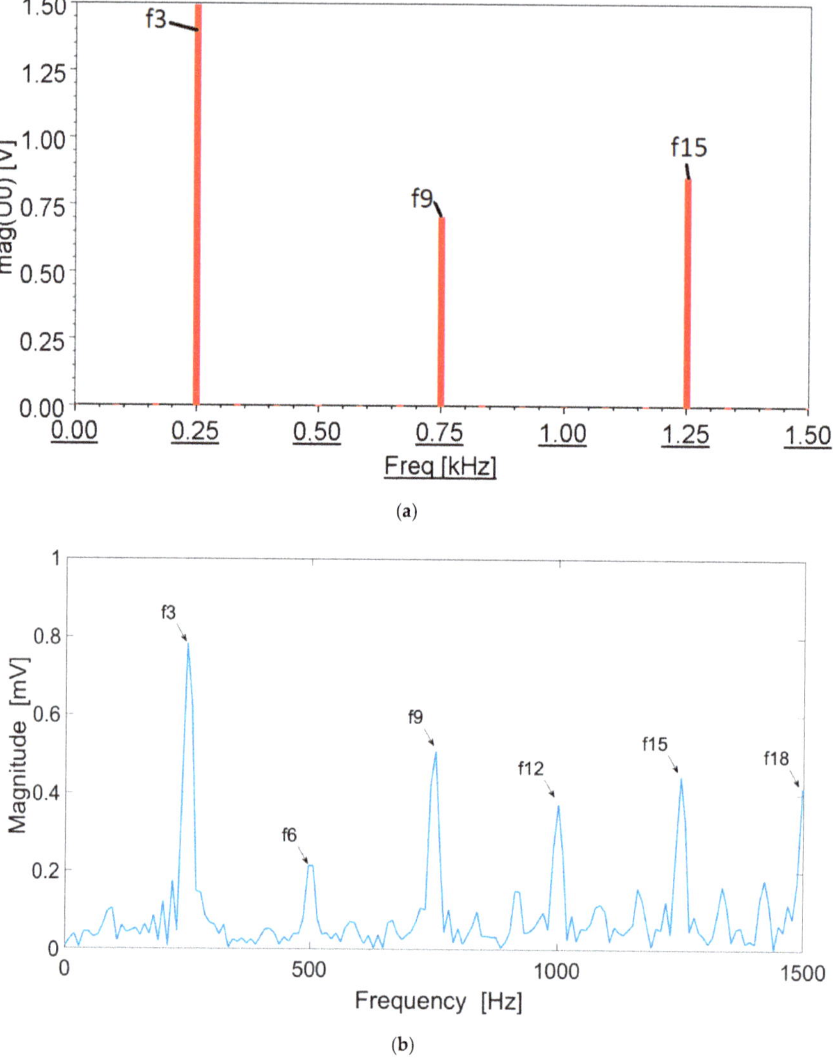

Figure 10. Fast Fourier Transform (FFT) of neutral voltage u_0 for star configuration in healthy mode: (**a**) numerical calculations; (**b**) measurements.

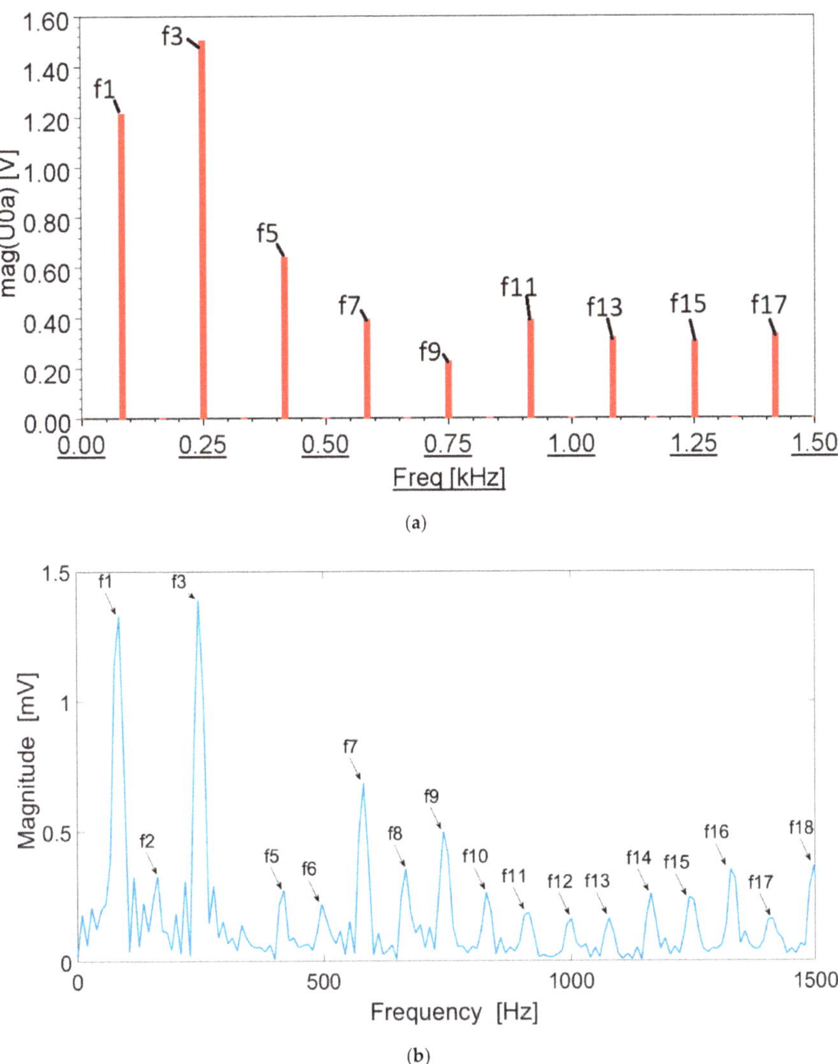

Figure 11. Star (Y)—FFT of neutral voltage star configurations of open circuits (OCs): (**a**) numerical calculations; (**b**) measurement.

4.2. Delta Configuration—Transient Analysis in Healthy Mode and Under Open Phase Fault

The delta winding configuration is shown in Figure 1c. As with the star configuration, an analysis of the symmetrical operation of the motor and its operation in a fault state was performed. Opening the switch OC1 caused a break in the conventional phase *Ph*1. However, unlike in the case of the star winding configuration, the motor still received a full six-step supply.

4.2.1. Delta (Δ) Configuration—Waveforms of Electromagnetic Torque and Currents

The waveforms of electromagnetic torque and line currents for the analysed states of operation are shown in Figures 12–14.

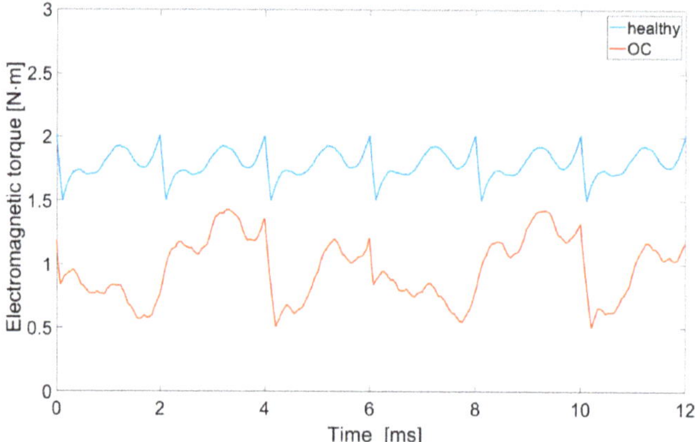

Figure 12. Delta (Δ) configuration—waveforms of electromagnetic torque.

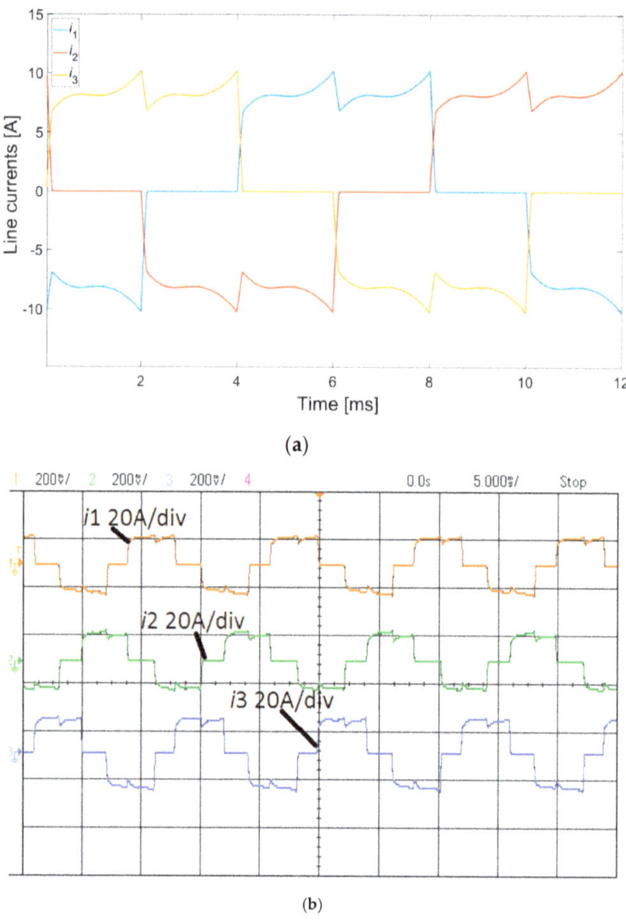

Figure 13. Delta (Δ) configuration—waveforms of line currents in healthy mode: (**a**) numerical calculations; (**b**) measurements.

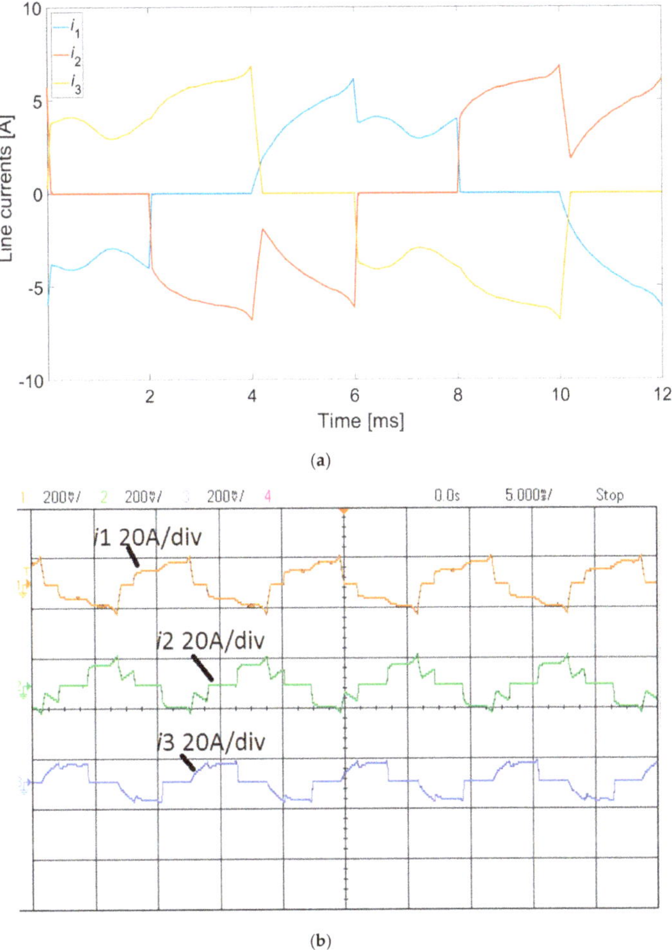

(b)

Figure 14. Delta (Δ) configuration—waveforms of line currents with open OC1: (**a**) numerical calculations; (**b**) measurements

The fault state OC1 caused by an open circuit in the conventional phase Ph1 had a negative impact on the value of the produced electromagnetic torque. The mean value of the torque decreased, simultaneous to a significant increase in torque ripple (Figure 12). There was a significant change in the waveforms of the line currents (Figure 14). The RMS values of line currents differed between phases. Greater differences were observed under laboratory conditions (Figure 14b). As with the star configuration, the motor could continue to operate with the load torque reduced by several tens of percent. The drive system could also be restarted after stopping.

4.2.2. FFT of u_0

For the delta configuration, based on harmonic analysis (FFT), the harmonics present in the voltage signal u_0 were identified. Analysis was performed for normal operation and for a fault state. The results of numerical calculations and laboratory measurements for normal operation are shown in Figure 15. Figure 16 shows the analysis of higher harmonics for the OC1 fault state.

As with the star configuration, the voltage signal u_0 for normal operation contains only the third harmonic and its odd multiples. The amplitude of the third harmonic was dominant in both the numerical calculations (Figure 15a) and the laboratory results (Figure 15b). In the practical setup, even multiples of the third harmonic also appeared (the 6th, 12th, 18th, etc.). This is a result of the electrical and magnetic asymmetry of the tested motor. The OC fault state caused a significant change in the content of harmonics of the voltage u_0. As in the star case, all odd harmonics appeared (Figure 16). In both the numerical calculations (Figure 16a) and the laboratory tests (Figure 16b), the appearance of the first harmonic was significant. Depending on the type of testing (numerical, experimental), the amplitude of this harmonic was 50% (or more) of the amplitude of the third harmonic. Under laboratory conditions, the ninth harmonic was practically absent in the fault state.

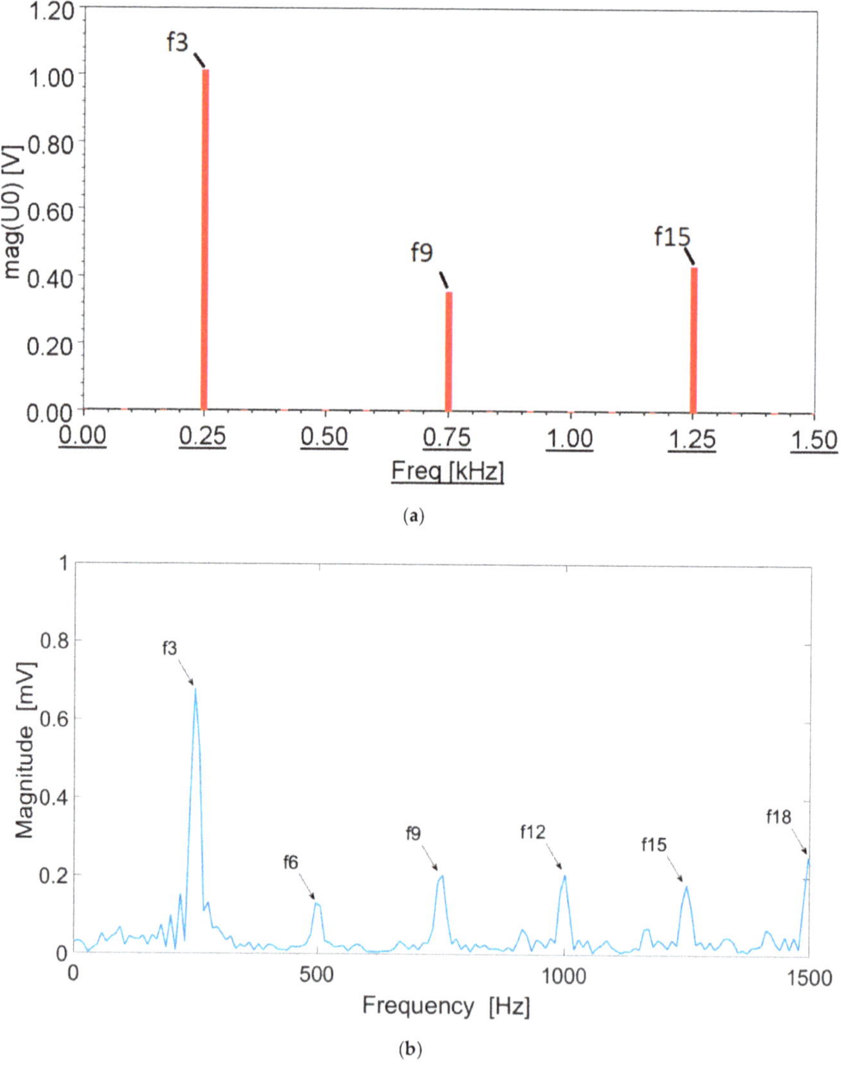

Figure 15. Delta (Δ) configuration—FFT of neutral voltage u_0 in healthy mode: (**a**) numerical calculations; (**b**) measurements.

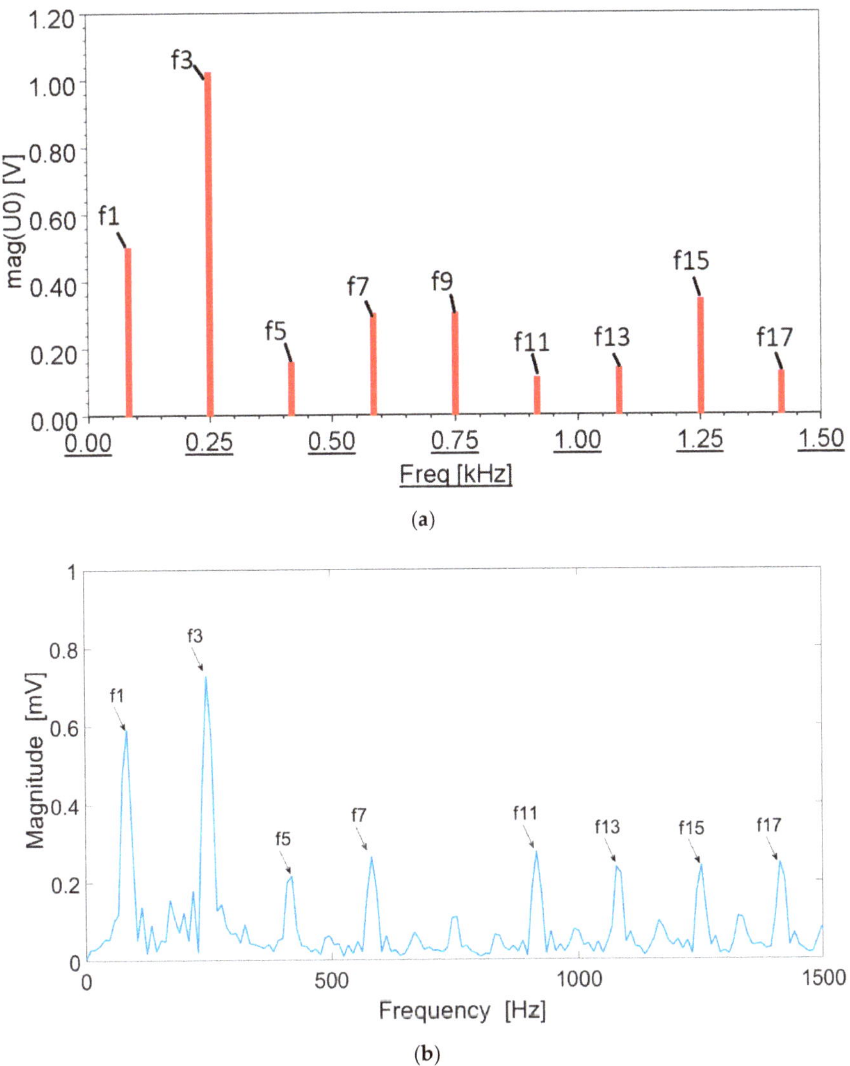

Figure 16. Delta (Δ) configuration—FFT of neutral voltage u_0 with open OC1: (**a**) numerical calculations; (**b**) measurements.

4.3. Star–Delta Configuration—Transient Analysis in Healthy Mode and Under Open Phase Fault

The YΔ type winding configuration is a combination of the well-known Y and Δ configurations (Figure 1d). In the case analysed, one half of the windings were connected in a delta arrangement. Analysis of the case of a fault state caused by an open circuit in one of the phases required consideration of two alternative scenarios. In the first scenario, the fault was produced by opening the key OC1. This was a situation very similar to the OC fault state for a star configuration. In the second scenario, when the key OC2 was opened (Figure 1d), the open circuit in the conventional phase Ph1 occurred in the delta part of the configuration. This case was similar, but not identical, to the case of an OC fault in a delta configuration.

4.3.1. Waveforms of Electromagnetic Torque and Currents

The waveforms of electromagnetic torque (Figure 17) show the effect of the analysed fault states. Figure 18 shows the waveforms of line currents under normal operation. The waveforms of line currents in the analysed fault states are shown in Figure 19 (OC1) and Figure 20 (OC2).

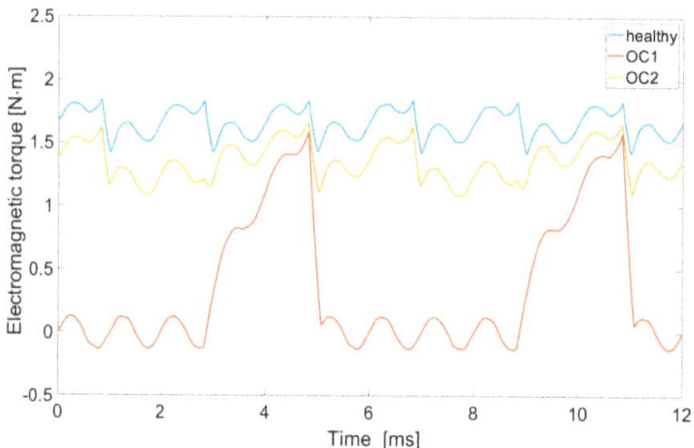

Figure 17. Star–delta (YΔ) configuration—waveforms of electromagnetic torque.

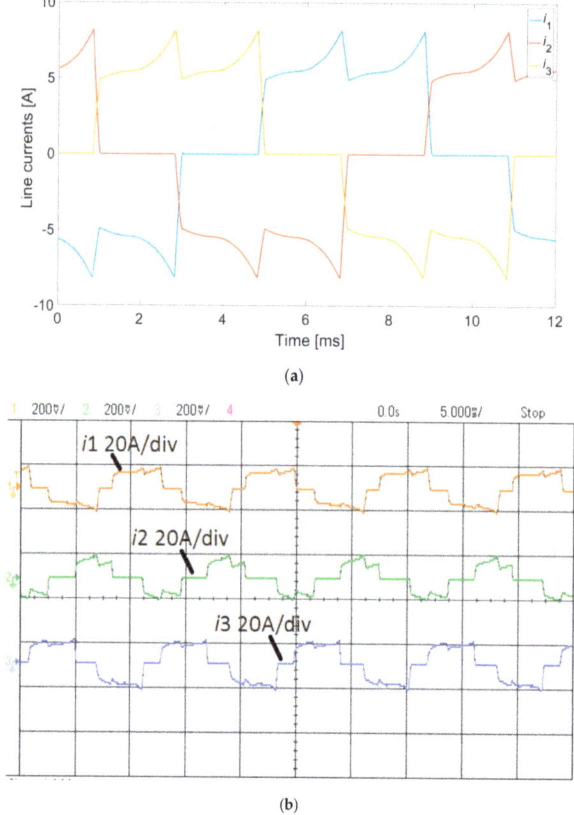

(b)

Figure 18. Star–delta (YΔ) configuration—waveforms of line currents in healthy mode: (**a**) numerical calculations; (**b**) measurements.

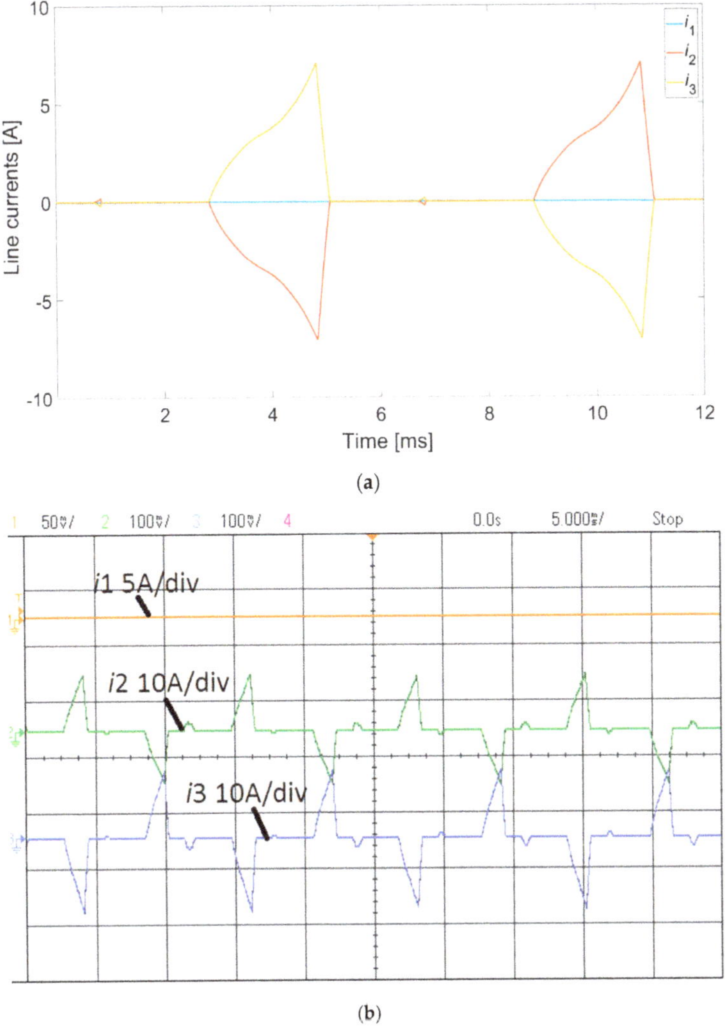

Figure 19. Star–delta (YΔ) configuration—waveforms of line currents in fault state OC1: (**a**) numerical calculations; (**b**) measurements.

Opening the switch OC1 (Figure 1d) caused an open circuit in the phase Ph1. The result of this, as in the case of the star configuration, was that only two steps of the available six were used. The consequences of the fault were also almost identical. The motor could operate in the fault state only with a significantly reduced load torque. The torque ripple increased by several hundred percent. If the open circuit was caused by the opening of the switch OC2, the consequences of the fault were less serious. There was a relatively small decrease in the electromagnetic torque generated (Figure 17) and a relatively small increase in ripple. The motor could continue to operate in all conditions. The changes in the waveforms of the line currents were also relatively minor (Figure 20). Relatively small differences were observed in the effective values of the particular line currents.

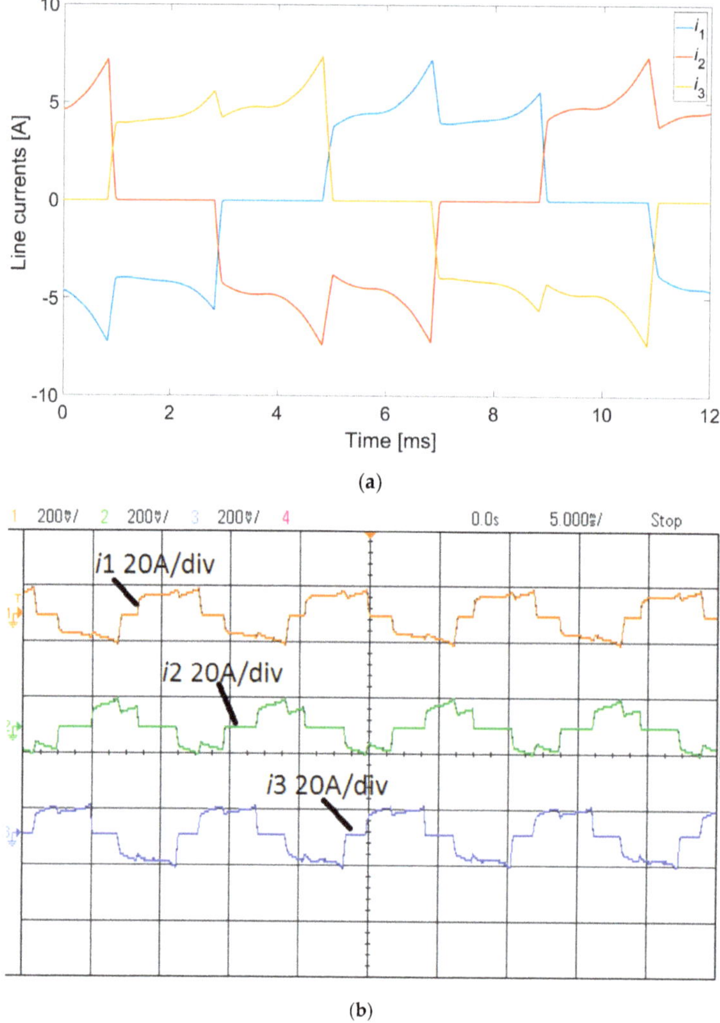

(b)

Figure 20. Star–delta (YΔ) configuration—waveforms of line currents in fault state OC2: (**a**) numerical calculations; (**b**) measurements.

4.3.2. FFT of u_0

For the analysed states, the waveform of the voltage u_0 was recorded for the YΔ configuration. Based on harmonic analysis (FFT), the significant harmonics present in the voltage signal u_0 were identified. The contents of higher harmonics are shown in Figure 21 for a normal operating state, in Figure 22 for fault state OC1, and in Figure 23 for fault state OC2.

In the healthy mode laboratory tests (Figure 21b), as in the case of the Y and Δ configurations, even multiples of the third harmonic appeared. Other harmonics also appeared, such as the 14th and 16th. However, in comparison with the cases of the Y and Δ configurations, the contribution of the third harmonic was dominant. This was not observed in the results of the numerical calculations (Figure 21a).

Fault state OC1 caused the appearance of the first harmonic, which is dominant for this state (Figure 22). Other odd harmonics also appeared—for example, the fifth (in the numerical calculations only) and the seventh. In the real system, the appearance of, for example, the second harmonic was also observed.

In fault state OC2, odd harmonics, such as the first, fifth and seventh appeared again. However, the appearance of the first harmonic, with an amplitude not exceeding 50% of the value of the third harmonic, is significant here.

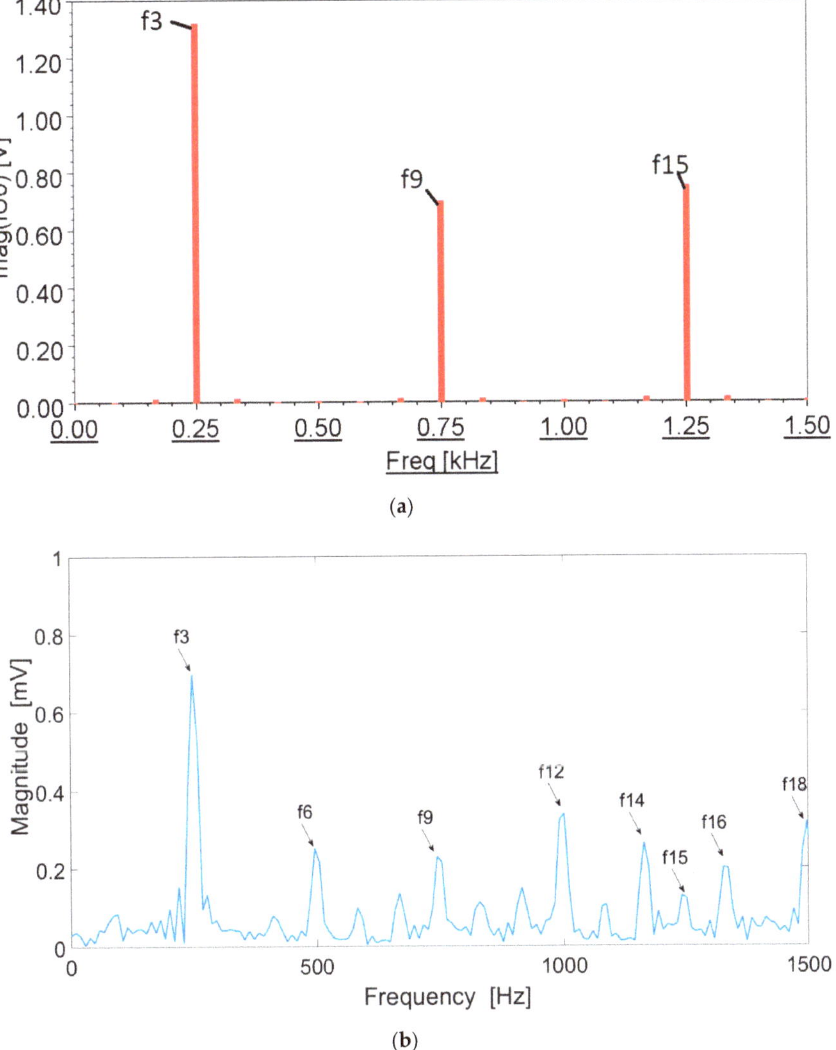

Figure 21. Star–delta (YΔ) configuration—FFT of neutral voltage u_0 in healthy mode: (**a**) numerical calculations; (**b**) measurements.

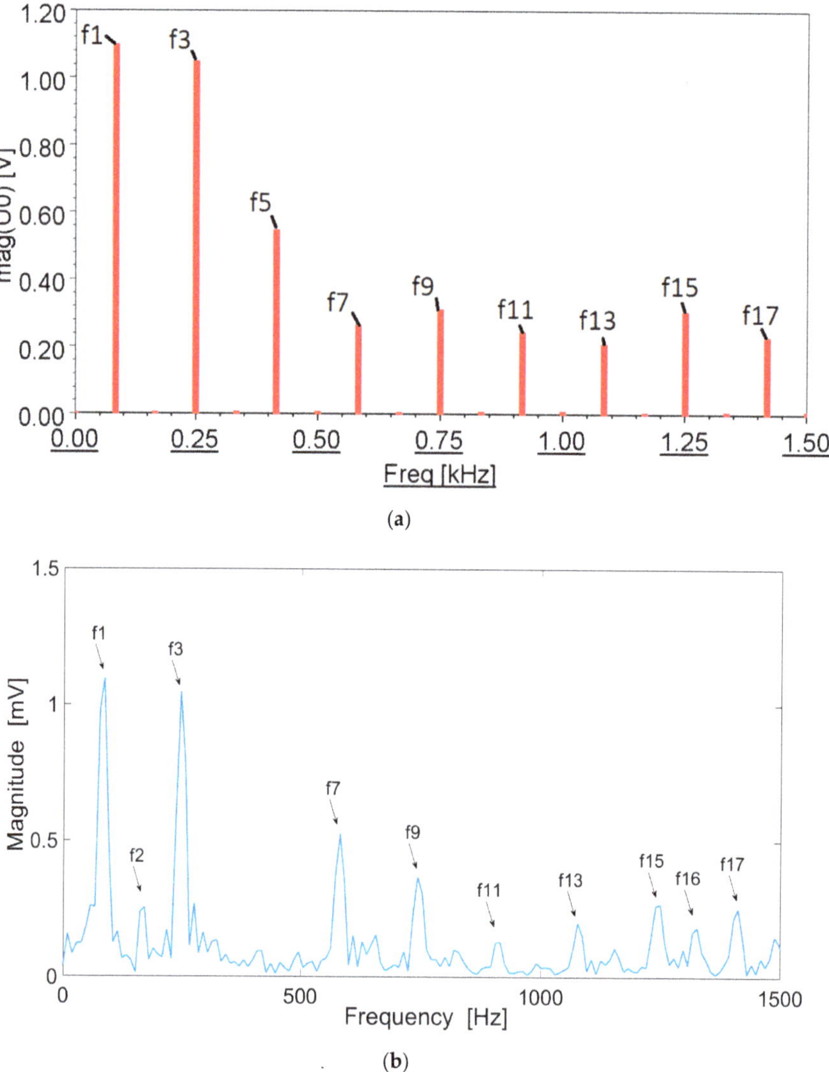

Figure 22. Star–delta (YΔ) configuration—FFT of neutral voltage u_0 in fault state OC1: (**a**) numerical calculations; (**b**) measurements.

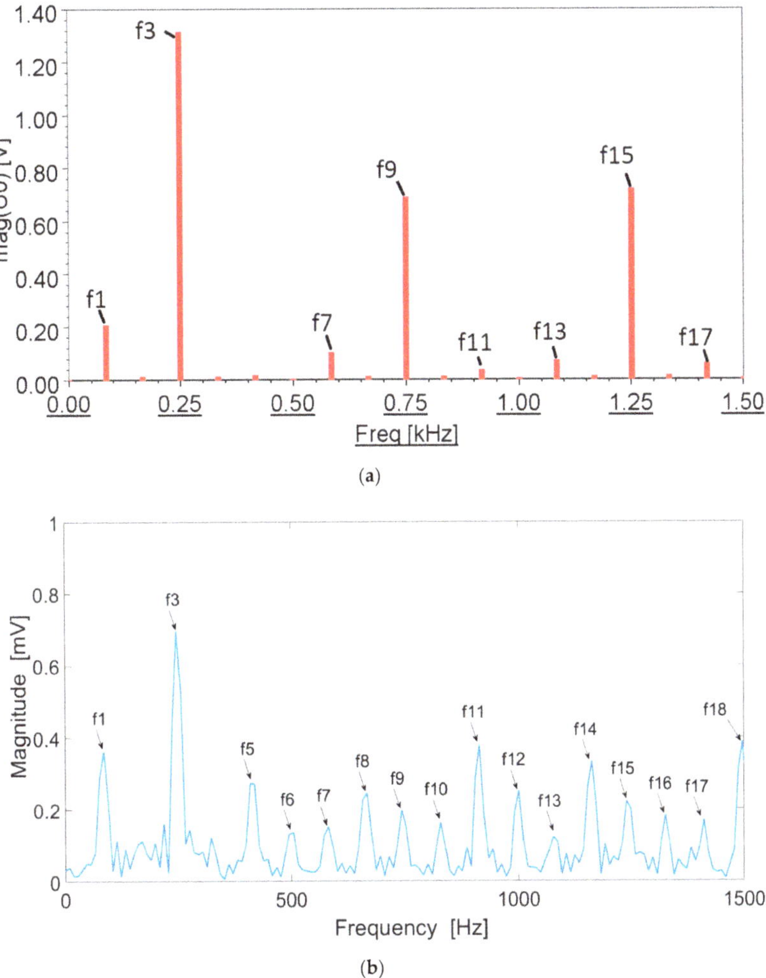

Figure 23. Star–delta (YΔ) configuration—FFT of neutral voltage u_0 in fault state OC2: (**a**) numerical calculations; (**b**) measurements.

4.4. Comparative Analysis in Healthy Mode and Under Open Phase Fault

To verify the state of symmetrical operation and fault states under laboratory conditions, the noise level and acceleration were recorded. This was done at the same load torque (0.5 Nm) and rotational speed (1000 r/min). The results are given in Table 3.

Table 3. Selected results for acceleration and noise level.

Configuration /State of Work	Symmetry		OC1		OC2	
	Acoustic Noise [dB]	Acceleration [m/s^2]	Acoustic Noise [dB]	Acceleration [m/s^2]	Acoustic Noise [dB]	Acceleration [m/s^2]
Y	1.2	64.6	6.8	72.1	-	-
Δ	1.2	64.4	1.1	64.3	-	-
YΔ	0.9	64.1	6.6	70.4	1.0	64.3

At the same operating point, the lowest noise level was produced by the YΔ configuration. The level of vibrations was also the lowest. However, the differences, relative to the Y and Δ configurations were, suboptimal. The situation was changed by an open circuit in one of the phases. Depending on the configuration, vibrations and noise may increase significantly (for the Y configuration and for OC1 in the YΔ configuration), although a lower noise level was recorded in the YΔ OC1 case (70.4 dB, compared with 72.1 dB for Y). In the case of an open circuit in the Δ configuration, or for OC2 in the YΔ configuration, the vibration and noise levels remained practically unchanged.

Torque–speed characteristics for the particular configurations under symmetrical operation and in fault states are shown in Figure 24. Overall efficiency, determined by a direct method, is shown in Figure 25.

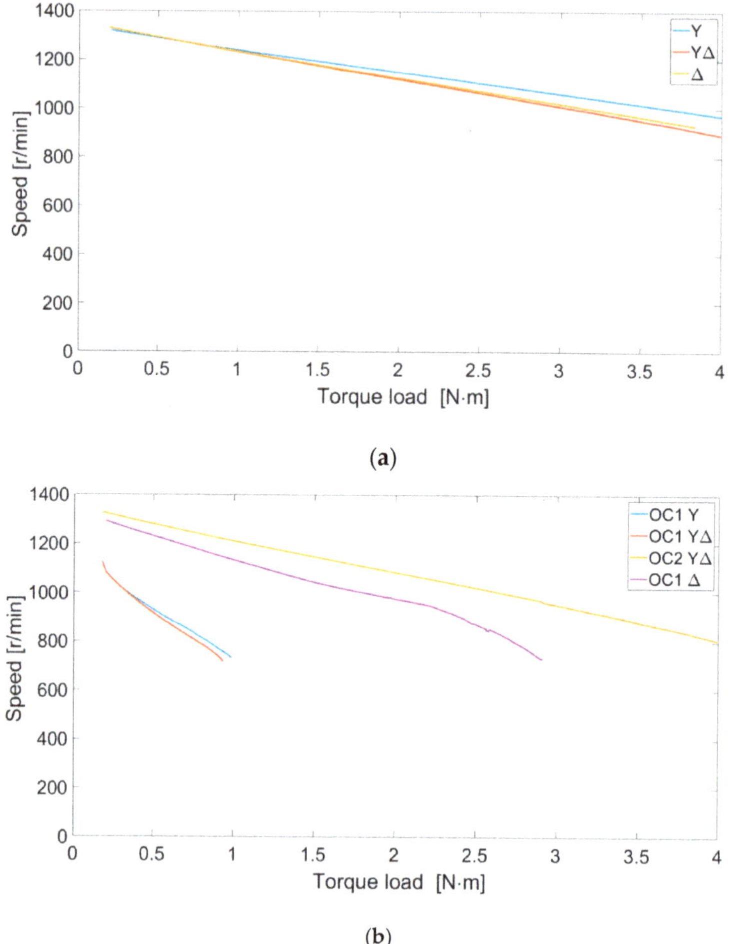

(**a**)

(**b**)

Figure 24. Speed vs. load torque for Y, YΔ and Δ configurations: (**a**) healthy mode; (**b**) OC fault state.

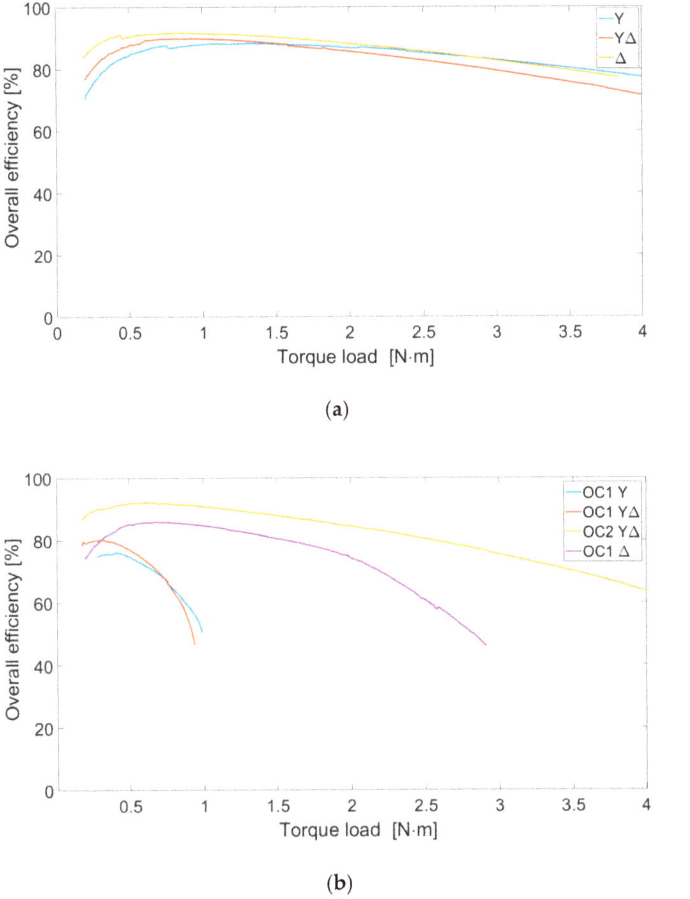

Figure 25. Overall efficiency vs. load torque for Y, YΔ and Δ configurations: (**a**) healthy mode; (**b**) OC fault state.

The torque speed characteristics for the YΔ and Δ configurations were less rigid. This is a result of the greater voltage drop between the power source and the system (higher current I_{dc}). The highest overall efficiency was provided by the delta configuration. The YΔ configuration had higher efficiency under partial load. For higher loads, its efficiency was the lowest of all those analysed.

5. Conclusions

The type of winding configuration used has an impact on the operation of a BLDC motor. A combined star–delta configuration does not produce significant advantages in normal operation. Under laboratory conditions, slightly lower levels of vibrations and noise were obtained, but the efficiency of the drive system was lower than in the case of the delta configuration. Differences appeared in the case of a fault caused by an open circuit in one of the phases. For the combined star–delta configuration, if the open circuit occurred in the delta part, this was the best tolerated fault case, with the load capacity of the motor reduced by less than 20% and the levels of vibrations and noise remaining almost unchanged. For the detection of fault states, the artificial neutral-point voltage can be used. In the voltage signal for normal operation, the third harmonic and its odd multiples were present. Any electrical or magnetic asymmetries or faults in the operation of the system produce changes in the content of higher harmonics in this signal. An open-circuit fault will cause all odd harmonics to appear in the voltage signal. The amplitude of the first harmonic is a key indicator. In the case of an open circuit in a star configuration or in the star part of a combined star–delta configuration, the amplitude of the first harmonic is comparable to that of the third harmonic. For a fault in a delta configuration or in the delta part of a combined configuration, the first harmonic does not exceed 50% of the value of the third harmonic. This is not the only fault state that may occur during operation of the motor.

Further research by the authors will include the analysis of other fault states of the analysed winding configurations, such as short circuits of turns or whole windings, and faults on the side of the supply converter.

Author Contributions: Conceptualization, M.K. and B.P.; methodology, J.P. and M.K.; software, M.K. and B.P.; validation, M.K. and J.P.; formal analysis, J.P.; investigation, M.K., J.P. and B.P.; resources, M.K.; data curation, B.P.; writing—original draft preparation, M.K.; writing—review and editing, M.K., B.P., J.P, P.B. and G.P.; visualization, M.K. and B.P.; supervision, J.P.; project administration, M.K.; funding acquisition, M.K. All authors have read and agreed to the published version of the manuscript.

Funding: This work is financed in part by the statutory funds (UPB) of the Department of Electrodynamics and Electrical Machine Systems, Rzeszow University of Technology and in part by Polish Ministry of Science and Higher Education under the program "Regional Initiative of Excellence" in 2019–2022. Project number 027/RID/2018/19, amount granted 11 999 900 PLN.

Conflicts of Interest: The authors declare no conflict of interest.

References

1. Zhou, R.; Li, G.; Wang, Q.; He, J. Torque Calculation of Permanent-Magnet Spherical Motor Based on Permanent-Magnet Surface Current and Lorentz Force. *IEEE Trans. Magn.* **2020**, *56*, 1–9. [CrossRef]
2. Onsal, M.; Cumhur, B.; Demir, Y.; Yolacan, E.; Aydin, M. Rotor Design Optimization of a New Flux-Assisted Consequent Pole Spoke-Type Permanent Magnet Torque Motor for Low-Speed Applications. *IEEE Trans. Magn.* **2018**, *54*, 1–5. [CrossRef]
3. Li, Z.; Chen, Q.; Wang, Q. Analysis of Multi-Physics Coupling Field of Multi-Degree-of-Freedom Permanent Magnet Spherical Motor. *IEEE Trans. Magn.* **2019**, *55*, 1–5. [CrossRef]
4. Wei, Y.; Xu, Y.; Zou, J.; Li, Y. Current Limit Strategy for BLDC Motor Drive with Minimized DC-Link Capacitor. *IEEE Trans. Ind. Appl.* **2015**, *51*, 3907–3913. [CrossRef]
5. Shao, J. An Improved Microcontroller-Based Sensorless Brushless DC (BLDC) Motor Drive for Automotive Applications. *IEEE Trans. Ind. Appl.* **2006**, *42*, 1216–1221. [CrossRef]
6. Zhu, J.; Bai, H.; Wang, X.; Li, X. Current Vector Control Strategy in a Dual-Winding Fault-Tolerant Permanent Magnet Motor Drive. *IEEE Trans. Energy Convers.* **2018**, *33*, 2191–2199. [CrossRef]
7. Niu, S.; Luo, Y.; Fu, W.; Zhang, X. An Indirect Reference Vector-Based Model Predictive Control for a Three-Phase PMSM Motor. *IEEE Access* **2020**, *8*, 29435–29445. [CrossRef]
8. Khan, K.R.; Miah, M.S. Fault-Tolerant BLDC Motor-Driven Pump for Fluids with Unknown Specific Gravity: An Experimental Approach. *IEEE Access* **2020**, *8*, 30160–30173. [CrossRef]
9. Chun, T.; Tran, Q.; Lee, H.; Kim, H. Sensorless Control of BLDC Motor Drive for an Automotive Fuel Pump Using a Hysteresis Comparator. *IEEE Trans. Power Electron.* **2014**, *29*, 1382–1391. [CrossRef]
10. Kumar, R.; Singh, B. Grid Interactive Solar PV-Based Water Pumping Using BLDC Motor Drive. *IEEE Trans. Ind. Appl.* **2019**, *55*, 5153–5165. [CrossRef]
11. Ferreira, F.J.T.E.; Baoming, G.; de Almeida, A.T. Stator Winding Connection-Mode Management in Line-Start Permanent Magnet Motors to Improve Their Efficiency and Power Factor. *IEEE Trans. Energy Convers.* **2013**, *28*, 523–534. [CrossRef]
12. Ibrahim, M.N.F.; Rashad, E.; Sergeant, P. Performance Comparison of Conventional Synchronous Reluctance Machines and PM-Assisted Types with Combined Star–DeltaWinding. *Energies* **2017**, *10*, 1500. [CrossRef]
13. Misir, O.; Morteza Raziee, S.; Hammouche, N.; Klaus, C.; Kluge, R.; Ponick, B. Prediction of Losses and Efficiency for Three-Phase Induction Machines Equipped with Combined Star–Delta Windings. *IEEE Trans. Ind. Appl.* **2017**, *53*, 3579–3587. [CrossRef]
14. Lei, Y.; Zhao, Z.; Wang, S.; Dorrell, D.G.; Xu, W. Design and Analysis of Star–Delta Hybrid Windings for High-Voltage Induction Motors. *IEEE Trans. Ind. Electron.* **2011**, *58*, 3758–3767. [CrossRef]

15. Song, X.; Han, B.; Zheng, S.; Fang, J. High-Precision Sensorless Drive for High-Speed BLDC Motors Based on the Virtual Third Harmonic Back-EMF. *IEEE Trans. Power Electron.* **2018**, *33*, 1528–1540. [CrossRef]

16. Ullah, Z.; Hur, J. A Comprehensive Review of Winding Short Circuit Fault and Irreversible Demagnetization Fault Detection in PM Type Machines. *Energies* **2018**, *11*, 3309. [CrossRef]

17. Yang, S. Online Stator Turn Fault Detection for Inverter-Fed Electric Machines Using Neutral Point Voltages Difference. *IEEE Trans. Ind. Appl.* **2016**, *52*, 4039–4049. [CrossRef]

18. Urresty, J.C.; Riba, J.R.; Romeral, L. A Back-emf Based Method to Detect Magnet Failures in PMSMs. *IEEE Trans. Mag.* **2013**, *49*, 591–598. [CrossRef]

19. Korkosz, M.; Prokop, J.; Bogusz, P. Modelling and experimental research of fault-tolerant dual-channel brushless DC motor. *IET Electr. Power Appl.* **2018**, *12*, 787–796. [CrossRef]

20. ANSYS. *Ansys Electronics Release 2019 R3*; ANSYS Inc.: Canonsburg, PA, USA, 2019.

Review

Modern Hybrid Excited Electric Machines

Marcin Wardach *, Ryszard Palka, Piotr Paplicki, Pawel Prajzendanc and Tomasz Zarebski

Faculty of Electrical Engineering, West Pomeranian University of Technology, Sikorskiego 37, 70-313 Szczecin, Poland; ryszard.palka@zut.edu.pl (R.P.); piotr.paplicki@zut.edu.pl (P.P.); pawel.prajzendanc@zut.edu.pl (P.P.); tomasz.zarebski@zut.edu.pl (T.Z.)
* Correspondence: marcin.wardach@zut.edu.pl; Tel.: +48-91-449-42-17

Received: 1 October 2020; Accepted: 10 November 2020; Published: 12 November 2020

Abstract: The paper deals with the overview of different designs of hybrid excited electrical machines, i.e., those with conventional permanent magnets excitation and additional DC-powered electromagnetic systems in the excitation circuit. The paper presents the most common topologies for this type of machines found in the literature—they were divided according to their electrical, mechanical and thermal properties. Against this background, the designs of hybrid excited machines that were the subject of scientific research of the authors are presented.

Keywords: permanent magnet machines; hybrid excitation; electric vehicles; wind power generator; finite element methods; variable speed machines

---

1. Introduction

Permanent magnets excited electrical machines, and in particular, hybrid excited, play nowadays an increasingly important role mainly due to their high efficiency and relatively high power-to-weight ratio. Their great popularity is related to dynamic developments in the field of modern materials, such as permanent magnets, magnetic composites, but also various types of power electronic components and fast and efficient microprocessor systems used to control the machine.

Over the past two decades, the development of technology has caused the properties of permanent magnets to improve rapidly. For this reason, permanent magnet machines have become competitive to those previously used.

Due to the development of machines of this type, manufacturers of various types of drive systems are increasingly turning to structures containing permanent magnets [1–5]. A very strong development in this area has been noticeable in the automotive industry. Currently, all major car companies offer cars powered by electric motors. Most of all, reliability and low production costs are required from motors in electric vehicles. Perhaps for this reason, until recently, the "strongest player" in this part of the automotive sector, which is the Tesla Company, used only induction motors in its designs. However, most manufacturers today use mainly permanent magnet motors [6].

Apart from the undoubted advantages mentioned earlier, permanent magnet machines have also some disadvantages. The disadvantages most frequently mentioned in the literature, include the higher cost of their production, compared to induction machines and limited and energy-costly ability to regulate the excitation flux. In variable speed drives, e.g., in electric vehicles, the drive is required to operate over the widest possible rpm range with the highest possible torque. Due to the constant flux of excitation from permanent magnets, to enable the motor to operate in the high-speed area—switching from constant torque to constant power characteristics—it is necessary to apply the d-axis current strategy [7]. This strategy is effective, but only to a limited extent. Too deep, field weakening can lead to permanent damage to the magnets, and also generates additional losses.

Another area by which electric machines, that can operate in a wide range of rotational speeds can be used, is wind energy that the electrical power that a wind turbine can generate depends on the strength of the wind. In the case of small (home) power plants, which usually do not have wind attack angle control systems on the turbine blades, the wind force will directly determine the rotational speed of the turbine, and therefore also the generator. There are several ways to regulate the rotational speed of a low-power wind turbine, from eccentric mounting of the turbine in relation to the axis of rotation of the entire nacelle, through more or less complicated mechanical systems causing the wind to press against the axis of rotation of the turbine, to electrical limiters (e.g., resistors) increasing the generator torque, which consequently brakes the turbine [8]. When a permanent magnet machine is used in a turbine of this type, the generation of energy depends on the occurrence of a minimum wind speed, as the generator voltage will be high enough to allow energy flow towards the receivers or energy storage systems.

In the above-mentioned areas (electric vehicle drives and generators in wind turbines), i.e., where it is necessary to regulate the excitation, rotational speed, induced voltage in the widest possible range and the efficiency of energy conversion is also important, electric machines with hybrid excitation can be successfully used.

The trend of minimizing the amount of permanent magnets in electric machines has been noticeable for many years. This is mainly due to economic aspects (PMs are relatively expensive), as well as limitations related to the temperature resistance, and permanent magnets may permanently lose their magnetic properties after exceeding a critical temperature. In addition, the development of electro-mobility forces, the use of electric machines with the widest possible range of rotational speed, at which the electric machine is able to generate the appropriate torque on the shaft. Otherwise, hybrid excitation allows for a much wider range of machine operation, compared to machines with only PM excitation. Research on hybrid-excited machines fits this trend perfectly, in which a part of the magnetic excitation flux is generated by the PM and the remainder by the corresponding electromagnetic system.

In this paper, in Section 2, the idea of controlled magnets is presented, and then in Section 3, an overview of the design solutions of hybrid-excited electric machines is presented. They are divided into several groups depending on the characteristic design properties. Then, based on this background, proprietary design solutions of hybrid excited machines are presented in Sections 4–7, which are implemented at the Department of Power Engineering and Electric Drives of the West Pomeranian University of Technology in Szczecin. The paper ends with a short summary.

2. Controlled Magnets

The idea of hybrid excitation of electrical machines comes from so-called controlled magnets which were initially used in different levitation systems and magnetic bearings [9–13]. In these systems, strong permanent magnets are used for force generation. Their flux is modulated by currents flying in properly placed coils. The principle of operation of such a system is shown in Figure 1. It is equipped with additional current coils, which demagnetize one part of the magnetic circuit and magnetize the other one, depending on changes in the size of the air gap. In this way, the system is stabilized in the tangential direction having negative stiffness values. Figure 1 shows a situation in which the moving part of the system was horizontally shifted in relation to the neutral position. In this case, it is necessary to reduce the magnetic flux on the side of the smaller air gap and increase it on the opposite side, in order to create reverse forces moving the permanent magnet system back to the central position. The magnetic flux produced by the coil current is directed against the main flux on the left side of the system, and in line with it on the right side. The levitation force in the configuration shown in Figure 1 is generated by a large, strong permanent magnet, which guarantees the creation of high lift forces, and to actuate the bearing in the tangential direction, small currents in the electromagnets are required. This means that the entire system works with a very low power consumption by the system measurement and control and executive elements. The operation of all levitation systems requires the optimization of the magnetic field distribution within the air gap by means of proper choice of geometry, material characteristics and power supply conditions [14,15]. Controlled magnets are used in various levitation systems, and their most spectacular use tangential directions the Transrapid levitation and side stabilization systems [9–13].

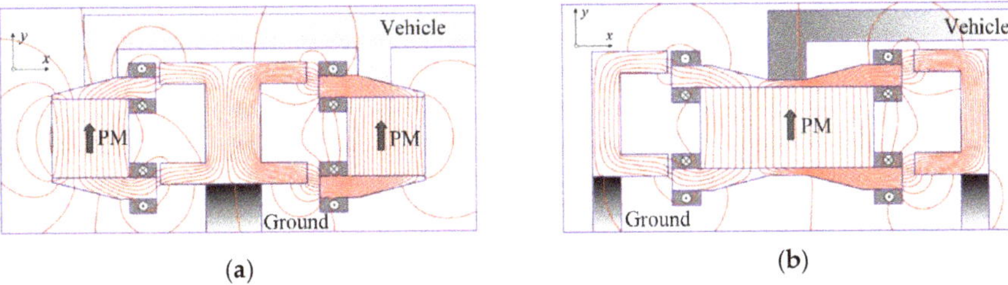

(a)

(b)

Figure 1. (**a**) Levitation systems with controlled permanent magnets; (**b**) Unbalanced position with current compensation.

Controlled magnets are the basic executive part of all magnetic bearings [16,17]. Figure 2 shows two examples of magnetic bearings: a conventional bearing with an external moving part and a hybrid bearing with the control coils placed in the stationary part of the bearing. Such bearings are stable in the axial direction, and the central position of the moving parts is maintained by appropriate control of the coils currents.

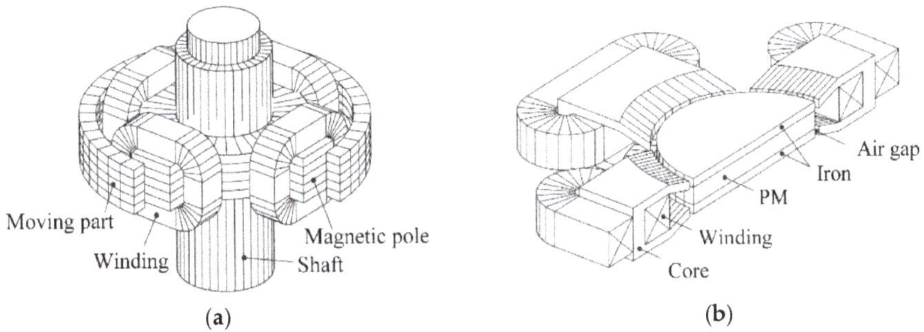

(a)

(b)

Figure 2. (**a**) Conventional magnetic bearing; (**b**) Hybrid magnetic bearing.

The mathematical description (voltage equations) of magnetic bearings is very similar to the description of hybrid-excited electric machines due to the appropriate selection of the coordinate system (rotating with the rotor). The main goal in magnetic bearings is to maintain a constant air gap, while in hybrid-excited machines, it is to obtain the required values of the magnetic flux.

3. Review of Hybrid Excited Machines

Hybrid excited electric machines can be divided into two groups. In the first group, the flux caused by the excitation winding passes through permanent magnets. The second group includes parallel excited hybrid machines. In these machines, the permanent magnet flux and the excitation winding flux have different trajectories. The magnetic permeability of PM is similar to air permeability. Therefore, for the first group machines, the coil magnetic reluctance is relatively high. This is a reason for the introduction of magnetic bridges into the machine in order to ensure lower reluctance for the excitation circuit.

For all machines with hybrid excitation the general mathematical model is described by Equations (1) and (2). They show that the induced voltage and magnetic flux related to the machine axes d and q – and, consequently, to the electromagnetic torque depend on the flux from permanent magnets and the current in the excitation coil [18,19]. Information about transformation from three-phase (L_1–L_2–L_3) system to two phase (α–β) system, and then to d–q axis, are described in details e.g., in [20],

$$
\begin{bmatrix} u_d \\ u_q \\ u_c \end{bmatrix} = \begin{bmatrix} R_s + sL_d & -\omega_e L_q & sM_{sc} \\ \omega_e L_d & R_s + L_q & \omega_e M_{sc} \\ sM_{sc} & 0 & R_c + sL_c \end{bmatrix} \begin{bmatrix} i_d \\ i_q \\ i_c \end{bmatrix} + \begin{bmatrix} 0 \\ \omega_e \Psi_{PM} \\ 0 \end{bmatrix} \tag{1}
$$

where u_d—d-axis voltage component, u_q—q-axis voltage component, u_c—voltage on the excitation coil, R_s—stator winding resistance, s—operator d/dt, L_d—d-axis inductance, ω_e—angular velocity, L_q—q-axis inductance, M_{sc}—mutual inductance between the stator winding and the excitation coil, R_c—excitation coil resistance, L_c—inductance of the excitation coil, i_d—d-axis stator current component, i_q—q-axis stator current component, i_c—excitation coil current, Ψ_{PM}—flux of permanent magnets,

$$
\begin{bmatrix} \Psi_d \\ \Psi_q \end{bmatrix} = \begin{bmatrix} L_d & 0 & M_{sc} \\ 0 & L_q & 0 \end{bmatrix} \begin{bmatrix} i_d \\ i_q \\ i_c \end{bmatrix} + \begin{bmatrix} \Psi_{PM} \\ 0 \end{bmatrix} \tag{2}
$$

where: Ψ_d—d-axis flux, Ψ_q—q-axis flux.

The comparison of different structures of hybrid excited electrical machines is very difficult because of their variety. They can be compared, e.g., in terms of their external characteristics (mechanical design) [1–3,21]. Hybrid excited electrical machines can also be categorized according to the path-determining design of the combined excitation flux. There is a huge number of design solutions for hybrid excited machines. This Chapter presents the most interesting ones found in the literature.

3.1. Synchronous Machines with Permanent Magnets

In order to regulate the excitation flux, in addition to permanent magnets, an additional source is used in the form of a winding. A different approach is presented [22]. The synchronous generator rotor has been modified by adding permanent magnets. In this way, it became independent to some extent from the failure of the sensitive part of the machine, which is the arrangement of brushes and slip rings. The machine can operate at high rotational speed with weakened excitation field. There is a high flux density between two adjacent PM poles, which can increase iron losses in the stator core.

Similar structures have been presented in [23,24]. Furthermore author of [23] also described a direct torque control strategy dedicated to hybrid excited permanent magnet machines.

3.2. Flux-Switching Machines

A novel hybrid excitation flux-switching motor (HEFS) presented in [25] is dedicated to hybrid vehicles. A new motor topology has been proposed, in which the dimensions of the magnets have been reduced to save space for additional excitation winding, while the rotor and stator lamination remain unchanged. It should be noted that this allows the machine stream to be adjusted by controlling the length of the magnets in the radial direction. This solution in its idea is to eliminate the disadvantages, even higher torque ripple due to the cogging torque, which has, for example, a permanent magnet motor with flux switching (FSPM). Similar design has been investigated in [26]. The paper presents numerical research as well as experimental tests on built machine prototype.

3.3. Doubly Salient Machines

The paper [27] presents design of hybrid excited doubly salient machine with parallel excitation system. Authors of the paper analyzed the regulation possibility of air gap flux in three types of the machine main poles. By performing simulation and then experimental tests, obtain very good control properties from approx. 30 V to approx. 220 V. Hybrid excited doubly salient machines are also presented in [28–30].

3.4. Axial Flux Machines

The text [31] discusses a synchronous, hybrid excitation, axial flux generator in autonomous mode, with a field winding powered by an armature winding. The proposed solution allows for very precise control of the magnetic flux, which allows to obtain the set value of the output voltage in cases where the load or speed changes, or both. Very interesting designs of hybrid excited axial flux machines are also described in various papers [32–35].

3.5. Axial-Radial Flux Machines

Structural optimization, which maximizes the flux control range of a dual excitation synchronous machine, is discussed in the paper [36]. The air gap flux in this type of machine can be regulated by controlling the field currents. A machine of this type is able to regulate the air gap flow more flexibly compared to conventional PM machines. This has been achieved at the expense of more volume and higher costs due to the presence of additional field windings. Both electromagnetic and thermal complexity has been well addressed through the use of equivalent circuit networks. It was also found that one of the analyzed configurations almost eliminates the PM flux. Similar construction of the machine is shown in [37]. On the other hand, in [38] the authors presented simulation studies of a machine with an excitation flux in both, radial and axial direction, while the rotor is similar to a rotor of a flux-switching machine. In addition, some parts of the machine are proposed to be made of SMC material.

3.6. Dual Rotor/Stator Machines

In [39], a new toroidal winding twin-rotor permanent magnet synchronous reluctance machine (PM-SynRM) is discussed, which is proposed for high electromagnetic torque taking full advantage of the permanent magnet torque and reluctance torque due to the special design of the mounting angles of the two rotors. Permanent magnet torque and reluctance torque of the proposed machine can obtain their maximum values near the same current phase angle due to the special configuration of the two rotors, which significantly increases the total torque. It turned out that, as a result of the FEM analysis, the proposed machine gives much better torque results. On the other hand, the proposed double rotor structure has excellent properties and resistance to irreversible magnetization. The paper [40] discusses the research on a hybrid excited machine with a double rotor, in which one part is a rotor with permanent magnets, and the other—a classic wound rotor.

Inverted structure is presented by the authors of the paper [41]. In the paper, a hybrid-excited PM machine, based on the flux modulation effect, has been proposed. The authors state that in the machine there is no risk of irreversible demagnetization of PMs. Moreover, the machine does not need slip rings and brushes, since the DC excitation coils are placed in the stator, which makes the structure simple and reliable. The paper shows FEA numerical model of the machine, its structure and the working principle. Similar design is presented in [42].

A very interesting, but at the same time very complicated structure, was presented by the authors [43]. The paper presents a machine with two stators (inner and outer) and two parts of the rotor, one of which was composed of alternately arranged N and S magnets, and the other part was a classic claw pole rotor with excitation from the coil inside it.

3.7. Hybrid Excited Machines with DC Winding on Stator

The concept of a machine with hybrid excitation with permanent magnets and excitation of the AC field winding is presented in [44]. In this machine, permanent magnets (PM) are placed on the rotor side and the AC windings on the stator side for flux control while ensuring high torque. Since the magnetic field PM rotates with the rotor, the alternating currents would have the same frequency as the motor speed. The obtained results and FEM analyzes show that, in the case of the HEPM machine, flux regulation and operation in a wide speed range can be realized, and the electrical parameters can be improved compared to the original IPMSM, which could verify the theoretical analysis presented above, expand the method of designing permanent magnet machines and control strategy and provide a reference to the design of machines with hybrid excitation of permanent magnets.

Similar design can be found in [45]. An example of a parallel hybrid excitation machine is hybrid excitation flux reversal machine (HEFRM). It has been designed for application in electric vehicles propulsion. It may not only show better overload and the possibility of weakening the flux, but also reduces the risk of PM demagnetization. The change in the air gap flux can be controlled by controlling the excitation winding current, which also improves the overload torque at low speeds.

3.8. Axial Flux SRM Machines

The issues of construction and modeling of a machine with some favorable features for wind energy conversion applications are presented in [46]. A machine with a double stator with an axial rotor and permanent magnet flux switching (AFSPM) was adopted for consideration. The developed model of this machine was verified by comparing its results with the results from the two-dimensional (2D) FEM model. The modeling approach adopted has proved to be effective and gives good results compared to FEM. The open-circuit AFSPM performance was compared to the previously developed SMPMAF prototype. This comparative study showed that the EMF waveform is very close to the sinusoidal signal for AFSPM, which is desirable for the intended applications. However, for SMPMAF the EMF wave contains more harmonics.

3.9. Consequent-Pole Permanent Magnet Machines

An innovative concept of a permanent magnet motor with sequence poles was proposed in [47]. The motor has a unique rotor configuration in which the actual PM pole pairs and image pole pairs are positioned every other pole pair. The pole pairs of the image are formed on parts of the iron core next to the actual solid parts of the rotor surface. One of the most important features of the proposed motor is the ability to run at high speed with a lower field-weakening current (negative d-axis current). This enables the operating range to be effectively extended at high speeds without increasing copper losses. One disadvantage is that the amount of effective magnetic flux is too low, which results in a lower torque in the low speed range. Therefore, optimization of the magnetic circuit is needed.

A slightly different construction is presented in [48]. The subject of the research was a machine, the rotor of which had alternating permanent magnets and iron poles, consequently one pole of the machine was a magnet and iron element. Between the two parts of the stator the excitation regulating coil was placed.

3.10. Claw Pole Machines

The rapid development of hybrid vehicles has prompted the need to develop a highly efficient source of electricity for this type of vehicle. One of the ways to achieve this goal is to equip the synchronous generator with claw poles.

The article [49] proposes a new design of a machine with a claw pole and compares its performance with a conventional machine. The new design features permanent magnets in the inter-claw region to reduce leakage flux and provide increased magnetic flux in the machine. It was observed that when adding permanent magnets weighing only a few grams, the machine output power increased significantly by over 22%. The geometrical dimensions of the magnets were also changed to verify their influence on the operation and it was observed that, as the mass of the magnet increases, there is a non-linear increase in the machine torque. The power-to-weight ratio of the machine has also been significantly improved, which is one of the main advantages for mild hybrid applications.

A similar variant of the rotor structure, only seven-phase, was considered in [50]. This solution is characterized by an unequal number of pairs of magnetic poles between the armature and the excitation circuit. In this case, it is necessary to model only a quarter of the structure, not a third as in the case of a classical generator, where the number of bars of the rotor and stator poles is the same $p = 6$. The main goal of the project was to increase the output power and reduce the core losses.

The paper [51] presents a hybrid excited machine, in which the sources of the excitation field are placed inside a claw rotor—toroidal permanent magnets are inside the toroidal core, on which the excitation coil is located.

4. Electric Controlled Permanent Magnet Synchronous Machine

Electric Controlled Permanent Magnet Synchronous machine (ECPMS-machine) is one of the hybrid excited machine concepts with good application prospects. The magnetic field excited by the DC coil current (I_{DC}) gives possibility to change a machine air gap flux and consequently the stator flux linkage Ψ_s. In this way, the output voltage of the machine is electrically controlled.

Figure 3 shows ECPMS-machine concept, in which an air gap flux is controlled by a DC excitation control coil (DC control coil), fixed on the stator or on the machine rotor. The presented 12-poles ECPMS-machine design has a winded double stator, which is separated by a stator DC control coil and is centrally placed between two stator laminations. The stator DC control coil is locked inside a toroidally-wound additional stator core. The machine rotor has two lamination stacks which are separated by toroidally-wound additional rotor core. The rotor lamination structure has multi-flux barriers and embedded flat magnets NdFeB type, that form six iron poles (IP) and six permanent magnets poles (PMP) for each of two rotor stacks. The main feature of the proposed rotor structure is related with proper machine air gap flux distribution and simultaneously required machine flux control (FC) ability. The presented ECPMS-machine concept has been presented in [7,52–54].

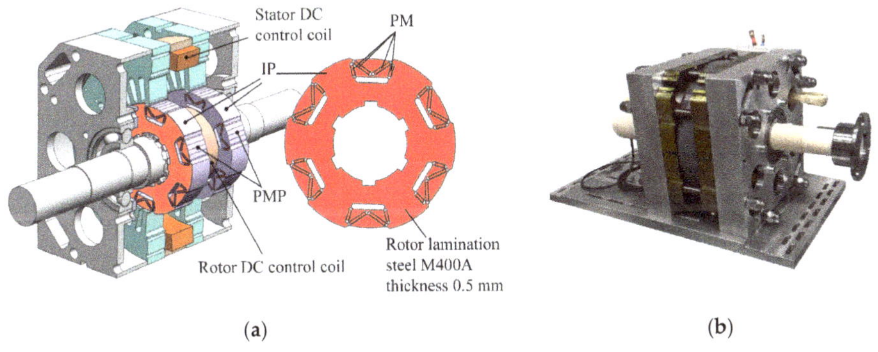

(a)

(b)

Figure 3. ECPMS-machine design: (a) Design; (b) Prototype [54].

4.1. Stator DC Control Coil of the ECPMS-Machine Concept

The concept of hybrid excitation with permanent magnets and the additional DC field winding locked on the stator machine is presented in Figure 4. The DC control coil placed on the stator side (Figure 4a) is used to control the machine air gap flux. The presented machine design concept has been widely analyzed in [52], where the influence of rotor structures on field regulation capability of the machine has been described. The main results of the study show a rotor structure which ensures the effective machine flux regulation. The results obtained during FEA carried out on the three-dimensional (3D) model of the machine (Figure 4b, where B is flux density expressed in tesla) confirmed effective flux control of the machine. The characteristics of magnetic flux linkage Ψ_s versus of DC coil magneto-motive force (MMF) θ_{DC}, which can be seen in the Figure 4c, is proof.

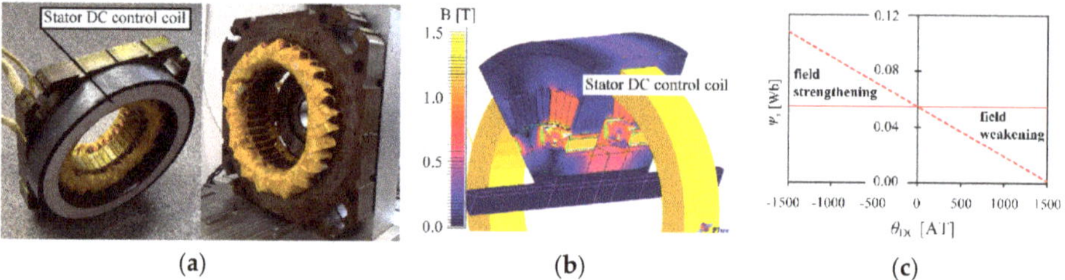

Figure 4. (**a**) The stator DC control coil location in the prototype; (**b**) 3D-FEA model; (**c**) Results of no-load magnetic flux linkage Ψ_s vs. DC coil MMF of ECPMS-machine design [54].

To validate the simulation results, a set of experimental tests have been carried out on a machine prototype. Figure 5 shows experimental validation results of no-load rotor speed obtained at constant DC-bus voltage of 100 V (Figure 5a), back-EMF waveforms recorded at 1000 rpm rotor speed (Figure 5b) and characteristics of no-load induced voltage in phase winding (Figure 5c) at different DC coil MMF of ECPMS-machine. The results have shown that the high (10:1) field control ratio of the presented hybrid excited machine concept is successfully achieved.

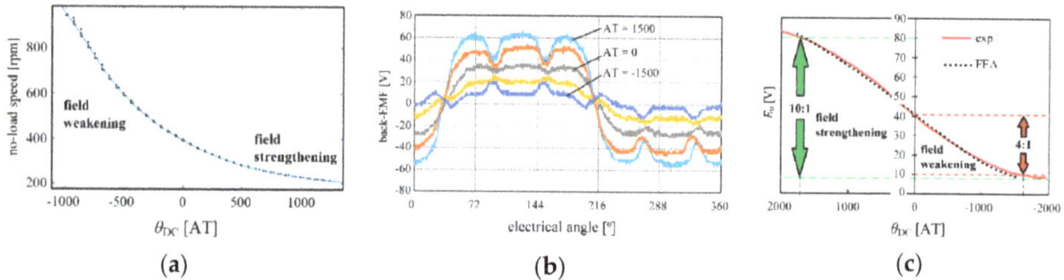

Figure 5. (**a**) Experimental results of no-load speed obtained at constant DC-bus voltage of 100 V; (**b**) Back-EMF waveforms at 1000 rpm rotor speed; (**c**) Characteristics of no-load phase output voltage at the different DC coil MMF of ECPMS-machine [54].

4.2. Rotor DC Control Coil of the ECPMS-Machine Concept

The DC excitation source can be placed in the rotor of ECPMS-machine. Figure 6a presents a 3D-FE model and rotor prototype of the machine where the placement of the DC control coil on the rotor is clearly shown. It should be noted that, depending of the presence of the stator DC control coil, the rotor DC control coil can be an additional or independent source of excitation.

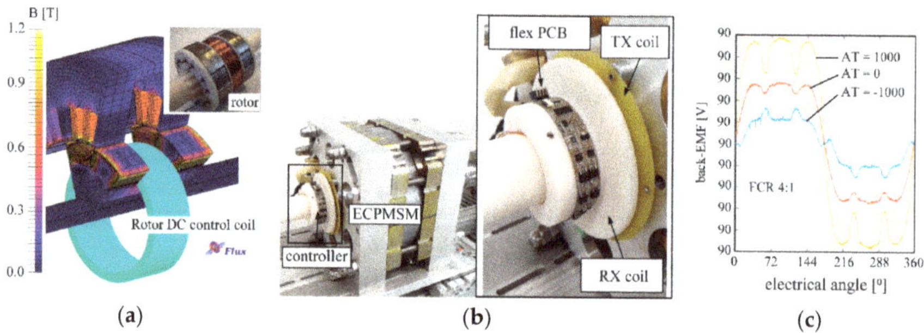

Figure 6. (**a**) 3D-FE model with rotor DC control coil location; (**b**) Machine prototype with contactless energy transfer system; (**c**) Experimental results of no-load induced voltage in phase winding at the different DC coil MMF of ECPMS-machine [54].

Commonly, in order to supply the windings placed on the rotor, it is necessary to use brushes and slip rings. Alternatively to this, modern contactless energy transfer (CET) system shown in Figure 6b have been performed and described in [54], and it has been successfully used, in this case.

Figure 6b shows construction details of the ECPMS-machine prototype with the rotor DC control coil and supply coils used for the wireless power transfer which have been designed and locked on the housing of the machine prototype. Such a solution includes a transformer whose windings are formed by a double-sided printed circuit board (PCB) (used 70 μm copper thickness) plates. On both parts of the secondary and primary plates, ferrite sheets have been used as the path of the magnetic flux (Wurth Electronic® WE-FSFS flexible ferrite sheet, number 344,003). Air gap between TX and RX-coil was approx. 1 mm, in this case.

Figure 6c shows experimental results of phase back-EMF waveforms recorded at constant rotor speed of 1000 rpm by three operating conditions at rotor DC control coil MMF (0 and ± 1000 AT). The results show that the field control ratio (FCR) up to 4:1 can be effectively obtained. Additionally, the result of FWR has been achieved at low losses and low power consumption of the rotor DC control coil. In this case, the power consumption of the DC control coil (at 1000 AT excitation) is approx. 20 W.

4.3. ECPMS-Machine Concept-Conclusion

The presented ECPMS-machine belongs to the concept of hybrid excited machines with excellent field-control capability, which can be used in wide adjustable speed drives. The experiment validations have shown that the field control ratio 10:1 of the presented machine can be effectively obtained. This property can be used in electric vehicle drives and other adjustable speed drive applications.

Advantages: wide flux control range, high starting torque, totally flux-weakening possibility, low demagnetization risk of rotor magnets, possibility to locate additional DC field sources on the stator or rotor machine, or both.

Drawbacks: additional components, complex machine structure and greater weight and dimensions compared to conventional machines.

5. Hybrid Excited Disk Type Machine

The hybrid-excited axial flux machine (HEAFM) is built on the basis of an internal double winding stator and two external 12-pole rotors connected by a ferromagnetic bushing. Additional electromagnetic excitation is placed in the stator circuit. The structure of the machine is shown in Figure 7.

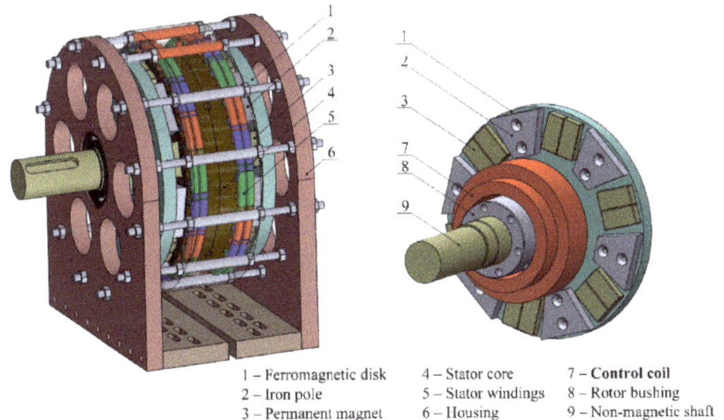

1 – Ferromagnetic disk	4 – Stator core	7 – **Control coil**
2 – Iron pole	5 – Stator windings	8 – Rotor bushing
3 – Permanent magnet	6 – Housing	9 – Non-magnetic shaft

Figure 7. HEAFM construction [55].

The machine's stator is made of two laminated toroidal cores, and in each of them 32 slots with a 3-phase winding. The DC coil used as an additional excitation source was mounted on the inside of the stator around the rotor bushing. The coil is stationary, so it has no brushes or slip rings. The rotor consists of two outer discs connected by a steel sleeve. On each disk are alternately mounted iron poles with magnets polarized in one direction. There are 6 pole pairs on each disc. The operating principle of HEAFM is shown in Figure 8. When no current flows through the DC coil, the main magnetic flux in the machine flows through the air gap between the magnets, and part of it between the magnet and the iron pole. Depending on the direction of the current in the DC coil, the iron poles magnetize, which in turn leads to a strengthening of the main flux (FS) or its weakening (FW).

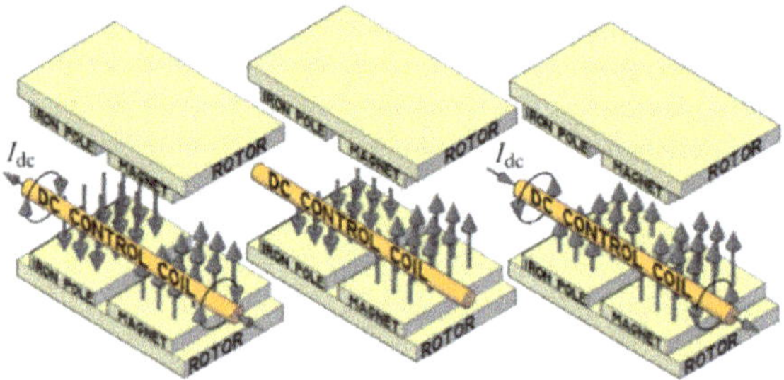

Figure 8. Main principle of the HEAFM.

The prototype of the machine (Figure 9) was made in accordance with the assumptions and tested on the experimental stand (Figure 10) in generator mode.

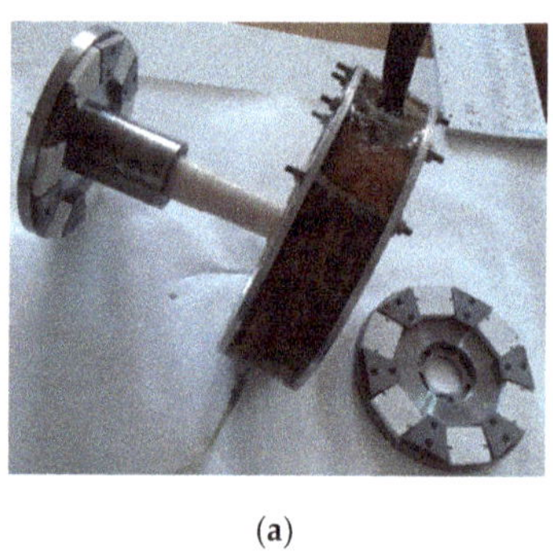

(**a**)

(**b**)

Figure 9. The prototype of the HEAFM machine: (**a**) A stator with a rotor, (**b**) A complex machine with a caliper 20 cm long [55].

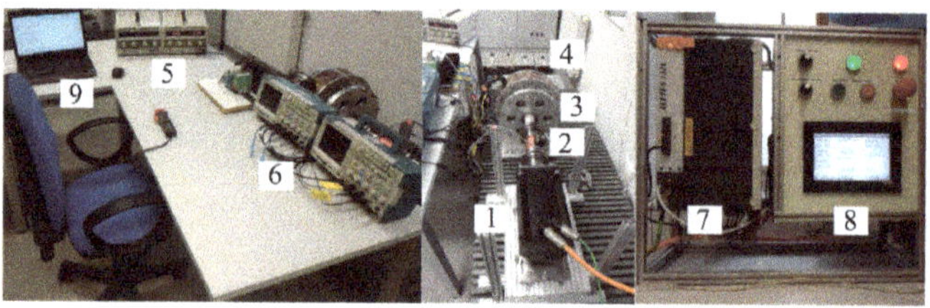

1 – motor,
2 – measuring shaft,
3 – HEAF generator,
4 – resistors,
5 – power supplies,
6 – oscilloscopes,
7 – inverter,
8 – inverter controller,
9 – laptop.

Figure 10. Experimental stand [55].

Waveforms induced in the machine without load at a speed of 600 rpm for different currents in the DC coil are shown in Figure 11 by a dotted line.

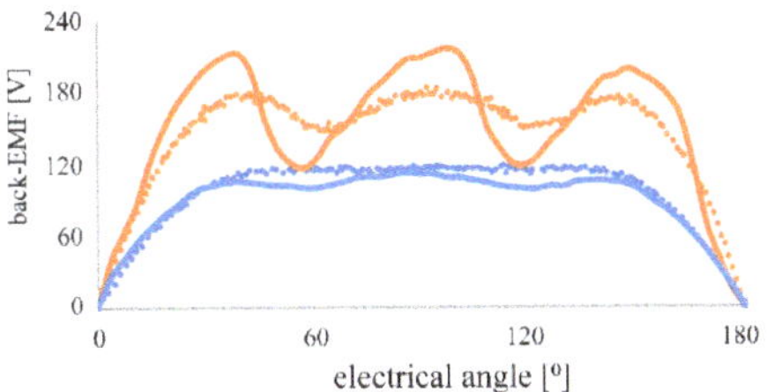

Figure 11. RMS of induced experimental (dotted line) and simulation voltages (solid line) in no-load machine for $I_{DC} = -5$ A (blue) and $I_{DC} = 5$ A (orange).

A 3D model of HEAFM was built and simulations were made using the finite element method (FEM). Figure 12 shows the distribution of magnetic induction in a magnetic circuit of the machine with the unpowered DC coil. Figure 13 shows the 2-D distribution of magnetic induction in an air gap for different currents in the DC coil on an arc, the radius of which is the average of the inner and outer radius of the machine active parts. This arc spans two adjacent poles: iron pole and PM pole.

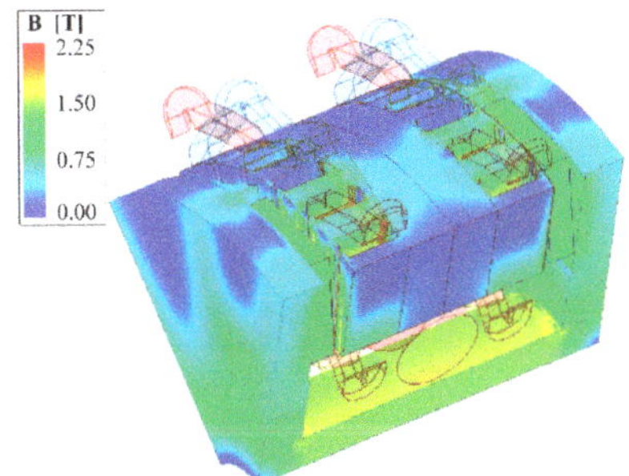

Figure 12. Distribution of magnetic induction in a machine for the unpowered DC coil.

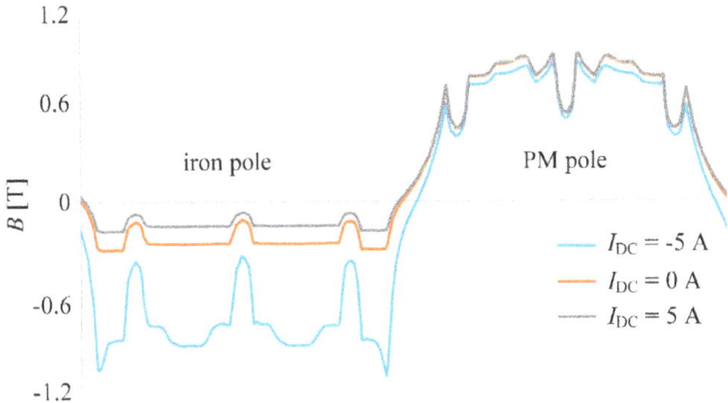

Figure 13. The distribution of magnetic induction in air gap for different currents in the DC coil [55].

The results show that the magnetic flux only changes under the iron pole, which is the advantage of this design, unfortunately the FS level is much higher than FW. Waveforms of magnetic induction in air gap for different currents in the DC coil are shown in Figure 13.

Subsequent simulation studies were to show the effect of modification of the machine's magnetic core on the magnetic field regulation (FCR) range. The height of the magnets was examined first. The base model was the machine in which the height of PM was 12 mm and it was changed in the range from 2 to 14 mm. ΔFCR is the ratio of the base machine FCR to the FCR of the machine with different heights of PM (Figure 14).

The results show that the amount of PM has an impact on FCR. With increasing PM height, the induced voltage increases, but at the same time the possibility of its regulation decreases.

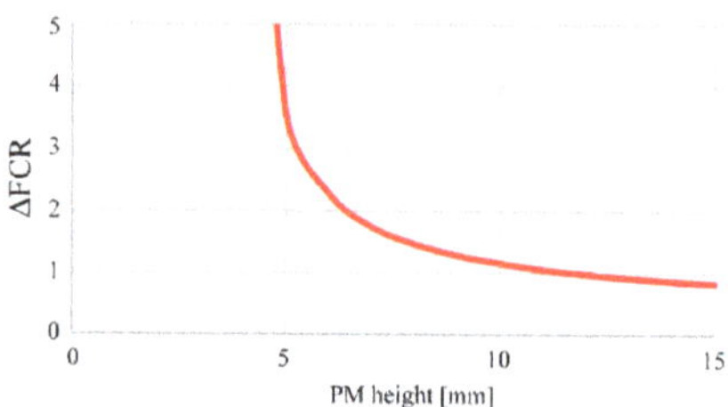

Figure 14. Relation ΔFCR to the height of PM based on FEA [55].

Further research aimed to demonstrate the impact of permanent magnet materials on FCR. For this purpose, various types of PM were shown, shown in Table 1.

Table 1. Specification of tested PMs.

Item (Unit)	Material Code	Remanent Magnetic Flux Density (T)	Coercive Force (kA/m)
Type A	N48H	1.388	1058.8
Type B	N38H	1.248	947.8
Type C	SmCo5	0.943	693.5
Type D	F30	0.399	226.5

FEM simulations have been conducted for no-load generator mode. Table 2 shows the distribution of induction throughout the entire air gap and the percentage ratio of magnetic flux flowing through the surface over the iron pole (IP) to pole with permanent magnet (PMP).

Table 2. Air gap magnetic flux density distribution for different types of PMs.

-	Type A	Type B	Type C	Type D
Magnetic flux near PMP Φ_{PMP}	2.05 mWb	1.87 mWb	1.63 mWb	0.58 mWb
Magnetic flux near IP Φ_{IP}	0.76 mWb	0.64 mWb	0.48 mWb	0.09 mWb
Percentage ratio of magnetic flux Φ_{IP}/Φ_{PMP}	37.10	34.22	29.45	15.52

The waveforms of induced voltages for selected materials of PM—neodymium (N38H) and ferrite (F30) depending on the currents in the additional DC coil at 200 rpm are also presented in Figure 15.

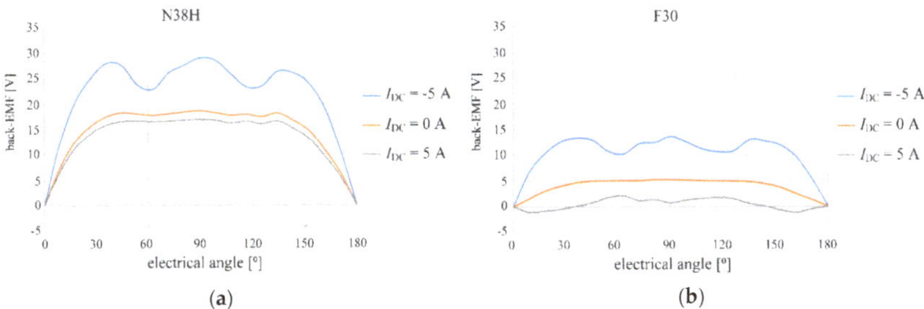

Figure 15. Induced voltage waveforms depending on different DC coil currents for: (**a**) N38H and (**b**) F30 PM types [55].

Based on the simulating results, which was made, it can be concluded that despite the N38H magnet is the strongest and has the largest coefficient $k_{\Phi\%}$, it allows the smallest possibility of flux control, and thus the induced voltage regulation.

The FCR coefficient of the analyzed machine is influenced, among other factors, by the magnetic circuit topology, including the shape, dimension and material of PM.

6. Hybrid Excited Claw Pole Machine (HECPM)

One of the innovative solutions is the use of hybrid excitation in claw pole machines by placing permanent magnets on or inside the claws. A construction of a hybrid excited claw machine with permanent magnets placed in milled areas on one [56] or both parts [57] of the rotor was proposed. Figure 16 shows these structures.

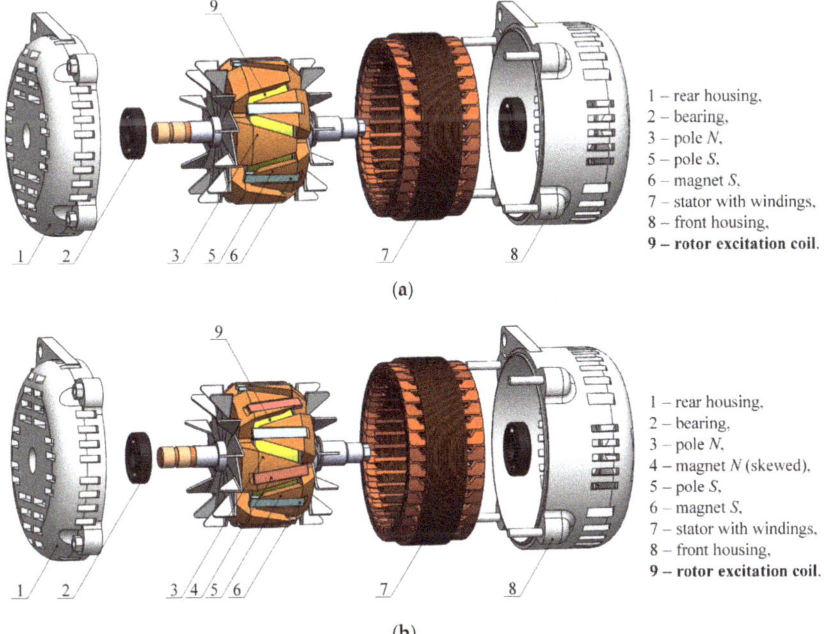

Figure 16. Visualization of a claw pole machine: (**a**) With permanent magnets on one part of the rotor [56]; (**b**) With permanent magnets on both parts of the rotor [57].

During the scientific research of the proposed solutions, in order to experimentally validate the results of numerical tests, a car alternator manufactured by Denso, with the nominal current $I_n = 100$ A and nominal voltage $U_n = 12$ V, was used and rebuilt, while maintaining the standard excitation regulator. As a result of the work, it was possible to develop a technical solution that made it possible to self-excite the machine without the use of an additional DC source. In the first solution [56], self-excitation took place at the rotational speed equal to 1300 rpm, while in the second solution [57]—at 850 rpm. This feature can be used in a home wind turbine—in the absence of wind, thanks to the use of one diode, the generator regulator would not get energy from the batteries, while when the wind of sufficient strength appears, the generator would be self-excited, and consequently the energy would be generated for the storage system.

The influence of the excitation current on the cogging torque of the machines was also investigated. The results are shown in Figure 17a which shows cogging torque waveforms for a machine with permanent magnets on one part of the rotor for five current values in the excitation coil-parameter α is an angular (mechanical angle) position between rotor and stator.

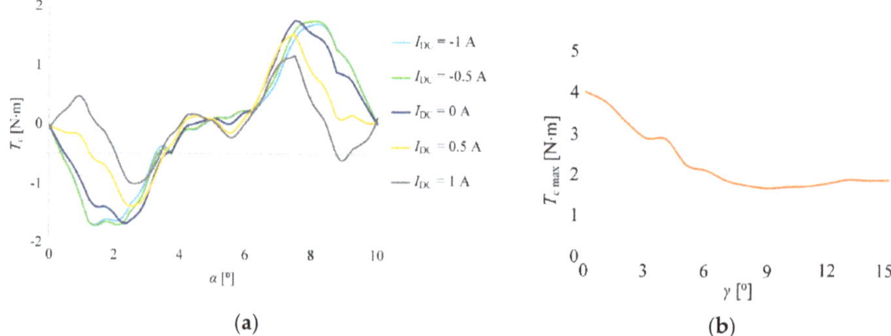

(a)

(b)

Figure 17. (**a**) Cogging torque waveforms for a machine with permanent magnets on one part of the rotor as a function of the current in the excitation coil (in kilo ampere-turns); (**b**) Maximum values of the cogging torque for a machine with permanent magnets on both parts of the rotor as a function of the angle γ, in a no-load excitation coil state [56,57].

In the second model, the magnets from the first model (on one part of the rotor) were retained and another 6 magnets were added to the second part of the rotor, however, simulation studies on the influence of the γ angle between the permanent magnets on this part of the rotor and the machine axis were previously carried out with using FEA. This angle varied from 0 to 15° in steps of 1°. It turned out that the angle at which the lowest cogging moment occurs is $\gamma = 9°$. The chosen results are presented in Figure 17b, which demonstrates the maximum values of the cogging torque as a function of the angle γ, in a no-load excitation coil state.

The test results showed that the induced voltage decreased with the increase of the angle γ, while the cogging torque reached the minimum for the angle $\gamma = 9°$. Additionally—thanks to the use of additional magnets set at an angle of $\gamma = 9°$—the machine was self-excited at speeds up to 450 rpm lower, moreover, despite the fact that twice as many sources of magnetomotive forces were installed, the cogging torque in the no-load state of the excitation coil did not increase in relation to the machine presented in [57]. Figure 18 presents experimental stand with HECPM.

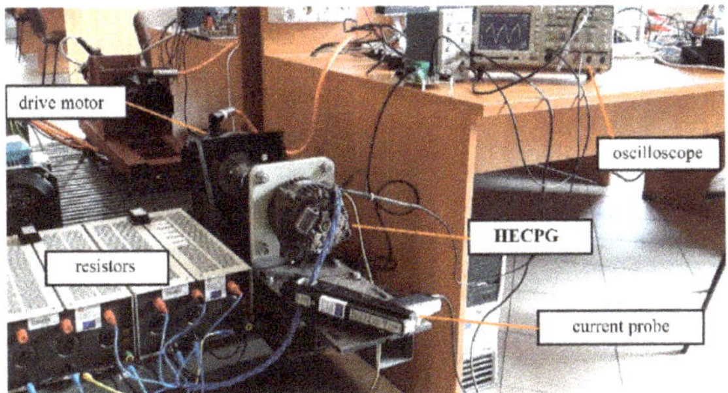

Figure 18. Test stand with HECPM [57].

Very important, from the technological point of view, is the simplicity and ease of production of all kinds of devices, including electromechanical energy converters. For this reason, a new approach to designing a hybrid excited claw machine has been proposed. The paper [58] presents the concept of building a machine with the use of a laminated rotor made by sheets of an appropriate shape (Hybrid Excited Claw Pole Machine with Laminated Rotor—HECPMLR). Figure 19 shows FEA model of the tested HECPMLR machine which is 1/6 of whole machine.

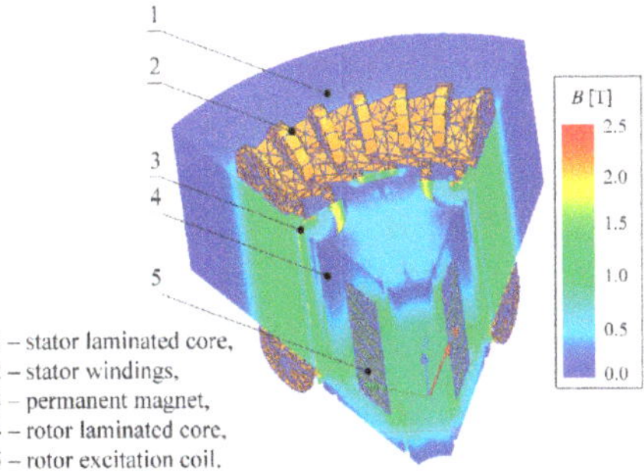

Figure 19. FEA model of HECPMLR [58].

This type of approach allows the construction of even the most complex electromagnetic structures. The paper [58] presents also preliminary results of simulation tests of the proposed structure and the relationship between the maximum cogging torque and the induced voltage distribution depending on the current in the excitation coil (Figure 20). Figure 20a shows the maximum value of the cogging torque T_{emax} depending on the current in the excitation coil I_{exc}, and Figure 20b—the distribution of the back-EMF depending on the current in the excitation coil, where α is a mechanical angle between rotor and stator.

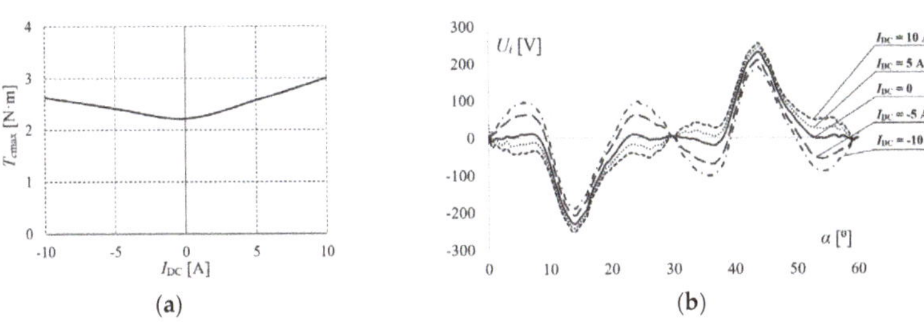

Figure 20. (a) Maximum value of the cogging torque depending on the current in the excitation coil; (b) Distribution of the back-EMF depending on the current in the excitation coil [58].

The research shows that the cogging torque always increases with the increase of the current in the excitation coil, regardless of its direction—Figure 20a. On the other hand, the induced voltage U_{imax} has the following adjustment range from 189.4 V to 253.8 V ($-18\% \div +10\%$).

7. PM Electric Machine with Magnetic Barriers and Excitation Coils in the Rotor (HESMFB)

The purpose of the work of a new design of HESMFB machine was to develop a construction with magnetic flux barriers, embedded PMs and additional electromagnetic excitation in the machine rotor. It should be added that in order to achieve a wide speed control range of PM machines a large inductance ratio L_q/L_d (L_q—inductance in q-axis, L_d—inductance in d-axis) of the machine is required. The magnetic flux density distribution in the FEA model has been presented in Figure 21.

As can be seen in Figure 21a, the large saturation in magnetic bridges in the rotor close to the air gap and permanent magnets is noticeably. Due to this, flux leakage is reduced because the most of the magnetic flux passes through the air gap. Figure 21b presents a novel conception of the machine rotor with barriers and hybrid excitation.

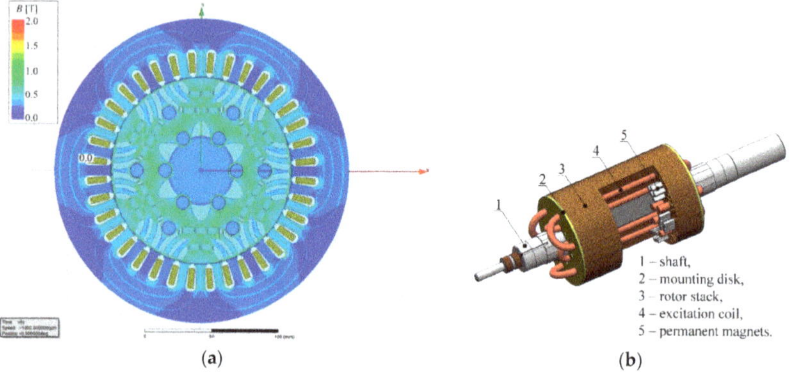

Figure 21. (**a**) Magnetic flux density distribution in the FEA model; (**b**) Concept of HESMFB machine rotor [59].

During FEA investigations induced voltage waveforms have been plotted—Figure 22a, where α is a mechanical angle between rotor and stator. Furthermore the influence of additional windings current density (j_{DC}) on the electromagnetic torque characteristics of the machine has been specified. These numerical tests have been conducted for three stator currents at $I_{s\ max} = 4$; 8 and 12 A, at various additional winding current density j_{DC} in the range from -8 A/mm^2 to $+8$ A/mm^2 and for whole load angle range (from 0 to 360$^{0\mathrm{el}}$). The maximum values of electromanetic torque T_e depending on j_{DC} and $I_{s\ max}$ have been presented in Figure 22b.

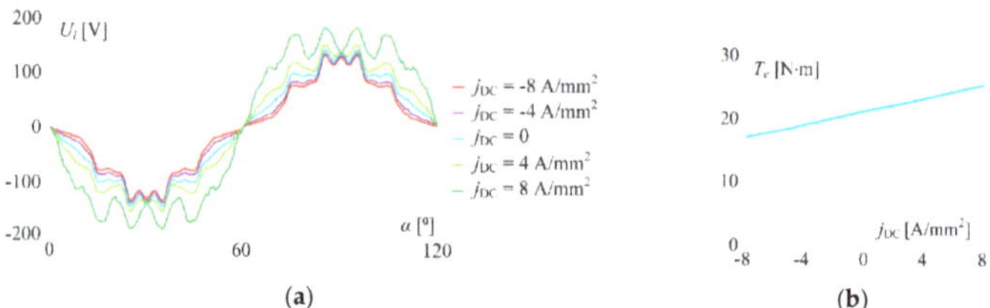

Figure 22. (**a**) No-load back-EMF waveforms at 1000 rpm under different current density levels; (**b**) Electromagnetic torque means values versus additional windings current density at the stator peak current $I_{s\ max} = 12$ A [59].

Next, the experimental tests have been conducted. Figure 23 presents chosen experimental results and comparison with simulation predictions.

Figure 23 show that in the proposed machine the induced voltage control range from 77.6 V to 129.8 V has been obtained. Whereas according to FEA results the back-EMF control range from 72.2 V to 129.3 V was reached. It follows from the above that the field control range (FCR) for the experiment is 1.67 but for FEA-1.79. This means a good representation of the real machine using the developed simulation model. Regarding the cogging torque, the results of the experiment differ slightly from the results obtained in the simulations. However, these differences are rather minor. These differences may result from the not perfect torque measuring, because of very small its values and the FEA model's mesh and accuracy.

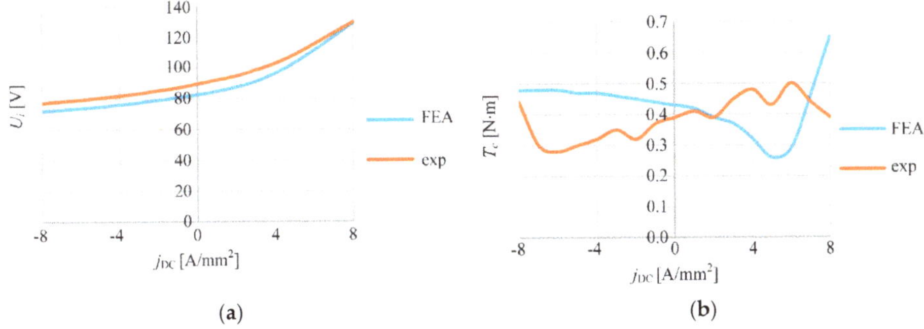

Figure 23. (**a**) RMS values of back-EMF; (**b**) Maximum values of cogging torque [59].

8. Conclusions

The paper provides an overview of various hybrid excited machine topologies. In the literature a lot of solutions for hybrid excited machines can be found. In this paper the most common ones have been presented. Against this background, some new designs, sometimes completely innovative, developed by the authors were presented. In Table 3, the advantages, disadvantages and characteristic features of these machines are summarized.

Table 3. Characteristic features of tested machines.

Machine Type	Advantages	Disadvantages	Features
ECPMS-machine	wide control range, FCR ~10, low PM demagnetization risk, possibility to locate DC coil on stator or rotor machine, or both	complicated structure, large dimensions	dedicated to high-speed drive, e.g., electromobility
HEAFM	middle control range useful FCR ~5 low PM demagnetization risk	very complicated structure, large dimensions	dedicated to low-speed applications, e.g., wind power
HECPM (HECPMLR)	middle control range, (easy implementation of complex rotor structures)	middle PM demagnetization risk, necessity to use brushes and slip rings or CET	dedicated to low and middle-speed applications, e.g., wind power
HESMFB	simple construction, middle control range, large inductance ratio L_q/L_d	middle PM demagnetization risk, necessity to use brushes and slip rings or CET	dedicated to high-speed drive, e.g., electromobility

Finally, we conclude that some of the presented solutions have very good flux control properties, but their complicated structure eliminates them from the possibility of practical application. Hence, the legitimacy of further search for such structures will be easy and inexpensive to manufacture, durable in operation, and at the same time will be characterized by a large range of control.

Author Contributions: All authors worked on this manuscript together. Conceptualization, M.W. and R.P.; investigation, resources, writing—original draft preparation and writing—review and editing, M.W., R.P., P.P. (Piotr Paplicki), P.P. (Pawel Prajzendanc), and T.Z. All authors have read and agreed to the published version of the manuscript.

Funding: This research received no external funding.

Acknowledgments: This work has been supported with the grant of the National Science Centre, Poland 2018/02/X/ST8/01112.

Conflicts of Interest: The authors declare no conflict of interest.

Nomenclature

2D	Two-dimensional
3D	Three-dimensional
AFSPM machine	Axial flux switching permanent magnet machine
AT	Ampere-turns
CET	Contactless energy transfer
DC	Direct current
ECPMS machine	Electric controlled permanent magnet synchronous machine
EMF	Electromotive force
FC	Flux control
FCR	Field control ratio
FEA	Finite element analysis
FEM	Finite element method
FS	Flux strengthening
FSPM machine	Flux switching permanent magnet machine
FW	Flux weakening
HEAFM	Hybrid excited axial flux machine
HECPMLR	Hybrid excited claw pole machine with laminated rotor
HEFRM	Hybrid excited flux reversal machine
HEFS machine	Hybrid excitation flux switching machine
HEPM machine	Hybrid excited permanent magnet machine
HESMFB	Hybrid excited synchronous machine with flux barriers
IP	Iron pole
IPMSM	Internal permanent magnet synchronous machine
MMF	Magnetomotive force
N	North (pole)
NdFeB	Neodymium, iron, boron (type of permanent magnets)
PM	Permanent magnet
PMP	Permanent magnets pole
PM-SynRM machine	Permanent magnet assisted synchronous reluctance machine
RMS (rms)	Root mean square
RX	Receiver
S	South (pole)
SMC	Soft magnetic composites
SMPMAF machine	Surface mounted permanent magnet axial flux machine
SRM	Switched reluctance machine
TX	Transmitter

References

1. Zhu, Z.Q.; Cai, S. Overview of Hybrid Excited Machines for Electric Vehicles. In Proceedings of the Fourteenth International Conference on Ecological Vehicles and Renewable Energies (EVER), Monte-Carlo, Monaco, 8–10 May 2019. [CrossRef]
2. Wang, Y.; Hao, W. General topology derivation methods and control strategies of field winding based flux adjustable pm machines for generator system in more electric aircraft. *IEEE Trans. Transp. Electrif.* **2020**, *6*, 1478–1496. [CrossRef]
3. Zhu, Z.Q.; Cai, S. Hybrid excited permanent magnet machines for electric and hybrid electric vehicles. *CES Trans. Electr. Mach. Syst.* **2019**, *3*, 233–247. [CrossRef]
4. May, H.; Palka, R.; Paplicki, P.; Szkolny, S.; Wardach, M. Comparative research of different structures of a permanent-magnet excited synchronous machine for electric vehicles. *Electr. Rev.* **2012**, *88*, 53–55.
5. Wardach, M.; Paplicki, P.; Palka, R.; Cierzniewski, P. Influence of the rotor construction on parameters of the electrical machine with permanent magnets. *Electr. Rev.* **2011**, *87*, 131–134.
6. Tesla Motor Designer Explains Model 3's Transition to Permanent Magnet Motor. Available online: https://electrek.co/2018/02/27/tesla-model-3-motor-designer-permanent-magnet-motor/ (accessed on 27 October 2020).
7. Di Barba, P.; Mognaschi, M.E.; Bonislawski, M.; Palka, R.; Paplicki, P.; Piotuch, R.; Wardach, M. Hybrid excited synchronous machine with flux control possibility. *Int. J. Appl. Electromagn. Mech.* **2016**, *52*, 1615–1622. [CrossRef]
8. Trzmiel, G. Analysis of Power Control Methods in Wind Power Plants. *Electr. Eng. Pozn. Univ. Technol. Acad. J.* **2017**, *89*, 395–404. [CrossRef]
9. May, H.; Shalaby, M.; Weh, H. Berechnung von geregelten permanent-magneten für tragen, führen und antriebsaufgaben. *Etz Arch.* **1979**, *2*, 63–67.
10. Weh, H. Linear electromagnetic drives in traffic systems and industry. In Proceedings of the LDIA'95, Nagasaki, Japan, 31 May–2 June 1995.
11. Weh, H.; May, H.; Hupe, H. Magnet concepts with one- and two-dimensional stable suspension characteristics. In Proceedings of the MAGLEV'89, Yokohama, Japan, 7–11 July 1989.
12. Weh, H.; May, H.; Hupe, H.; Steingröver, A. Improvements to the attractive force levitation concept. In Proceedings of the MAGLEV'93, Argonne, IL, USA, 19–21 May 1993.
13. Weh, H.; Steingröver, A.; Hupe, H. Design and simulation of a controlled permanent magnet. *Int. J. Appl. Electromagn. Mater.* **1992**, *3*, 199–203.
14. Palka, R. Synthesis of magnetic fields by optimization of the shape of areas and source distributions. *Arch. Elektrotechnik* **1991**, *75*, 1–7. [CrossRef]
15. Sikora, R.; Palka, R. Synthesis of Magnetic-Fields. *IEEE Trans. Magn.* **1982**, *18*, 385–390. [CrossRef]
16. Schweitzer, G.; Traxler, A.; Bleuler, H. *Magnetlager*; Springer: Berlin/Heidelberg, Germany, 1993.
17. Meins, J. *Elektromechanik*; Teubner: Stuttgart, Germany, 1997.
18. Mbayed, R.; Vido, L.; Salloum, G.; Monmasson, E.; Gabsi, M. Effect of iron losses on the model and control of a hybrid excitation synchronous generator. In Proceedings of the IEEE International Electric Machines & Drives Conference, Niagara Falls, ON, Canada, 14–17 May 2011; pp. 971–976.
19. Huang, M.; Huang, Q.; Zhang, Y.; Guo, X. A comprehensive optimization control method for hybrid excitation synchronous motor. *Math. Probl. Eng.* **2020**. [CrossRef]
20. Faleh, R.; Rahi, M.S. Improved performance of three phase motor with high power quality based on dual boost converter. *Int. J. Innov. Res. Electr. Electron. Instrum. Control Eng.* **2016**, *4*, 128–135. [CrossRef]
21. Geng, H.; Zhang, X.; Zhang, Y.; Hu, W.; Lei, Y.; Xu, X.; Wang, A.; Wang, S.; Shi, L. Development of brushless claw pole electrical excitation and combined permanent magnet hybrid excitation generator for vehicles. *Energies* **2020**, *13*, 4723. [CrossRef]
22. Luo, X.; Lipo, T.A. A synchronous/permanent magnet hybrid AC machine. *IEEE Trans. Energy Convers.* **2000**, *15*, 203–210.
23. Xia, Y.; Yi, X.; Wen, Z.; Chen, Y.; Zhang, J. Direct Torque control of hybrid excitation permanent magnet synchronous motor. In Proceedings of the 22nd International Conference on Electrical Machines and Systems (ICEMS), Harbin, China, 11–14 August 2019.

24. Fodorean, D.; Djerdir, A.; Viorel, I.A.; Miraoui, A. A double excited synchronous machine for direct drive application—Design and prototype tests. *IEEE Trans. Energy Convers.* **2007**, *22*, 656–665. [CrossRef]

25. Hua, W.; Cheng, M.; Zhang, G. A novel hybrid excitation flux-switching motor for hybrid vehicles. *IEEE Trans. Magn.* **2009**, *45*, 4728–4731. [CrossRef]

26. Sun, X.Y.; Zhu, Z.Q. Investigation of DC winding induced voltage in hybrid-excited switched-flux permanent magnet machine. *IEEE Trans. Ind. Appl.* **2020**, *56*, 3594–3603. [CrossRef]

27. Chen, Z.; Zhou, N. Flux regulation ability of a hybrid excitation doubly salient machine. *IET Electr. Power Appl.* **2011**, *5*, 224–229. [CrossRef]

28. Leonardi, F.; Matsuo, T.; Li, Y.; Lipo, T.A.; McCleer, P. Design considerations and test results for a doubly salient PM motor with flux control. In Proceedings of the IAS'96, Conference Record of the 1996 IEEE Industry Applications Conference Thirty-First IAS Annual Meeting, San Diego, CA, USA, 6–10 October 1996. [CrossRef]

29. He, M.; Xu, W.; Zhu, J.; Ning, L.; Du, G.; Ye, C. A novel hybrid excited doubly salient machine with asymmetric stator poles. *IEEE Trans. Ind. Appl.* **2019**, *55*, 4723–4732. [CrossRef]

30. Chen, Z.; Sun, Y.; Yan, Y. Static characteristics of a novel hybrid excitation doubly salient machine. In 'Proceedings of the Eighth International Conference on Electrical Machines and Systems, Tokyo, Japan, 27–29 September 2005; pp. 718–721.

31. Ostroverkhov, N.; Chumack, V.; Monakhov, E. Synchronous axial-flux generator with hybrid excitation in stand alone mode. In Proceedings of the 2nd Ukraine Conference on Electrical and Computer Engineering, Lviv, Ukraine, 2–6 July 2019.

32. Ostroverkhov, M.; Chumack, V.; Monakhov, E. Ouput voltage stabilization process simulation in generator with hybrid excitation at variable drive speed. In Proceedings of the 2nd Ukraine Conference on Electrical and Computer Engineering (UKRCON), Lviv, Ukraine, 2–6 July 2019. [CrossRef]

33. Geng, W.; Zhang, Z.; Li, Q. Concept and electromagnetic design of a new axial flux hybrid excitation motor for in-wheel motor driven electric vehicle. In Proceedings of the 22nd International Conference on Electrical Machines and Systems (ICEMS), Harbin, China, 11–14 August 2019. [CrossRef]

34. Hsu, J.S. Direct control of air-gap flux in permanent-magnet machines. *IEEE Trans. Energy Convers.* **2000**, *15*, 361–365. [CrossRef]

35. Aydin, M.; Huang, S.; Lipo, T.A. A new axial flux surface mounted permanent magnet machine capable of field control. In Proceedings of the International Conference Record of the IEEE IAS Annual Meeting, Pittsburgh, PA, USA, 13–18 October 2002; pp. 1250–1257.

36. Hoan, T.K.; Vido, L.; Gillon, F.; Gabsi, M. Structural optimization to maximize the flux control range of a double excitation synchronous machine. *Math. Comput. Simul.* **2019**, *158*, 235–247. [CrossRef]

37. Mohammadi, A.S.; Pedro, J.; Trovão, F.; Antunes, C.H. Component-Level Optimization of Hybrid Excitation Synchronous Machines for a Specified Hybridization Ratio Using NSGA-II. *IEEE Trans. Energy Convers.* **2020**, *35*, 1596–1605. [CrossRef]

38. Kosaka, T.; Matsui, N. Hybrid excitation machines with powdered iron core for electrical traction drive applications. In Proceedings of the ICEMS 2008 International Conference on Electrical Machines and Systems, Wuhan, China, 17–20 October 2008; pp. 2974–2979.

39. Zhang, Z.; Liu, Y.; Zhao, W.; Wang, X.; Kwon, B. A novel dual-rotor permanent magnet synchronous reluctance machine with high electromagnetic performance. In Proceedings of the 22nd International Conference on Electrical Machines and Systems (ICEMS), Harbin, China, 11–14 August 2019.

40. Naoeand, N.; Fukami, T. Trial production of a hybrid excitation type synchronous machine. In Proceedings of the IEMDC 2001 IEEE International Electric Machines and Drives Conference, Cambridge, MA, USA, 17–20 June 2001; pp. 545–547.

41. Liu, Y.; Zhang, X.; Niu, S.; Chan, W.L.; Fu, W.N. Novel hybrid-excited permanent magnet machine based on the flux modulation effect. In Proceedings of the 22nd International Conference on Electrical Machines and Systems (ICEMS), Harbin, China, 11–14 August 2019. [CrossRef]

42. Wei, L.; Nakamura, T. Optimization design of a dual-stator switched flux consequent pole permanent magnet machine with unequal length teeth. *IEEE Trans. Magn.* **2020**, *56*, 1–5. [CrossRef]

43. Liu, X.; Lin, H.; Zhu, Z.Q.; Yang, C.; Fang, S.; Guo, J. A Novel dual-stator hybrid excited synchronous wind generator. *IEEE Trans. Ind. Appl.* **2009**, *45*, 947–953. [CrossRef]

44. Liu, L.; Li, H.; Chen, Z.; Yu, T.; Liu, R.; Mao, J. Design of a hybrid excited permanent magnet machine with AC field winding excitation. In Proceedings of the 22nd International Conference on Electrical Machines and Systems (ICEMS), Harbin, China, 11–14 August 2019.
45. Gao, Y.; Qu, R.; Li, D.; Li, J. A novel hybrid excitation flux reversal machine for electric vehicle propulsion. In Proceedings of the IEEE Vehicle Power and Propulsion Conference (VPPC), Hangzhou, China, 17–20 October 2016. [CrossRef]
46. Diab, H.; Amara, Y.; Barakat, G. Open circuit performance of axial air gap flux switching permanent magnet synchronous machine for wind energy conversion: Modeling and experimental study. *Energies* **2020**, *13*, 912. [CrossRef]
47. Noguchi, T. Proposal and operation characteristics of novel consequent pole permanent magnet motor. *J. Phys. Conf. Ser.* **2019**, *1367*, 012032. [CrossRef]
48. Tapia, J.A.; Leonardi, F.; Lipo, T.A. Consequent pole permanent magnet machine with extended field weakening capability. *IEEE Trans. Ind. Appl.* **2003**, *39*, 1704–1709. [CrossRef]
49. Upadhayay, P.; Kedous-Lebouc, A.; Garbuio, L.; Mipo, J.C.; Dubus, J.M. Design & comparison of a conventional and permanent magnet based claw pole machine for automotive application. In Proceedings of the 15-th International Conference on Electrical Machines, Drives and Power Systems (ELMA), Sofia, Bulgaria, 1–3 June 2017.
50. Bruyere, A.; Henneron, T.; Locment, F.; Semail, E.; Bouscayrol, A.; Dubus, J.M.; Mipo, J.C. Identification of a 7-phase claw-pole starter-alternator for a micro-hybrid automotive application. In Proceedings of the 18th International Conference on Electrical Machines, Vilamoura, Portugal, 6–9 September 2008. [CrossRef]
51. Popenda, A.; Chwalba, S. The synchronous generator based on a hybrid excitation with the extended range of voltage adjustment. *Electr. Rev.* **2019**. [CrossRef]
52. Paplicki, P. A novel rotor design for a hybrid excited synchronous machine. *Arch. Electr. Eng.* **2017**, *66*, 29–40. [CrossRef]
53. Wardach, M.; Palka, R.; Paplicki, P.; Bonislawski, M. Novel hybrid excited machine with flux barriers in rotor structure. *COMPEL* **2018**. [CrossRef]
54. Wardach, M.; Bonislawski, M.; Palka, R.; Paplicki, P.; Prajzendanc, P. Hybrid excited synchronous machine with wireless supply control system. *Energies* **2019**, *12*, 3153. [CrossRef]
55. Paplicki, P.; Prajzendanc, P. The influence of permanent magnet length and magnet type on flux-control of axial flux hybrid excited electrical machine. In Proceedings of the 14th Selected Issues of Electrical Engineering and Electronics (WZEE), Szczecin, Poland, 19–21 November 2018. [CrossRef]
56. Wardach, M. Torque and back-emf in hybrid excited claw pole generator. *COMPEL* **2018**. [CrossRef]
57. Wardach, M. Hybrid excited claw pole generator with skewed and non-skewed permanent magnets. *Open Phys.* **2017**, *15*, 902–906. [CrossRef]
58. Wardach, M. Design of hybrid excited claw pole machine with laminated rotor structure. In Proceedings of the Innovative Materials and Technologies in Electrical Engineering (i-MITEL 2018), Sulecin, Poland, 18–20 April 2018. [CrossRef]
59. Wardach, M.; Paplicki, P.; Grochocki, P.; Piatkowska, H. Influence of rotor design on field regulation capability of hybrid excited electric machines. In Proceedings of the 14th Conference on Selected Issues of Electrical Engineering and Electronics (WZEE'2018), Szczecin, Poland, 19–21 November 2018. [CrossRef]

Publisher's Note: MDPI stays neutral with regard to jurisdictional claims in published maps and institutional affiliations.

Creative

Development and Verification of a Simulation Model for 120 kW Class Electric AWD (All-Wheel-Drive) Tractor during Driving Operation

Seung-Yun Baek [1,2], Yeon-Soo Kim [1,3], Wan-Soo Kim [1,2], Seung-Min Baek [1,2,*] and Yong-Joo Kim [1,2,*]

[1] Department of Biosystems Machinery Engineering, Chungnam National University, Daejeon 34134, Korea; kelpie0037@gmail.com (S.-Y.B.); kimtech612@gmail.com (Y.-S.K.); wskim0726@gmail.com (W.-S.K.)

[2] Department of Smart Agriculture Systems, Chungnam National University, Daejeon 34134, Korea

[3] Convergence Agricultural Machinery Group, Korea Institute of Industrial Technology (KITECH), Gimje 54325, Korea

* Correspondence: bsm1104@naver.com (S.-M.B.); babina@cnu.ac.kr (Y.-J.K.); Tel.: +82-42-821-7870 (S.-M.B.); +82-42-821-6716 (Y.-J.K.)

Received: 18 April 2020; Accepted: 7 May 2020; Published: 12 May 2020

Abstract: This study was conducted to develop a simulation model of a 120 kW class electric all-wheel-drive (AWD) tractor and verify the model by comparing the measurement and simulation results. The platform was developed based on the power transmission system, including batteries, electric motors, reducers, wheels, and a charging system composed of a generator, an AC/DC converter, and chargers on each axle. The data measurement system was installed on the platform, consisting of an analog (current) and a digital part (rotational speed of electric motors and voltage and SOC (state of charge) level of batteries) by a CAN (controller area network) bus. The axle torque was calculated using the current and torque curves of the electric motor. The simulation model was developed by 1D simulation software and used axle torque and vehicle velocity data to create the simulation conditions. To compare the results of the simulation, a driving test using the platform was performed at a ground speed of 10 km/h in off- and on-road conditions. The similarities between the results were analyzed using statistical software and we found no significant difference in axle torque data. The simulation model was considered to be highly reliable given the change rate and average value of the SOC level. Using the simulation model, the workable time of driving operation was estimated to be about six hours and the workable time of plow tillage was estimated to be about 2.4 h. The results showed that the capacity of the battery is slightly low for plow tillage. However, in future studies, the electric AWD tractor performance could be improved through battery optimization through simulation under various conditions.

Keywords: electric AWD tractor; SOC level; simulation model; load measurement system; driving test

1. Introduction

Due to the increasing oil consumption of agricultural machinery and the seriousness of the pollution caused by diesel engines [1,2], the field of agricultural machinery has been actively researching electric drive type power transmission systems [3–6]. Electric tractors can reduce CO_2 emissions by up to about 70% compared to engine-driven tractors [7]. Electric drive power transmission systems can be classified into series hybrid, parallel hybrid, and electric drive [8]. In series hybrid, the engine is used only to charge the battery. Instead of the engine, the electric motor receives energy from the battery and drives the vehicle [9]. Parallel hybrids are equipped with an additional internal combustion engine to distribute power optimally by driving the vehicle with an electric motor at low loads, with an engine at heavy loads, and simultaneously driving the vehicle at high loads [10]. The electric drive type is a method of removing the engine for battery charging and using regenerative braking for battery charging and driving the vehicle [11]. For automobiles, series hybrid and parallel hybrid vehicles have been commercialized for a long time; recently, an electric drive type has also been commercialized. However, in the field of agricultural tractors, the electric drive power transmission method is less commercialized compared to the conventional diesel engine tractor. High-torque driving motor technology is not yet suitable for application in series hybrid to a tractor. The energy conversion efficiency of an electric motor varies with load, making it difficult to maintain optimum efficiency [12]. The parallel hybrid type struggles to properly distribute power because the driving motor must perform power generation and torque assist according to load fluctuations generated during operation to efficiently distribute power [13]. In addition, the electric drive type requires expensive and large-capacity batteries [14] because agricultural operations must be performed using only a battery without an engine.

To overcome the limitations of the high-torque electric motor, electric drive technology has been studied for electric all-wheel-drive (AWD) vehicles in which four motors are mounted on the drive shaft [15]. Electric AWD can easily be applied using an electric motor with a small torque capacity. A separate engine can be installed for charging the battery, which minimizes the battery capacity. Therefore, AWD technology can replace a high-power engine with four electric motors, which can continuously perform high-torque operations by charging the battery through the additional engine.

The simulation model for the AWD platform was verified to improve tracking performance [16], energy efficiency, and steering performance [17]. The AWD platform controller algorithm was developed to improve steering performance [18] and ride comfort [19]. Research on electric AWD technology is active in the automotive field, but research on agricultural machinery such as tractors is lacking. Most AWD platforms use one motor to divert power to each axle; research is needed that applies four motors to a tractor. Since the electric AWD tractor directly drives the axle using a motor, the capacities of the electric motor and the battery must be verified. In the automotive field, actual vehicle tests were conducted to verify the design of electric drive AWD systems and component parts. However, repeating the actual vehicle test is expensive and time-consuming [20]. To solve this, the research trend is to verify the platform using 1D simulation programs such as Simcenter AMESim (ver. 16, Siemens, Munich, Germany) and SimulationX (ver. 4.0, ESI ITI GmbH, Dresden, Germany).

A dynamic model of a hybrid vehicle in AMESim was used to develop a coupling device [21], predict the driving performance [22], optimize the torque control strategies [23] and the design for a dual clutch transmission [24], and predict the shock during mode shifting and riding [25]. Hybrid vehicle models were developed using SimulationX to evaluate component and hybrid control strategies for optimizing the engine power of the hybrid system [26], to review the dynamic characteristics of the power coupling device [27], and to verify and calibrate the control algorithms for electric and hybrid vehicles [20]. In previous studies, electric drive systems were simulated using programs for specification review and performance prediction before actual vehicle development. This is because accurate performance evaluation is difficult using only simulation models. Therefore, we aimed to develop a prototype of an electric AWD tractor and simulation model and verified the model by comparing the results of the simulation with those of the actual vehicle test. To verify the simulation model, we compared the axle torque and the charging/discharging performance of the battery according to load conditions. The rate of change for SOC (state of charge) level can be predicted through the torque generated on the axle, and the available time for agricultural operation of the electric AWD tractor can be estimated through the rate of change for the SOC level.

This study was basic research on the development of an electric AWD tractor, and the simulation model was developed and verified through an actual platform. The goals of the research were to (1) develop an electric AWD tractor based on power transmission systems; (2) develop a simulation model that reflects the specifications of the parts of an electric AWD tractor; (3) verify the simulation model through comparison of measured data with simulation results by conducting a driving test; and (4) develop a reliable simulation model that can be used to develop and supplement electric AWD tractors.

2. Materials and Methods

2.1. Concept of Electric AWD Tractor

Figure 1 depicts the composition of an electric AWD tractor and the power flow of an AWD system. Power transmission systems of electric AWD tractors consist of batteries, electric motors, reducers, wheels, and a charging system, composed of generators, AC/DC converters, and chargers. The platform was developed based on power transmission systems, and the axle must be composed of a combination of a motor and a speed reducer to cope with the high loads generated during work [15]. The capacities of the battery and generator were selected so that the motor could be supplied with constant power by considering the electric tractor's farming time of 6 h [28]. The four AWD systems are driven independently, which reduces the turning radius and improves the ease of slip control [29]. The electric motor operates by being supplied with energy from a battery and a generator, and the torque output from the motor is transmitted to the wheel through two reducers. Each control unit is configured to communicate via CAN (controller area network). The function of the AWD was developed by installing electric motors, reducers, and wheels with the same specifications on each axle. The electric systems basically consisted of a plug-in hybrid electric vehicle (PHEV) system that uses rechargeable batteries to supply electrical energy continuously.

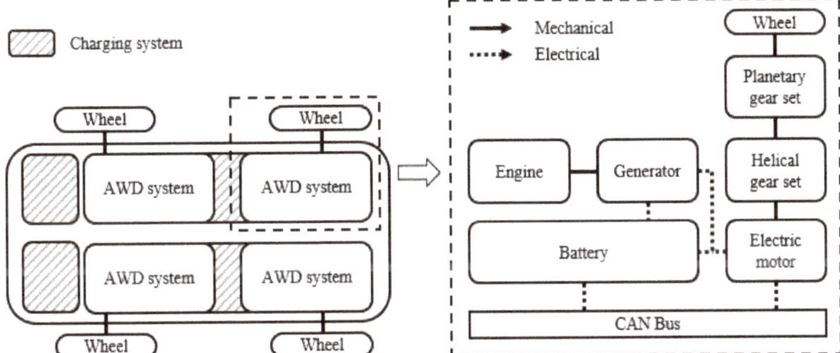

Figure 1. Schematic diagram of the power flow for the electric all-wheel-drive (AWD) tractor.

2.2. Development of Electric AWD Tractor

The electric AWD tractor was developed by combining selected electric and mechanical components for each axle, as shown in Figure 2. The total weight and dimensions were 7430 kg and 5500 × 2500 × 1950 mm (length × width × height), respectively. The weight and dimensions of a conventional tractor of the same class are 7125 kg and 4855 × 2450 × 3095 mm, respectively, so the electric AWD tractor has specifications similar to a conventional tractor. Table 1 shows the components and specifications of the electric AWD tractor. The battery type adopted was LiFePO$_4$ because it can cope with the high power output of the electric motor during operation and has a long life cycle [30]. To stably cope with the high torque of the axle generated in the harsh farming environment, the battery had a capacity of 14.6 kWh, a rated voltage of 70.4 V, and a of 2 C. It was connected to each electric motor and generator. The C-rate was selected to ensure no damage occurred to the battery when a load higher than the rated power of the motor occurred. The generators (GENEX ST15000, Honda, Tokyo, Japan) with a gasoline engine and AC/DC converter were added to supply electrical energy to batteries for continuous charging during operation. The selected rated output of the generator was 13.5 kW, considering overcharging of the battery. The battery was charged with the current generated from the generator through the converter and charger. The selected rated power of the converter was 15 kW, which was similar to the output of the generator to increase efficiency. The maximum output current of the charger was selected as 50 A considering the safety of battery charging. The converter and charger were selected considering the input current for charging the battery and the safety of the output of the generator. Four electric motors were used to replace the engine of a conventional tractor to cope with the high load during the agricultural work. Electric motors (1238-6501 AC-34, Curtis, New York, NY, USA) were selected as an easy-to-obtain product in Korea; the maximum torque and rotational speed were 119.7 Nm and 8000 rpm, respectively. We selected the motor considering the maximum power and the rated power to develop a 120 kW class tractor. The maximum and rated powers of each electric motor were about 37 and 30 kW, respectively. To increase the torque generated on the axles during tractor operation, reducers were installed on each motor output shaft. The planetary gear reducer used an existing knuckle arm to facilitate tire mounting for the tractor, and the gear ratio was about 12.05. The helical gear reducer was selected as 4.3 to increase the torque with the planetary gear reducer, and the gear ratio considered the maximum torque generated during agricultural operations. For the wheels, we selected 380/85R24 agricultural wheels (AGRIMAX RT 855, BKT, Gujarat, India), mainly used for front wheels for a large tractor to ensure the traction and ground speed of the platform.

Figure 2. Photo of the configuration of the electric AWD tractor.

Table 1. Specifications of the electric AWD tractor and charging system.

	Item		Specification
	Length × Width × Height (mm)		5500 × 2500 × 1950
	Weight (kg)		7429
		Max. torque (Nm)	119.7
	Electric motor	Max. rotational speed (rpm)	8000
		Max. power (kW)	37
Electric AWD tractor		Capacity (kWh)	14.6
	Battery	Type	LiFePO$_4$
		Voltage (V)/C-Rate (C)	70.4/2
	Reducer	Planetary gear ratio	12.05
		Helical gear ratio	4.3
	Tire		380/85R24
	Generator	Rated power (kW)	13.5
Charging system	Converter	Rated power (kW)	15
	Charger	Max. output current (A)	50

2.3. Simulation Model

The power transmission system and the control system of the electric AWD tractor were developed using SimulationX (ver. 4.0, ESI ITI GmbH, Dresden, Germany). The simulation model consisted of a mechanical model, electrical model, and control model, as shown in Figure 3. Each model reflected the actual specifications of the components. The motor model reflected the torque and rotational speed (T-N) curve provided by the manufacturer, and the battery model reflected the specifications of the capacity and voltage. The output according to the rated rotational speed was input to the generator model, and each gear ratio was input to the reducer models. The tire model was configured to be the same as the actual driving condition of the platform through input of the slip ratio for each soil condition [31]. By inputting the load data between the reducer and the tire model, the output power of the electric motor was determined. The battery SOC level changed according to the power of the electric motor during simulation.

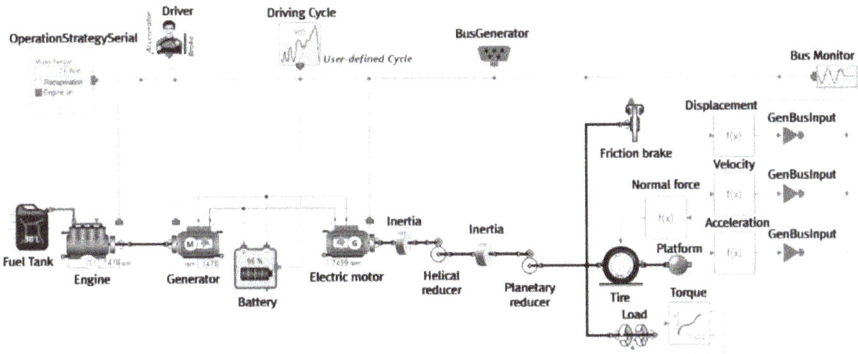

Figure 3. Simulation model for power transmission system of the electric AWD tractor using SimulationX (reproduced from [7]).

2.3.1. Mechanical Model

The mechanical model consisted of an electric motor, reducers, and tire. The axle load was input from the input torque model configured between the reducer and the tire, and the electric motor model was configured to output the motor torque considering the input load and the gear ratio of reducers. The motor torque is shown in Equation (1), which considers the air-gap torque, acceleration torque, and friction torque. Under constant voltage conditions, the current of the motor that has the most influence on the SOC is shown in Equation (2), and the torque proportional to the current is shown in Equation (3) using the gear ratio. The tire model represents wheel–road and estimates slip values at two speeds, and assigns coefficients of friction according to the prescribed slip characteristics [32].

$$T_{motor} = T_e - T_a - T_l, \tag{1}$$

where T_{motor} is motor torque (Nm), T_e is air-gap torque (Nm), T_a is acceleration torque (Nm), and T_l is motor friction torque (Nm).

$$P_{motor} = V_{in}I_{in} = \frac{2\pi T_{motor}N_{motor}}{60,000}, \tag{2}$$

where P_{motor} is motor power (kW), V_{in} is supply voltage (V), I_{in} is supply current (mA), T_l is motor friction torque (Nm), and N_{motor} is rotational speed of motor (rpm).

$$T_o = T_{in}\gamma_{reducer}, \tag{3}$$

where T_o is the output torque of the reducer (Nm), T_{in} is the input torque of the reducer (Nm), $\gamma_{reducer}$ is the gear ratio of the reducer.

2.3.2. Electrical Model

The electrical model consisted of a generator with engine and a battery. The generator was always driven so that it supplied 13.5 kW of output to the battery and electric motor, and the output of the generator was determined as shown in Equation (4). The battery applied a capacity of 14.6 kWh, and the initial SOC level was set in accordance with the test conditions. SOC level was calculated from the current output from the motor and generator [30,33], as shown in Equation (5).

$$P_{generator} = V_{in}I_{in} \; (P_{generator} \geq 0),\tag{4}$$

where $P_{generator}$ is motor power (kW).

$$\mathrm{SOC}(k) = \mathrm{SOC}(0) - \frac{T}{C_n} \int_0^k (\eta I(t) - S_d)dt,\tag{5}$$

where $SOC(0)$ is the battery initial SOC, $I(t)$ is the current at time t, T is the sampling period, C_n is the nominal capacity of the battery, η is the coulombic efficiency, and S_d is the self-discharging rate. For a LiFePO$_4$ battery, $\eta > 0.994$ at room temperature and the self-discharging rate is about 5% per month.

2.3.3. Simulation Condition

The driving cycle model represents the preset vehicle velocity and vehicle acceleration. Vehicle velocity was calculated based on motor rotational speed and tire diameter, as shown in Equation (6) [34]. The operation strategy serial model operates the generator according to the setting range of the SOC level of the battery. The charging range of the SOC level was set to the condition that the generator is always driven. All models were connected via a bus system, and the models interacted according to input conditions. The time steps of the simulation were set to 35 s and 40 s in the off-road and the on-road condition, respectively, to match the load measurement time during the driving test.

$$V_{platform} = \frac{N_{motor}}{\gamma_{reducer}} \times \frac{60}{1000} \times d_{tire} \times \pi,\tag{6}$$

where $V_{platform}$ is velocity (km/h), N_{motor} is the rotational speed of the motor (rpm), $\gamma_{reducer}$ is the gear ratio of the reducer, and d_{tire} is the tire diameter (m).

2.4. Verification of Electric AWD Tractor

2.4.1. Load Measurement System

The load measurement system was developed and installed on an electric AWD tractor for measuring real-time data during operation, as shown in Figure 4. The current sensors (LF-1005S, LEM USA Inc., Milwaukee, WI, USA) and CAN logging system were installed to measure current and rotational speed of electric motors and voltage of batteries for calculating axle torque. The current sensors used in this study were installed between each battery and controller of the electric motor, generating a coil magnetic field according to the current flow and recognizing the current generated by the magnetic field detecting of the hall sensor up to 1000 mA. The SOC level was measured using the CAN logging system connected to a battery management system (BMS). A data acquisition device (QuantumX MX840, HBM, Darmstadt, Germany) was installed to acquire and transfer the signals from the current sensors and the CAN logging system to the laptop computer. The accuracy classes of the data acquisition device ranged from 0.05% to 0.1% depending on the sensor technology. Two 24-bit analog input channels with a sampling rate of 300 Hz per channel were used to acquire the current, and four digital input channels with a sampling rate of 300 Hz per channel were used to measure the CAN data. The measured signals were transmitted and saved on the laptop computer through the signal processing program CATMAN (ver. 3.1, HBM, Darmstadt, Germany).

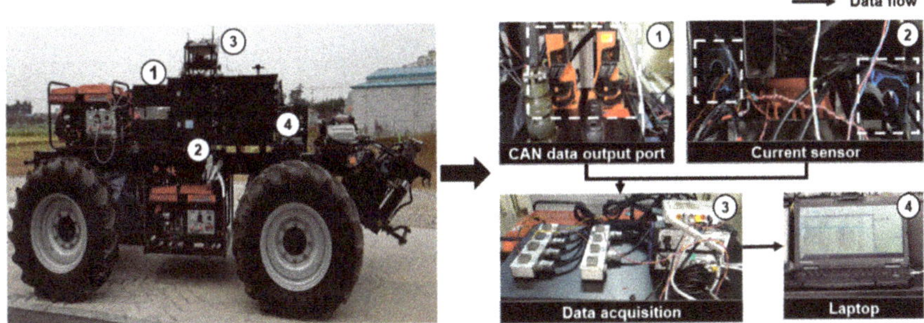

Figure 4. Load measurement system and data flow of electric AWD tractor.

2.4.2. Test Procedure

We conducted driving tests to measure data for calculating axle torque on a test track (Latitude: 35°56′08.5″ N, Longitude: 126°55′33.1″ E) in Iksan-si, Jeollabuk-do province, Republic of Korea. Driving tests were conducted at a ground speed of 10 km/h considering the working speed of a tractor, which is the standard used in the tractor design [35] under off-road (field) and on-road (asphalt) conditions, as shown in Figure 5. The driving test was conducted with the generator always running during all operations and the generator's gasoline fuel tank was filled with 30 L. The driving test was conducted through a wireless controller that controlled the speed of the four motors. The platform traveled approximately 45 m for 25 s in off-road conditions and 80 m for 35 s in on-road conditions at 10 km/h. Data measurement was performed in off-road and on-road conditions for 35 s and 40 s, respectively, and the measurement time included a stop section of about 10 s before and after the driving test.

(**a**) (**b**)

Figure 5. Photos of driving operation of electric AWD tractor under (**a**) off-road and (**b**) on-road conditions.

Figure 6 shows the overall test procedure as a flow chart. The current and rotational speed of electric motors and voltage of batteries measured during the test were converted into torque using the current–torque characteristic curve. The current–torque characteristic curve was provided by the motor manufacturer, and the output torque of the motor was calculated according to voltage, motor rotation speed, and current. The calculated torque was converted to axle torque by applying gear ratios of 4.3 and 12.05. The calculated torque and velocity of the vehicle were used as input conditions for the simulation. The measured data and simulation results of the axle torque and SOC level of the battery were compared through simulation, and the similarities between the two results were verified using t-test calculated with IBM SPSS Statistics software (SPSS 25, IBM Corp., Armonk, NY, USA).

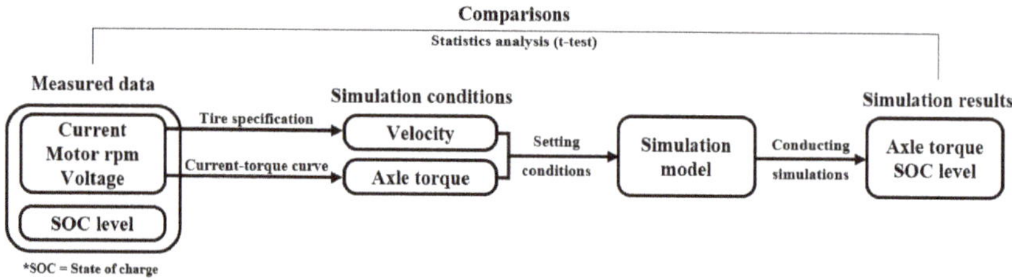

Figure 6. Flow chart of the test procedure for development and verification of electric AWD tractor.

3. Results

3.1. Axle Torque

The axles of the electric AWD tractors had the same size of wheels and same weight distribution ratio, so the output torque of the motors did not vary significantly from axle to axle. However, when the steering of the vehicle adopts skid steering [16], a difference occurs in the output torque of the motors located on the left and right sides, so the simulation model was verified using data from each axle. Figure 7 provides a comparison of the measured and simulated results of the averaged axle torque at a ground speed of 10 km/h in off-road conditions. The axle torque was considerably affected by the ground speed in off-road conditions. In addition, the axle torque has been shown to vary not only by soil irregularities, but also by steering for straight driving. The axle torque peaked as the electric motor accelerated, and then changed due to repeated deceleration and acceleration for steering. The maximum and average measured torques of the left axle were 6847.4 and 3771.8 Nm, respectively, and 6345.5 and 2580.1 Nm, respectively, for the right axle. The simulation results showed similar trends in all sections, but the largest differences for each axle were 1425.2 and 1850.2 Nm in the sections where the platform starts to operate and where the maximum torque occurred, respectively.

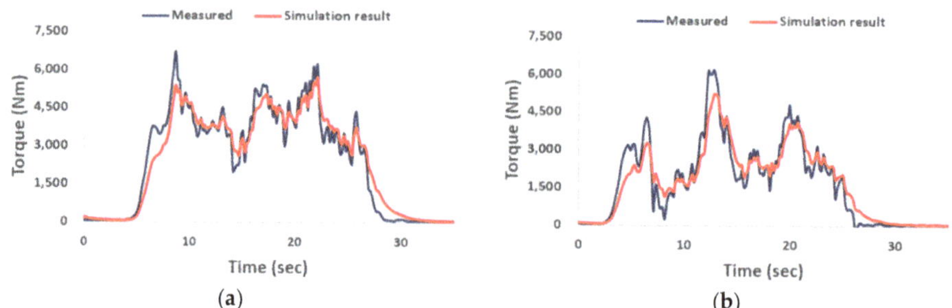

Figure 7. Comparison of averaged axle torque of (**a**) left and (**b**) right sides of the tractor for measured and simulated results at a ground speed of 10 km/h in off-road conditions.

Figure 8 provides a comparison of the measured and simulated results of the averaged axle torque at a ground speed of 10 km/h in on-road conditions. The axle torque has been shown to vary with steering for straight driving. The driving test performed in the on-road condition had a small load fluctuation compared to the off-road condition because the soil had little influence on asphalt driving. The maximum and average measured torques of the left axle were 1905.9 and 689.1 Nm, and 1644.9 and 161.7 Nm for the right axle, respectively. The minimum torques of the left and right axles were −625.0 and −732.1 Nm, respectively, by decelerating the electric AWD tractor for steering.

The simulation results showed similar trends in all sections, but the highest differences for the left and right axle were 572.9 and 581.3 Nm, respectively, in the section where the maximum torque occurred.

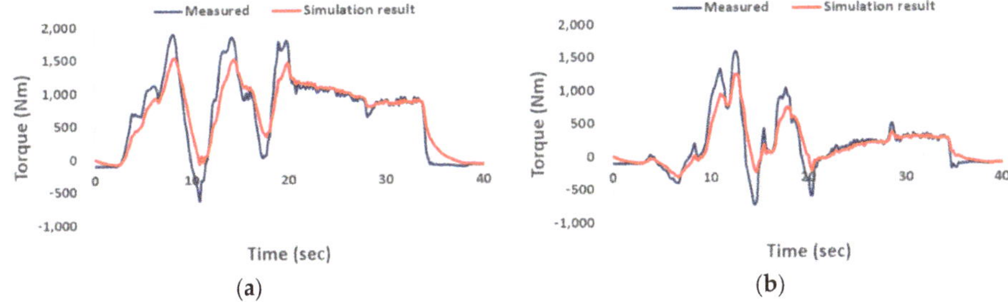

Figure 8. Comparison of averaged tractor axle torque of (**a**) left and (**b**) right sides for measured and simulated results at ground speed of 10 km/h in on-road conditions.

The simulated axle torque was smaller than the measured value in most sections. This is because the simulation analysis was performed in different conditions compared to the actual vehicle test. For example, the inertia models connected to the motor and reducer model were set to a default value because it is difficult to measure the exact value during a driving test. As the input value of the inertia model decreases, it is judged that the simulated torque of the electric motor is lower than the measured value because a lower force was required. However, it is possible to obtain more similar results by selecting the proper value of the inertia model using the inertial measurement unit (IMU). Table 2 lists the axle torque differences between the measured and simulated results based on the operation conditions. The average torques of the left and right motors were about 1800 and 1900 Nm higher in the off-road than on-road conditions, respectively. The maximum average torques of the motors were similar on the right and left sides, but the average value was about 700 Nm higher on the left side on the off-road condition. This difference was caused by the load according to the difference in the rotational speed of the left and right sides by the skid steering. The results of the t-test showed that the *p*-value was higher than 0.05 under all conditions, indicating no significant differences between the measured and simulated results of axle torque. Therefore, the simulation model was able to evaluate the axle torque of the electric AWD tractor during operation.

Table 2. Comparison of axle torque between measured and simulated results based on the operation conditions at 10 km/h.

Field Conditions	Electric Motor	Measured Axle Torque (Nm)	Simulated Axle Torque (Nm)	*p*-Value
Off-road	Left	2490.8 ± 2005.3 [1]	2497.4 ± 1865.5	0.846
	Right	1702.6 ± 1596.4	1705.3 ± 1402.5	0.941
On-road	Left	689.1 ± 601.1	687.8 ± 484.9	0.934
	Right	161.7 ± 413.0	166.3 ± 310.1	0.696

Note: [1] Average ± standard deviation.

3.2. SOC Level of Battery

The measured and simulated SOC levels were compared when the vehicle was traveling at 10 km/h. Figure 9 depicts the results in off-road conditions. The initial and final SOC levels of the battery connected to the left motor were measured at 96.30% and 96.10%, respectively, and 80.00% and 79.90% for the right motor, respectively. Simulation analysis showed that the initial and final SOC levels of the battery connected to the left motor were 96.30% and 96.11%, respectively, and 80.00% and 79.90% for the right motor, respectively. The rate of change of the SOC level during operations by driving the left motor for measured and simulated results were about 0.20% and 0.18%, respectively, with a 0.02% difference between results. The rate of change of the SOC level by the right driving motor for measured and simulated results was similar, with a 0.10% difference between results.

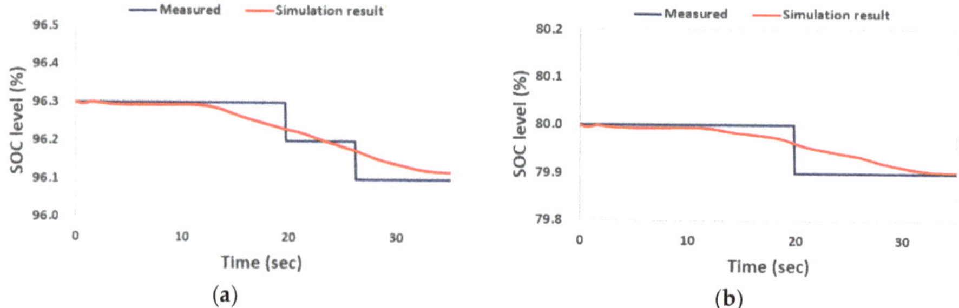

Figure 9. Comparison of averaged SOC level of (**a**) left battery and (**b**) right battery for measured and simulated results at a ground speed of 10 km/h under off-road conditions.

Figure 10 depicts the measured and simulated results of the averaged SOC level at a ground speed of 10 km/h under on-road conditions. The initial and final SOC levels of the battery connected to the left motor were measured at 97.20% and 97.00%, respectively; for the battery connected to the right motor, the measurements were 80.00% and 79.90%, respectively. Simulation analysis showed that the initial and final SOC levels of the battery connected to the left motor were 97.20% and 96.99%, respectively. For the battery connected to the right motor, these values were 80.00% and 79.89, respectively. The rate of change in the SOC level during operations by driving the left motor for the measured and simulated results were about 0.20% and 0.21%, respectively, a 0.01% difference between results. The results for the right driving motor showed a similar difference of 0.10%.

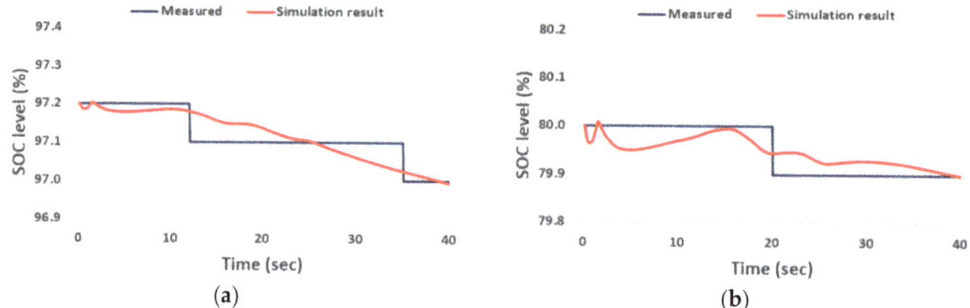

Figure 10. Comparison of averaged SOC level of (**a**) left battery and (**b**) right battery for measured and simulated results at a ground speed of 10 km/h under on-road conditions.

The SOC level is changed by the motor output, but we found that the SOC level of the battery connected to each motor did not change dramatically by continuously supplying power to the battery and motor by always driving the generator during driving operation. Table 3 provides a comparison of the SOC level between the measured and simulated results based on the operation conditions. The t-test results showed that there was no significant difference between measured and simulated results of the SOC level for the all battery except right battery at a ground speed of 10 km/h in off-road conditions. The p-value was less than 0.05 only for the right battery under off-road conditions, and there were no significant differences except for this one condition. The final variation in SOC (SOC_{fv}) was calculated using the initial and final values of the battery SOC level [3], and the SOC_{fv} was compared through measurement and simulation analysis results. Similarity was not verified in the one item in the t-test, but the difference in SOC_{fv} showed a similar trend. According to the low resolution power of the measurement system, the results of measurement and simulation did not match in all sections. Due to the low resolution power, the SOC level measured through the measurement system can be checked only up to 0.1% change. Otherwise, the simulation results of the SOC level are different from the measurement results because the reduction rate appears in real time. The SOC level is expected to be more accurate compared with the higher resolution power of the measurement system. Besides, the SOC_{fv} and the average and standard deviation values of the two results showed slight differences. Therefore, the simulation model was able to evaluate the SOC level of the electric AWD tractor during operation according to the axle torque.

Table 3. Comparison of SOC level and final variation in SOC (SOC_{fv}) between measured and simulated result based on the operation conditions at 10 km/h.

Field Conditions	Battery	*t*-Test			SOC_{fv} (%)	
		SOC Level (%)				
		Measured	Simulated	*p*-Value	Measured	Simulated
Off-road	Left	96.20 ± 0.08 [1]	96.23 ± 0.06	0.400	0.20 [2]	0.18
	Right	79.95 ± 0.05	79.96 ± 0.03	2×10^{-9} *	0.10	0.10
On-road	Left	97.12 ± 0.06	97.12 ± 0.06	0.719	0.20	0.21
	Right	79.95 ± 0.05	79.95 ± 0.03	0.326	0.10	0.11

Note: * Significant difference at $p < 0.05$; [1] Average ± standard deviation; [2] $SOC_{fv} = SOC_{initial} - SOC_{final}$.

4. Discussion

The axle torque of the AWD tractor showed similar trends in all sections except when the platform started to operate or maximum torque was reached. The measured torque data tended to be similar to the simulation data, but the measured values were higher in most sections. The selection of motors and reducers for electric-powered AWD tractors needs to account for these differences, especially in the development and complementary phases. The SOC level rate of change values were similar, with a maximum error of 0.02%, but the form of decline in the graph was different given the resolution of the measurement system. The decrease rate of SOC level showed slight changes due to the continuously operated generator during driving operation. The decrease rate per minute of SOC level was about 0.3% in off-road and on-road conditions. The SOC level was not significantly different between off-road and on-road as regenerative braking occurred in on-road driving conditions [11,36]. In addition, as the driving test was conducted for a short period of time, the SOC level of the battery was determined to have little influence on the output torque of the motor according to the field conditions, because it mainly used the current supplied from the generator. However, the electric AWD tractor cannot be operated only with the current supplied from the generator if the working time is prolonged.

We determined that the reduction rate of the SOC battery level changes according to current consumption by increasing the output torque of the motor during long working hours. The SOC level reduction rate is expected to be greater in off-road than on-road conditions when the driving test is conducted without regenerative braking. The workable time of the electric AWD tractor was estimated to be about 6 h based on the axle torque generated during driving operation. The platform can travel for about 6 h at a speed of 10 km/h, and estimated driving range is calculated as approximately 60 km according to the reduction rate of the SOC level and workable time. Estimated driving range is expected to be longer due to battery charging by regenerative braking in the actual driving environment. Since the tractor was designed based on the results of research on driving operation [37–39], available time for work was predicted through torque comparison by plow tillage using the existing research. The torque data were compared with off-road driving data because it is difficult to compare the average value given the negative value generated by regenerative braking during driving operation under on-road conditions. Tillage is the agricultural preparation of soil by mechanical agitation of various types, such as digging, stirring, and overturning [40], and it is the harshest operation, accounting for over 60% of engine power [41]. The average axle torques of the driving operation and plow tillage were reported to be 1452.1 and 3680.3 Nm, respectively [15]. The estimated torque of each axle for plow tillage was 3221.5 Nm and the workable time was predicted to be 2.4 h, as summarized in Table 4. The average torque generated on the axle during plow tillage is about 2.5 times higher than the averaged torque during driving operation [7], and the workable time for plow tillage using electric AWD tractors is estimated to be about 40% lower than workable time for driving operation. The average working time of the tractor is six hours in Korea, and it can be assumed that the morning and afternoon work are performed for three hours each. The battery performance of the electric AWD tractor developed in this study was lower than three hours; however, the working time can be extended by increasing the capacity of the battery and by fast charging during the break in the middle of the workday.

Table 4. Estimated torque and workable time for electric AWD tractor according to torque ratio of operations.

Agricultural Operation	Torque Ratio	Estimated Torque (Nm)	Estimated Time (Hours)
Plow tillage	2.5 [1]	3221.5 [2]	2.4

Note: [1] Ratio of axle torque during driving operation; [2] Averaged torque applying torque ratio to driving data.

The electric AWD tractor was developed considering the load generated during operation, the required output, and the workable time. The developed system has the advantage of being applicable to various fields including construction machinery, depending on the selection of electric motor and reducer according to the output and torque, and the capacity of the battery and generator considering the usage time and discharge rate. The platform has the required output according to the combination of the electric motor and the reducer, and the usage time and the size of the platform are determined by selecting the proper capacity of the battery and the generator. The developed simulation model consists of all components including electric motors, reducers, batteries, and generators, and it is also possible to simulate and analyze the AWD platforms with different outputs and sizes by modifying each performance map and specifications. However, the simulation model can be utilized through partial modification of the simulation model in the case of a similar AWD platform, and it is judged that there are limitations in the simulation and analysis of the AWD platforms with different systems. In future studies, simulations and field tests including tillage should be performed and verified under a variety of conditions to optimize the electric-powered tractor components.

5. Conclusions

In this study, an electric AWD tractor was developed based on a power transmission system. A simulation model reflecting the specifications of this electric AWD tractor was developed and verified using measured data from driving tests conducted under off-road and on-road conditions. The measured data were converted to torque using equations and were used for simulation conditions. A comparison of the simulation analysis results with the measured data showed that the torque generated on the axle was similar in value and shape, and we found no significant differences in the statistical analysis results. Although the SOC level showed a significant difference in the statistical analysis results, the rate of change per minute, and the SOC_{fv}, the simulation results were considered to be reliable. The axle torque is closely related to the SOC level because it is proportional to the current supplied from the battery to the electric motor. As the measured data for both factors matched the simulation results, we determined that the operating time of the platform can be estimated through simulation. The workable time of the electric AWD tractor was estimated through simulation models and existing research data. As a result of simulation, the workable time for plow tillage using the electric AWD tractor was estimated to be about 2.4 h. The results are less than the target hours (three hours) of work. In future studies, performance could be improved through battery optimization through a simulation.

Author Contributions: Conceptualization, S.-Y.B. and Y.-J.K.; methodology, S.-Y.B. and W.-S.K.; software, S.-Y.B. and Y.-S.K.; validation, S.-Y.B.; writing—original draft preparation, S.-Y.B.; writing—review and editing, S.-Y.B., S.-M.B. and Y.-J.K.; supervision, Y.-J.K.; project administration, Y.-J.K. All authors have read and agreed to the published version of the manuscript.

Funding: This work was supported by the Industrial Strategic Technology Development Program (10083590, Development of construction and agricultural robot platform technology with autonomous system for all-wheel drive of 100kW class) funded by the Ministry of Trade, Industry and Energy (MI, Korea).

Conflicts of Interest: The authors declare no conflict of interest.

References

1. Lee, D.H.; Choi, C.H.; Chung, S.O.; Kim, Y.J.; Inoue, E.; Okayasu, T. Evaluation of tractor fuel efficiency using dynamometer and baler operation cycle. *J. Fac. Agric. Kyushu Univ.* **2016**, *61*, 173–182.
2. Chen, Y.; Xie, B.; Mao, E. Electric Tractor Motor Drive Control Based on FPGA. *IFAC-PapersOnLine* **2016**, *49*, 271–276. [CrossRef]
3. Lee, D.H.; Kim, Y.J.; Choi, C.H.; Chung, S.O.; Inoue, E.; Okayasu, T. Development of a parallel hybrid system for agricultural tractors. *J. Fac. Agric. Kyushu Univ.* **2017**, *62*, 137–144.
4. Kang, E.; Pratama, P.S.; Byun, J.; Supeno, D.; Chung, S.; Choi, W. Development of Super-capacitor Battery Charger System based on Photovoltaic Module for Agricultural Electric Carriers. *J. Biosyst. Eng.* **2018**, *43*, 94–102.
5. Azwan, M.B.; Norasikin, A.L.; Sopian, K.; Abd Rahim, S.; Norman, K.; Ramdhan, K.; Solah, D. Assessment of electric vehicle and photovoltaic integration for oil palm mechanisation practise. *J. Clean. Prod.* **2017**, *140*, 1365–1375. [CrossRef]
6. Moreda, G.P.; Muñoz-García, M.A.; Barreiro, P. High voltage electrification of tractor and agricultural machinery—A review. *Energy Convers. Manag.* **2016**, *115*, 117–131. [CrossRef]
7. Baek, S.Y.; Kim, W.S.; Kim, Y.S.; Kim, Y.J.; Park, C.G.; An, S.C.; Moon, H.C.; Kim, B.S. Development of a Simulation Model for an 80 kW-class Electric All-Wheel-Drive (AWD) Tractor using Agricultural Workload. *J. Drive Control* **2020**, *17*, 27–36.
8. Ko, J.W.; Ko, S.; Park, Y.C. A Study on Battery Performance of a Motor Driven Local Transportation Vehicle. *J. Korean Soc. Mar. Eng.* **2012**, *36*, 430–436. [CrossRef]
9. Liu, W.; He, H.; Wang, Z. A Comparison Study of Energy Management for A Plug-in Serial Hybrid Electric Vehicle. *Energy Procedia* **2016**, *88*, 854–859. [CrossRef]
10. Park, J.K. *Design of Parallel Hybrid Tractor and Evaluation of Charge/Discharge Performance*; Sungkyunkwan University: Seoul, Korea, 2013.
11. Zhao, W.; Wu, G.; Wang, C.; Yu, L.; Li, Y. Energy transfer and utilization efficiency of regenerative braking with hybrid energy storage system. *J. Power Sources* **2019**, *427*, 174–183. [CrossRef]
12. Liu, M.; Xu, L.; Zhou, Z. Design of a Load Torque Based Control Strategy for Improving Electric Tractor Motor Energy Conversion Efficiency. *Math. Probl. Eng* **2016**, *2016*, 2548967. [CrossRef]
13. Park, Y., II; Kim, Y.J. Technology for the Agricultural Hybrid Tractor. *Auto J.* **2014**, *36*, 30–34.
14. Mousazadeh, H.; Keyhani, A.; Javadi, A.; Mobli, H.; Abrinia, K.; Sharifi, A. Evaluation of alternative battery technologies for a solar assist plug-in hybrid electric tractor. *Transp. Res. Part D Transp. Environ.* **2010**, *15*, 507–512. [CrossRef]
15. Kim, W.S.; Baek, S.Y.; Kim, T.J.; Kim, Y.S.; Park, S.U.; Choi, C.H.; Hong, S.J.; Kim, Y.J. Work load analysis for determination of the reduction gear ratio for a 78 kW all wheel drive electric tractor design. *Korean J. Agric. Sci.* **2019**, *46*, 613–627.
16. Kim, B.S. *Slip Detection and Control Algorithm to Improve Path Tracking Performance of Four-Wheel Independently Actuated Vehicles*; Hongik University: Seoul, Korea, 2019.
17. Song, H.W. *A Study on the Improvement of Energy Efficiency and Cornering Stability Performance for 4WD in-Wheel Electric Vehicle*; Sungkyunkwan University: Seoul, Korea, 2013.
18. Jeon, D.S. *Study about Skid Steer of Independent 4 Wheel Drive Vehicle*; Kookmin University: Seoul, Korea, 2008.

19. Song, C.H. *Development of Motor Control Algorithm of an All-Wheel Drive In-Wheel Electric Vehicle to Improve Ride Comfort*; Sungkyunkwan University: Seoul, Korea, 2015.

20. Klein, S.; Xia, F.; Etzold, K.; Andert, J.; Amringer, N.; Walter, S.; Blochwitz, T.; Bellanger, C. Electric-Motor-in-the-Loop: Efficient Testing and Calibration of Hybrid Power Trains. *IFAC-PapersOnLine* **2018**, *51*, 240–245. [CrossRef]

21. Zhang, D.X.; Zeng, X.H.; Wang, P.Y.; Wang, Q.N. Co-simulation with AMESim and MATLAB for differential dynamic coupling of hybrid electric vehicle. In Proceedings of the IEEE Intelligent Vehicles Symposium, Xi'an, China, 3–5 June 2009; pp. 761–765.

22. Hong, J.; Kim, S.; Min, B. Drivability development based on CoSimulation of AMESim vehicle model and simulink HCU model for parallel hybrid electric vehicle. SAE Technical Papers. In Proceedings of the SAE World Congress & Exhibition, Detroit, MI, USA, 20–23 April 2009.

23. Hwang, H.S.; Yang, D.H.; Choi, H.K.; Kim, H.S.; Hwang, S.H. Torque control of engine clutch to improve the driving quality of hybrid electric vehicles. *Int. J. Automot. Technol.* **2011**, *12*, 763–768. [CrossRef]

24. Cao, X.; Du, C.; Yan, F.; Xu, H.; He, B.; Sui, Y. Parameter Optimization of Dual Clutch Transmission for an Axle-split Hybrid Electric Vehicle. In Proceedings of the IFAC-PapersOnLine, Vienna, Austria, 4–6 December 2019; Elsevier B.V.: Amsterdam, The Netherlands, 2019; Volume 52, pp. 423–430.

25. Zeng, X.; Yang, N.; Wang, J.; Song, D.; Zhang, N.; Shang, M.; Liu, J. Predictive-model-based dynamic coordination control strategy for power-split hybrid electric bus. *Mech. Syst. Signal Process.* **2015**, *60*, 785–798. [CrossRef]

26. Wenyong, L.; Abel, A.; Todtermuschke, K.; Tong, Z. Hybrid vehicle power transmission modeling and simulation with SimulationX. In Proceedings of the 2007 IEEE International Conference on Mechatronics and Automation, ICMA 2007, Heilongjiang, China, 5–9 August 2007; pp. 1710–1717.

27. Tiantian, Z.; Sihong, Z.; Xiaoting, D. Research of Dynamic Characteristics on Power Coupling Equipment of Hybrid Electric Tractor. *Tract. Farm Transp.* **2013**, *4*, 7.

28. Korea Agricultural Machinery Industry Cooperative (KAMICO); Korean Society for Agricultural Machinery (KSAM). *Agricultural Machinery Yearbook in Korea*; Korea Agricultural Machinery Industry Cooperative and Korean Society for Agricultural Machinery: Suwon, Korea, 2019.

29. Jayaram, P.; Bharath, N.; Ahmed, R. A Novel Design for an all Terrain Custom-Built Vehicle Dynamics. *Int. J. Adv. Sci. Res. Eng.* **2018**, *4*, 186–194. [CrossRef]

30. Zhang, R.; Xia, B.; Li, B.; Cao, L.; Lai, Y.; Zheng, W.; Wang, H.; Wang, W. State of the art of lithium-ion battery SOC estimation for electrical vehicles. *Energies* **2018**, *11*, 1820. [CrossRef]

31. Kim, Y.S.; Lee, S.D.; Kim, Y.J.; Kim, Y.J.; Choi, C.H. Effect of tractor travelling speed on a tire slip. *Korean J. Agric. Sci.* **2018**, *45*, 120–127.

32. Wohnhaas, A.; Hötzer, D.; Sailer, U. Modulares Simulationsmodell eines KFZ-Antriebstrangs unter Berücksichtigung von Nichtlinearitäten und Kupplungsvorgängen.pdf. *VDI Ber.* **1995**, *1220*, 97–117.

33. Zhang, C.; Li, K.; Pei, L.; Zhu, C. An integrated approach for real-time model-based state-of-charge estimation of lithium-ion batteries. *J. Power Sources* **2015**, *283*, 24–36. [CrossRef]

34. Lee, N.G.; Kim, Y.J.; Kim, W.S.; Kim, Y.S.; Kim, T.J.; Baek, S.M.; Choi, Y.; Kim, Y.K.; Choi, I.S. Study on the Improvement of Transmission Error and Tooth Load Distribution using Micro-geometry of Compound Planetary Gear Reducer for Tractor Final Driving Shaft. *J. Drive Control* **2020**, *17*, 1–12.

35. Park, S.U. *Fatigue Life Evaluation of Spiral Bevel Gear of Transmission Using Agricultural Workload of Tractor*; Chungnam National University: Daejeon, Korea, 2019.

36. Xu, G.; Li, W.; Xu, K.; Song, Z. An Intelligent Regenerative Braking Strategy for Electric Vehicles. *Energies* **2011**, *4*, 1461–1477. [CrossRef]

37. Kim, Y.-J.; Chung, S.-O.; Choi, C.-H.; Lee, D.-H. Evaluation of Tractor PTO Severeness during Rotary Tillage Operation. *J. Biosyst. Eng.* **2011**, *36*, 163–170. [CrossRef]

38. Lee, D.H.; Kim, Y.J.; Chung, S.O.; Choi, C.H.; Lee, K.H.; Shin, B.S. Analysis of the PTO load of a 75 kW agricultural tractor during rotary tillage and baler operation in Korean upland fields. *J. Terramech.* **2015**, *60*, 75–83. [CrossRef]

39. Kim, D.C.; Ryu, I.H.; Kim, K.U. Analysis of tractor transmission and driving axle loads. *Trans. Am. Soc. Agric. Eng.* **2001**, *44*, 751–757.

Energies **2020**, *13*, 2422

40. Reddy, G.S.; Narsaiah, J.; Shashikala, G. Dynamic Analysis on Tillage Equipment Used in Agriculture Using Ansys Software. *Int. J. Sci. Res. Sci. Technol.* **2017**, *3*, 890–897.
41. Kim, W.S.; Kim, Y.S.; Kim, T.J.; Park, S.U.; Choi, Y.; Choi, I.S.; Kim, Y.K.; Kim, Y.J. Analysis of power requirement of 78 kW class agricultural tractor according to the major field operation. *Trans. Korean Soc. Mech. Eng. A* **2019**, *43*, 911–922. [CrossRef]

MDPI

St. Alban-Anlage 66

4052 Basel

Switzerland

Tel. +41 61 683 77 34

Fax +41 61 302 89 18

www.mdpi.com

Energies Editorial Office

E-mail: energies@mdpi.com

www.mdpi.com/journal/energies